Solar
Energy

Solar
Energy

DONALD RAPP

Jet Propulsion Laboratory
Pasadena, California

University of Texas at Dallas
Richardson, Texas

Prentice-Hall, Inc.
Englewood Cliffs, NJ   07632

Library of Congress Cataloging in Publication Data

RAPP, DONALD, 1934–
    Solar energy.

    Includes bibliographical references and index.
    1. Solar energy.   I. Title.
TJ810.R26   621.47   80-15175
ISBN   0-13-822213-4

Editorial/production supervision
    and interior design by Karen Skrable
Cover design by Judith A. Matz
Manufacturing buyer: Anthony Caruso
Cover photograph courtesy of NASA

Printed in the United States of America

10 9 8 7 6 5 4 3 2 1

Prentice-Hall International, Inc., *London*
Prentice-Hall of Australia Pty. Limited, *Sydney*
Prentice-Hall of Canada, Ltd., *Toronto*
Prentice-Hall of India Private Limited, *New Delhi*
Prentice-Hall of Japan, Inc., *Tokyo*
Prentice-Hall of Southeast Asia Pte. Ltd., *Singapore*
Whitehall Books Limited, *Wellington, New Zealand*

Contents

Preface

Solar energy engineering is a field that spans many basic disciplines, including chemistry; physics; mechanical, chemical and electrical engineering; mathematics; and economics. Solar energy engineering is therefore a difficult subject to teach, since it requires a broad understanding of many subjects. For the past four years, I have been teaching a two-semester course in solar energy to first-year graduate students majoring in "environmental sciences." The course is probably equivalent to a junior or senior level undergraduate course in an engineering school. The lecture notes from the course formed the basis of this book. *Solar Energy* is intended as a textbook for undergraduate engineers or first-year graduate students. It would also be useful to engineers who wish to learn about the field of solar energy on their own and to "do it yourselfers" who want to learn about solar energy applications. Such readers may not wish to grapple with the derivations given in the book, but would only want to make use of the results.

The first chapter reviews the patterns of energy use and resources in the United States, and explores the potential for solar energy to supply some of this energy in the future. The second chapter is concerned with the apparent position of the sun to an observer on the earth at any latitude at any time of any day of the year. Some of the geometry becomes rather complex, and some readers may wish to merely glance at the derivations. The contrast between the apparent path of the sun across the sky in winter and in summer is emphasized. Equations are derived for the angles that define the position of the sun at any hour of any day for an observer at any latitude.

The third chapter deals with methods of measurement and expected levels of solar intensities. A chapter on flat plate collectors follows, emphasizing efficiency as a function of design parameters. The fifth chapter discusses the economics of new energy saving technologies for which a large initial investment is made; annual savings in fuel are accrued for each year of operation. Simple procedures are given for evaluating the value and pay back period of such an installation.

Chapters 2 to 5 provide a basic background for the student in solar geometry, solar intensities, flat plate collectors, and economics. These fundamentals are applied to solar energy systems in Chapters 6 and 7. Chapter 6 gives a detailed account of the design and operation of a solar domestic hot water system, and utilizes computer simulation to illustrate the performance of alternate designs. Chapter 7 discusses space heating loads and solar space heating systems. Computer simulation is used to show the dependence of system performance on design parameters. Economic analyses are made of the designs discussed in Chapters 6 and 7.

In order to have an adequate background for dealing with solar assisted heat pumps and solar cooling systems, a chapter on heat engines and heat pumps is provided. Students with good backgrounds in engineering thermodynamics may choose to bypass Chapter 8.

The combination of flat plate solar collector technology and heat pumps is discussed in Chapter 9. Design, performance, and economics are emphasized.

Solar air cooling schemes generally require elevated temperatures for which flat plate collectors are usually inadequate or marginal. Therefore, Chapter 10 is given on "intermediate temperature collectors" which are designed to operate in the range 200°F to 330°F. A lengthy discussion is given of the relation between acceptance aperture and concentration. Modes of operation involving fixed, periodically adjusted, and continuous tracking collectors are discussed. A number of specific collectors are discussed with special emphasis given to the "CPC" collector. Many of the derivations are rather lengthy, and the casual reader may wish to skip these derivations.

Solar space cooling systems are discussed and illustrated with computer simulation in Chapter 11.

Chapter 12 gives a brief treatment of highly concentrating solar collectors for temperatures greater than $\sim 300°F$. Chapter 13 gives a brief synopsis of storage techniques.

Chapter 14 deals with solar total energy systems in which solar energy is used to drive an on-site electrical power plant, with the turbine exhaust heat used for space heating, space cooling, and hot water.

Chapter 15, the final chapter, briefly reports on rights of solar access. This chapter was written by Ian Whitlock, a graduate student at the University of Texas at Dallas.

Solar Energy can be used as the text for a full year course, or, by eliminating certain topics, for a one semester course in solar energy. The following sections should be retained in a one semester course:

Chapter	Sections Retained
1	1.2, 1.3
2	all
3	3.1 to 3.6
4	4.1, 4.2
5	5.1 to 5.3
6	6.1 to 6.3, 6.5, 6.7, 6.8
7	7.1 to 7.3, 7.5, 7.7
8	8.8
10	10.6 to 10.8, 10.11
11	11.1, 11.2, 11.6

A student who has never had FORTRAN programming should take a course in FORTRAN concurrent with the solar energy course.

This book has been written in English engineering units. At many key points, SI units have been provided in parentheses. During the period of transition to general acceptance of SI units, it is hoped that this will be sufficient.

DONALD RAPP

1

Potential
Impact of
Solar Energy

This chapter provides a brief outline of how solar energy could play a major role in the energy supply of the United States. We begin by reviewing the basic patterns of energy usage that have evolved in the United States and by estimating conventional energy resources. Projections of future energy demands are given, and the difficulties of supplying those demands are described. The various solar energy technologies and approaches are categorized and described briefly. The potential for these technologies to displace conventional nonrenewable energy sources is discussed, with special reference to domestic hot water, space heating, space cooling, industrial process heat, and electricity.

1.1 Energy Use, Consumption, and Resources

1.1.1 Energy Use Patterns

The historical percentage mix of various fuels used in the United States is shown in Figure 1.1. The periods of rapid industrial growth in the United States can be identified with the large-scale availability of cheap fuels: first coal, then oil, and then natural gas. During the late 1970s, the United States depended on petroleum and natural gas for about 70% of its total energy supply. However, the fuel mix for the world places a considerably larger emphasis on coal and less emphasis on natural gas.

The history of total energy usage in the United States is plotted in Figure 1.2, along with projected usage based on 2% or 3% annual compounded growth. The difference in usage in the year 2000 between 2% and 3%

1

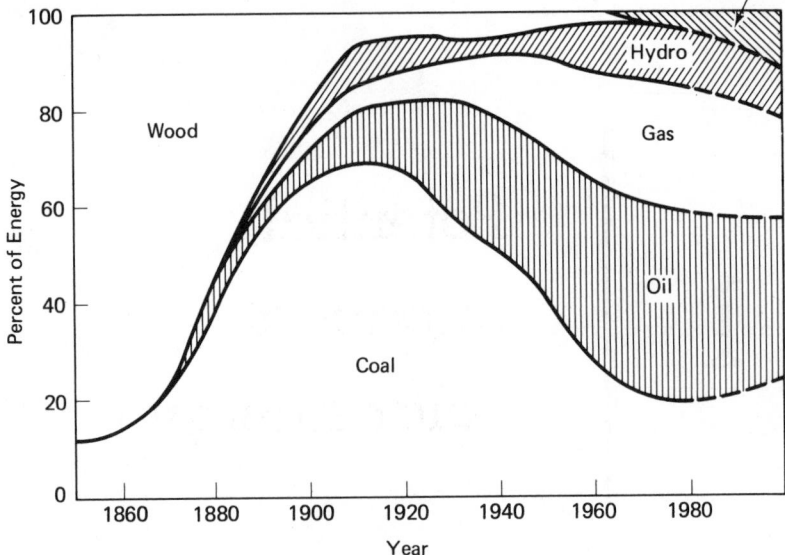

Figure 1.1 Percent of total energy used in the United States as supplied by various sources. Adapted from *Energy in the United States: Sources, Uses and Policy Issues* (1963) by H. H. Landsburg and S. H. Schurr, by permission of Random House, Inc.

Figure 1.2 History of total energy usage in the United States.

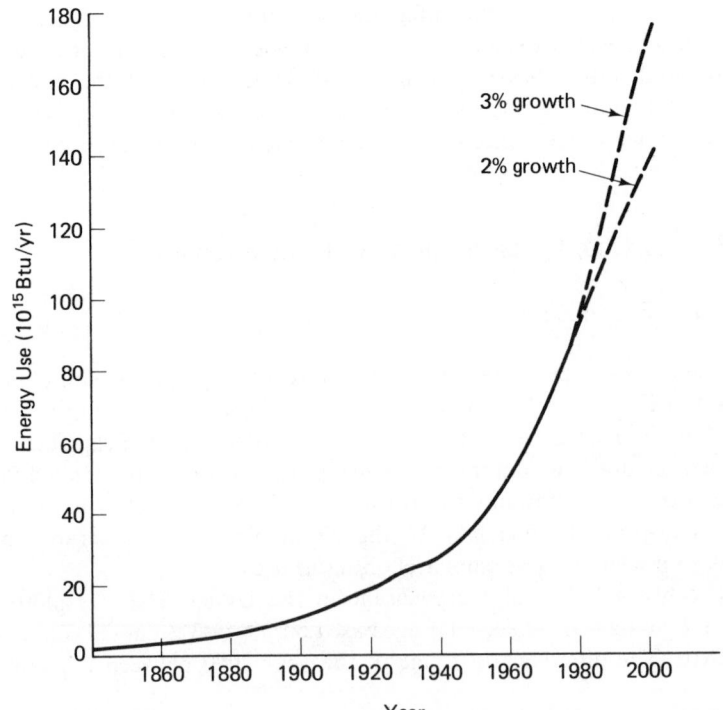

growth from 1978 to 2000 is 143×10^{15} Btu per year vs. 178×10^{15} Btu per year. This difference of 35×10^{15} Btu per year is equal to the entire energy consumption of the United States in 1950. It is clear that continued growth at historic levels will lead to very large demands of energy by the year 2000.

1.1.2 Energy Conversion Patterns

The energy conversion patterns in the United States are drawn roughly to scale in Figure 1.3. It can be seen that four areas, electrical energy generation, residential and commercial, industrial, and transportation, each consume roughly one-quarter of the total energy used. Energy losses and rejected energy represent over 50% of the energy expended. The most rapidly growing segment of the energy consuming sectors is electric energy generation. Projections indicate that by the year 2000 close to 50% of the raw energy consumed in the United States may be used for electric power generation. When the electrical power generated is fed back into the economy, we may identify four end-use sectors: industrial, transportation, residential, and commercial. The division of total energy used in these areas is shown in Figure 1.4. Summaries of the energy used for various specific end uses in these categories are given in Tables 1.1 through 1.3. Some of the areas that can be immediately identified as being appropriate for augmentation with solar energy are listed in Table 1.4.

1.1.3 Energy Resources

The major developed energy resources of the United States include coal, oil, natural gas, hydroelectric power, and nuclear energy (^{235}U). In addition, other sources such as oil shale, solar energy, wind, and geothermal energy may be developed further during the coming decades.

TABLE 1.1 Energy use in the commercial area, 1968*

Type of Use	Energy Used (trillions of Btu)	Percent of Area Total	Percent of National Total
Space heating	4182	47.7	6.9
Air conditioning	1113	12.7	1.8
Feedstock	984	11.2	1.6
Refrigeration	670	7.6	1.1
Water heating	653	7.5	1.1
Cooking	139	1.6	0.2
Other	1025	11.7	1.7
Area total	8766		14.4

*Data for energy used from *Patterns of Energy Consumption in the United States*, Office of Science and Technology, Executive Office of the President, Jan. 1972, p. 68.

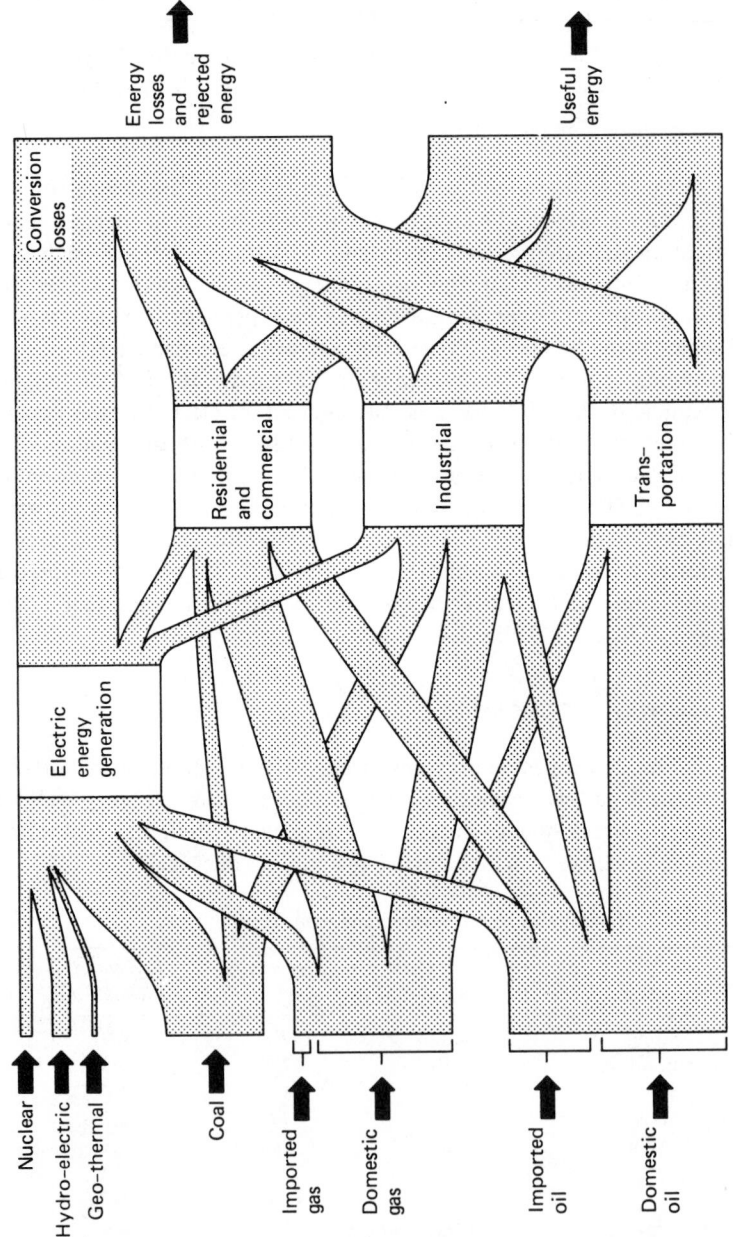

Figure 1.3 Energy conversion patterns in the United States. Adapted from *The Energy Source-book* (1977) by permission of the Center for Compliance Information, The Aspen Corporation.

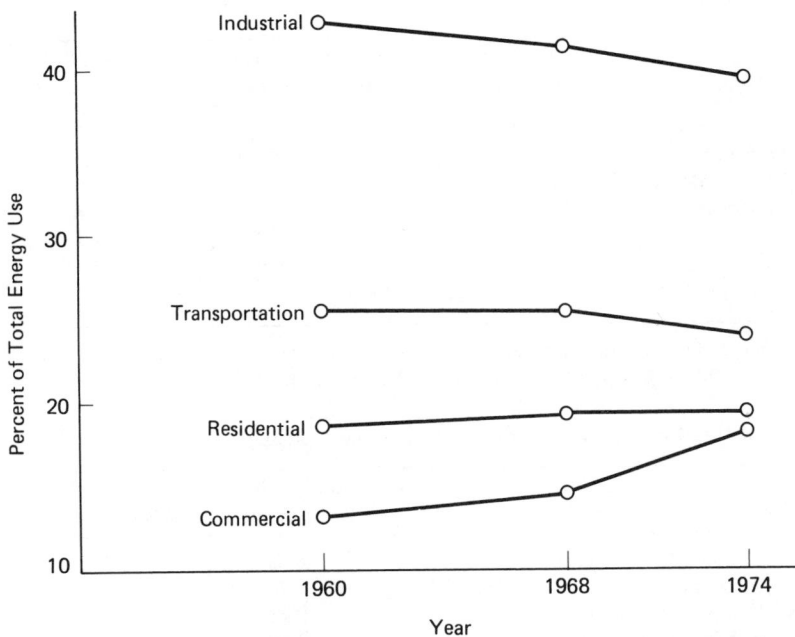

Figure 1.4 Energy use distribution among areas.

TABLE 1.2 Energy use in the residential area, 1968*

Type of Use	Energy Used (trillions of Btu)	Percent of Area Total	Percent of National Total
Space heating	6,675	57.5	11.0
Water heating	1,736	14.9	2.9
Refrigeration	692	6.0	1.1
Cooking	637	5.5	1.1
Air conditioning	427	3.7	0.7
Lighting	412	3.5	0.7
Television operation	352	3.0	0.6
Food freezing	220	1.9	0.4
Clothes drying	208	1.8	0.3
Clothes washing	41	0.4	0.1
Dish washing	36	0.3	0.1
Other	180	1.5	0.3
Area total	11,616		19.3

*Data from *Patterns of Energy Consumption in the United States*, Office of Science and Technology, Executive Office of the President, Jan. 1972, pp. 38, 62.

TABLE 1.3   Energy use in the industrial area, 1968*

Type of Use	Energy Used (trillions of Btu)	Percent of Area Total	Percent of National Total
Process steam	10,132	40.6	16.7
Direct heat	6,929	27.8	11.5
Electric motors	4,794	19.2	7.9
Feedstock	2,202	8.8	3.6
Electrolytic processes	705	2.8	1.2
Other	198	0.8	0.3
Area total	24,960		41.2

*Data from *Patterns of Energy Consumption in the United States*, Office of Science and Technology, Executive Office of the President, Jan. 1972, p. 6.

TABLE 1.4   Energy use sectors for possible solar augmentation, 1968

Type of Use	Energy Used (10^{15} Btu/yr)	Percent of National Total
1. Space heating, residential	6.68	11.0
2. Domestic hot water, residential	1.74	2.9
3. Space cooling, residential	0.43	0.7
4. Space heating, commercial	4.18	6.9
5. Space cooling, commercial	1.11	1.8
6. Hot water, commercial	0.65	1.1
7. Process steam, industrial	10.13	16.7
Total	24.92	41.1

1.1.4 Coal Resources

The coal resources of the United States can be estimated with moderate accuracy because coal occurs in layers that are easily mapped. The estimates of Averitt[1] are generally accepted as reliable by most investigators in the field. If the sheer number of tons of coal in the ground is taken at face value, the resource is quite large—approximately 1600×10^9 tons. At the present rate of utilization of coal in the United States (600×10^6 tons/year), our coal would last for a very long time. Indeed, some have argued that coal could be used as the sole fuel for the United States as oil and natural gas gradually diminish in importance during the coming decades. There are several problems encountered in the extensive utilization of coal. These problems include:

[1]P. Averitt, "Coal Resources of the United States," *U.S. Geol. Survey Bull. 1275*, January 1, 1967.

1. Extensive mining requires expensive, large-scale land reclamation.
2. Much of the coal is used in the eastern/central part of the United States, where most of the coal deposits have high sulfur content. Burning this coal would lead to great pollution control problems.
3. Use of low sulfur western coals would require expensive procedures for transportation by railroad or slurry.
4. Waste materials from coal burning must be disposed of.
5. Effective processes for liquefaction and gasification of coal must be developed.

Even if all these problems could be solved, the vastness of our coal resources are somewhat illusory. Not all the coal in place can actually be removed. If, say, 35% could be removed from the ground, that leaves 560×10^9 tons to be utilized. If the entire 1978 U.S. energy usage were obtained from coal, it would have required about 3×10^9 tons. If U.S. energy usage continues to grow at 3% per year, annual usage would grow to 5.75×10^9 tons by the year 2000, 12.0×10^9 tons by 2025, and 25.2×10^9 tons by 2050. Cumulative usage of coal from 1978 through 2040 would then be approximately 560×10^9 tons. This would use up the entire domestic supply.

1.1.5 Petroleum and Natural Gas Resources

Petroleum and natural gas resources are difficult to estimate. Hubbert[2] devised a number of interesting approaches for making crude estimates. Rapp[3] updated the estimates of petroleum resources in 1974. One of the best methods is based on an analysis of the requirements for saturation drilling of all the sedimentary basins in the United States. If it is assumed that, in order to find essentially all of the petroleum in the United States, an exploratory well of approximately 20,000 ft depth must be drilled every 2 mi^2 in all the sedimentary basins, then it follows that about 4.5×10^9 linear feet must be drilled. By the year 1977, more than 2×10^9 ft had already been drilled. Data are available over the years on the number of barrels of oil discovered per foot drilled. The recovery curve is of declining exponential form, with about 200 bbl/ft in the early days of petroleum exploration, dropping to about 30 to 40 bbl/ft in the late 1970s. By extrapolating the recovery

[2]M. King Hubbert, "Energy Resources," *National Research Council Pub. 10000*, National Academy of Sciences, 1962.

[3]D. Rapp, "Estimation of the Degree of Petroleum Advancement in the United States," *Energy Sources*, 2, 125, 1975.

curve into the future and integrating the bbl/ft over feet of drilling to 4.5×10^9, estimates of recoverable petroleum can be made. The best estimates for the 48 conterminous states is that there were originally 175×10^9 bbl recoverable. As of 1978, roughly 120×10^9 bbl had been removed from the ground, leaving about 55×10^9 bbl recoverable. Alaskan and offshore oil will amplify this figure considerably. However, at the 1978 rate of annual consumption of over 6×10^9 bbl per year,[4] there is clear cause for alarm. In 1978, U.S. oil imports were nearly 50% of annual consumption. The consequences of this in terms of balance of payments problems, weakening of the dollar, and weakening of our international political influence are disastrous. The situation in natural gas is also rapidly growing worse. The rate of growth in natural gas usage in the United States from 1950 to 1970 was over 6% per year.[5] This rate of growth is not sustainable for many decades. Indeed, serious shortages had already developed by 1975. These shortages might be partially alleviated for a few years by temporary exploration encouraged by price increases, but there is little doubt that, regardless of pricing, fundamental shortages will develop in natural gas during the 1980s.

1.1.6 Other Energy Resources

Many of the best natural sites for hydropower have already been developed. It is believed that the potential for hydropower to make major new contributions to the U.S. energy supply does not exist.

Nuclear power via ^{235}U fission spread rapidly in the United States during the 1960s and 1970s. The number of nuclear electric power plants has grown[6] from the first plant in 1957 to approximately 60 existing plants at present, and about 100 plants are either planned or under construction. However, the radioactive waste problem has not been settled, and the wastes are accumulating in temporary storage sites. Furthermore, it is estimated that the domestic supply of uranium at moderate prices is quite limited and may become severely strained by 2000.

It must be concluded that the development of renewable energy resources is necessary for survival of the future generations. The major renewable resources appear to be nuclear fusion, solar energy, and wind energy. Technologically, none of these energy sources is even close to being readily available as a major prime source of energy. Further development will be long and costly—but is absolutely essential.

[4]A. McRae, J. L. Dudes, and H. Rowland, eds., *The Energy Source Book*, Aspen Systems Corp., Germantown, MD. 1977.

[5]Ibid.

[6]Ibid.

1.1.7 Future Energy Demand

The rate of energy consumption by the United States, and by the world, is growing rapidly. It was pointed out in Sec. 1.1.1 that the difference between 2% growth and 3% growth in energy usage in the United States from 1978 to 2000 results in a difference in annual consumption in 2000 of 143×10^{15} vs. 178×10^{15} Btu per year. Some observers predict energy growth in the United States at around 3% per year. This implies that energy consumption in 2000 will be triple the energy consumption in 1960. Furthermore, it appears that there will be a gradual shift to more usage in the form of electricity, with raw energy use for electricity approaching 50% of the U.S. total by 2000. This will greatly increase the amount of heat rejected into the environment. If these changes are allowed to take place, it will make the task of supplying energy from solar sources more difficult. It appears that our consumption is tending to run away from our development of alternative energy resources. The Energy Act of 1978 was a small first step in the direction of moderation and regulation of energy consumption. The Energy Act actually is a set of five acts:

(1) *The energy tax act:*
   a) Provides residential credit for energy conservation (up to 15% of $2000).
   b) Provides residential credit for renewable energy resources (up to 30% of $2000, 20% of next $8000).
   c) A stepped gas guzzler tax on automobiles.
(2) *The national energy conservation act:*
   A hodgepodge of miscellaneous programs for energy conservation.
(3) *The power plant and industrial fuel use act:*
   a) Bans use of natural gas or oil in new large installations (but has many exemptions).
   b) Bans use of natural gas or oil in existing installations after 1990 (with exemptions).
(4) *The public utility regulatory policies act*
(5) *The natural gas act:*
   Regulates prices of natural gas.

1.1.8 Effect of Population Growth

One of the major impediments to control of growth of world energy usage is the increase in world population. The history of world population is shown in Figure 1.5.[7] The world population in 1978 was over 4×10^9 and is

---

[7]N. Keyfitz, "Population Growth: Causes and Consequences" in *Environment*, ed. W. W. Murdoch, Sinauer Assoc., Sunderland, MA, 1975.

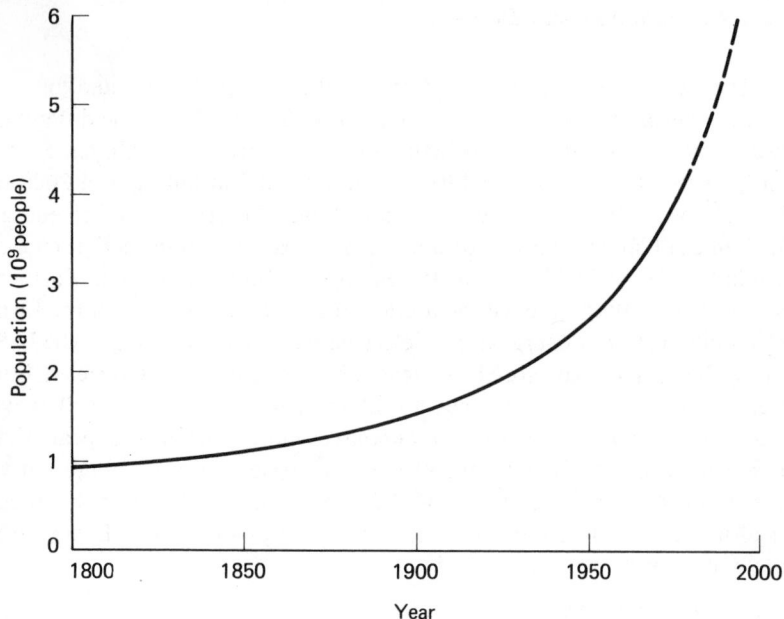

Figure 1.5 History of world population growth.

expected to exceed 6.5 × 10⁹ by the year 2000. The United States, with 5%
of the world's people, consumes nearly 30% of the world's energy. To bring
the rest of the world up to 1978 U.S. energy standards by 2000 would require
a 6-fold increase in per capita usage. When this is coupled with the projected
60% increase in population, the result is a growth factor of about a 10-fold
increase in world energy consumption. The unbridled growth in world population may be the greatest impediment to the success of alternative energy
sources.

1.2 Hierarchies of Solar Technology

1.2.1 Photovoltaic Cells

Solar energy can be utilized in a variety of ways. The solar spectrum
is a distribution of wavelengths from the ultraviolet to the infrared characterized by an average solar surface temperature of approximately 6000°K.
In the ultraviolet and visible parts of the spectrum, the light quanta have
sufficiently high energy to induce chemical transitions in absorbing materials.
This is the basis for solar photochemical processes and photovoltaic conversion to electricity. Solar photochemical conversion is still in a conceptual

stage, and it is difficult to predict its future. Photovoltaic conversion uses solid-state semiconductor crystals that are doped with positive and negative atoms to produce an electric field across the crystals. When sunlight passes through such a crystal, some of the light is absorbed by atomic electrons, which become excited up to conduction levels. The electric field drives the electrons through the cell and the outer circuit, producing a flow of electrical power. Although the voltage attainable with one cell is small (approximately 0.5 volt), large numbers of cells can be wired in series to achieve high voltages. The solar cells originally produced during the 1970s were obtained by taking slices out of a long crystal of silicon, much like delicatessen-sliced salami. The processing was batch and resulted in high costs for cells. The cells have other deficiencies, including decreased performance at elevated temperatures and gradual dimunition of power over several years. The cost of electricity from these batch-produced cells was about 40 times that of electricity from conventional sources in 1975. In order to bring the cost of solar cells down to competitive levels, a number of companies are developing procedures for continuous production of silicon ribbon or of other solar cell materials. The Department of Energy is highly optimistic that costs of solar cells will be reduced to competitive levels by the mid-1980s.

1.2.2 Photovoltaic Installations

Since cells are very expensive, attempts have been made to operate cells behind Fresnel lenses on tracking mounts, the theory being that lenses plus tracking ought to be cheaper than cells. However, since solar cells only convert approximately 10–15% of the solar radiant power to electricity, and much of the remainder ends up as heat, effective cooling must be provided for the cells. It is not clear whether the combination of lenses, tracking mounts, and cooling system is cheaper than arrays of cells without lenses and cooling. If the cells can be operated at elevated temperatures ($130°F$ to $300°F$) it may be possible to produce electricity *and* collect thermal energy for low-level heat applications.

One of the major problems encountered in electricity generated by solar cells is the difficulty of storing the electricity. A terrestrial solar cell system can only provide electricity for a limited number of hours and, even then, not on all days. One method for providing storage is to use the electricity to electrolyze water and store hydrogen as a potential fuel. The hydrogen can be burned to produce heat or used in fuel cells to produce electricity for subsequent use. The economics of such a system are uncertain. Another approach[8] is to place the solar cells on large satellite arrays in a synchronous orbit, where the satellite remains above a fixed point on the

---

[8]P. Glaser, "Solar Power from Satellites," *Physics Today*, February, 1977, p. 30.

earth. Such solar-generated electricity is converted on the satellite to micro-waves which are beamed to a receiving station on the earth below. The energy is then reconverted into conventional electricity and conveyed to users by power lines. The satellite would be outside of the earth's shadow for about 23 hr per day. Some suggestions have been made that costs of such systems could be reduced by mining materials such as silicon on the moon and doing vacuum metallurgy on the surface of the moon. The energy required to "blast off" from the moon is also less than that for earth. It is too early to evaluate the environmental hazards and other technical problems associated with such schemes. Immense cost reductions would be required to make them competitive.

Solar cells use the high energy part of the solar spectrum to induce chemical transitions in matter with large energy gain per absorption. As a result, only a limited portion of the spectrum is available, and it is doubtful whether conversion of power in the solar spectrum to electrical power will exceed 15–20%. Another way to utilize solar energy is to attempt to absorb as much as possible of the solar spectrum with a black surface, and utilize solar energy in the form of heat. Such solar thermal conversion processes are the main topic of this book.

1.2.3 Solar Thermal Conversion

When a black surface is exposed to solar radiation, it heats up. As its temperature increases, the surface loses heat at an increasing rate to its surroundings. A steady state is reached when the rate of solar heat gain is balanced by the rate of heat loss to the ambient. This approach has the advantage that the entire solar spectrum is available for absorption by the surface. There are a number of ways to utilize solar thermal conversion. *Passive systems* are systems where the solar energy is absorbed directly into the living space where it is to be used. *Active systems* utilize external solar collectors with a heat transfer fluid to convey collected heat to the living area or to storage. Almost all solar thermal conversion systems utilize heat storage to extend the period of operation beyond the hours when the sun is shining. In most cases, the storage simply consists of a tank of heated water. However, other more sophisticated storage techniques can be employed in many cases with some advantages.

1.2.4 Flat Plate Collectors

The simplest solar collector for active solar conversion is the flat plate collector. The flat plate collector is simply a black sheet of metal with tubes attached for heating a fluid, enclosed in a box with a transparent cover. Most residential applications are built around the use of flat plate collectors. Flat plate collectors are adequate for domestic hot water heating and space heating. They are probably only marginally useful in solar space cooling applications.

To achieve the higher temperatures required for space cooling, industrial process heat, mechanical power, or electrical generation, concentrating solar collectors are required.

1.2.5 Concentrating Collectors

Concentrating collectors employ lenses or reflectors to concentrate the radiant power from the sun passing through a large aperture onto a smaller absorbing surface. Since the rate of heat loss from a surface is proportional to its area, roughly the same amount of solar gain as a flat plate collector can be achieved, but with a much smaller heat loss. Thus, adequate collection efficiencies (approximately 50%) can be achieved at elevated temperatures. One of the difficulties with concentrating collectors is the requirement that they be aligned with the sun. There are three categories of operation. For very low amounts of concentration (generally less than 2:1) it is possible to have a fixed-orientation concentrating collector. For intermediate concentration (generally less than 5:1) it is possible to utilize a periodic adjustment scheme whereby the collector tilt is adjusted perhaps once a month. For higher concentrations, continuous tracking schemes must be used. In some cases, movement through one angle is adequate. For very high temperatures, movement through two orthogonal axes is required.

1.3 The Potential for Solar Energy Applications

1.3.1 Domestic Hot Water

The first area that is most appropriate for large-scale application of solar energy is domestic hot water heating. Hot water in the residential sector accounts for about 3% of the nation's energy consumption. Taking into account the number of situations where solar energy would not be appropriate, such as high-rise apartments, wooded areas, houses with poor orientation, and regions of low solar availability, it appears that only about half of the energy required for hot water could be displaced by "retrofitting" to existing structures. A large part of the remainder could eventually be displaced by solar energy over a period of years as new dwellings are built. The total cost of providing most of the residential hot water in this country from solar energy will probably be about 60×10^6 dwellings $\times$ $2000 per dwelling $=$ $120,000,000. The energy displaced per year would be about 2.5×10^{15} Btu per year. It is shown in a later section that when electricity is the conventional source of energy for domestic hot water, the investment is well worth it—the pay-back period being about 5 or 6 years. The pay-back period is considerably longer when natural gas is used for hot water, but this is due to the artificially low price of natural gas due to government control. The technology for solar heating of domestic hot water is well developed. Standard flat plate collectors, storage tanks, and control systems are commercially available. Although the potential here is only a few percent of total annual energy consumption, it is an important first step.

1.3.2 Space Heating

Space heat in the residential and commercial sectors accounts for about 18% of the annual energy consumption in the United States. This is one of the largest single pools of energy usage that is conceptually displaceable by existing technologies in renewable energy resources. However, the task is not simple. Many existing structures are not very suitable for solar energy application. Unlike domestic hot water systems, which are fairly uniform in design, a wide variety of space heat designs are used in various applications. Integration with solar energy may occur through a variety of approaches, including hot air systems, hot water systems, water/air systems, and passive systems. Solar space heating systems tend to be less cost competitive than solar domestic hot water systems because space heat systems do not displace conventional energy for all 12 months of the year. The technology for solar space heating is not very different in principle than for solar hot water heating. Flat plate collectors can be used, but they should have better performance characteristics. This can be achieved with extra insulation, extra glazing, and selective absorber surfaces.

Solar space heating systems may have application in conjunction with the use of heat pumps. This may be of growing importance for the future as more of our energy becomes available as electricity generated from nuclear or coal resources. If solar energy can be collected at low temperature levels (50–80°F) (283–300°K), and a heat pump can boost this heat to the ∼130°F (328°K) level for interior space heating, advantageous results may be obtained. As electricity is used more for space heating, integration with heat pump technology will become increasingly important.

The possibility that solar energy might eventually displace a large part of the 20% of the nation's energy supply used for hot water and space heat is of critical importance for future planning of energy supplies for the United States. The government has attempted to support this endeavor by authorizing tax credits for such installations. However, the major impediment to wide acceptance of this technology is the unfamiliarity of life cycle economics to the general public. Solar energy is capital intensive, requiring large initial investments with variable pay-back periods.

1.3.3 Space Cooling

Solar space heating/space cooling/hot water systems for residential and commercial applications represent, for two reasons, another level of higher sophistication in technology. One is that a relatively sophisticated technology is required for producing a cooling effect from a heat source. Three approaches were in development during the late 1970s. One involves utilizing a small heat engine to drive a conventional compressor refrigeration machine for cooling. Another involves evaporative cooling by humidification of desiccant dried air, with solar heat used for regeneration of the desiccant. The third method utilizes an absorption cycle chiller. All three of these approaches require a heat source at a somewhat more elevated temperature [minimum of $\sim 190°F$ (361°K), with advantages if as high as 300°F (422°K)] than is required for space heating or hot water ($\sim 120°F$) (322°K). Therefore, not only does the cooling effect require more sophistication, but the solar collectors also require more sophisticated design. The best flat plate collectors are barely marginal for achieving the minimum temperatures required for space cooling systems. In many applications, concentrating collectors will be required for achieving reasonable efficiencies at the elevated temperatures needed for space cooling. It is found upon analysis that as the degree of concentration is increased, the achievable working temperatures also increase. However, the aperture for acceptance of light decreases in a reciprocal relationship. Therefore, collectors with high concentration require accurate sophisticated tracking techniques to keep them normal to the sun. This increases the expense and the probability of failure modes of such collectors. For space cooling applications, collectors with moderate concentrations and simple periodic pointing adjustment (as opposed to continuous tracking) will probably be preferred. Such collectors can be designed in various ways, using reflectors or lenses. Although the total amount of energy used in the United States for space cooling (probably about 3–4% of the total in 1978) is not as large as that used for space heating, space cooling is one of the most rapidly expanding areas of energy consumption. As the population movement to the "sunbelt" continues, the disproportionate growth in energy used for space cooling will continue.

1.3.4 Industrial Process Heat

Another major area of energy consumption is the use of process steam in the industrial sector. This accounted for about 17% of the nation's energy consumption in 1970. Such process steam is typically utilized in the 300°F–500°F (422–533°K) temperature range. Concentrating solar collector systems can supply heat at these temperatures. However, there are a variety of approaches possible, and it is not clear which methods offer the greatest advantages. Energy storage remains a major problem for such systems. No simple inexpensive method for storing heat in the 300–500°F (422–533°K) temperature range exists. Further work is required in this important area.⌉

1.3.5 Interface with Utilities

An important aspect of almost all solar energy systems is the fact that optimum design usually involves solar energy supplying perhaps 60–80% of the energy requirements of a user, with conventional "back-up" energy supplying the remainder. If one attempted a design in which solar energy provided all the energy required, it would be found that the collector and storage requirements would be so great that the cost would be prohibitively expensive. Therefore, the interface between individually installed solar energy systems and public utilities is an important matter. If the solar energy systems in a region tend to require back-up conventional energy at the same times due to outages of solar availability, then the potential demands from the utility during peak periods could be nearly as high as if there were no solar installations. Nevertheless, the average rate of utility usage would be severely reduced if solar energy installations were widespread in an area. The economic impact on a utility could be devastating if it had to provide very high levels of plant capacity for peak periods, which operate with relatively low average load factors. Therefore, there are complications involved in the widespread introduction of solar energy in any region. Economic mechanisms must be found for making solar energy a favorable technology for utilities.

1.3.6 Production of Electricity

In the late 1970s, more than 25% of the nation's raw energy consumption was used to produce electricity. Electricity production is also a very rapidly growing area of energy consumption. Solar energy could conceivably make major contributions to this consumption area. This could take place either through direct conversion to electricity in solar cells or via solar thermal conversion in heat engines.

1.3.7 Electricity from Solar Cells

The use of solar cells in terrestrial fields is technically possible but requires great cost reduction. The cells could be emplaced in fixed or periodically adjusted mountings or could be the receivers of concentrating collectors. Since it is difficult to store electricity, such systems might be best used in conjunction with pumped hydro-storage sites. It has been suggested[9] that solar cell systems be mounted on satellites in synchronous orbit about the earth, where they would be outside the earth's shadow for about 23 hr per day. The electricity generated on the satellite would be converted to microwaves, which would be transmitted to a receiving station on the earth directly beneath the satellite. The microwaves would then be reconverted to electrical power at the receiving station. This approach has the advantage that the energy storage problem would be solved through use of several satellites with staggered time zones. However, such an approach will require cost reductions and technology development of heroic proportions, and there are many environmental dangers that require further analysis.

1.3.8 Electricity from Solar Thermal Conversion

Solar thermal conversion can also be used for electrical power genera-tion by using solar heat to drive a heat engine connected to a generator. Some of the approaches that are being considered include:

1. Distributed fields of concentrating collectors supplying hot fluid to a conventional power plant. If the power plant is not too large and on-site, the heat from the engine exhaust can be used for low-level applications such as hot water and space heat. Such a system is called a *total energy system*.
2. Distributed fields of concentrating collectors (probably of "dish" con-figuration) with a small Stirling cycle engine and electrical generator attached to each collector module. The electricity from each module is wired to a central distribution center.
3. Central receiver systems in which a receiver is placed at the top of a tower surrounded by a field of reflectors mounted on pedestals (helio-stats). A fluid is pumped up from the ground to the receiver, where it is heated to a high temperature. The heated fluid can be used to drive a heat engine system at ground level. This might be a Brayton cycle, a Rankine cycle, or a hybrid of the two. The advantage of this system is that all the plumbing is confined to a single area.

[9]P. Glaser, "Solar Power from Satellites," *Physics Today*, February, 1977, p. 30; "Report of the Solar Power Satellite Task Group," Universities Space Research Assn., P. O. Box 1892, Houston, TX, November, 1977; P. Glaser, G. M. Hanley, R. H. Nansen, and Richard L. Kline, "First Steps to the Solar Power Satellite," *IEEE Spectrum*, May, 1979, p. 52; W. C. Brown, "Solar Power Satellites," *IEEE Spectrum*, June, 1979, p. 36.

1.3.9 The Hydrogen Economy

The problem of energy storage remains an obstacle to all three of these approaches. Heat storage at high temperatures [$\sim 1000°F$ ($811°K$)] does not appear viable. A long-term ultimate goal would be to use collected solar energy to dissociate water. This could either be done catalytically in high temperature reactors or by electrolysis. The hydrogen produced[10] could be stored for extended periods of time and transported by pipeline. Fuel cells could be used to reconvert the hydrogen to electricity with high efficiency.

It is possible that solar energy may become an important contributor to the total energy supply of the United States. This will require considerable effort in the development of technology and cost reduction in production.

[10]T. N. Veziroglu, *Hydrogen Energy*, Plenum, New York, 1975.

2 | Solar Geometry

This chapter concerns the geometry of the earth relative to the sun. The basic equations are derived that determine the angles that solar rays make with various surfaces in various locations during various parts of the year. The earth rotates about an axis that is tilted relative to its plane of motion, and this gives rise to the seasonal and diurnal variations in the position of the sun to an observer on the earth. It is shown that in the northern hemisphere the sun appears low in the southern sky during the winter but is more overhead during the summer. This allows simple passive solar heating to be effected with south-facing windows with overhangs. Some of the derivations in this chapter are rather difficult; some readers may wish to skip the detailed derivations.

2.1 Motion of the Earth About the Sun

The earth is approximately spherical, with a diameter of about 7900 mi (12,700 km). It makes one rotation about its axis every 24 hr and completes a rotation about the sun in approximately $365\frac{1}{4}$ days. The earth moves about the sun in an approximately circular path, with the sun located slightly off center of the circle. This offset is such that the earth is closest to the sun near January 1 and furthest from the sun near July 1. The distance in January is about 3.3 % closer, and since the solar intensity is proportional to the inverse square of the distance, the intensity is about 7 % higher in January than in July. The earth's axis of rotation is tilted at 23.45° with respect to its orbit

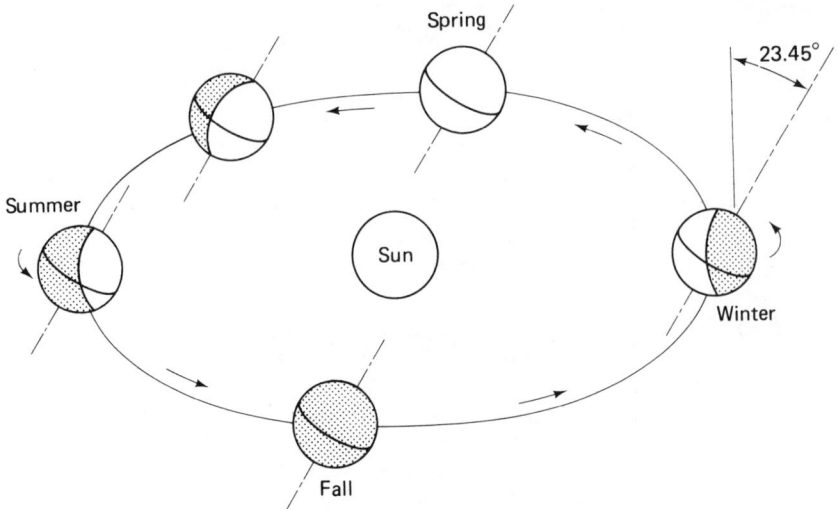

Figure 2.1 Movement of the earth around the sun.

plane about the sun. This tilt remains fixed in space and is the cause of the seasons (see Figure 2.1). In this figure, the earth's axis is drawn with the north pole at the top, tilted at 23.45° to the orbit plane. When it is winter in the northern hemisphere, the north end of the axis is tilted away from the sun, and when it is summer in the northern hemisphere, the north end of the earth's axis is tilted toward the sun. The days when the axis is tilted exactly toward or away from the sun are June 21 and December 21. On March 21 and September 21 the tilt of the earth's axis is along the orbit and is neither toward nor away from the sun. Clearly, in the southern hemisphere the seasons are reversed, and the winter occurs south of the equator when it is summer north of the equator. Since the earth is closer to the sun when it is winter in the northern hemisphere (and summer in the southern hemisphere), there is a tendency for the seasonal differences in temperature between winter and summer to be greater in the southern hemisphere than in the northern hemisphere.

2.2 Angle of Declination

The following discussion is limited to the northern hemisphere but can be easily extended to the southern hemisphere. The apparent position of the sun to an observer on the earth's surface is such that the direct solar rays come from a direction above (or below) the equatorial plane in summer (winter). Solar noon at any locality is defined as the time when a line from the center of the earth to the sun passes through the meridian of longitude that contains the locality. The angle that the sun's rays make with the equatorial plane at solar noon is called the *angle of declination*. The angle of declination varies from 23.45° on June 21, to 0 on September 21, to $-23.45°$ on December 21, to 0 on March 21. It is illustrated for winter and summer conditions in Figure 2.2.

An approximate model for estimating the angle of declination on any day of the year is given next. Consider the diagram in Figure 2.3. An x-y-z Cartesian coordinate system is arranged with the x and y directions in the orbit plane and with the z direction perpendicular to the orbit plane. The

Figure 2.2 Angle of declination.

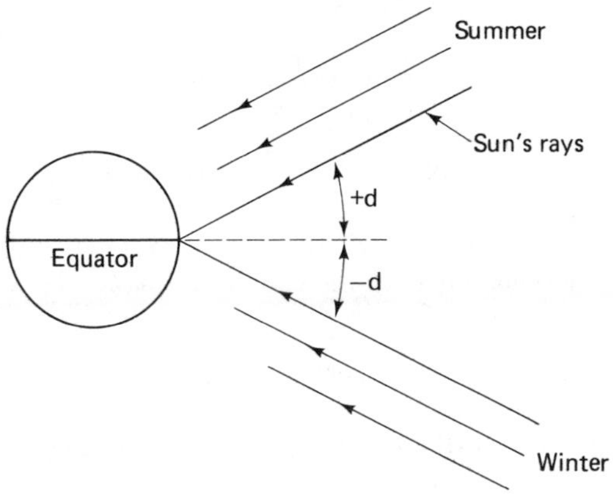

Figure 2.3 Vectors for calculating the angle of declination.

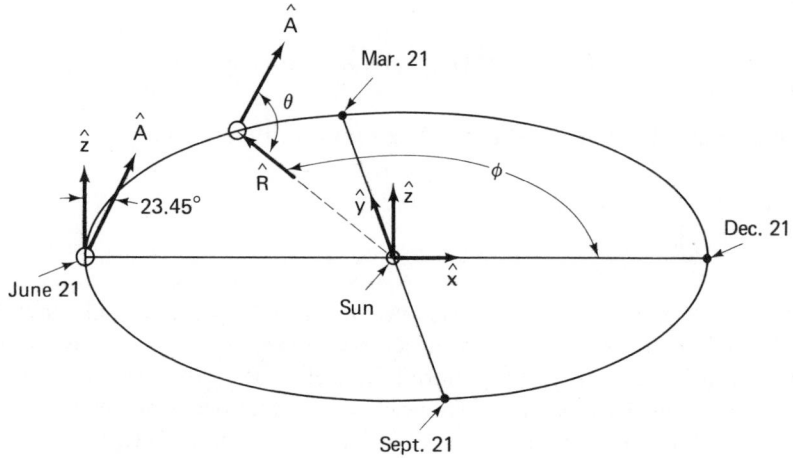

vectors with carets over them are of unit length. The $\hat{A}$ vector is along the earth's axis pointed from the south pole toward the north pole. The vector $\hat{R}$ is in the earth orbit plane from the sun toward the earth. The angle θ is the angle between $\hat{R}$ and $\hat{A}$; and angle ϕ in the earth's orbit plane defines the day of the year.

Let the day of the solar year be D_s, where $D_s = 1$ on December 21 and $D_s = 365$ on December 20. Then, since ϕ advances by 180° in 182.6 days,

$$\phi = (D_s - 1)\frac{180}{182.6} \quad \text{(degrees)} \tag{2.1}$$

The vector $\hat{R}$ can be resolved into $\hat{x}$ and $\hat{y}$ components,

$$\hat{R} = -(\cos \phi)\hat{x} + (\sin \phi)\hat{y} \tag{2.2}$$

The $\hat{A}$ vector can most easily be resolved by noting its position on June 21. It is clear that

$$\hat{A} = \cos (23.45°)\hat{z} + \sin (23.45°)\hat{x} \tag{2.3}$$

The angle of declination is the angle the sun's rays make with the equatorial plane. Since $\hat{A}$ is perpendicular to the equatorial plane, the angle of declination is

$$d = 90° - \theta \tag{2.4}$$

The angle θ is found from the relation

$$\cos \theta = -\hat{R} \cdot \hat{A} = \cos \phi \sin (23.45°) \tag{2.5}$$

Combining Eqs. (2.1), (2.4), and (2.5), we find that

$$\sin d = -\cos \left[(D_s - 1)\frac{180°}{182.6} \right] \sin (23.45°) \tag{2.6}$$

A plot of this dependence of d on D_s is shown in Figure 2.4.

2.3 Solar Time

Time on the earth's surface is reckoned from midnight at the Greenwich meridian (zero longitude) as Greenwich civil time (GCT). Local civil time (LCT) at any place on earth is more advanced at the same instant by 1 hr for each 15° further east than Greenwich, and is retarded 1 hr for each 15° further west than Greenwich. Thus, at a point 180° in longitude, the hour is

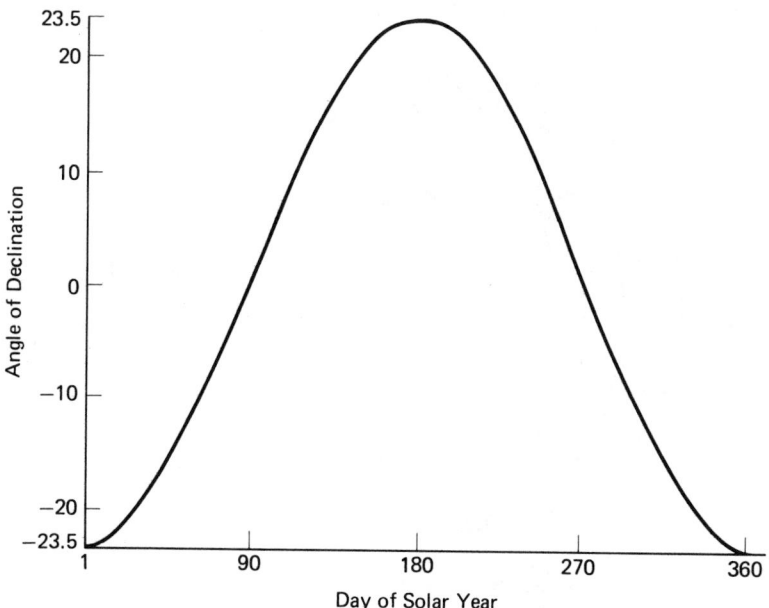

Figure 2.4 Angle of declination vs. day of solar year.

12 hr different than at Greenwich. An international date line separates a change in day as one goes across the 180° longitudinal line.

Ordinarily, we arbitrarily divide regions between each 15° of longitude into areas of local standard time. These correspond to the civil times at the east boundaries of the region. In the United States, for example, we have Eastern Standard Time (EST), Central Standard Time (CST), Mountain Standard Time (MST), and Pacific Standard Time (PST), defined as the civil times at the longitudes 75°W, 90°W, 105°W, and 120°W, respectively. The local civil time at any local point within any of these zones is obtained by subtracting 4 min for each degree of longitude that the local point is west of the reference longitude of the time region. For example, at a point of longitude 110°W, when it is some time (GCT) in Greenwich, the local standard time is (MST), and

$$(MST) = (GCT) - 7 \ hr$$

The local civil time is

$$(LCT) = (MST) - (5°) \ (4 \ min/°)$$
$$(LCT) = (MST) - 20 \ min$$

This is illustrated in Figure 2.5.

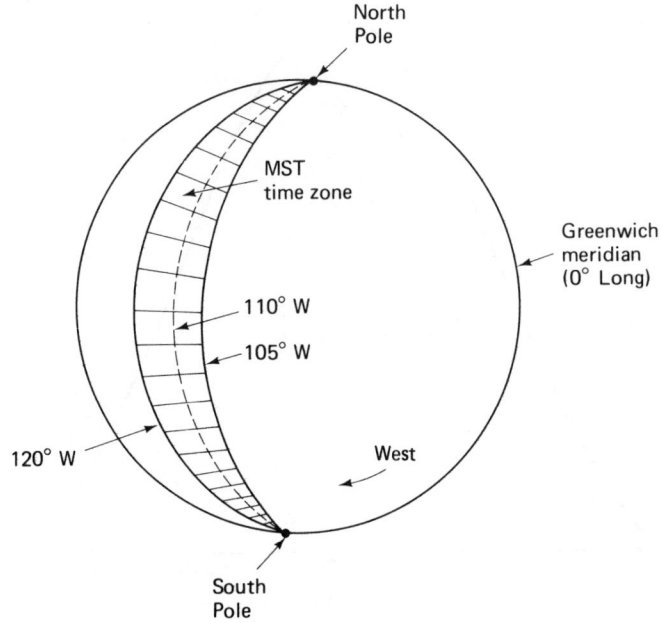

Figure 2.5 Illustration of local civil time in the mountain standard time zone (MST = GCT − 7 hr.) [Local civil time at 110°W is MST − 20 min.]

If the earth's rotation were perfectly regular and if the earth's orbit were a perfectly circular plane, the local civil time would be a measure of solar time. However, because the earth's motion contains irregularities, local solar time, LST, (defined by the apparent diurnal motion of the sun) is slightly different than local civil time, LCT. Astronomers have worked out the necessary corrections, and they are listed in Table 2.1, where corrections

TABLE 2.1   The equation of time (min : sec)

Day:	1	8	15	22
Month:				
January	−(3:16)	−(6:26)	−(9:12)	−(11:27)
February	−(13:34)	−(14:14)	−(14:15)	−(13:41)
March	−(12:36)	−(11:04)	−(9:14)	(−7:12)
April	−(4:11)	−(2:07)	−(0:15)	1:19
May	2:50	3:31	3:44	3:30
June	2:25	1:15	−(0:09)	−(1:40)
July	−(3:33)	−(4:48)	−(5:45)	−(6:19)
August	−(6:17)	−(5:40)	−(4:35)	−(3:04)
September	−(0:15)	2:03	4:29	6:58
October	10:02	12:11	13:59	15:20
November	16:20	16:16	15:29	14:02
December	11:14	8:26	5:13	1:47

are given in the form

$$LST = LCT + \text{Eq. of time} \tag{2.7}$$

Values for the (Eq. of time) can be obtained by interpolation in Table 2.1. For example, at a point of 110°W longitude on January 4, the table indicates that the local solar time is between 3 min 16 sec and 6 min 26 sec earlier than local civil time.

2.4 Location of the Sun Relative to a Horizontal Plane

At any point on the earth, at any time, a line may be drawn toward the center of the sun as shown in Figure 2.6. The direction of this line may be specified by spherical polar angles (θ_h, ϕ_h), where θ_h is the angle between the line to the sun and the vertical, and ϕ_h is the angle between the projection of the line to the sun in the horizontal plane and the line pointing north in the horizontal plane. The purpose of this section is to derive expressions for these angles which may be applied at any time to any point on the earth's surface. Two methods will be used. In the first method, only θ_h is derived, but the spatial visualization is relatively easy. In the second method, both θ_h and ϕ_h are obtained, but the spatial relations are more difficult to comprehend.

Figure 2.6 Angles locating the sun relative to a horizontal plane.

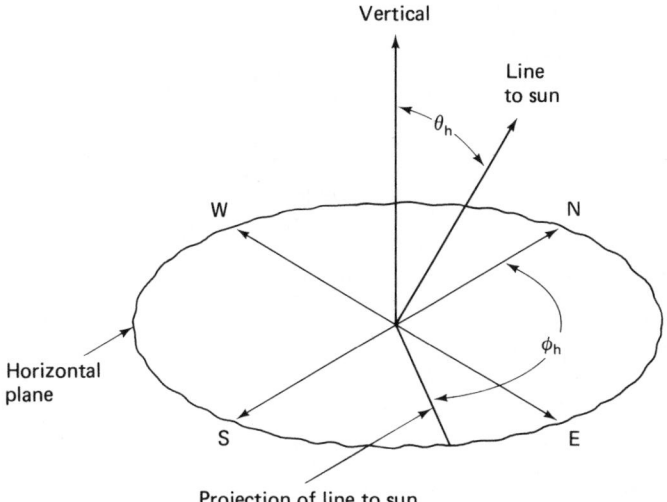

26

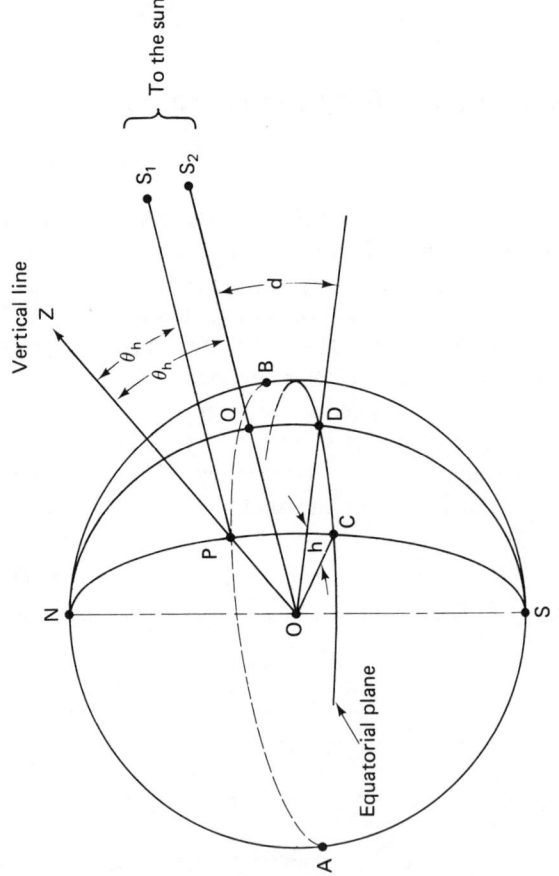

Figure 2.7 Diagram for calculation of θ_h.

We begin with a method for obtaining θ_h based on the diagram of Figure 2.7. For this diagram, we use the following definitions:

$\quad\quad O =$ center of the earth

$\quad\quad P =$ point on the earth where θ_h is to be determined

$\quad\quad N =$ north pole

$\quad\quad S =$ south pole

$\quad NPCS =$ longitudinal great circle containing P

$\quad OQS_2 =$ line from center of earth to sun

$\quad NQDS =$ longitudinal great circle containing Q

$\quad OCD =$ triangle in the equatorial plane

$\quad APQB =$ great circle containing P and Q

$\quad PS_1 =$ line from P to sun

$\quad OPZ =$ vertical line at P

It is desired to determine θ_h, the angle between a line to the sun and a vertical line at point P. A line from the center of the earth to the sun passes through point Q. The arc PQ is part of a great circle, and since angle POQ is equal to θ_h, arc PQ is also equal to θ_h. The hour angle h is defined as the angle of arc along the equator between the meridian containing point P and the meridian containing line OQ from the center of the earth to the sun. Thus, arc CD equals h, and angle COD equals h. The hour angle can be converted to time units at the rate of 1 hr per $15°$. Arc PC is equal to the latitude, by definition.

Consider the spherical triangle NPQ, which is made up of segments of great circles as shown in Figure 2.8(a). For any general spherical triangle made up of segments of great circles, as in Figure 2.8(b), it can be shown that the law of cosines is

$$\cos a = \cos b \cos c + \sin b \sin c \cos \alpha \tag{2.8}$$

Figure 2.8 Spherical triangles: (a) general triangle, (b) triangle from Figure 2.7.

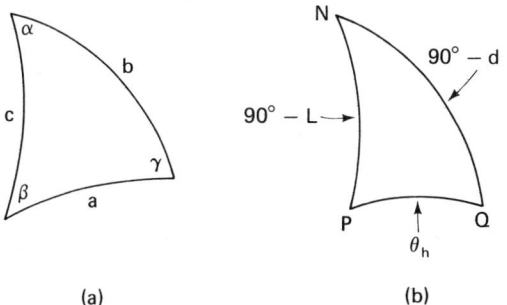

(a)   (b)

This may be identified with Figure 2.8(a) by relating $a \rightarrow \theta_h$, $b \rightarrow 90° - d$, $c \rightarrow 90° - L$, and $\alpha \rightarrow h$. Then we obtain

$$\cos \theta_h = \cos d \cos L \cos h + \sin L \sin d \qquad (2.9)$$

as the desired expression. For any point P, L is presumably known. The angle of declination can be estimated from Eq. (2.6), and the hour angle is given by

$$h = (H - 12)15° \qquad (2.10)$$

where H is local solar time. Thus θ_h is obtained for any point, at any time of day, for any day of the year.

In order to calculate the angle ϕ_h it is necessary to explicitly draw a horizontal plane at point P as shown in Figure 2.9. In this figure, the declination plane is the locus of points from the center of the earth to the sun (i.e., point Q in Figure 2.7) as the earth revolves over 1 day. It is drawn for a summer day. From Figure 2.1 it can be seen that the declination plane is parallel to the equatorial plane but is offset by an arc equal to the declination angle, above the equator in summer and below the equator in winter. A line is drawn from P toward the sun, and the projection of this line in the horizontal plane (tangent to the earth at P) is shown as a dashed line in Figure 2.9. In the horizontal plane, north-south and east-west lines are drawn, the north-south line being tangent to the longitudinal meridian passing through P. The angle between the north line and the projection of the sun line is defined as ϕ_h. Figure 2.9 may now be redrawn by (1) moving the horizontal

Figure 2.9 Horizontal and declination planes.

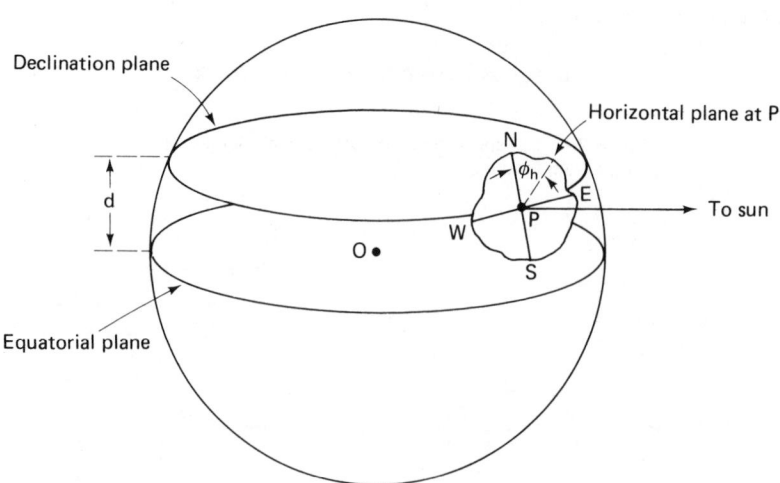

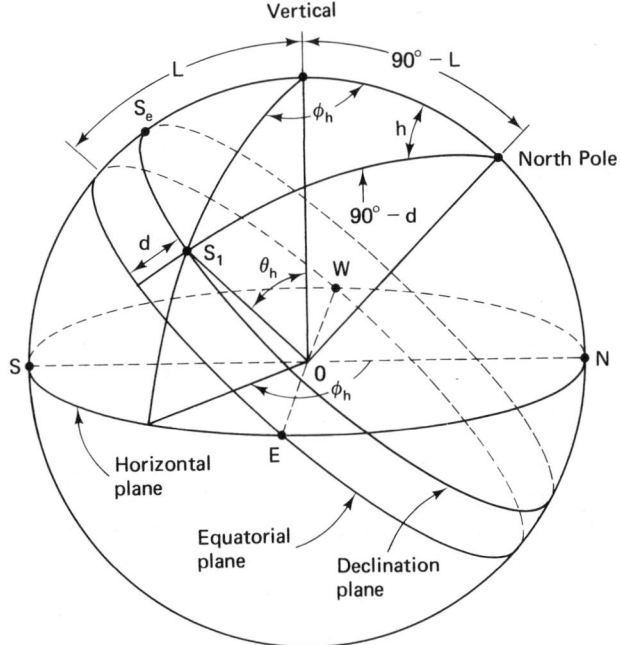

Figure 2.10 Diagram for obtaining θ_h and ϕ_h.

plane in a parallel direction so that it passes through the center of the earth O, and (2) tilting the figure so that the vertical line OP is pointed upward on the page. This is done in Figure 2.10. Line OS_1 is the line to the sun at any instant. Point S_1 lies on the intersection of the declination plane with the earth's surface, which is the locus of points of intersection of lines to the sun with the earth's surface. When, at some later time, the line to the sun reaches S_2, solar time is 12:00 noon at the point considered. Since it takes 24 hr for point S_1 to traverse $360°$ and return to its original position, spherical angle S_2-N-S_1 is the hour angle h. The angles θ_h, ϕ_h, d, and L are as given in Figure 2.10. Next, consider the spherical triangle in Figure 2.10 as shown in detail in Figure 2.11. Applying Eq. (2.8) to the spherical triangle, with α chosen as h, we find that

$$\cos \theta_h = \cos L \cos d \cos h + \sin L \sin d \qquad (2.11)$$

which is the same as Eq. (2.9). Next, if Eq. (2.8) is applied to a spherical triangle with α chosen as ϕ_h, we obtain

$$\cos \phi_h = \frac{\sin d - \sin L \cos \theta_h}{\cos L \sin \theta_h} \qquad (2.12)$$

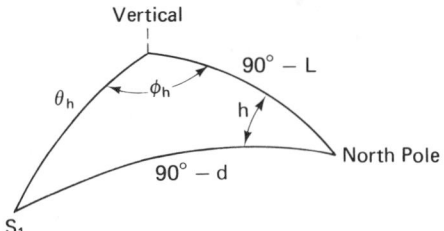

Figure 2.11 Spherical triangle from Figure 2.10.

which determines the angle ϕ_h. Equations (2.11) and (2.12) are the solutions to the problem of finding θ_h and ϕ_h in terms of d, L, and H.

A number of conclusions can be drawn from these results. The hour angle is symmetric about solar noon, and therefore θ_h is also symmetric about solar noon. Thus, for example, θ_h has the same value at solar time 14:00 as it does at 10:00. The angle ϕ_h can be shown to be symmetric in time about the south direction. Since ϕ_h in Eq. (2.12) is measured from the north direction, the angle $(\phi_h - 180°)$ is symmetric about solar noon. It can readily be shown that, at noon, $\theta_h = L - d$, and therefore $\cos \phi_h = -1$.

It is a simple matter to examine the angles at solar noon, since $h = 0$ and $\cos h = 1$. At solar noon,

$$\cos \theta_h = \cos (L - d) \qquad (2.13)$$

and therefore

$$\theta_h = |L - d| \qquad (2.14)$$

According to Eq. (2.14), the solar elevation at noon can be traced out for any latitude or any day of the year. At the equator, $L = 0$, and therefore θ_h varies from 23.45° on December 21, to 0° on March 21, to 23.45° on June 21, to 0° on September 21. At any other latitude, θ_h varies in a similar way, except that $|L - d|$ replaces d. Thus, at solar noon in the northern hemisphere at latitudes greater than 23.45°, the sun always lies in the southern sky, away from vertical. The elevation of the sun above the southern horizon at noon is $90° - L + d$. Since d is negative in winter and positive in summer, the sun lies 46.90° lower in the southern sky on December 21 than it does on June 21, at noon.

On March 21 and September 21, $d = 0$, and Eq. (2.11) reduces to

$$\cos \theta_h = \cos L \cos h \qquad (2.15)$$

On these days, the motion of the sun moves in an apparent plane containing the horizontal EW line and tilted up from the southern horizon at the angle $90° - L$.

The times at which sunrise and sunset occur are determined by the values of the hour angle when $\theta_h \rightarrow 90°$ (or $\cos \theta_h \rightarrow 0$). This occurs when

$$\cos h = - \tan L \tan d \qquad (2.16)$$

On March 21 and September 21, sunrise and sunset occur 6 hr from noon, and thus the day is 12 hr long on these days. At the equator, all days are 12 hr long. However, at any other latitude, the days are longer in summer than in winter, the gain in length of day in summer just equalling the loss in day length in winter. In Figure 2.12, the day lengths on December 21 and June 21 are plotted vs. (north) latitude. The results show that the disparity between summer and winter increases with latitude until, at 66.55°N. latitude, the day length on June 21 becomes 24 hr, and the day length on December 21 goes to zero. Further north of this latitude, there is a period of days in winter with no daylight and a period of days in summer when the sun does not set.

Figure 2.12 Length of the day as a function of latitude.

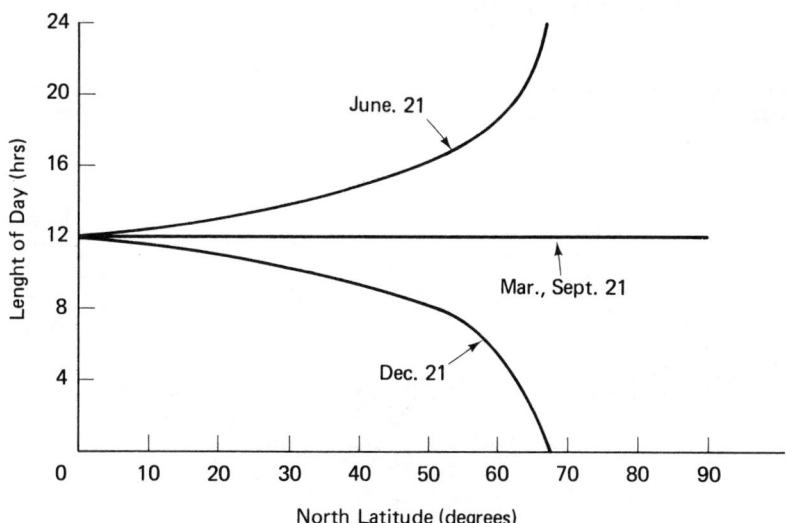

In general, the angle θ_h is determined mostly by the day of the year, and ϕ_h is mainly determined by the hour of the day. The apparent path of the sun across the sky at the equator is illustrated in Figure 2.13. On March 21 and September 21, the sun moves in a vertical plane through the horizontal EW line, with θ_h changing by 15° for each hour of change in time. On December 21, the plane is shifted 23.45° to the south.

A good way to present the motion of the sun across the sky is to plot θ_h and ϕ_h on circular polar plots as shown in Figure 2.14. In these plots, con-

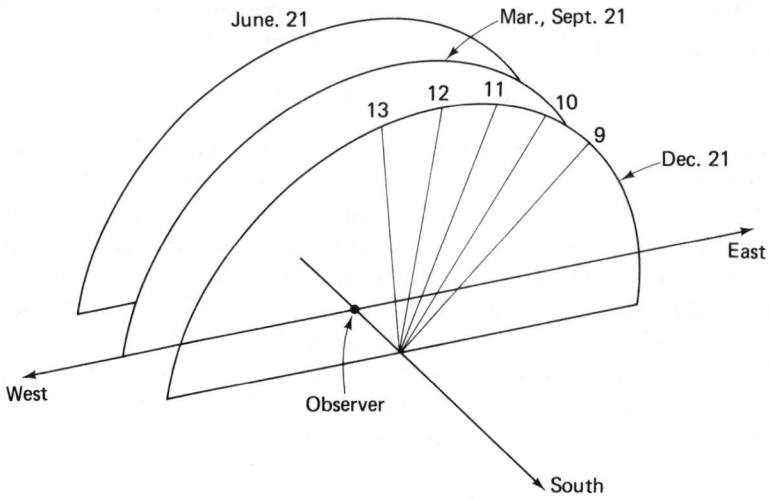

Figure 2.13 Position of the sun to an observer at the equator.

Figure 2.14 Variation of angles θ_h and ϕ_h with hour on the 21st day of each month at 32°N latitude.

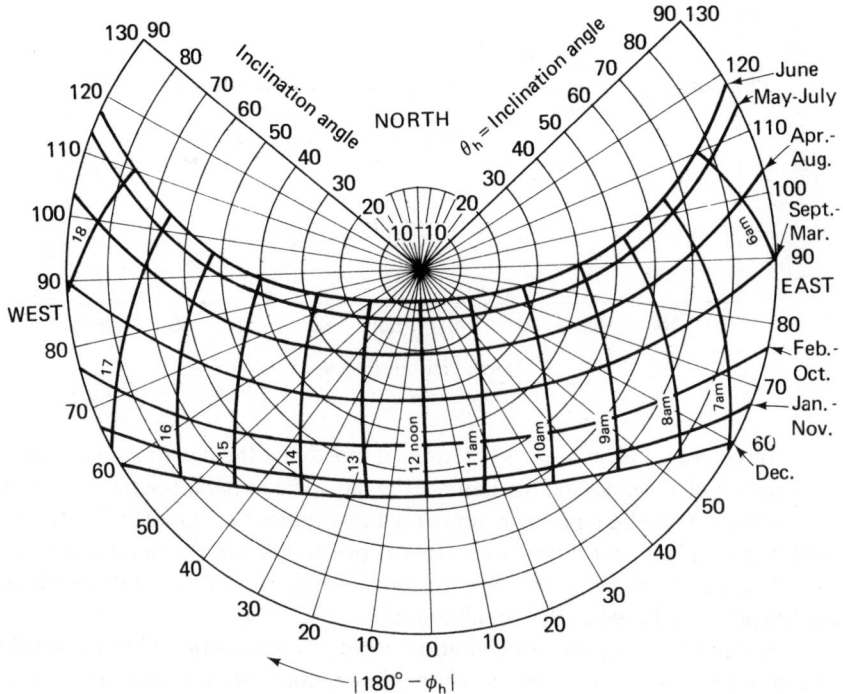

centric circles refer to changing values of θ_h, and rotation of a radius refers to changes in ϕ_h. The solar paths on the 21st day of each month are given by approximately horizontal curved lines, whereas lines of constant hour are drawn as approximately vertical curved lines. As one moves north from the equator to higher latitudes, the family of curves of apparent motion shifts south.

2.5 Effect of Tilting the Reference Plane

The sun is so far away that, for most practical purposes, it is adequate to assume that the sun's rays are parallel and independent of the position on the earth where they are received. Then any two surfaces tilted at the same angle relative to the line from the center of the sun to the center of the earth will have solar reference angles θ, ϕ which are the same. Consider the diagram shown in Figure 2.15. The surface A is horizontal at point P_1 at latitude $L_1°$.

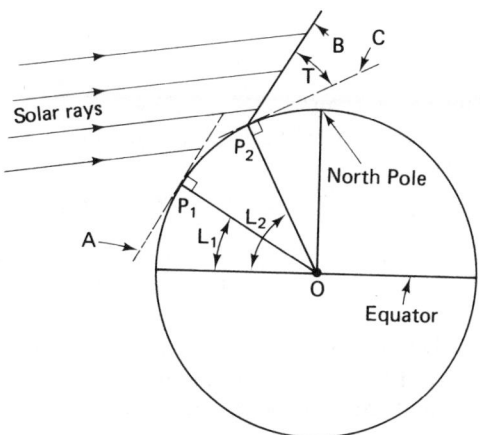

Figure 2.15 Equivalence of a tilted plane to a horizontal plane of lower latitude.

At a higher latitude, $L_2°$, the horizontal surface is C. If at point P_2 a tilted surface B is formed by tipping the horizontal plane up toward the south at tilt angle T, then when T is set equal to $L_2 - L_1$, the solar rays impinge on surface B at the same angles as they do on surface A. Thus, by tilting at the latitude angle, the surface becomes parallel to a horizontal surface at the equator. The only difference between a surface tilted at the latitude and a horizontal plane at the equator is the difference in path length of atmosphere traversed by the rays in the two cases. However, the solar angles are the same. Thus, the tilted surface shown is spatially equivalent to a horizontal surface at an effective latitude $L - T$. Therefore, if a set of reference angles are

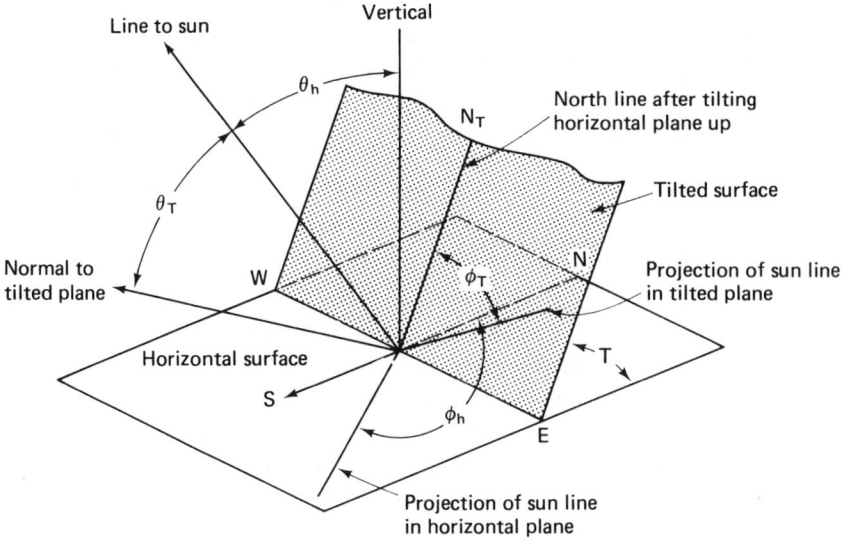

Figure 2.16 Effect of tilting a surface toward the south.

defined as in Figure 2.16, it follows that Eqs. (2.11) and (2.12) may be applied
to θ_T and ϕ_T (instead of θ_h and ϕ_h) but with L replaced by $(L - T)$. Thus,

$$\cos \theta_T = \cos (L - T) \cos d \cos h + \sin (L - T) \sin d \qquad (2.17)$$

$$\cos \phi_T = \frac{\sin d - \sin (L - T) \cos \theta_T}{\cos (L - T) \sin \theta_T} \qquad (2.18)$$

Tilted planes are extremely important for solar energy collection
because they provide surfaces which are much more normal to the sun than
horizontal surfaces. A flat plate solar collector tilted at the latitude receives
sunlight at the same angles as a horizontal surface at the equator. The effect
of tilting a surface on the angles θ_T and ϕ_T on January 21, at $L = 32°N$, is
shown in Figure 2.17. It can be seen that tilting at about 53° makes the sun
line essentially normal to the sun at midday.

A general tilted surface, which is not simply tilted up toward the south
with the east-west line remaining horizontal, may be treated in terms of the
diagram shown in Figure 2.18. The actual tilted plane is not shown, but
the normal to the tilted plane is given, and this uniquely determines the
attitude of the tilted plane. The normal is described by spherical polar angles
(θ_n, ϕ_n) relative to the horizontal plane. The line to the sun is described by
angles (θ_h, ϕ_h) relative to the horizontal plane. According to a theorem in
analytic geometry, the cosine of the angle θ_T between the sun line and the

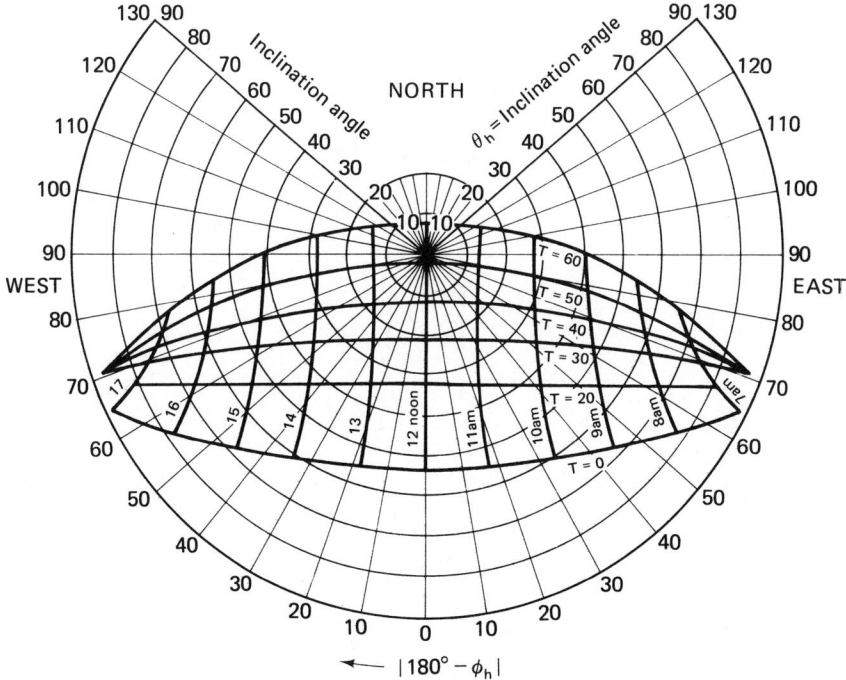

Figure 2.17 Variation of θ_h and ϕ_h with hour on January 21 at 32°N latitude for various tilt angles T.

Figure 2.18 Diagram for determination of the angle θ_t between the line to the sun and the normal to an arbitrarily tilted surface. (Tilted surface not shown).

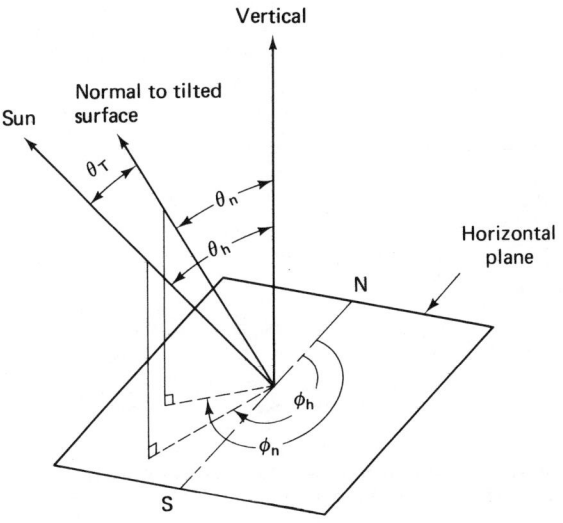

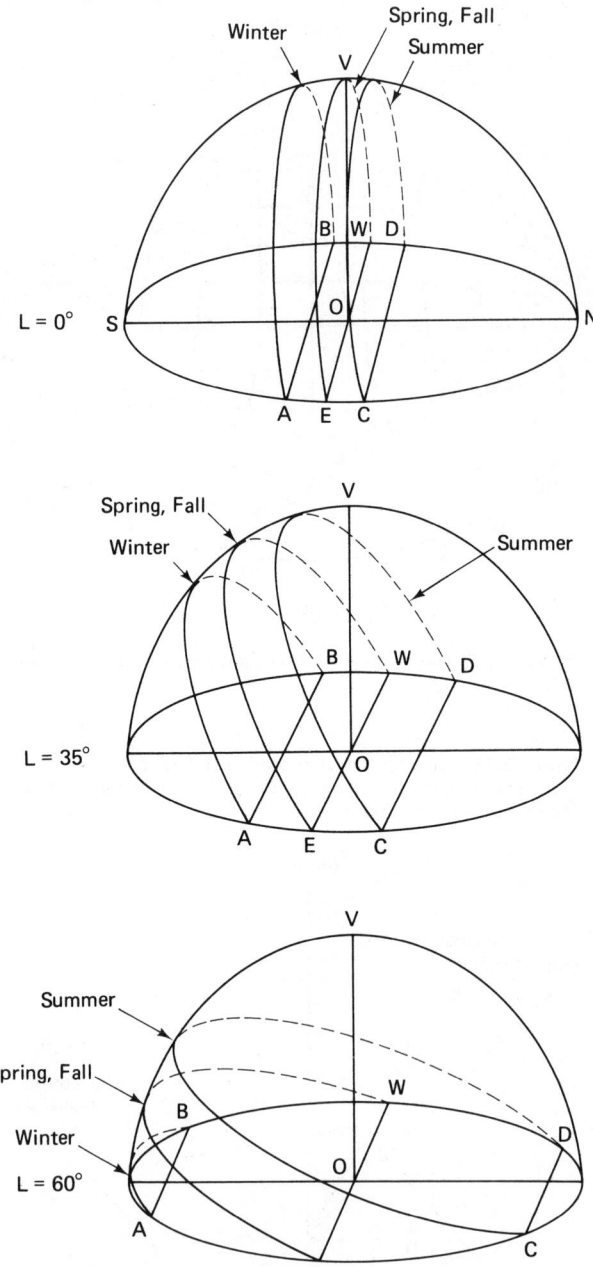

Figure 2.19 Apparent paths of the sun to an observer at point *O* for various latitudes and seasons.

normal to the tilted surface is given by

$$\cos \theta_T = \cos \theta_n \cos \theta_h + \sin \theta_n \sin \theta_h \cos (\phi_h - \phi_n) \qquad (2.19)$$

2.6 Apparent Trajectory of the Sun to an Observer on the Earth

The apparent path of the sun across the sky to an observer on the earth's equator is shown in Figure 2.13. Each day is 12 hr long at the equator, the only difference between seasons being the small change in tilt of the solar plane. An observer at the equator can tell what time it is, if he knows which direction is south, by simply noting the position of the sun east or west of the south line.

As the latitude of the observer is increased northerly, the solar planes tilt over to the south. One way of depicting the apparent solar motion is to draw a hemisphere above the observer and to trace the apparent solar path as a line on this hemisphere: this path would be observed by a person at the center of the hemisphere looking up toward the sky. In Figure 2.19, the curves are sketched for December 21, March 21 or September 21, and June 21 for three latitudes.

2.7 Radiation Flux Intercepted by a Tilted Plane

Consider a flux of solar rays as shown in Figure 2.20. The intensity of the solar rays may be specified by placing an absorbing surface perpendicular to the rays and measuring the heat received by the surface per unit area per unit time. Solar intensities are usually given in units of either Btu/ft²-hr,

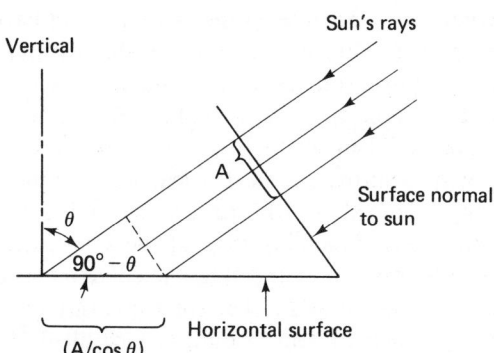

Figure 2.20 Diagram of planes normal and oblique to the sun. The same amount of solar energy falls on areas A and $A/\cos \theta$.

Langleys/hr (cal/cm²-hr), or W/m². Solar intensities measured at normal incidence give a proper indication of direct solar intensity. However, this requires a receiver plane which moves continuously to be perpendicular to the sun's rays. In many cases, this is not convenient, and solar intensities are monitored with a surface that remains horizontal at all times. Such a surface receives direct solar rays at an oblique angle as shown in Figure 2.20. If an area A of the surface normal to the sun receives a certain solar flux, this flux falls upon an area ($A/\cos \theta$) of the horizontal surface. Thus, the intensity recorded by the horizontal surface is $\cos \theta$ times the intensity recorded by the surface perpendicular to the sun. In general, the intensity recorded by a surface with a normal making an angle θ with the solar rays is $\cos \theta$ times the solar intensity recorded by a surface with its normal along the solar rays. In actual practice, some light reaches a horizontal surface after scattering of direct rays from the sun by clouds on the atmosphere, and these "diffuse" rays are not properly described by this formulation, since they impinge upon the horizontal surface at angles different from θ.

It is obvious that in any scheme to collect solar energy with planar collectors or planar mirrors, the planes should be as perpendicular to the solar rays as possible because of the $\cos \theta$ factor described above. At the equator, the ideal attitude of a fixed planar collector is horizontal, whereas at northerly latitudes it should be tilted toward the south. Most solar intensity monitoring stations use horizontal plane collectors which receive light from the entire sky. It is important to be able to estimate from these data what the intensity would be on a surface tilted up toward the south as in Figure 2.16. With this information, the designer of solar energy systems can orient the collectors at an optimum angle for collection of the greatest amount of solar energy when it is needed. To convert the intensities measured on a horizontal surface to those that would be received on a tilted surface, one should separate the light received on the horizontal surface into direct and diffuse components. The direct component is not difficult to correct for tilt angle with the cosine law given above. It is more difficult to correct the diffuse component, because cosine factors would have to be integrated over a range of incidence angles from all over the sky. Since the fraction of light received from diffuse sources is generally unknown anyway, a useful approximation is to assume that all the light is direct. If most of the diffuse light comes from a solid angle centered on the general region near the sun, this is probably a good approximation. In clear weather the fraction of light that is diffuse ranges from about 10 % to 30 % depending on latitude, time, and atmospheric conditions. Assuming that all the solar intensity is produced by direct co-parallel rays from the sun, the solar intensity on a surface normal to the sun is denoted as I_N. The solar intensity on a horizontal surface is denoted as I_h, and the intensity on a surface tilted $T°$ toward the south is I_T. Then, neglecting diffuse light, an approximation to the tilt correction factor based on only direct rays is

$$R = \frac{I_T}{I_h} = \frac{I_N \cos \theta_T}{I_N \cos \theta_h} = \frac{\cos \theta_T}{\cos \theta_h} \qquad (2.20)$$

where $\cos \theta_h$ is given in Eq. (2.11), and $\cos \theta_T$ is given in Eq. (2.17). The tilt correction factor is particularly easy to evaluate at solar noon. At this time, $\theta_h = L - d$, and $\theta_T = L - T - d$. The tilt correction factor at $12:00$ noon is illustrated in Figure 2.21 for various tilt angles at latitude 33°N. It can be seen that the correction factor increases in winter and decreases in summer for surfaces tilted toward the south. As the tilt is increased from zero to the latitude, there is a large gain in winter effectiveness at the cost of a

Figure 2.21 Tilt correction factors at 33°N latitude as a function of tilt angle.

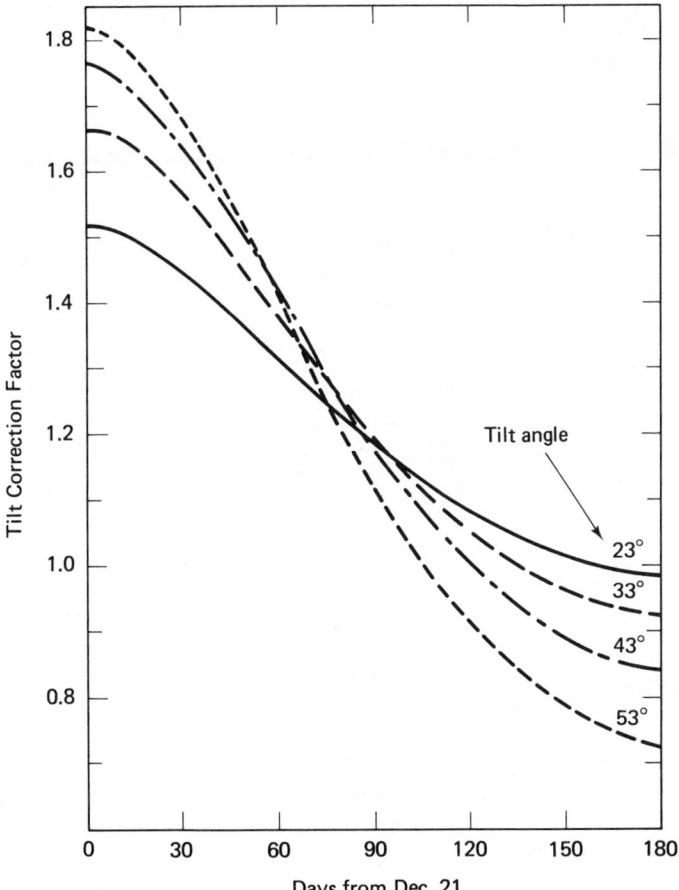

small decrease in summer effectiveness. However, as the tilt angle is increased past the latitude, the loss in summer begins to exceed the gain in winter. For systems that work mainly in winter, the optimum tilt will be greater than the latitude. However, for systems that function all year, the optimum tilt may be equal to or less than the latitude. The tilt correction factors for hours other than noon are shown for 33°N latitude in Figures 2.22, 2.23, and 2.24.

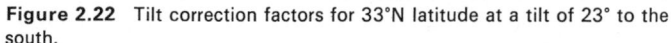

Figure 2.22 Tilt correction factors for 33°N latitude at a tilt of 23° to the south.

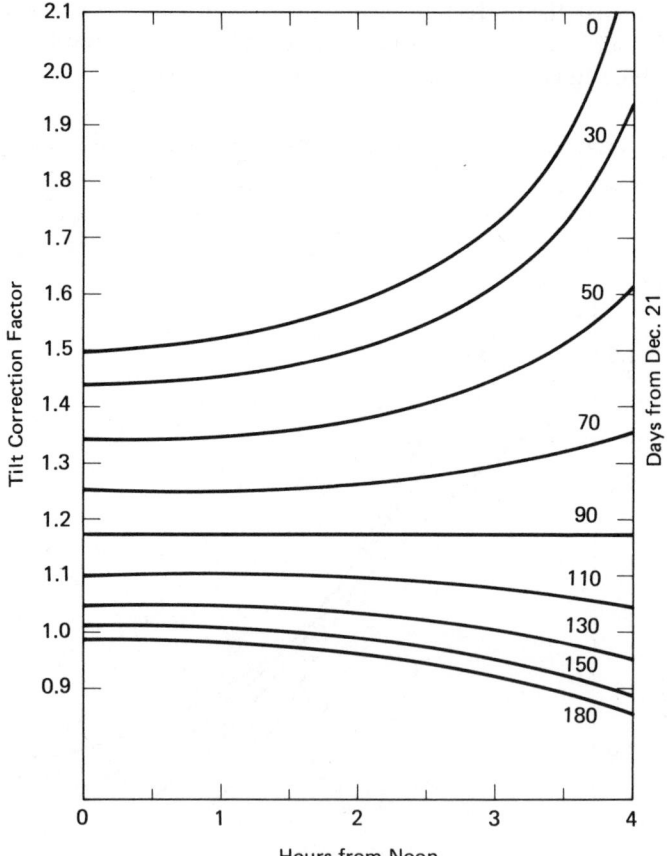

Since the elevation of the sun above the southern horizon monotonically decreases in winter as the time moves away from solar noon, the tilt correction factor must increase as the time moves away from noon. In the summer, just the opposite takes place, and the tilt correction factor is a maximum at noon.

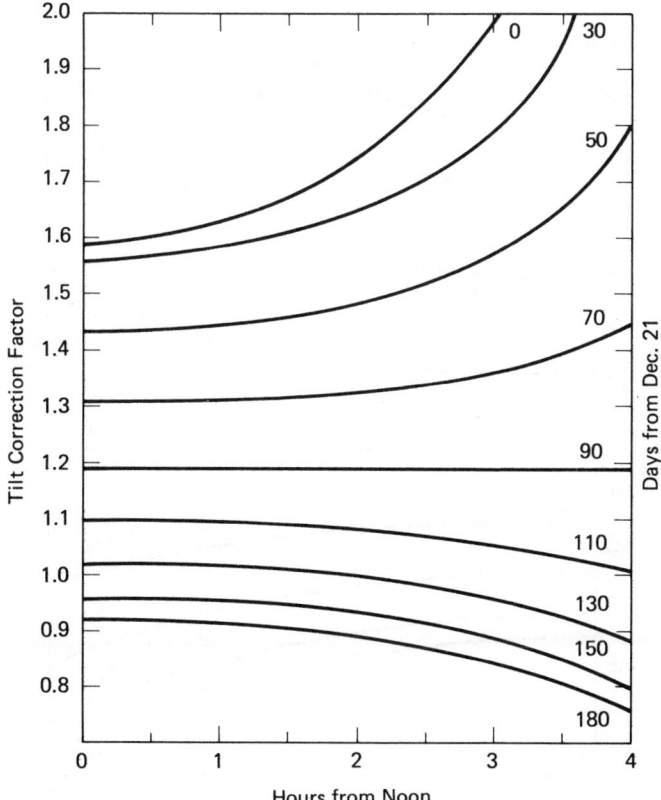

Figure 2.23 Tilt correction factors for 33°N latitude and a tilt angle of 33°.

The junctures between these regions occur on March 21 and September 21, when the tilt correction factor remains constant all day.

2.8 Angles Appropriate to Linear Arrays

In some solar energy applications, the geometric symmetry of the solar collectors is that of linear arrays in the east-west direction. For example, a parabolic trough reflector with a linear receiver suspended above the reflector would be of this type. Such collectors would track the movement of the sun only in elevation above the southern horizon but would not track the diurnal motion of the sun across the sky from east to west. At each time of day, the mirrors would be rotated about the receiver so that a plane containing the horizontal east-west line and the line to the sun would bisect the parabolic

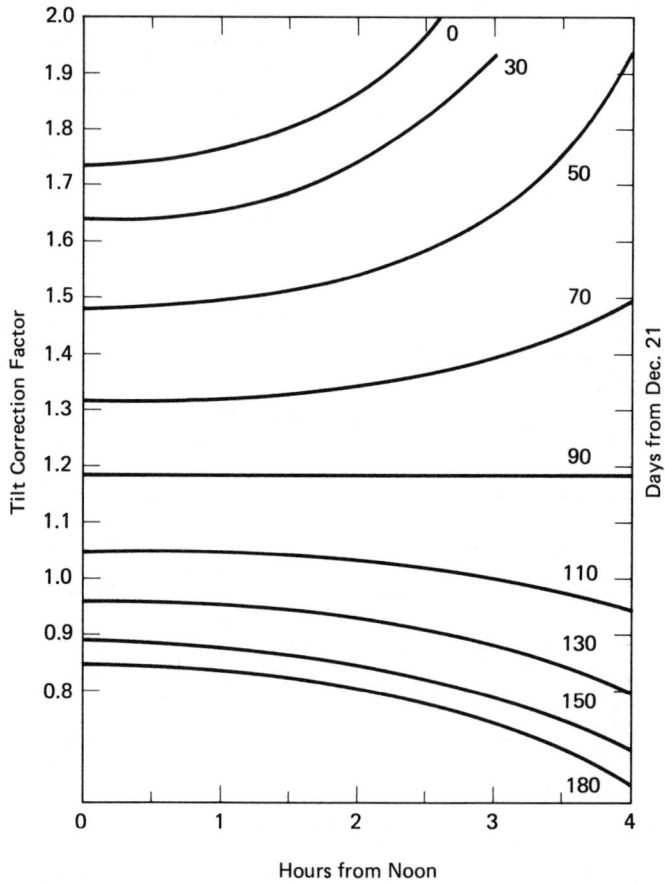

Figure 2.24 Tilt correction factors for 33°N latitude and 43° tilt angle.

Figure 2.25 Line focus of EW parabolic trough at off-noon hours.

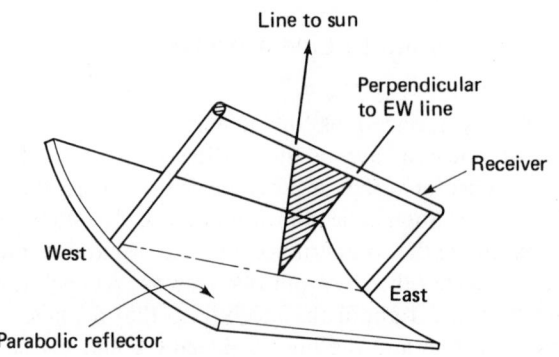

trough as shown in Figure 2.25. To determine the angle at which to orient
the trough, it is necessary to know the angle the above defined plane makes
with the horizontal plane. The angle between the two planes is denoted χ
and is illustrated in Figures 2.26 and 2.27. Point O is a point on the earth,

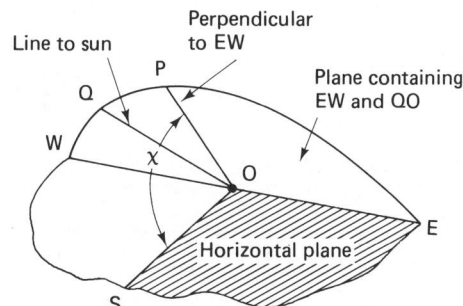

Figure 2.26 Definition of angle χ.

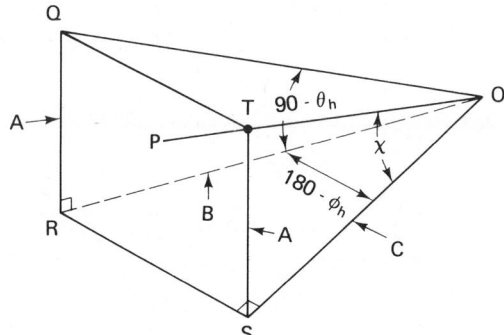

Figure 2.27 Diagram for calculation of angle χ.

and lines OE, OW, and OS are drawn in the horizontal plane in the east,
west, and south directions, respectively. Line OQ is the line to the sun, and
OP is the line in the tilted plane that would become OS if the tilted plane
were rotated down to horizontal. The angle χ between the planes is the angle
between OP and OS. In Figure 2.27, QR is dropped perpendicular to the
horizontal plane from OQ, and RS is drawn from R perpendicular to OS.
Point T is located on OP by drawing ST vertically from S to the intersection
with OP. The lengths of TS and QR are equal and are denoted as A. The
length OR is B, and the length OS is C.

Since ϕ_h is measured clockwise from the north direction (see Figure
2.6), angle SOR is equal to $(180° - \phi_h)$. Angle QOR is $(90° - \theta_h)$. From
triangles QOR and TOS,

$$\tan (90° - \theta_h) = \frac{A}{B}$$

$$\tan \chi = \frac{A}{C}$$

But $C = B \cos (180° - \phi_h) = -B \cos \phi_h$. Therefore,

$$\tan \chi = \frac{\tan (90° - \theta_h)}{(-\cos \phi_h)}$$

$$= \frac{-\cot \theta_h}{\cos \phi_h} \tag{2.21}$$

Figure 2.28   Dependence of χ on hour for the 21st day of each month at the equator. At any north latitude L, subtract L from the vertical scale.

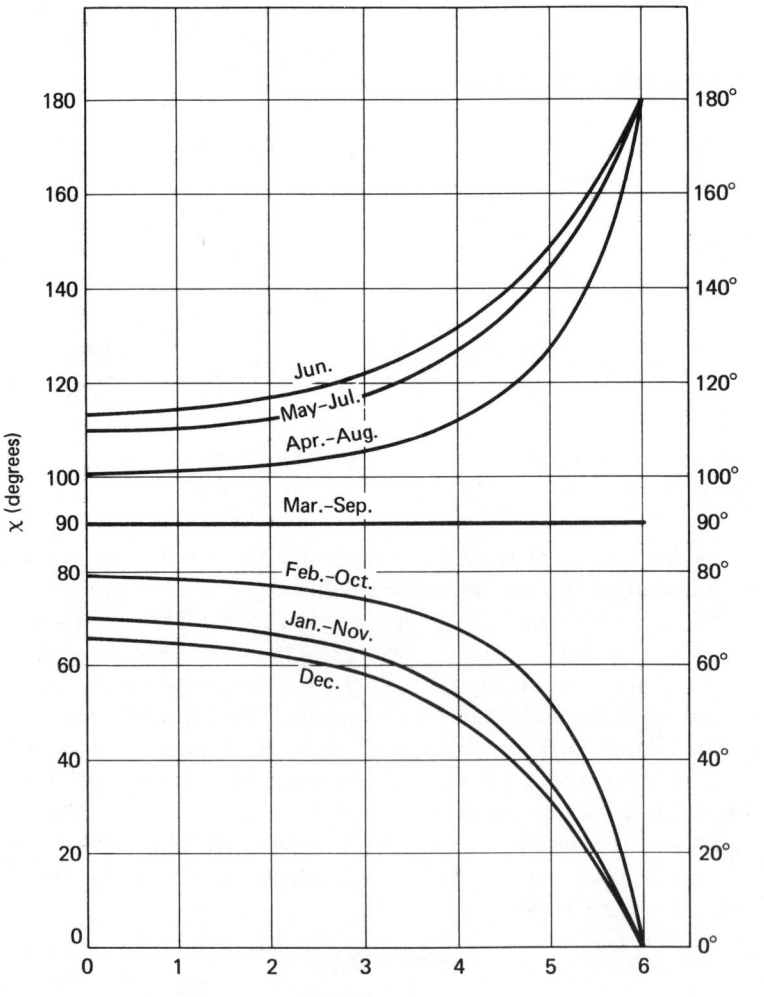

Thus, since θ_h and ϕ_h can be calculated at any hour of any day, χ can also be so calculated. At solar noon, points Q and T are the same, and $\chi = 90° - \theta_h$. A plot of χ vs. hour for the 21st day of each month is shown in Figure 2.28 for an observer at the equator. This same plot can be used for any observer at L degrees north latitude by subtracting L from the values of χ on the vertical scale. When the value of χ goes to zero, the sun sets in the southern sky, and when it goes to 180°, it sets in the northern sky. On March and September 21, when the sun sets in the east and west, χ is constant all day, and the day lasts for 6 hr on either side of noon.

2.9 Simple Passive Systems for Solar Transmission[1]

A passive system is sometimes defined as a system that does not collect heat in a transfer fluid for circulation but which generally admits sunlight directly to the region of a building for heating of the space occupied by users. Although a wide variety of passive systems can be imagined for various applications, the main consideration here is the proper design of south-facing windows with overhangs for winter transmission of solar rays directly into a residence. For the vast majority of locations in the United States, the latitudes range from 30°N to 45°N. As an example, consider a location at 38°N latitude. During the middle of winter, the sun rises in the southeast and sets in the southwest, reaching a maximum elevation above the southern horizon of about 30°. The shortest day is December 21, when the maximum elevation is 28.5°. In the summer time, however, the sun lies lowest in the southern sky at solar noon when it lies about 70° to 75° above the southern horizon. At times other than noon, the sun is tilted further toward the northern sky. Therefore, it is clear that a south-facing window can be arranged with a suitable overhang so that the low-lying winter sun is admitted but the high summer sun is shaded from reaching the window. This is illustrated in Figure 2.29.

Consider the diagram shown in Figure 2.30. The width of the overhang is R, the length of the top wall is T, and the window height is W. The angles for rays arriving at the top and bottom of the window are a and b, respectively. The problem is to select values for R, T, and W to guarantee full winter illumination, but with suitable cut-off dates for the overhang to shield the window between spring and fall. For most common home designs, $T + W = 8$ ft (2.44 m), which we shall adopt here. To determine the dates at which full, partial, and zero illumination of the window occur, it is generally sufficient to consider the solar angles at 12:00 noon. This is because,

[1]For a much more elaborate treatment of passive systems, consult E. Mazria, *The Passive Solar Energy Book* (expanded professional book with 13 appendices), Rodale Press, Emmaus, PA 1979.

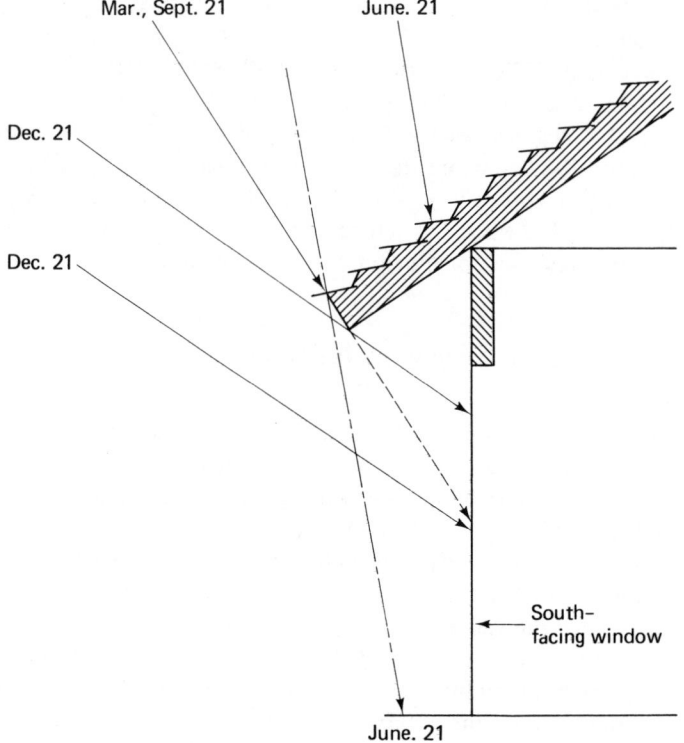

Figure 2.29   Diagram for passive systems.

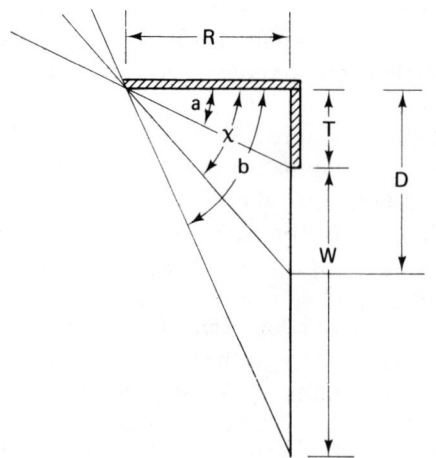

Figure 2.30   Angles for passive transmission.

46

between March 21 and September 21, the sun always lies further above the southern horizon at hours away from noon than at noon, and therefore an overhang that will shade the window at noon will shade it at hours other than noon. Conversely, between September 21 and March 21, the sun always lies lower in the southern sky at off-noon hours than it does at noon, and therefore a design that provides for full illumination at noon must provide for full illumination at off-noon hours during that half of the year.

At any hour of any day of the year, the line to the sun lies in a plane which makes an angle χ with the horizontal plane. The variation of χ with hour, day, and latitude was explained in Sec. 2.8. At solar noon, this angle reduces to $90° - L + d$, where L is the latitude and d is the declination. The angles a and b are defined by the equations

$$\tan a = \frac{T}{R} \tag{2.22}$$

$$\tan b = \frac{T + W}{R} \tag{2.23}$$

For complete illumination of the window, we require $\chi < a$, and for zero illumination, $\chi > b$. The days of the year when complete illumination first occur are determined by

$$\tan a = \frac{T}{R} = \tan \chi$$

and the days when zero illumination first occur are determined by

$$\tan b = \frac{T + W}{R} = \tan \chi$$

The fraction of the window illuminated is defined as

$$f = \frac{W - (D - T)}{W} \tag{2.24}$$

It follows that

$$f = \frac{\tan b - \tan \chi}{\tan b - \tan a} \tag{2.25}$$

The above equations will now be applied to an example where the latitude is 30°N, and it is desired to have complete illumination from October 21 through February 21 and no illumination from April 21 to August 21. By calculating the angle of declination on February 21 and on April 21, using

Eq. (2.6), it is found that

$$\text{On } 2/21: \quad \tan \chi = 1.106 = \frac{T}{R}$$

$$\text{On } 4/21: \quad \tan \chi = 2.830 = \frac{T + W}{R}$$

From these results, it follows that

$$T = 1.106R$$

$$W = 1.724R$$

Since $T + W = 8$ ft (2.44 m), one may then show that $R = 2.83$ ft (0.862 m), $T = 3.13$ ft (0.954 m), and $W = 4.87$ ft (1.48 m). With this design, the fractional illumination is as shown in Figure 2.31 as Case I. If the calculations are now repeated for full illumination between November 10 and January 30 and zero illumination between April 1 and September 11, the results are as shown in Figure 2.31 for Case II. The design for Case II has $T = 3.58$ ft

Figure 2.31 Variation of transmission factor f with season at 30° latitude.

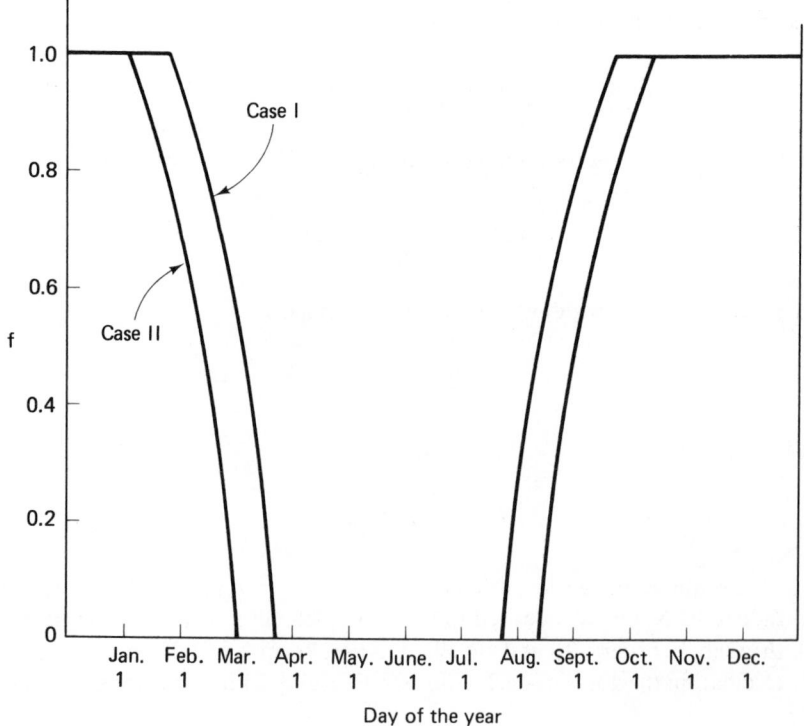

(1.09 m), $R = 4.01$ ft (1.22 m), and $W = 4.42$ ft (1.35 m). In any climate, one can adjust the design to provide illumination only when it is needed. However, the illumination characteristics are symmetric in the solar year, and this does not always coincide with climatic variations. For example, in Case I above, it probably would have been desirable to extend the period of no illumination further into the autumn than August 21 by increasing the overhang, but then one would lose illumination in the spring. The window and overhang measurements should be adjusted to coincide as much as possible with local climatic conditions.

Worked Examples

1. What is the day length on December 21, March 21, and June 21 at 35°N latitude?

    The angle of declination varies from $-23.45°$ to $0°$ to $+23.45°$ on these three days. The day lengths are found from Eq. (2.16). The results are

    December 21:

    $$\cos h = -\tan 35° \tan (-23.45°)$$
    $$\cos h = +0.700 \times 0.434 = 0.304$$
    $$h = 72.3° = \frac{72.3}{15} = 4.82 \text{ hr}$$
    $$\text{day length} = 9.64 \text{ hr}$$

    March 21:

    $$\cos h = 0, \qquad h = 90° = 6 \text{ hr}$$
    $$\text{day length} = 12 \text{ hr}$$

    June 21:

    $$\cos h = -0.304, \qquad h = 107.7°$$
    $$\text{day length} = 14.4 \text{ hr}$$

2. What angles do the solar rays make with a horizontal surface and a surface tilted at 45° to the south on March 21 at 3 pm solar time at 35°N latitude?

    Equations (2.11) and (2.17) are appropriate here.

    $$\cos \theta_h = \cos L \cos d \cos h + \sin L \sin d$$
    $$\cos \theta_h = \cos 35° \cos 0° \cos 45° + \sin 35° \sin 0°$$
    $$\cos \theta_h = 0.819 \times 0.707 = 0.579$$
    $$\theta_h = 54.6°$$

$$\cos \theta_t = \cos (-10°) \cos 45° = 0.696$$
$$\theta_t = 45.9°$$

3. What is the solar elevation χ at 3 pm on March 21 at 35°N latitude?

According to Eq. (2.21),

$$\tan \chi = \frac{-\cot \theta_h}{\cos \phi_h}$$

From Eq. (2.12),

$$\cos \phi_h = \frac{\sin d - \sin L \cos \theta_h}{\cos L \sin \theta_h}$$

$$\cos \phi_h = \frac{0 - 0.574 \times 0.576}{0.819 \times 0.815} = -0.495$$

$$\phi_h = 119.7° \text{ measured clockwise from north}$$

$$\tan \chi = \frac{-0.711}{-0.495} = 1.436$$

$$\chi = 55.1°$$

4. At what local time does solar noon occur at 123°W longitude on February 15?

Local solar time = local civil time + eq. of time

Local civil time = local standard time $- \frac{3}{15}$ hr

$12:00:00 =$ (local standard time $- 0:12:00) + (-0:14:15)$

Local standard time $= 12:00:00 + 0:12:00 + 0:14:15$

$$= 12:26:15$$

5. What angles are made by direct solar rays with a plate tilted toward the south at the latitude angle at solar noon on December 21, March 21, and June 21?

The plate is parallel to a horizontal plane at the equator, and therefore at solar noon the angles are 23.45°, 0°, and 23.45°.

6. A south-facing window at 40°N latitude has $W = 1.43$ m and $T = 1.00$ m as in Figure 2.30. What overhang R is required to make March 21 the last day of full illumination in the spring?

For full illumination,

$$\tan \chi = \frac{T}{R} = \frac{1.00}{R}$$

On March 21, $d = 0$, and $\chi = 90° - L = 50°$ (all day).

$$\tan \chi = 1.19$$
$$R = 0.839 \text{ m}$$

Problems

2.1. What are the times of sunrise and sunset at 45°N latitude on December 21?

2.2. If a surface is tilted toward the south at 55° at a site with latitude = 45°N on December 21, what is θ_t at 0, 1, 2, 3, and 4 hr from solar noon?

2.3. What is the solar elevation (angle between the horizontal plane and the plane formed by a horizontal east-west line and a line to the sun) at each of the times in Problem 2.2?

2.4. A stranded sailor on a deserted island keeps a calendar and knows that it is December 21. At midday, he observes that the highest elevation of the sun above the northern horizon is 35°. What is his latitude?

2.5. Consider a south-facing window as shown in Figure 2.30, with $W = 5$ ft (1.52 m) and $T = 3$ ft (0.914 m) at a latitude of 35°N. How long should the overhang be to allow 50% transmission at solar noon, 60 days on either side of June 21?

2.6. Calculate the number of hours between sunrise and sunset on December 21, March 21, and June 21 at latitudes 30°N, 40°N, and 50°N.

2.7. At what local time does solar noon occur at a location at 125°W longitude on December 1?

2.8. What is the angle that solar rays make with a surface at latitude 35°N on March 21:
(a) At 8 am with an east-facing wall?
(b) At 4 pm with a west-facing wall?

2.9. What is the solar elevation (χ) at a latitude of 40°N at noon, 2 hr from noon, and 4 hr from noon on December 21?

2.10. A building is located south of a flat plate collector array. The building is much wider than the collectors and is 60 ft (18.3 m) high. If the sides of the building are in the east-west and north-south directions and the collector is at ground level, what is the minimum distance required from the front of the collectors to the north face of the building to assure no shading of the collectors on December 21, 3 hr from noon, at the latitude where you live?

2.11. Make a diagram of the illuminated area of the floor of a house, produced by a south-facing window, at 0, 1, 2, 3, and 4 hr from noon on December 21, at your latitude. The window is 1 ft × 1 ft and the lower edge is 3 ft above the floor.

3 | Solar Intensities

This chapter concerns the variation in levels of solar intensity, both above the earth's atmosphere and at ground level. The methods for measuring solar intensity are described, and typical results are presented for the various components of solar intensity. The national network for monitoring solar intensity is described. It is shown that substantial uncertainties exist in the levels of solar intensities at many localities in the United States. The new network for improved measurements is described. Methods are given for estimating solar intensities when measurements are not available.

3.1 Solar Spectrum and Intensities above the Atmosphere

The sun emits energy in the form of electromagnetic radiation distributed over a wide range of wavelengths. Wavelengths shorter than about 0.4 μm (micrometer) are described as ultraviolet, whereas wavelengths longer than about 0.7 μm are infrared. The visible region lies in between the ultraviolet and infrared. Rocket- and satellite-borne instruments have measured the solar spectrum above the earth's atmosphere; the results are shown in Figure 3.1. In this figure, the observed irradiance is shown to compare closely with the expected irradiance from a perfectly black body at 5489°C (5762°K), especially at longer wavelengths. The area under the curve in Figure 3.1

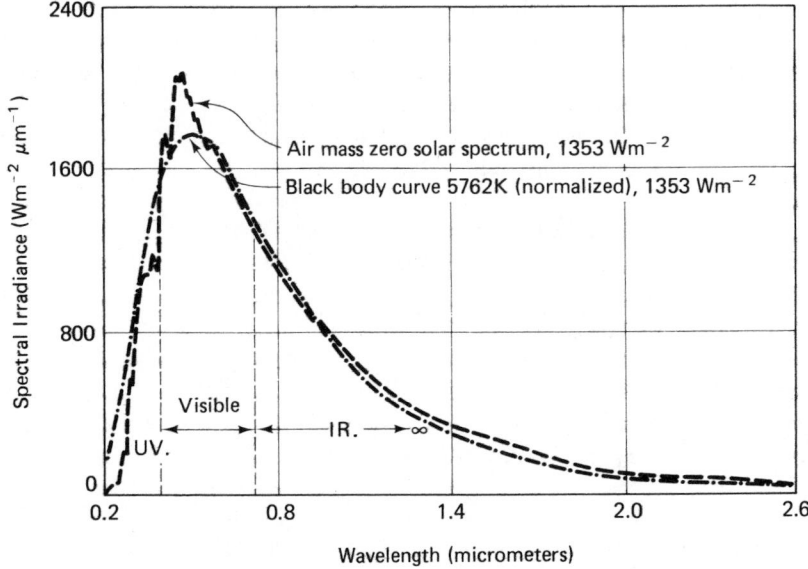

Figure 3.1 Spectral distribution of sunlight compared with a black body at 5762°K. By permission of the author and the American Section of the International Solar Energy Society, Inc. These curves are based on a solar constant of 1353 W/m², whereas recent measurements indicate a better value is 1377 W/m². A. P. Thomas and M. P. Thekaekara, "Experimental and Theoretical Studies on Solar Energy for Energy Conversion," *Proc. 1976 Annual Mtg. of the American Section of the Int'l Solar Energy Soc.,* Winnipeg, Canada, 1976.

represents the solar power per unit area falling on a surface normal to the line to the sun above the earth's atmosphere. This intensity varies with the day of the year due to variations in the earth-sun distance as shown in Figure 3.2. The "solar constant" is defined as the extraterrestrial solar intensity at the mean distance of the earth from the sun. The accepted value for the solar

Figure 3.2 Ratio of extraterrestrial radiation at normal incidence to the solar constant.

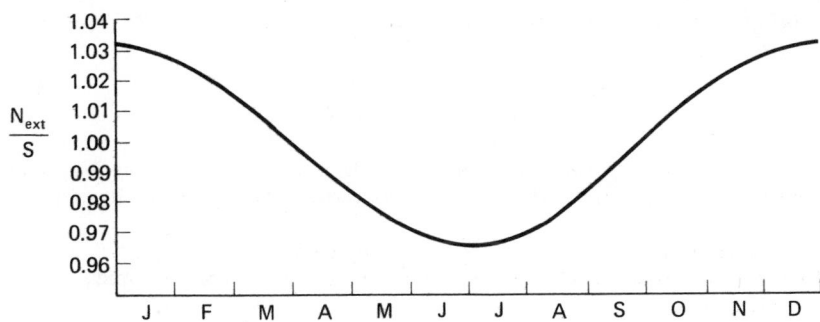

constant was 1353 W/m² for many years. Recent measurements indicate, however, that the best value[1] appears to be 1377 W/m². This value is now generally accepted. The distance between the earth and the sun is at a minimum around January 2 and at a maximum around July 2. The estimated extraterrestrial solar intensities on a surface normal to the sun on the 21st day of each month are shown in Table 3.1.

TABLE 3.1 Extraterrestrial solar intensities normal to the sun on the 21st day of each month

Month	Watts/meter²	Btu/hr-ft²
January	1422	450.9
February	1401	446.7
March	1387	439.8
April	1364	432.3
May	1344	426.0
June	1333	422.8
July	1334	423.0
August	1345	426.6
September	1367	433.4
October	1390	440.6
November	1411	447.4
December	1423	451.2

An analytical formula can be derived for estimating the extraterrestrial irradiance normal to the sun on any day.[2] The solar constant S is defined as the intensity normal to the sun when the earth is at the mean distance $\bar{r}$ from the sun. Thus, the extraterrestrial irradiance on any arbitrary day is

$$N_{\text{ext}} = S \left(\frac{\bar{r}}{r} \right)^2 \tag{3.1}$$

since the intensity varies inversely with the square of the distance. The earth's orbit around the sun can be approximated by an ellipse with the equation

$$r = \frac{A(1 - \epsilon^2)}{1 + \epsilon \cos \phi} \tag{3.2}$$

[1]C. Frohlich, "Contemporary Measures of the Solar Constant," in *The Solar Output and Its Variation*, ed. O. R. White, Colorado Associated University Press, Boulder, 1977, pp. 93–109.

[2]N. B. Guttmann and J. D. Matthews, Appendix II of SOLMET, Vol. 2, *Hourly Solar Radiation Rept. TD-9724*, National Oceanographic and Atmospheric Administration, Asheville, NC.

where r is the distance from the sun, ϵ is the eccentricity, A is half the length of the major axis, and ϕ is the angular displacement from the major axis. The mean distance is found by integrating r over the entire ellipse. The result is

$$\bar{r} = \frac{1}{2\pi} \int_0^{2\pi} r(\phi)d\phi = A(1 - \epsilon^2)^{1/2} \tag{3.3}$$

Thus, the instantaneous value of N_{ext} is

$$N_{ext} = \frac{S(1 + \epsilon \cos \phi)^2}{(1 - \epsilon^2)} \tag{3.4}$$

The eccentricity of the earth's orbit is 0.01672, and $\phi = 0$ on January 2. For any day of the calendar year,

$$\phi = (D - 2)\frac{360°}{365.2} \tag{3.5}$$

From Eqs (3.4) and (3.5), N_{ext} can be calculated. The extraterrestrial irradiance on a surface parallel to a horizontal surface at any locality is given by

$$I_{ext} = N_{ext} \cos \theta \tag{3.6}$$

where θ is the angle between the solar rays and the normal to the horizontal surface. An expression for $\cos \theta$ is given in Chapter 2. The hourly variation of I_{ext} on several days of the year for $L = 30°$ and $L = 40°$ is shown in Figure 3.3.

The solar intensity received at ground level is the residual intensity after absorption and scattering processes in the atmosphere. It is common to refer to the spectrum received above the atmosphere as *air mass zero* and the spectrum received at ground level when the sun is directly overhead and the atmosphere is clear as *air mass one*. As the zenith angle of the sun dips toward the horizon the path length of rays through the atmosphere increases. When the rays have to pass through a path of atmosphere equivalent to 2, 3, 4, . . . clear atmospheres, the spectra received at ground level are referred to as "air mass 2, 3, 4, ..." In Figure 3.4, the estimated ground level irradiance is given as a function of air mass. The variation of air mass with time of day for various days of the year is illustrated in Figure 3.5 for 40°N latitude.

The effective path length through the atmosphere at any time is given by

$$l = h \sec \theta_h \tag{3.7}$$

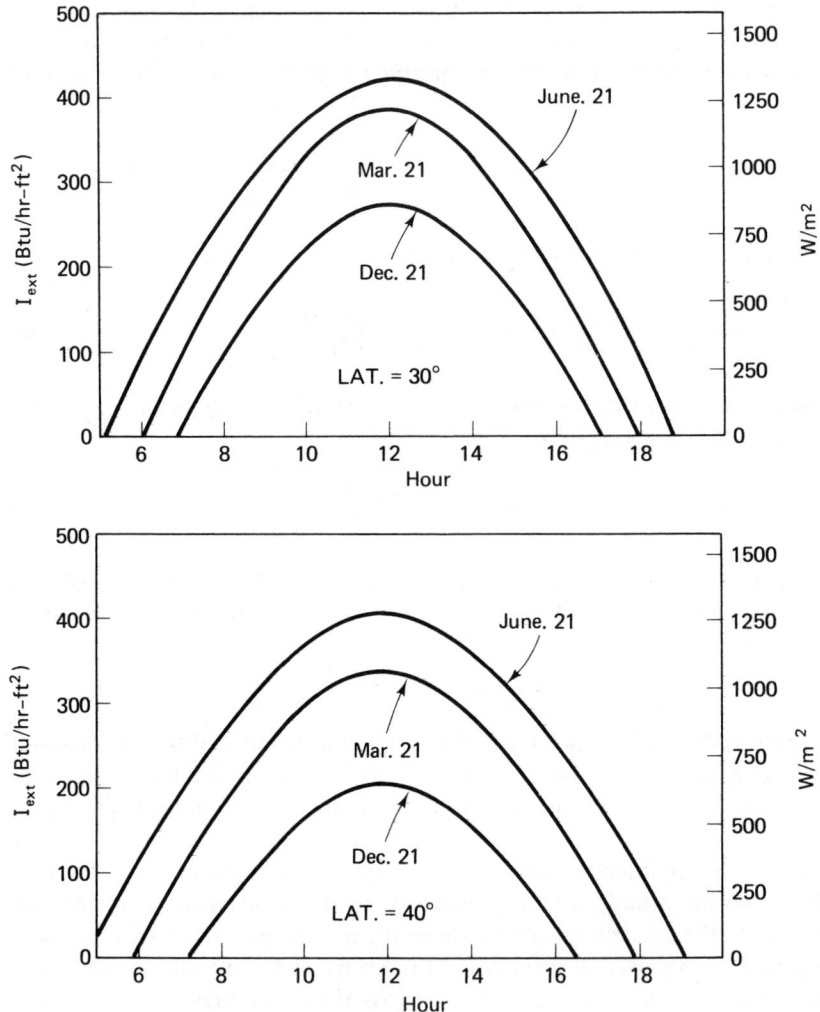

Figure 3.3 Extraterrestrial radiation on a horizontal plane.

where θ_h is defined in Chapter 2, and h is the effective vertical distance through the atmosphere at any locality. Although h will vary somewhat with location, it will be taken here as simply one "air mass." The fraction of light absorbed or scattered by the atmosphere at any wavelength λ over any path length l is exp $[-\beta(\lambda)l]$, where $\beta(\lambda)$ is the extinction coefficient at λ. The transmission of sunlight through the atmosphere is an average of this function over the irradiance as a function of wavelength. This can be approximated by use of an average extinction coefficient $\bar{\beta}$, such that the transmission is exp $[-\bar{\beta}l]$. Typical values for exp $[-\bar{\beta}l]$, when l is of the order of one air

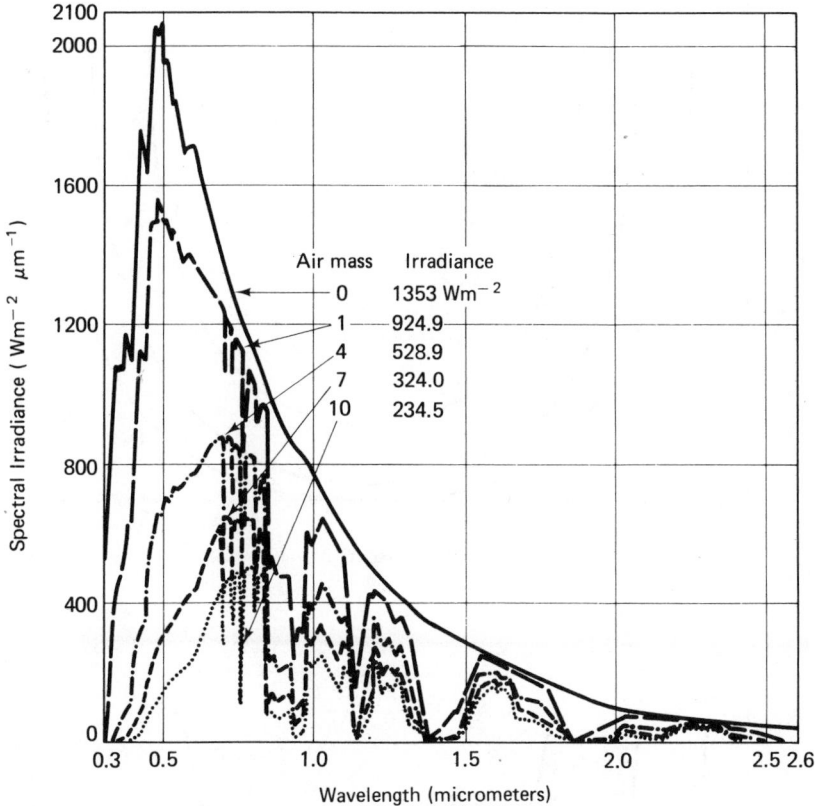

Figure 3.4 Spectral irradiance as a function of air mass. By permission of the author and the American Section of the International Solar Energy Society, Inc. A. P. Thomas and M. P. Thekaekara, "Experimental and Theoretical Studies on Solar Energy for Energy Conversion," *Proc. 1976 Annual Mtg. of the American Section of the Int'l Solar Energy Soc.,* Winnipeg, Canada, 1976.

mass, range around 0.65. Thus, a reasonable approximation for a typical β is determined by

$$\exp\left[-\bar{\beta}\right] = 0.65$$
$$\bar{\beta} = 0.431$$

3.2 Instrumentation for Measuring Solar Intensities

The solar intensities received at ground level can be measured in many ways. The most common measurement is the total radiant flux of sunlight falling on a horizontal absorbing surface, including light from the entire hemisphere above the surface. This quantity is commonly referred to as *insolation, total*

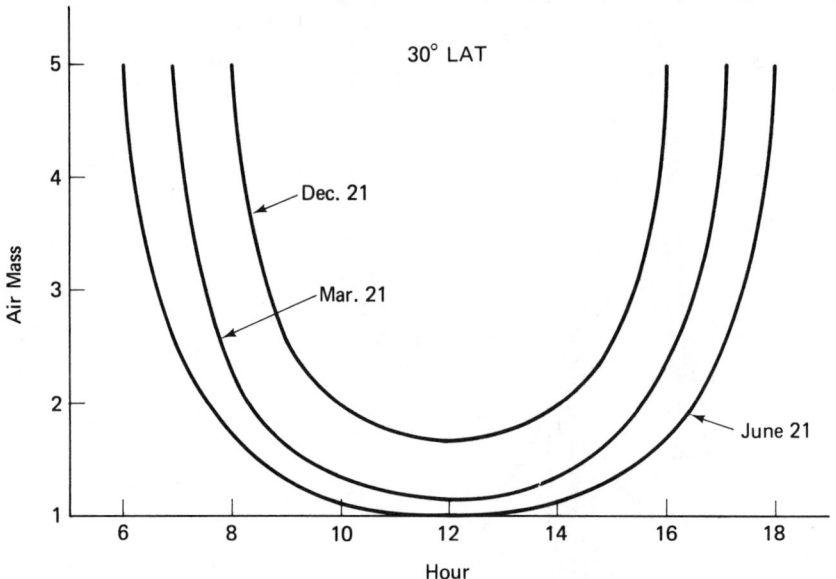

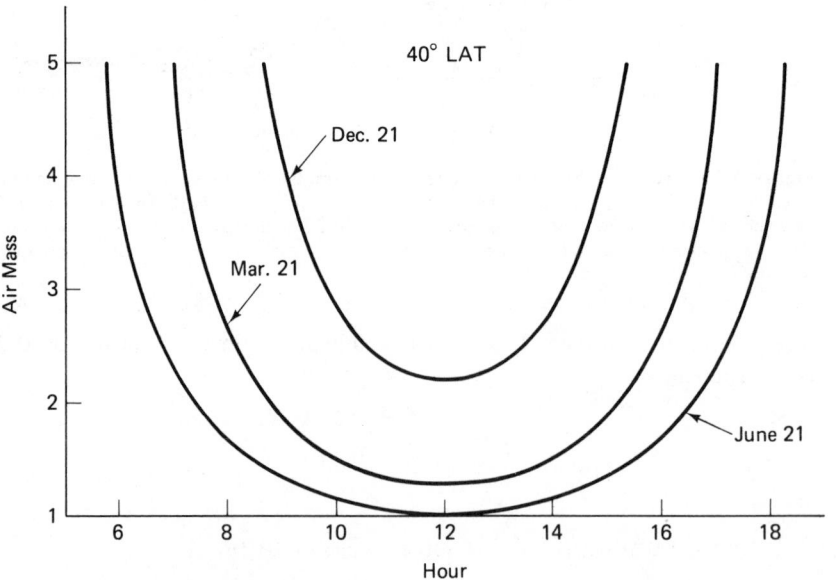

Figure 3.5 Air mass as a function of hour at 30° and 40° latitudes.

insolation, or *global insolation.* Similar measurements can be made on a tilted surface, usually tilted up toward the south. The total insolation received on a horizontal surface is denoted I_h, and that on a surface tilted up to T degrees to the south is I_T. The units most often used are

$$1 \text{ Langley/hr} = 1 \text{ cal/cm}^2\text{-hr} = 3.687 \text{ Btu/ft}^2\text{-hr} = 11.63 \text{ W/m}^2$$

Another measure of solar intensity is the rate at which heat energy is collected by a black surface normal to the sun's rays, with a relatively small entrance aperture (usually 5.8°). This is called the *direct normal* intensity or *normal incidence solar intensity.* It is given the symbol N. The direct normal intensity represents, mainly, direct rays from the sun, although there are scattered rays in a narrow range of angles within the entrance aperture. Assuming that these scattered rays can be neglected and that the solar rays can be treated as parallel, N represents the attenuated direct solar intensity after passing through the atmosphere.

The most common instrument for measuring global insolation is a device referred to as a *pyranometer, global pyranometer,* or *total pyranometer.* In essence, it consists of a black metal surface placed horizontally under a single or double glass hemisphere, with a set of thermocouples with the hot junctions attached to the underside of the black surface. The cold junctions of the thermocouples are either connected to a white surface adjacent to the black surface or to an electronic cold junction inside the instrument. Electronic circuits can be used to compensate for the variation in instrument sensitivity with ambient temperature. The basic concept is that the black surface will absorb most of the incident sunlight and will reach an equilibrium temperature higher than ambient or a white (reflective) surface reference, and the temperature elevation can be related to the rate of heat gain from the sun. Since the black surface is covered by a hemisphere of glass, it measures the total global solar input, including light scattered by the atmosphere, clouds, the ground, and nearby structures. A schematic illustration of such a device is shown in Figure 3.6, and an actual instrument is illustrated in Figure 3.7. In actuality, a thermopile is constructed by placing many thermocouples in series to increase the sensitivity of the device. The output voltage from this device can be calibrated against standard light sources to produce a scale of light intensity vs. output voltage. In most applications, the instrument is mounted horizontally and measures the total global solar input of both direct and diffuse rays from the sky. In a few instances, instruments have been mounted in a position tilted toward the south to simulate a tilted flat plate collector.

A second instrument used for monitoring solar intensity is the *normal incidence pyrheliometer,* or simply *pyrheliometer,* for measuring solar intensities normal to the sun. The principle of operation is the same as for the

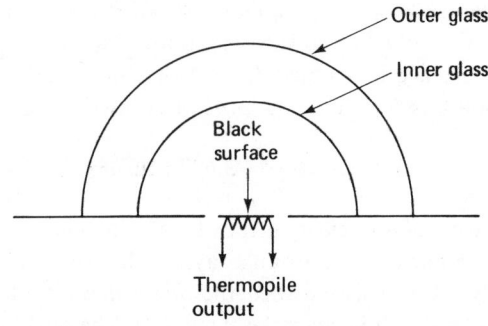

Figure 3.6 Construction of a pyranometer.

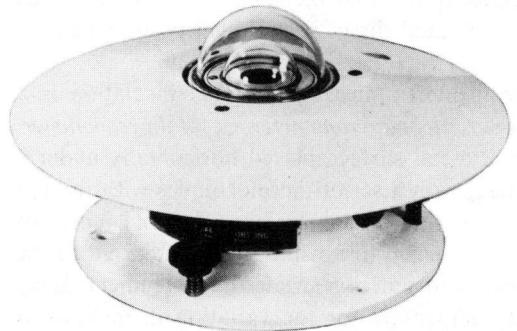

Figure 3.7 The Eppley PSP pyranometer. By courtesy of the Eppley Laboratory, Inc.

global pyranometer, except that the black surface is mounted behind a tube with apertures arranged so that the light entering the tube must have entered within an acceptance cone of 5.8° as shown in Figures 3.8 and 3.9. This device is usually mounted on an *equatorial mount* so that it can track the movement of the sun across the sky and stay normal to the sun. An equatorial mount is built on a basic reference plane that is tilted toward the south at an angle equal to the latitude so that it is parallel to a horizontal plane at the

Figure 3.8 Construction of pyrheliometer.

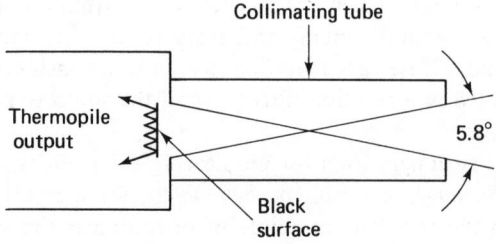

Figure 3.9 The Eppley normal incidence pyrheliometer. By courtesy
of the Eppley Laboratory, Inc.

equator. From Figure 2.13, it is clear that the sun moves in a plane which is
shifted by the angle of declination from the normal to the equatorial reference
plane, and the motion in the plane is at the rate of 15° per hr. Therefore,
the arrangement shown in Figures 3.10 and 3.11 constitutes an equatorial
mount for a pyrheliometer. The reference plane, tilted at the latitude, acts
like a horizontal plane at the equator, and a motor rotates the long axis of
the pyrheliometer at the rate of 15° per hr. The motor axis is tilted at the
angle of declination relative to the normal to the reference plane. In actual
operation, the angle of declination is usually adjusted every day or two,
and a clutch is provided on the instrument for unwinding the leads and
resetting the alignment.

For some purposes, it is desirable to determine the angular distribution
of light entering a normal incidence pyrheliometer. Instruments for doing
this are not commercially available. However, a few hand-built research
tools have been constructed and employed for these purposes. One approach
is a variant of the normal incidence pyrheliometer with a lens in front of the
thermopile as shown in Figure 3.12. By adjusting the diameter of an iris in
the focal plane of the lens, the effective acceptance aperture of the lens can
be varied. One can then take measurements of normal incidence solar intensity as a function of angular aperture. Only a few such measurements have
been reported.[3] Another approach, which is more accurate but considerably

[3]B. J. Petterson, "A Normal Incidence Pyrheliometer with Selective Fields of View,"
Rept. No. SLA-73-7001, Sandia Laboratories, Albuquerque, NM, 1973; J. R. Hickey and
A. R. Karoli, "A Variable Field-of-View Pyrheliometer," *Proc. 1977 Annual Mtg. Amer.
Sec. Int. Solar Energy Society*, Orlando, FL, June, 1977.

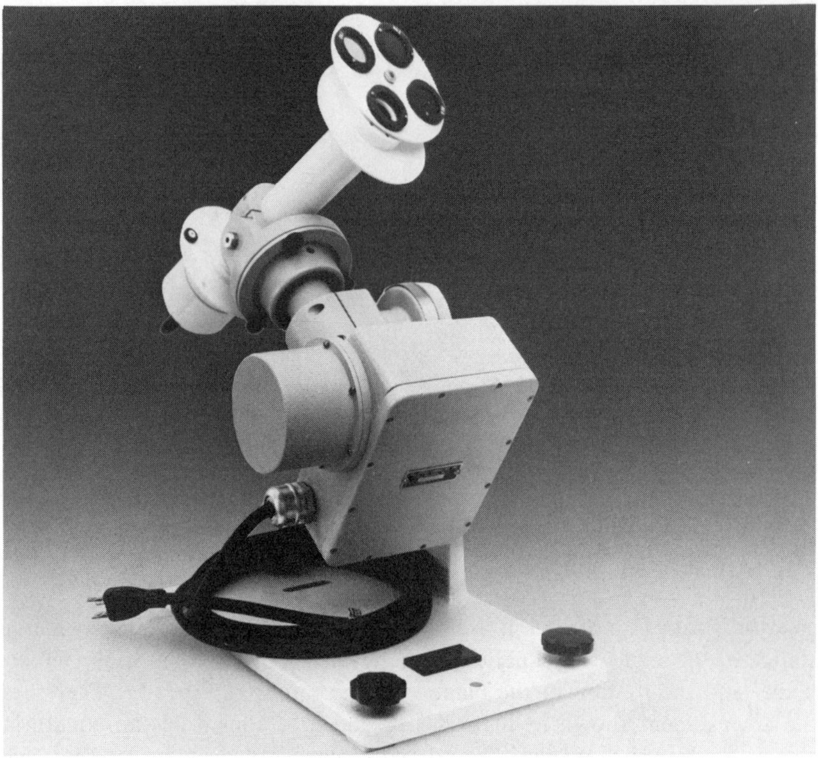

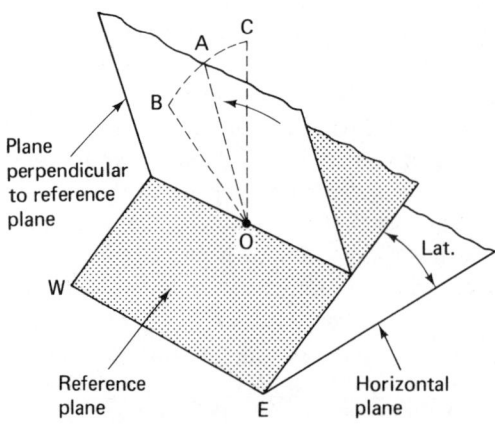

Figure 3.10 The Eppley solar tracker equatorial mount. By courtesy of the Eppley Laboratory, Inc.

Figure 3.11 Equatorial mount. Line *OA* rotates 15° per hr in the perpendicular plane. The pyrheliometer axis is *OB* in winter, *OC* in summer, and *OA* on March 21 and September 21 (at 12 noon).

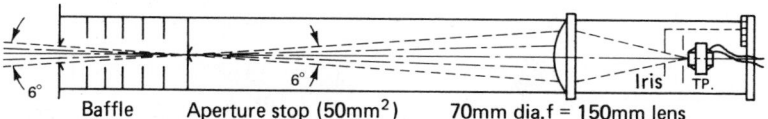

Figure 3.12 Optical schematic of variable field pyrheliometer.[3] By permission of the author and the American Section of the International Solar Energy Society, Inc.

more expensive, is the circumsolar telescope with a very small aperture compared to the apparent diameter of the sun (which is about 0.5°). By tracking the telescope across the sun's disc, it is possible to measure the amount of solar intensity at any angle to the line to the sun. The data taken so far indicate that the circumsolar radiation (radiation coming from angles outside the solar disc) is generally small compared to that coming from near the solar disc.[4]

The output from the various solar sensors consists of a voltage (usually in millivolts) which must be recorded. In some cases, this output is fed into a strip-chart recorder which produces a continuous trace of solar intensity vs. time. Solar intensities at hourly intervals or daily integrated totals are usually reported. In other applications, the solar intensities are integrated electronically over preset time intervals and printed on a digital printer.

3.3 Solar Intensities at Earth Level Normal to the Sun

The pattern of normal incidence solar intensity during clear weather is illustrated in Figure 3.13 for a summer day and a winter day in Fort Hood, Texas.[5] It can be seen that the day is longer in summer and the peak intensity is higher in the winter. The higher peak intensity in the winter is due partly to the earth being closer to the sun and partly to the lower absolute humidity in the atmosphere, which produces less atmospheric absorption. Once the sun rises above the horizon, the normal incidence pyrheliometer (NIP) receives the solar rays at normal incidence. The intensity recorded by the NIP is the extraterrestrial intensity, less the amount of radiation absorbed and scattered by the atmosphere. Near sunrise and sunset, the path length of the solar rays through the atmosphere is very long (large air mass), and as the sun rises in the sky the absorptive path length decreases. This is why the shape of the NIP curves are not rectangular step functions. For the cases

[4]See Sec. 3.11 of this book.

[5]D. Rapp and A. A. J. Hoffman, "On the Relation between Insolation and Climatological Variables-III. The Relation between Normal Incidence Solar Intensity, Total Insolation, and Weather at Fort Hood, Texas," *Energy Conversion*, *17*, 163, 1977.

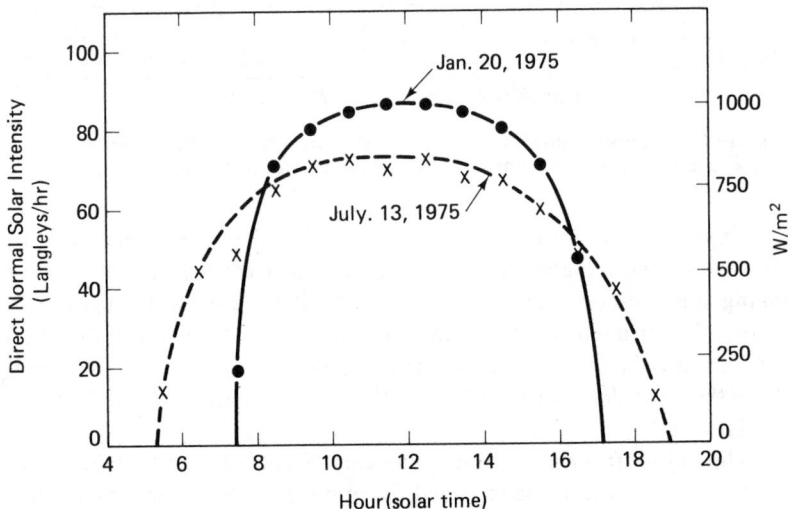

Figure 3.13 Direct normal solar intensity as a function of hour on clear summer and clear winter days.[5] Note the longer day and lower peak intensity in the summer. By permission of Pergamon Press Ltd.

illustrated in Figure 3.13 the actual peak NIP values are less than 70% of the extraterrestrial values. This is typical of many locations in the United States. Even in very clear weather, less than two-thirds of the extraterrestrial radiation penetrates the atmosphere and reaches ground level as direct rays. Of course, in cloudy weather the NIP output can go to zero. For clear weather in North Central Texas, it has been found that the hourly patterns of normal incidence solar intensity are quite repetitive. In Figures 3.14 to 3.16 typical data are given for several clear days in Central Texas.[6] An arbitrary mathematical function that roughly fits these hourly patterns is*

$$N(H, D) = N_0\left[1 - \left(\frac{|H - 12|}{4.7 + 0.0126D}\right)^{3.5}\right] \qquad (3.8)$$

where

$$N_0 = 77 - 6\cos\left[\frac{(D_s - 220)\pi}{182.6}\right] \qquad (3.9)$$

and N is the normal incidence solar intensity in Langleys/hr (1 Langley/hr = 11.63 W/m²) at solar time H on a day which is D days from December 21 in the shortest direction, and D_s is the day of the solar year (December

[6]D. Rapp and A. A. J. Hoffman, "On the Relation between Insolation and Climatological Variables-III. The Relation between Normal Incidence Solar Intensity, Total Insolation, and Weather at Fort Hood, Texas." *Energy Conversion*, *17*, 163, 1977.

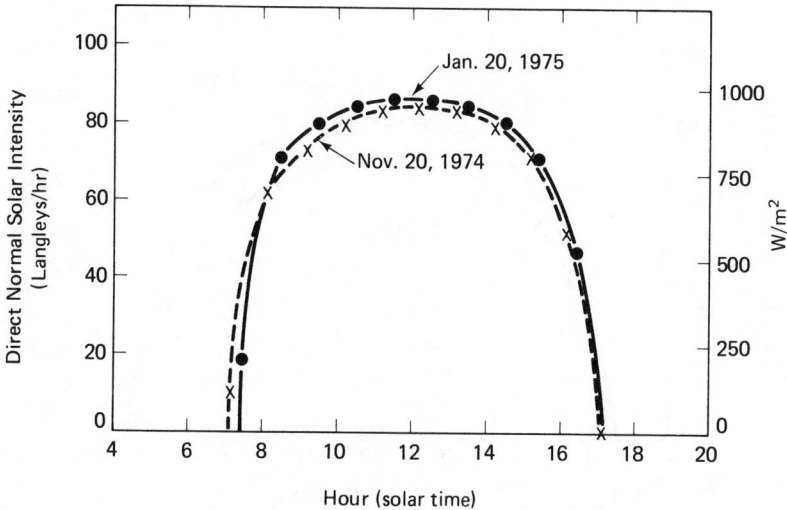

Figure 3.14 Variation of direct normal solar intensity with hour for clear winter days equally spaced about December 21.[5] The two curves are nearly identical, but the peak in January is slightly higher than the peak in November. By permission of Pergamon Press Ltd.

Figure 3.15 Variation of direct normal solar intensity with hour for clear days equally spaced about December 21.[5] Note the higher peak intensity in February than in October. By permission of Pergamon Press Ltd.

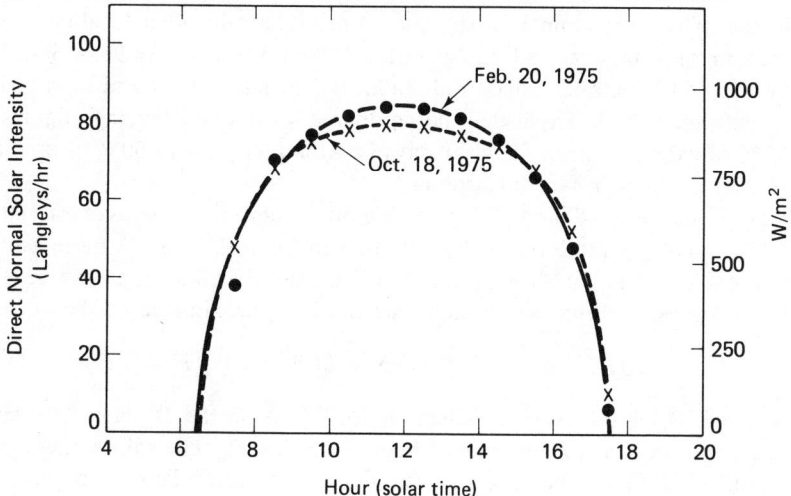

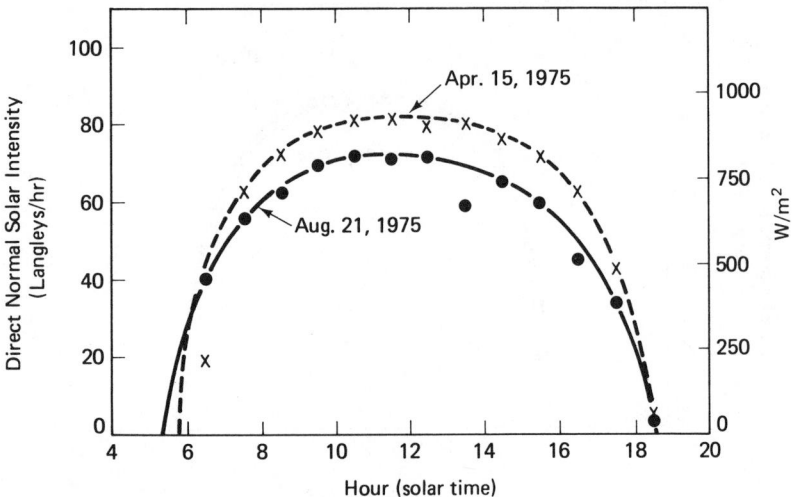

Figure 3.16 Variation of direct normal solar intensity with hour for clear days equally spaced about December 21.[5] Note the higher peak intensity in April than in August. By permission of Pergamon Press Ltd.

$21 = 1$, June $21 = 183$, and December $20 = 365$). It can be seen that the peak intensity, N_0, reached at solar noon, varies from a high of 83 Langleys/hr (965.3 W/m²) on $D_s = 37$ (January 26) to a low of 71 Langleys/hr (825.7 W/m²) on $D_s = 220$ (July 30). These results represent a location in Central Texas. It is not certain how similar the results are in other localities. Note that in late January, 83 Langleys/hr = 306.0 Btu/hr-ft² = 965.3 W/m² is 69.1 % of the extraterrestrial radiation, and in late July the ratio is 63.0 % of extraterrestrial. These may be compared with yearly average figures of 65 % of extraterrestrial for clear days in Albuquerque and 60 % of extraterrestrial in Raleigh, North Carolina.[7]

Equations (3.8) and (3.9) provide a purely empirical means of estimating $N(H, D_s)$ at ground level in clear weather in Central Texas. A more general approach that might allow transference to other localities is to assume that the direct normal intensity at any time can be approximated by the function

$$N(H, D_s) = N_{ext}(D_s) \exp \{-\beta(D_s) \sec [\theta_h(H, D_s)]\} \qquad (3.10)$$

where $\beta(D_s)$ is the average extinction coefficient on the D_s day of the solar year, and $\sec \theta_h$ is the air mass at hour H on day D_s. The values of N_0 from from Eq. (3.9) can be used to infer $\beta(D_s)$ in Central Texas. The result is shown in Figure 3.17.

[7]C. M. Randall and M. E. Whitson, Jr., "Hourly Insolation and Meteorological Data Base," *Aerospace Corp. Report*, El Segundo, CA, 1978.

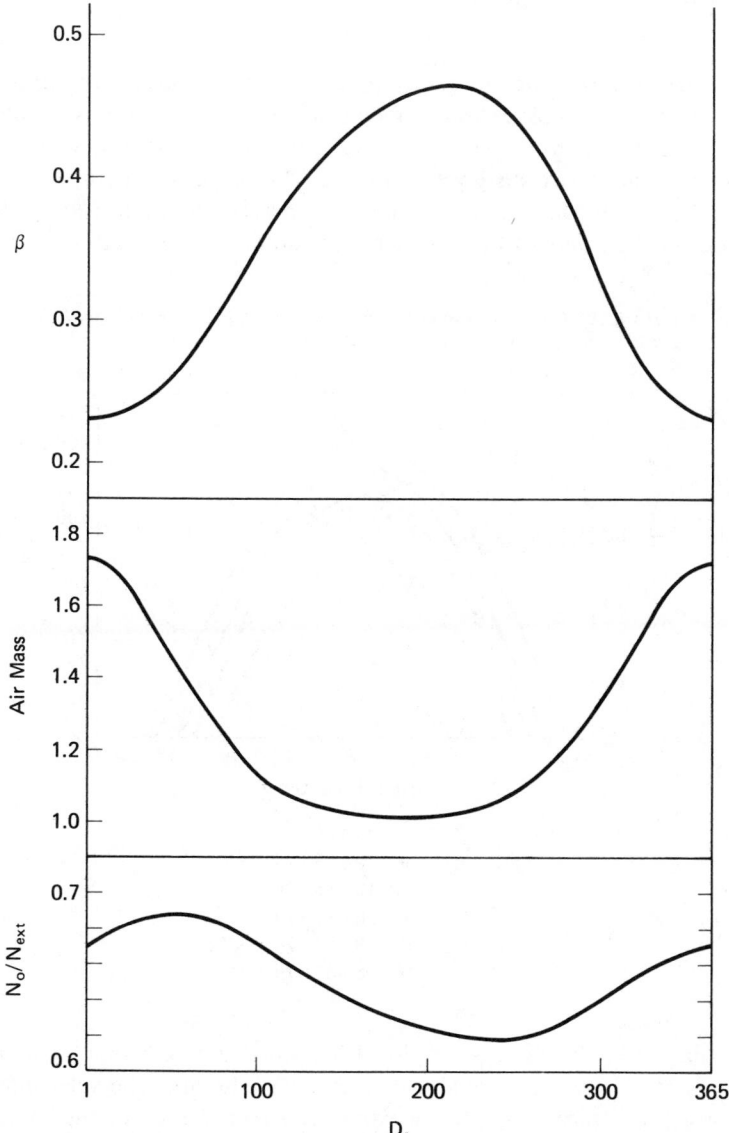

Figure 3.17 Variation of the noon extinction coefficient β, the air mass at noon, and the ratio of noon direct normal intensity to extraterrestrial intensity with day of the solar year at Fort Hood, Texas, in clear weather.

3.4 Insolation on Surfaces

The expected forms for the hourly patterns of global insolation on a horizontal surface are considerably different than the hourly patterns for normal incidence solar intensity. An example of patterns of insolation on clear days in North Central Texas is given in Figure 3.18, based on data taken at Fort Hood, Texas.[5] It can be seen that there is a regular variation from winter to summer as the length of the day increases and the peak insolation increases.

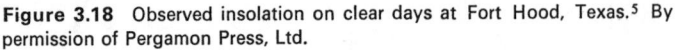

Figure 3.18 Observed insolation on clear days at Fort Hood, Texas.[5] By permission of Pergamon Press, Ltd.

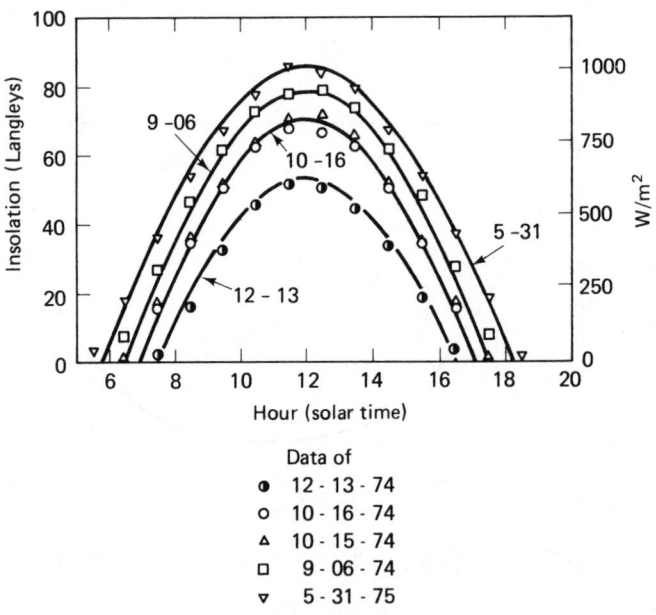

The analysis of the patterns of global insolation is more complex than the analysis of patterns of normal incidence solar intensity. The solar intensity at normal incidence is simply the extraterrestrial radiation minus losses due to atmospheric absorption and scattering, and the peculiar patterns that are obtained are due to the variation in path length of light through the atmosphere as the sun's elevation changes during the day. The global insolation, however, depends on the direct normal solar intensity, the solar elevation angle, and the diffuse scattered light from the sky and from clouds. In the middle latitudes, such as are encountered in the United States, the major factor determining the form of the curves in Figure 3.18 is the solar elevation angle. For the latitude (31°) of the data in Figure 3.18, the summer midday sun is within about 10° of vertical, whereas the winter midday sun is 47° lower in the southern sky. This results in a cosine factor which greatly reduces the winter intensity on a horizontal surface.

Data from the same station[5] on a tilted surface on clear days are shown in Figure 3.19. The latitude is 31°N, and the tilted surface is tilted at 30° up toward the south. It can be seen that roughly the same peak intensities near noon are found in clear weather at all times of the year on the surface tilted near the latitude angle.

Figure 3.19 Insolation on horizontal and tilted (30°) surfaces at Fort Hood, Texas, on January 4 and July 28.[5]

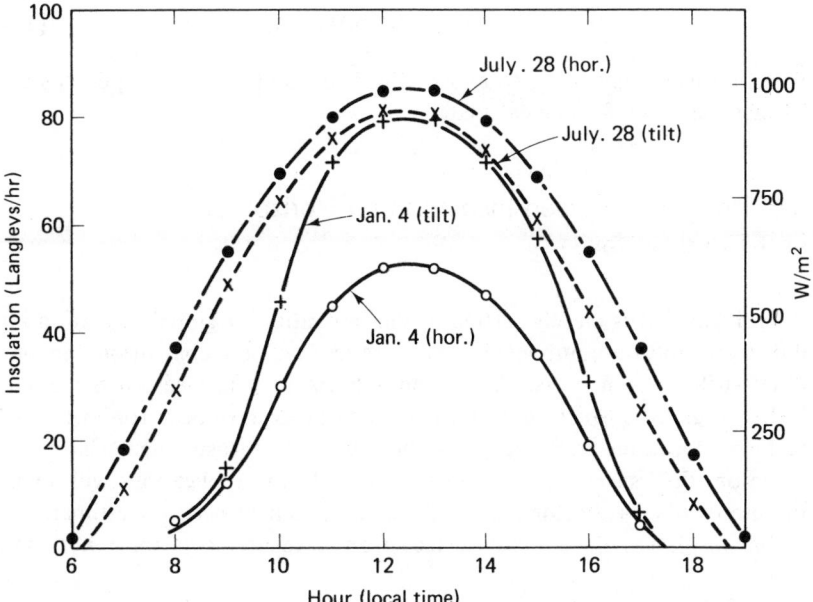

3.5 Direct and Diffuse Radiation

The direct normal solar intensity N can be measured with a narrow aperture pyrheliometer. The global insolation on surfaces that are horizontal (I_0) and tilted up toward the south at angle T (I_T) can also be measured. The question now arises as to what fraction of the global insolation is direct and what fraction is diffuse. The direct normal intensity represents the power incident on a surface normal to the sun from direct rays. The component of direct rays on any other surface can be found by multiplying N by the cosine of the angle between the normal to the other surface and the line to the sun. But, according to Chapter 2, this cosine is given in Equations (2.11) and (2.17). Thus, the direct intensities on horizontal and tilted surfaces are

$$D_0 = N \cos \theta_0 \tag{3.11}$$

$$D_T = N \cos \theta_T \tag{3.12}$$

where Eq. (2.17) is used for the cosine, and $T = 0$ for horizontal surfaces. If these direct ray intensities are subtracted from the observed global intensities, the remainders must be the diffuse intensities on the surfaces. Thus, the diffuse intensities are given by

$$F_0 = I_0 - D_0 \tag{3.13}$$

$$F_T = I_T - D_T \tag{3.14}$$

From measurement of I_0, I_T, and N, the direct and diffuse components on a surface may therefore be calculated.

3.6 Relation between Insolation on Horizontal and Tilted Surfaces

A problem that is closely related to the separation of global insolation into direct and diffuse components is the estimation of global insolation on a tilted surface from measurements on a horizontal surface. Such measurements of global insolation on horizontal surfaces have been routinely made at many locations over the years, but very few measurements have been made on tilted surfaces. Since most solar collector applications will involve the use of tilted collector planes, there is a practical need for estimation of global insolation on a tilted surface from measurements on a horizontal surface.

If the rays falling on a horizontal surface were all direct, one could calculate the intensity on a tilted surface from that on a horizontal surface by multiplying it by $(\cos \theta_T / \cos \theta_0)$. It is a simple matter to calculate this ratio at any hour in any location. Unfortunately, the rays falling on a horizontal surface are partly direct and partly diffuse. It is incorrect to multiply the diffuse portion of radiation on a horizontal surface by this ratio to calculate diffuse radiation on a tilted surface. The relation between diffuse radiation on horizontal and tilted surfaces can be determined empirically. The measured ratio is called Q and is given by

$$Q = \frac{F_T}{F_0} \tag{3.15}$$

The ratio of direct radiation on a tilted surface to the direct radiation on a horizontal surface is

$$R = \frac{\cos \theta_T}{\cos \theta_0} \qquad (3.15a)$$

The measured overall ratio is

$$S = \frac{I_T}{I_0} = \left(\frac{F_0}{I_0}\right)Q + \left(1 - \frac{F_0}{I_0}\right)R \qquad (3.16)$$

For stations where I_0 is measured but I_T is not, the question arises as to how to estimate S. The ratio R can always be calculated. If the fraction of diffuse radiation (F_0/I_0) is small, S will be closely approximated by R. The question arises as to what degree R may be used to approximate S. The answer to this question cannot be given, in general, but it is instructive to review a short study made in North Central Texas.[8]

3.7 Case Study: Central Texas

All the data considered here were taken at Fort Hood, Texas, by the U.S. Army Electronics Command. Only data for the months from March, 1976, to February, 1977, are considered here. The latitude at the site is 31°0' N, and the tilted pyranometer was mounted at 30° to the horizontal.

To explore the variation of diffuse radiation during clear periods, the days during the 12 month period, March, 1976–February, 1977, were separated into "clear" and "cloudy" categories. A day was designated as "clear" if it followed the expected clear day patterns from 10:00 to 15:00 and had a daily total insolation of at least 85% of the expected total for cloudless days. An example of a clear day is January 4, 1977, for which data are presented in Table 3.2 and Figures 3.20, 3.21, and 3.22. It can be seen that the fraction of diffuse radiation on a horizontal surface falls from nearly 100% at sunrise and sunset to about 11% during the midday period. The ratios of insolation on tilted and horizontal surfaces are given in Figure 3.22. The observed ratio $I_T/I_0 = S$ is nearly equal to the calculated ratio for direct rays, $R = \cos \theta_T/\cos \theta_0$, during the midday period.

Each of the clear days were analyzed in the same way, and average values for the clear days in each month were computed. For the 2 month period in midwinter (December, 1976, and January, 1977) the results are shown in Figure 3.23. It can be seen that the fraction of diffuse radiation on a horizontal surface averages around 12% during the midday period in clear

[8]D. Rapp and D. Oxley, "On the Relation between Global Insolation on Horizontal and Tilted Surfaces," *Energy Conversion*, *18*, 39, 1978.

TABLE 3.2   Solar intensities on a typical clear day (1-4-77)*

Hour	N	I_0	I_T	D_0	D_T	F_0/I_0	F_T/I_T	Q	R	S
8	2	5	5	0.2	0.7	0.96	0.86	0.90	3.31	1.00
9	8	12	14	2.3	4.4	0.81	0.68	0.98	1.96	1.17
10	36	30	46	15.5	26.1	0.49	0.43	1.37	1.69	1.53
11	76	45	72	40.5	64.3	0.10	0.11	1.71	1.59	1.60
12	79	52	80	46.3	71.8	0.11	0.10	1.45	1.55	1.54
13	79	52	80	46.3	71.8	0.11	0.10	1.45	1.55	1.54
14	77	47	72	41.0	65.1	0.13	0.10	1.15	1.59	1.53
15	73	36	58	31.4	53.0	0.13	0.09	1.09	1.69	1.61
16	47	19	31	13.3	26.1	0.30	0.16	0.87	1.96	1.63
17	8	4	6	0.8	2.8	0.79	0.54	1.02	3.31	1.50

*All intensities are in Langleys/hr (1 Langley/hr $=$ 11.63 W/m²)

Figure 3.20   Solar intensity as function of time[8] on a clear day (1-4-77) at Fort Hood, Texas. N is the direct normal intensity, I_0 is the global insolation on a horizontal surface, and I_T is the global insolation on a surface tilted at 30° toward the south. By permission of Pergamon Press Ltd.

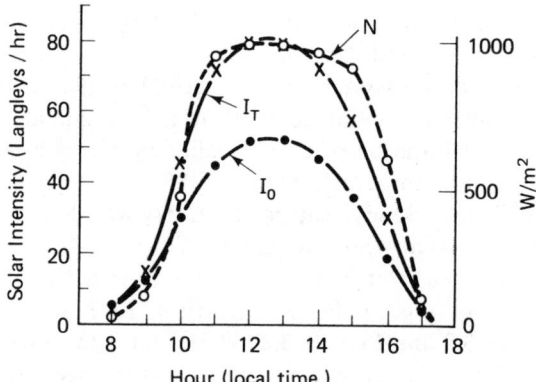

weather. The calculated ratio R for direct rays closely approximates the actual ratio of radiation on tilted and horizontal surfaces S during the middle of the day.

Values of R and S on several clear days are shown in Figure 3.24. Average values of R and S for each of the months of the year were found from an analysis of 12 months of data. The results indicated that R closely approximates S during the middle of the day in clear weather. The average percentage of diffuse radiation on a horizontal surface was also evaluated from 12 months of data. The results indicate a range of 12–15% during the

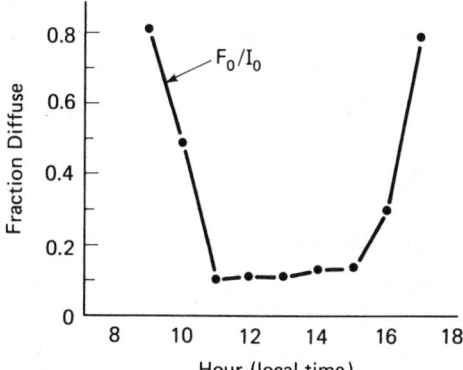

Figure 3.21 Fraction of global insolation on a horizontal surface that is diffuse for 1-4-77 at Fort Hood, Texas.[8] By permission of Pergamon Press Ltd.

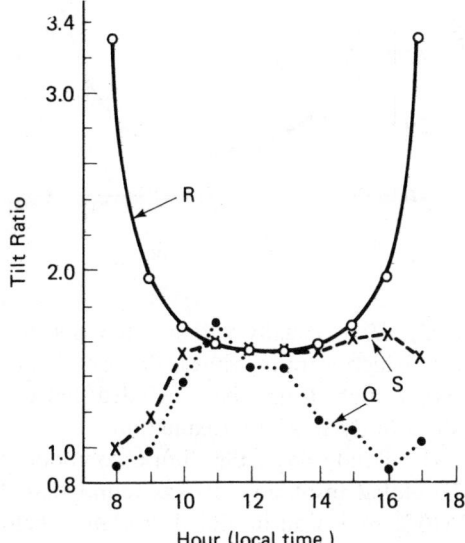

Figure 3.22 Ratio of global insolation on a tilted surface (30° toward the south) to global insolation on a horizontal surface for 1-4-77 at Fort Hood, Texas.[8] R is calculated from solar geometry for direct rays, S is observed, and Q is estimated from the data for diffuse rays. By permission of Pergamon Press Ltd.

midday period in clear weather. Because the diffuse component is small during midday, it is good approximation to use R instead of S to correct I_0 to I_T. This approximation breaks down during early morning or late afternoon hours, but these hours are not too important for solar collection.

In testing the hypothesis that R might be used to correct I_0 to I_T for the tilted surface on clear days, only hours for which $I_T > 22$ Langleys/hr were included, because this is an estimated threshold for collection by flat plate collectors. Daily sums were calculated for a 12 month period that do not include periods where I_T was below 22. For periods of clear days when

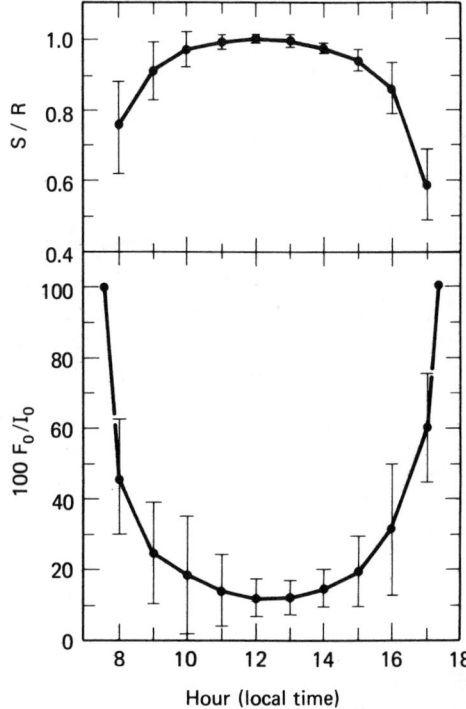

Figure 3.23 Statistical summary of clear days[8] in December, 1976, and January, 1977, showing average values and ranges (±S.D.) of S/R and 100 F_0/T_0. By permission of Pergamon Press Ltd.

Hour (local time)

$I_T > 22$, RI_0 is usually within a few percent of I_T. For 58 clear days, the bulk of which were in winter, the use of the simple approximation RI_0 for I_T is 3.3% high. It may be concluded that use of RI_0 to estimate I_T on clear days is a fairly good approximation.

On cloudy days, the diffuse component is a much higher percentage of the global insolation. It was found that during the midday period, for each 10% reduction in global insolation below clear day values, there was an approximately 15% addition to the percentage of diffuse radiation. When the global insolation drops to 60% (or less) of clear day values, almost all the insolation is diffuse. The ratios Q tend to be 15–30% below R during such periods. A comparison of daily total insolation on a tilted surface (for hours when $I_T > 22$) with those calculated from RI_0 for cloudy days was performed for a 12 month period. The errors can be as large as 18% on a monthly basis. When summed over all days, clear and cloudy, the values of I_T and RI_0 are within 5% of one another. It may be concluded that RI_0 can be a useful approximation to I_T in Central Texas when the intensities are high enough to warrant reasonable performance by solar collectors.

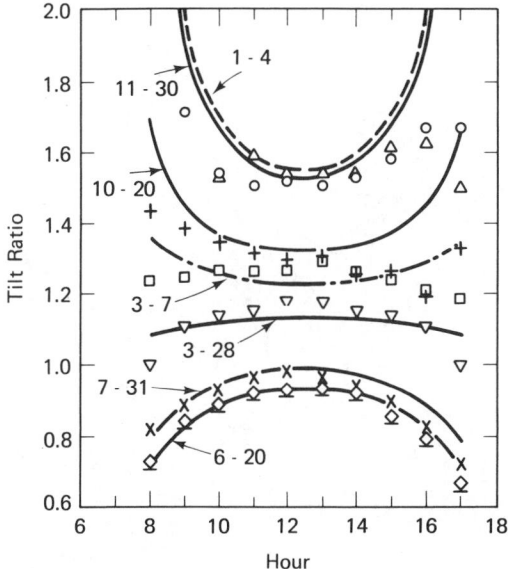

Figure 3.24 Tilt correction ratio on clear days at various timesof the year.[8] The lines are calculated values from $R = \cos \theta_T / \cos \theta_0$, while the data points are actual values of $S = I_T / I_0$. The data points refer to dates as follows: $\triangle$ = 1/4, $\bigcirc$ = 11/30, $+$ = 10/20, $\square$ = 3/7, $\triangledown$ = 3/28, $\times$ = 7/31, and $\diamondsuit$ = 6/20. By permission of Pergamon Press Ltd.

3.8 Networks for Monitoring Solar Intensities

3.8.1 History

The National Weather Service operated a nationwide network for monitoring solar intensities since the early 1950s. Most of the measurements utilized horizontally mounted total hemispheric pyranometers, but a few sites were equipped with normal incidence pyrheliometers. Some of the stations only reported data on a daily total basis while others reported on an hourly basis. The stations in the new network, begun in 1978, are illustrated in Figure 3.25. The data from the old network have been available for many years from the National Climatic Center in Asheville, North Carolina, in the form of computer tapes or data books. The hourly data on computer tapes were integrated with climatic data so thet all standard meteorological

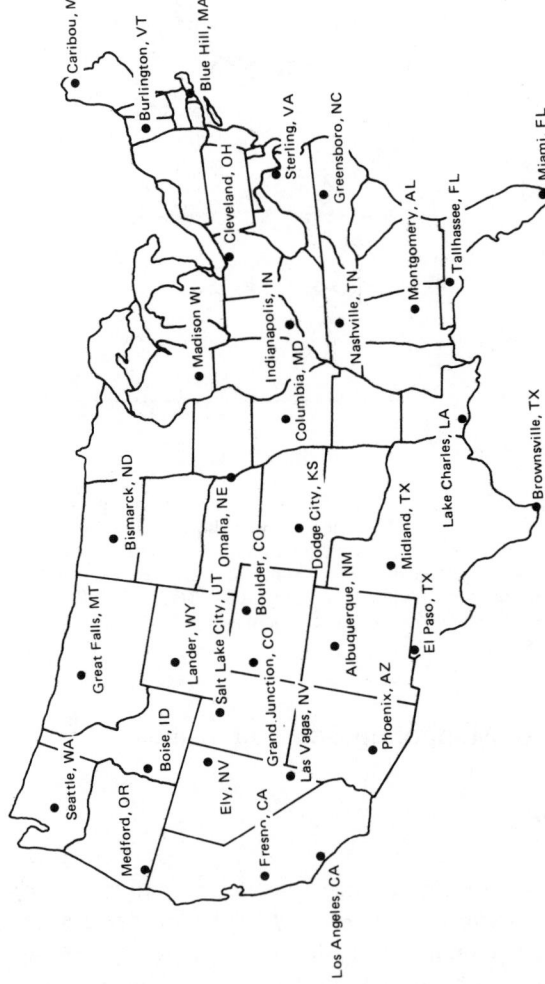

Figure 3.25 New NOAA network for recording total and direct solar radiation. Courtesy National Oceanic and Atmospheric Administration, Asheville, NC.

data were available along with solar intensity at each hour. From averages of this data, rough contours of variation of total insolation have been plotted.[9]

3.8.2 Errors in the Raw Data

Some of the inadequacies in this program for monitoring solar intensity include:

1. Lack of pyrheliometer data at most stations.
2. Inadequate density of stations.
3. Total absence of pyranometer data on tilted surfaces.
4. Lack of statistical analysis of data to report such things as range of fluctuations about average, probabilities of repeated good or poor solar days in succession, and breakdown of radiation into direct and diffuse.
5. Lack of correlation of solar intensities with climatological variables.
6. Lack of methodology for interpolating between measurement stations.
7. Lack of adequate instrument calibration.

One might excuse these inadequacies on the grounds that the National Weather Service was not prepared, either psychologically or financially, to do much more during the period 1950–1972 when cheap fossil fuels were available. Budgetary allocations for monitoring solar intensities were minimal, and the resources of the National Weather Service were severely strained to maintain the network.

If all that was wrong with the solar intensity data base was that it simply was incomplete, then at least one could rely on the data, sparse as it is. However, it is now generally accepted that the accuracy of the data are highly questionable, and that the reliability of all the data are subject to grave doubt. Durrenberger and Brazel[10] have given some of the history of the realization that errors exist in the data and have presented further proof of that hypothesis. This history shows that Bennett[11] analyzed solar radiation data in 1965 and warned about problems in using the data. Despite these warnings, workers in the field of solar energy applications have used the data at face value, sometimes reaching dubious conclusions. A case in

[9] *Climatic Atlas of the United States*, National Oceanic and Atmospheric Administration, Washington, DC, 1968.

[10] R. W. Durrenberger and A. J. Brazel, "Need for a Better Solar Radiation Data Base," *Science*, **193**, 1154, 1976.

[11] I. Bennett, *Solar Energy* **9**, 145, 1965.

point is the station at Inyokern, California. According to the insolation
tables and maps prepared by the National Climatic Center in the 1962 *Climatic Atlas of the United States*, the station at Inyokern receives far more
solar radiation than any other location in the United States. This was accepted by climatologists.[12] However, in 1966, the National Weather Service made
calibration checks on a number of instruments at various localities in the
United States. The results were circulated in an unpublished National
Weather Service report, and they indicated errors ranging from 1 to 16.7%
at the various sites. The largest error was at Inyokern, where the instrument
in place was high by 16.7%. Further evidence was afforded by comparing
average levels of solar intensity at the same location over different periods.
For Inyokern, readings taken from 1967–1974 tended about 20% lower than
readings from 1951–1966, upon which the *Climatic Atlas* was based. It is now
quite certain that Inyokern has solar levels that are not higher than any of the
surrounding desert areas such as Las Vegas and Phoenix. Yet, because of the
mistaken data, Inyokern has been suggested as a site for a large-scale solar
power project by several contractors.[13]

Further evidence for errors in the solar intensity data base was presented at the 1977 annual meeting of International Solar Energy Society.
Bahm[14] analyzed all the days during a span of many years at several localities.
For each day, he estimated the extraterrestrial radiation (ETR) falling on a
surface above the earth's atmosphere but parallel to a horizontal surface.
He then divided the actual total insolation for the day by (ETR)/100, and
called this "% of ETR." He found that on clear days typical values of %
of ETR ranged from 70 to 85. If fluctuations in solar intensity from year to
year at a location are due to variations in incidence of clouds, the peak values
of % of ETR for clear days in any part of the year should not vary much
from year to year. Changes in these patterns indicate either long-term changes
in atmospheric clarity or instrument drift due to changes in instrument
sensitivity. Plots of % of ETR for clear days at Albuquerque, New Mexico,
and Inyokern, California, are shown in Figures 3.26 and 3.27. It seems evident
from these figures that instrument drifts occurred at times in these measurements. Further information is obtained by preparing plots of the percentage
of clear days corresponding to any region of % of ETR. Such plots show the

[12]W. Sellers, *Physical Climatology*, University of Chicago Press, Chicago, 1965, p.
24.

[13]J. C. Powell, *et al.*, "Dynamic Conversion of Solar Generated Heat to Electricity,"
NASA Rept. CR-134724, Honeywell, Inc., and Black and Veatch, August, 1974.

[14]R. J. Bahm, "Instrument Errors in National Weather Service Solar Radiation
Data," *Proc. 1977 International Solar Energy Society Meeting*, Orlando, FL, June, 1977,
p. 14–17.

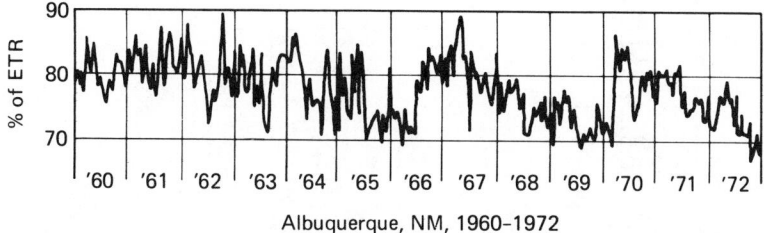

Albuquerque, NM, 1960-1972

Figure 3.26 A typical clear day plot for Albuquerque, New Mexico.[14] By permission of the author and the American Section of the International Solar Energy Society, Inc.

distribution of various values of % of ETR among the clear days during any period. For all clear days in Albuquerque from 1952 to 1973, the plot is as shown in Figure 3.28. The vertical dashed line is at 80%, and the most probable value of % of ETR is made for each of the periods in Albuquerque over which a single instrument was in operation, and results are shown in Figure 3.29. Evidence is clear for downward drift of several of the instruments. In each case, the sensitivity of the instrument appears to fade after a year or two of exposure. Similar plots for the periods 1954–1966 and 1967–1971 are given for Inyokern, California, in Figure 3.30. It is clear that the earlier data are repeatedly higher than the later data. The % of ETR values in the upper 80s from 1953 to 1966 appear unrealistically high. The bimodal pattern found for the entire 20 year period 1952–1971 indicates the variation in instrument sensitivity.

Martin, Berdahl, Grether, and Wahlig[15] found drifts in instrument sensitivities at California measurement stations by similar techniques.

Further evidence of instrumental drift and errors in calibration have recently been provided by the NOAA.[16] For each location where hourly insolation has been measured, a "clear day plot" has been prepared. This is a 15 day moving average of measured percentage of extraterrestrial radiation at solar noon on clear days. Clear day plots for three localities are shown as illustrations in Figures 3.31 to 3.33. The data show many spurious variations. Measurements made with early instruments appear higher than more recent measurements. The data available on past calibration procedures are sparse. In general, these plots do not inspire much confidence in the early data.

[15]M. Martin, P. Berdahl, D. Grether, and M. Wahlig, "Rehabilitation Techniques for Daily Solar Radiation Data," *ibid*, p. 14–22.

[16]SOLMET, Vol. 2, *Final Rept TD-9724*, "Hourly Solar Radiation—Surface Meteorological Observations," NOAA, Asheville, NC, February, 1979.

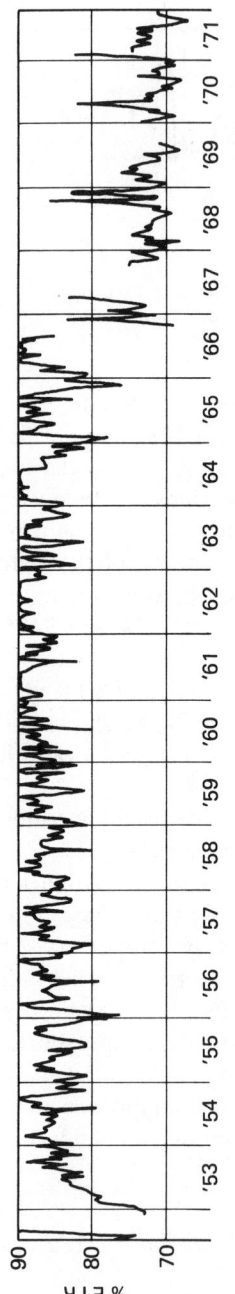

Figure 3.27 The clear day plot for Inyokern.[14] By permission of the author and the American Section of the International Solar Energy Society, Inc.

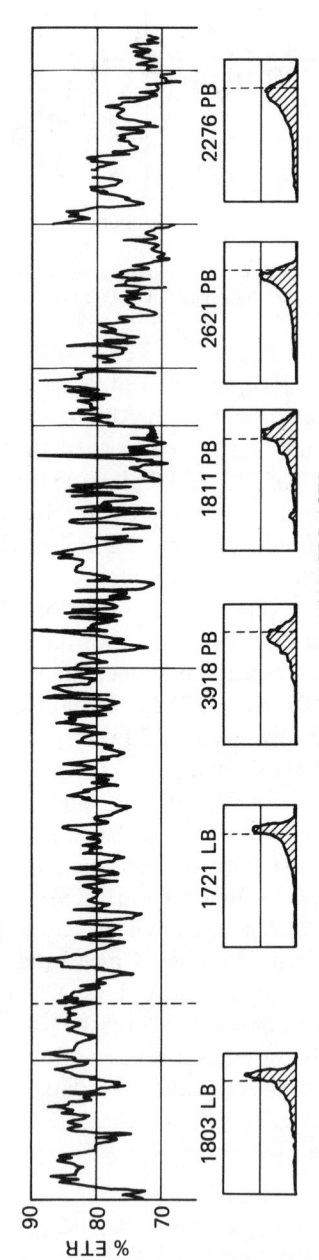

Figure 3.28 The clear day plot for all the Albuquerque data with histograms for each instrument.[14] By permission of the author and the American Section of the International Solar Energy Society, Inc.

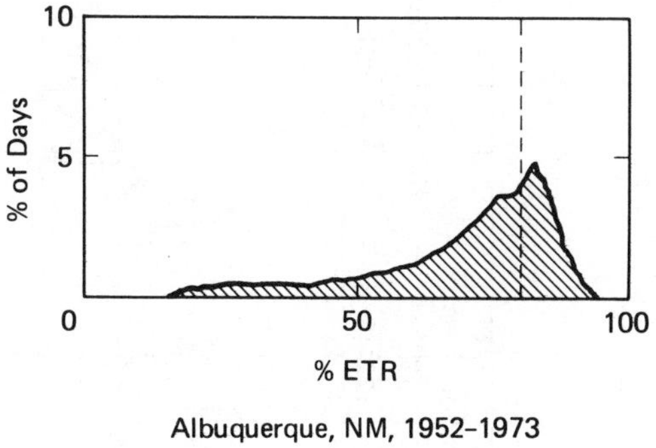

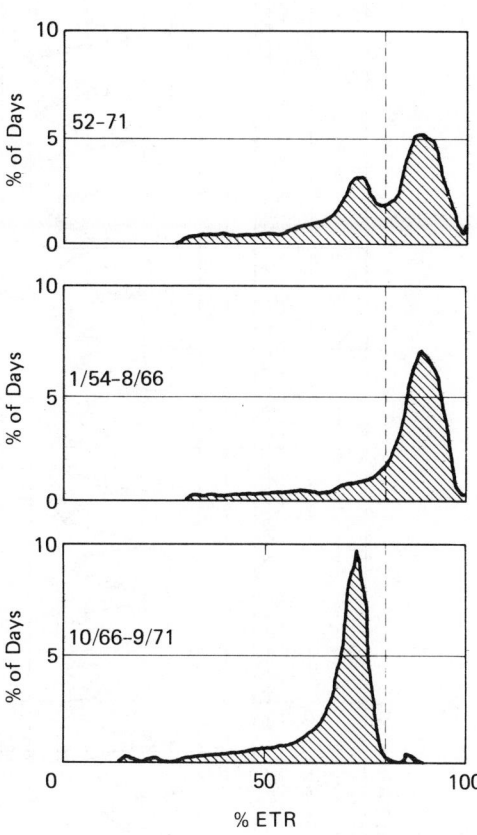

Figure 3.29 The histogram for all the Albuquerque data.[14] By permission of the author and the American Section of the International Solar Energy Society, Inc.

% ETR

Albuquerque, NM, 1952–1973

52–71

1/54–8/66

10/66–9/71

% ETR

Figure 3.30 Histograms for the entire data for Inyokern (top), for the notorious instrument No. 1620 (middle), and for a good instrument (bottom).[14] By permission of the author and the American Section of the International Solar Energy Society, Inc.

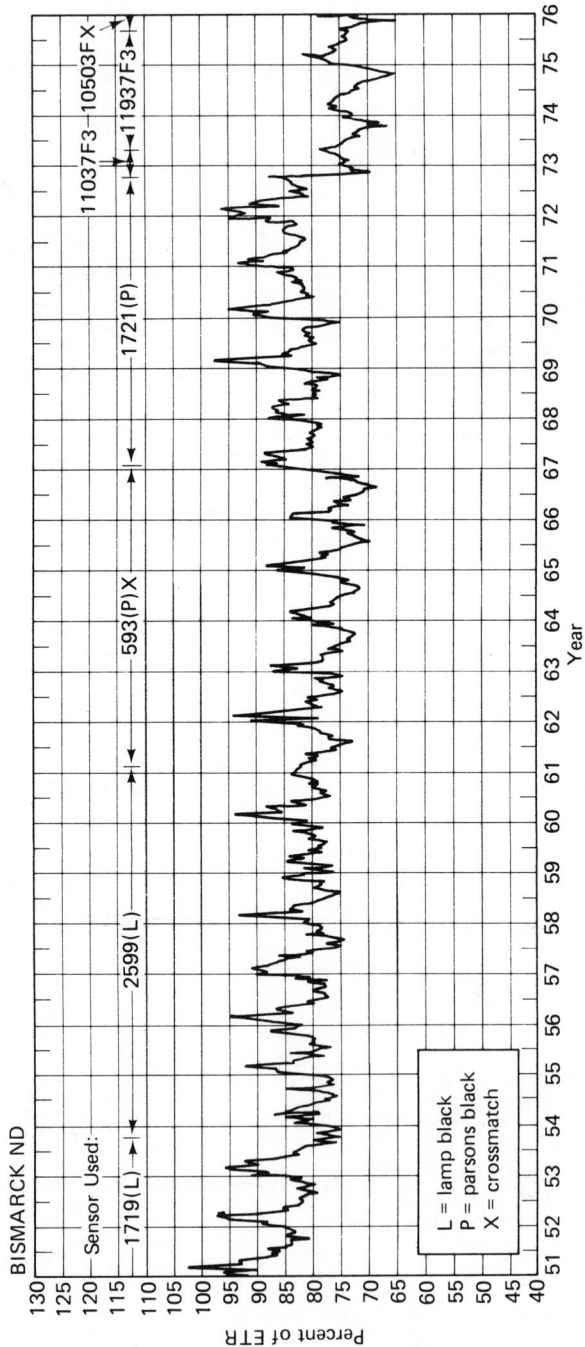

Figure 3.31 Clear solar noon transmission plot (15 day average) at Bismarck, North Dakota.[16] Courtesy National Oceanic and Atmospheric Administration, Asheville, NC.

82

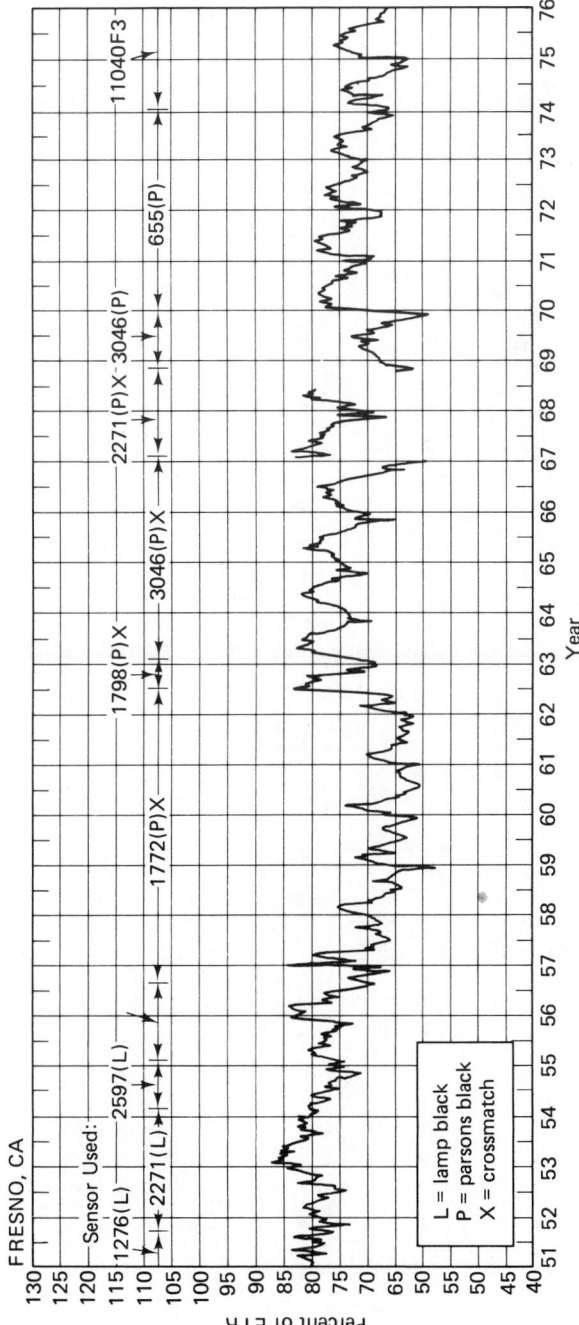

Figure 3.32 Clear solar noon transmission plot (15 day average) at Fresno, California.[16] Courtesy National Oceanic and Atmospheric Administration, Asheville, NC.

83

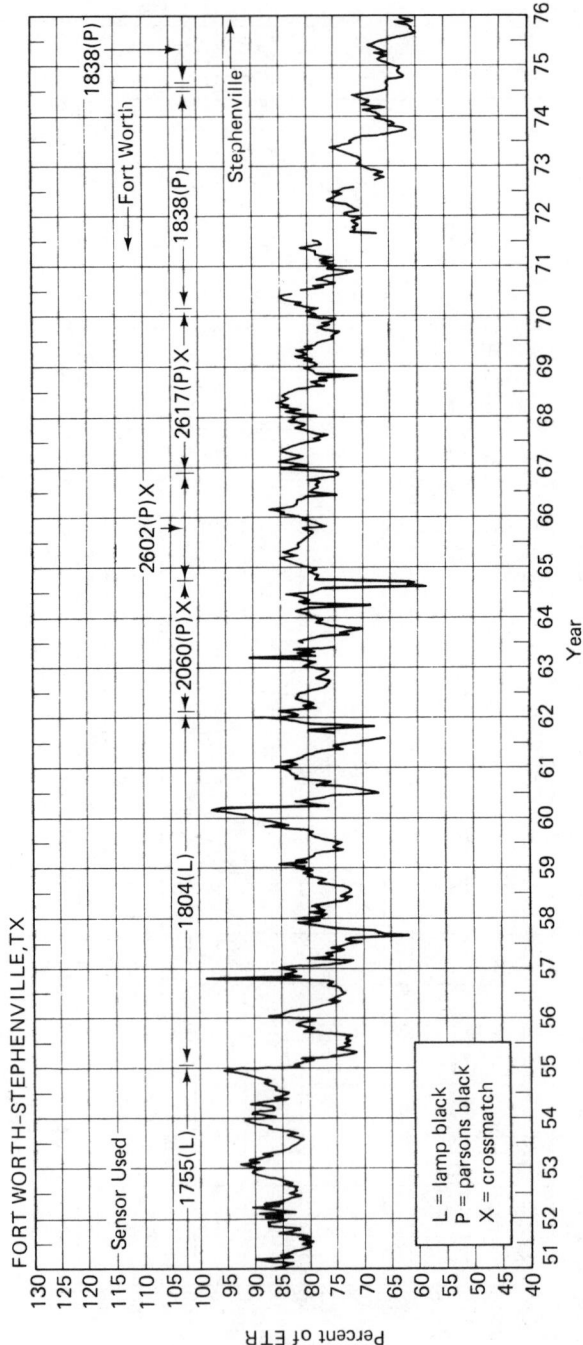

Figure 3.33 Clear solar noon transmission plot (15 day average) at Fort Worth-Stephenville, Texas.[16] Courtesy National Oceanic and Atmospheric Administration, Asheville, NC.

3.8.3 Rehabilitation of Data

The Department of Energy has been very slow to act to correct this situation. After several years of preliminary fumbling, the new solar intensity program was finally making progress in 1978. The new program can be divided into three categories:

1. "Rehabilitation" of global solar data at 26 stations on an hourly basis. Theoretical estimation of direct normal intensities are also included.
2. Installation of new instruments (both pyranometers and pyrheliometers) at 38 sites. Automatic data acquisition and instrument calibration are also provided for.
3. Theoretical analyses of trends in solar intensity and climatology. Estimates of general levels of solar intensity at approximately 200 sites are being prepared.

The rehabilitation procedure for the 26 stations where hourly global insolation are available is described next. NOAA has prepared computer tapes containing the raw data and corrected data. One set of data is called *engineering corrected*, in which an attempt has been made to correct for instrumental effects indicated by sensor, station, and recorder histories. The corrections that have been identified include calibration changes, solar radiation absolute scale variations, degradation of black paint in sensors, and variation of sensitivity with temperature. Unfortunately, calibrations were rarely performed. All instruments were calibrated before field installation. When instruments were withdrawn from the field, it was intended that they be recalibrated. In many cases the instruments were broken, or the recalibration was not carried out for some other reason. In such cases, the initial calibration was used for the entire time the instrument was in the field. When the final calibration was available, a linear variation of sensitivity with time was assumed for the period the instrument was in the field. The major cause for variation in instrument sensitivity over a period of years is thought to be chemical and physical changes in the black absorbing surface in a pyranometer. Visual observation of early pyranometers has indicated coloration changes in the black surface after exposure. Typical periods of installation in the field were several years. Corrections for variation in the accepted absolute radiation scale were made over the years. Modern new pyranometers are temperature-compensated. Most of the data taken in the years 1950–1975 was taken with noncompensated instruments. The manufacturer's recommended variation of sensitivity with temperature was used to apply a further correction to the raw data. The result of applying these various corrections to the raw data are labelled the *engineering corrected data*.

The major sources of error in the data lie in calibration errors and calibration variations. These were performed at such rare intervals that great uncertainty remains over wide intervals in the data. It might be thought that the engineering corrected data should be fairly accurate for about a year or more after the initial introduction of a new instrument. It is doubtful that large drifts in sensitivity would occur within a year. However, the accuracy of the initial calibration is difficult to evaluate.

In those sites for which new Eppley PSP pyranometers have been installed in the past several years, the absolute levels of global insolation generally lie lower than the levels corresponding to the first year of operation of the older pyranometers. It is generally believed that the newer data taken with Eppley PSP pyranometers are reliable and that the data taken with older pyranometers may have had calibration errors, even when first placed in the field. Therefore, even the engineering corrected data do not appear trustworthy on an absolute basis in any locality.

In order to put the solar data on a common basis independent of the fluctuations in sensitivity of instruments, NOAA has carried out a second rehabilitation procedure known as the *standard year irradiance* model, yielding *standard year corrected* data. This model is based on the key postulate that all engineering corrected data represent numbers that are proportional to the actual solar intensities. It is assumed that gradual (of the order of weeks or months) variations in instrument sensitivity change the proportionality constant between engineering corrected data and true solar intensity. A theoretical model is used to estimate how the solar intensity should vary with day of the year during clear weather at solar noon. Then, the instrument sensitivity on any day (cloudy or clear) is inferred by comparing the solar noon engineering corrected intensity on the nearest clear day with the theoretical model. All engineering corrected data for a day are then scaled up or down in accordance with this inferred instrument sensitivity, to obtain the standard year corrected data. A major part of the difficulty in doing this is estimating the theoretical solar noon global insolation in clear weather for any day of the year. This can only be done in an average sense, because of atmospheric fluctuations from year to year, and because the critical period when summer humidity levels move into some areas varies from year to year. A rather detailed sophisticated calculation of atmospheric transmission was performed by Hoyt.[17] His calculation estimates losses of the direct beam from the sun in passing through the atmosphere due to absorption and scattering at each location based on a model atmosphere based on empirical correlations. Hanson[18] modified Hoyt's calculations somewhat by comparing

[17]D. Hoyt, Appendix V of "Hourly Solar Radiation—Surface Meteorological Observations," *Report No. TD 9724*, Vol. 2, NOAA, Asheville, NC, February, 1979.

[18]K. Hanson, Appendix IV, *ibid. loc. cit.*

the calculations with recent Eppley PSP data at three locations. He found that the theory was about a few percent high and therefore empirically modified the theory to reduce predicted levels of insolation in proportion to air mass and water vapor content. Hanson's modification of Hoyt's theory was used to generate the standard year irradiance data that are distributed by the NOAA. Unfortunately, this modification underestimates solar intensities at certain northern localities such as Seattle.[18] An evaluation of the data has been prepared by Rapp.[19]

Fortunately, the new NOAA network is now generating high quality data on both total and direct normal solar intensities at most sites. These data are being published monthly as data books of hourly integrated intensities.

3.9 Correlation of Solar Intensities with Energy Demand

In designing solar energy systems, it is not only desirable to have separate estimates of levels of solar intensities and energy demand for space heating and cooling but also to have the correlation between these variables. Relatively little information has been amassed on the degree of correlation between days of high and low levels of solar intensity and days of high and low levels of energy demand. A very brief study for one locality (Killeen, Texas) was reported by Rapp and Hoffman.[20] Additional unpublished work in Fort Worth, Texas, corroborates the findings in that work. The daily space heating and cooling demand for a building can be correlated with variations in heating and cooling degree-days (see Sect. 7.1). The studies in North Central Texas indicate that for such locations there is essentially no correlation between solar availability and heating degree-days in midwinter. That is, there is no more probability of a bright sunny day being unusually hot or cold in winter than a dull overcast day. On the other hand, in midsummer there is a definite direct correlation between level of solar intensity for a day and cooling degree-days. The daily direct normal intensity is correlated with degree-days for Killeen, Texas, in Figures 3.34, 3.35, and 3.36. It can be seen that there is a considerable difference between summer and winter. It is not known how the corresponding curves vary at other localities.

[19]D. Rapp, "A Critique on NOAA Procedures for Rehabilitation of Solar Data," *Energy Conversion*; **19**, 101, 1979.

[20]D. Rapp and A. A. J. Hoffman, "On the Relation between Insolation and Climatological Variables-IV. Construction of a Model Year of Solar Intensity and Climate," *Energy Conversion*, **17**, 173, 1977.

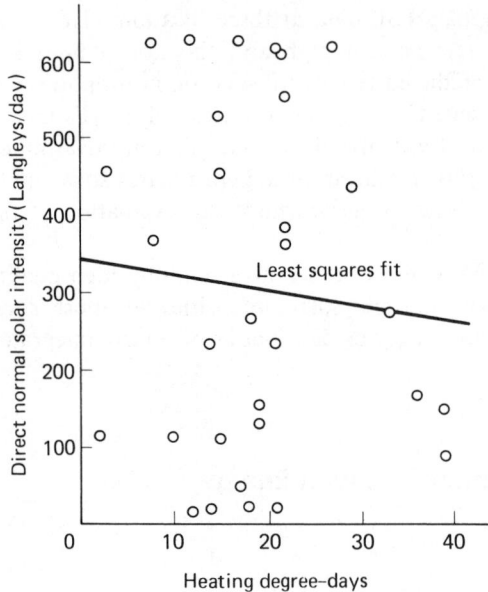

Figure 3.34 Correlation of direct normal intensity with heating degree-days for the month of January, 1961.[20] By permission of Pergamon Press Ltd.

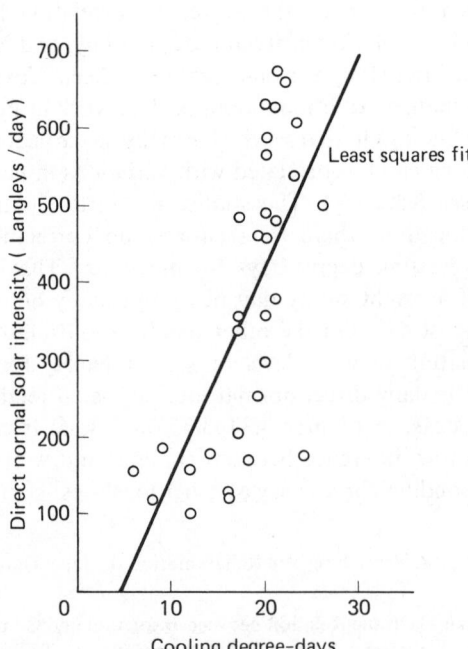

Figure 3.35 Correlation of direct normal intensity with cooling degree-days for August, 1966.[20] By permission of Pergamon Press Ltd.

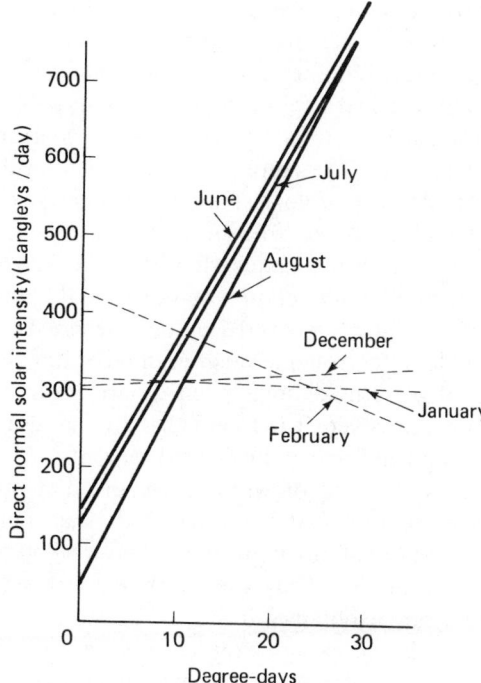

Figure 3.36 Lines representing average dependence of direct normal intensity on degree-days over the 1961–1969 period for summer and winter months.[20] By permission of Pergamon Press Ltd.

3.10 Estimation of Solar Intensities

There are a number of occasions when solar intensities must be estimated for some locality. These include:

1. Cases where local data are not available.
2. Cases where a model year is desired to test conceptual designs of solar energy systems via computer simulation.

The appropriate methods for estimating solar intensities depend on the kind of data required. In cases where local data are not available, one must usually depend on some form of interpolation between stations. A situation that arises in some instances is where solar data are available for a short period (perhaps a year or two), and it is desired to estimate long-term patterns of solar availability from long-term data on climatological variables. Another situation is where global insolation data are available and it is desired to estimate direct normal intensities.

In some cases, data are available for a period of a year or two, and it is desired to approximately infer long-term patterns. This is important because proper design of solar energy installations requires a knowledge of average levels of intensity over a long period, as well as expected variations about the average. Long-term data are available at many stations on climatological variables, and therefore if a relationship between solar intensity and climatological variables can be found, the solar intensity can be estimated over the long term. The simplest approach is to deal with the single variable cloud cover. At meteorological monitoring sites, observations are made hourly on the cloud cover (fraction of the sky covered by clouds). Unfortunately, the estimation of opaque cloud cover is a somewhat subjective operation. Different station operators appear to make rather different judgments of the fraction of the sky covered by opaque clouds. All correlations based on this variable suffer from the vagaries of recording this variable. When an examination of patterns of solar intensity are made under conditions of essentially zero cloud cover, it is found that the patterns tend to be quite repeatable. Rapp and Hoffman[21] divided the year into 26 two week periods and examined patterns of insolation within each period at various locations over long time periods (approximately 20 years). It was found that the insolation during clear weather at any hour of any two week period tended to be repeatable to within about 5% Thus, the hourly patterns on clear days during any 2 week period can be obtained from 1 or 2 years of data by examining clear periods. The insolation pattern expected for clear weather acts as an "envelope" for the actual insolation on any day under actual weather conditions. One may write[22]

$$I(H, W) = f(c, v)I_c(H, W)$$

where c is the cloud cover (0 to 10), H is the hour, W is the two week period (1 to 26), v is the surface visibility in miles, I_c is the clear weather insolation expected at hour H for period W, and I is the actual insolation. It is assumed here that f is determined by c and v. A statistical study can be made for any locality of the dependence of f on c and v. It was found[22] that when the visibility is restricted there are very few occurrences when c is not 10 (100% cloud cover). Thus, there tends to be a curve of f vs. c for high v and a single value of f for $c = 10$ and low v. An example is shown in Figure 3.37. On a statistical basis,[23] substantial reductions in I do not tend to occur until c

[21]D. Rapp and A. A. J. Hoffman, "On the Relation between Insolation and Climatological Variables-V. Estimation of Availability of Solar Energy," *Energy Conversion*, **18**, 31, 1978.

[22]D. Rapp and A. A. J. Hoffman, "On the Relation between Insolation and Climatological Variables-II. Prediction of Insolation at Fort Hood, Texas," *Energy Conversion*, **17**, 31, 1977.

[23]*Ibid.*

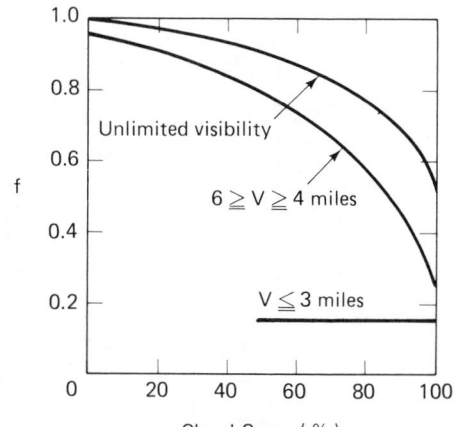

Figure 3.37 Coefficient for correcting clear weather insolation for the effect of weather.[22] The coefficient is given as a function of cloud cover for various ranges of visibility. By permission of Pergamon Press Ltd.

rises to about 0.6. Having found the function $f(c, v)$ based on radiation data for a year or two, one may then use long-term meteorological data on hourly values of c and v to estimate long-term variations in I. In actual fact, there are a range of values of f that occur for each (c, v) combination. The correlation described above only leads to the average value of f corresponding to any (c, v) combination. One could study the range of variations about the average and use this function to generate randomly selected variations about the average. For purposes of generating approximate model years of data, this does not seem to be necessary.

When it is desired to estimate the direct normal solar intensity from available hourly data on insolation, a relationship must be sought between the direct normal intensity and the total insolation. A good way to do this is to define clear weather envelopes for direct normal intensity $N_{cl}(H, W)$, and to let

$$N(H, W) = g(f)N_{cl}(H, W)$$

where N is the actual intensity, and g is a factor (less than unity) that depends on f. The clear weather patterns $N_{cl}(H, W)$ are known at only a few locations and would have to be approximated at other localities. The function $g(f)$ has been studied at Fort Hood, Texas,[24] and has the form shown in Figure 3.38. Randall and Whitson[25] have prepared a report giving a more elegant procedure for estimating N at any hour for a wider range of locations.

[24]D. Rapp and A. A. J. Hoffman, "On the Relation between Insolation and Climatological Variables-III. The Relation between Normal Incidence Solar Intensity, Total Insolation, and Weather at Fort Hood, Texas," *Energy Conversion*, **17**, 163, 1977.

[25]C. M. Randall and M. E. Whitson, Jr., "Hourly Insolation and Meteorological Data Base," *Aerospace Corp. Report*, El Segundo, CA, 1978.

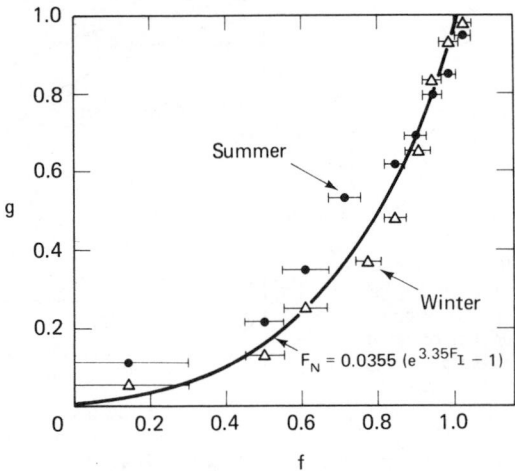

Figure 3.38 Dependence of g on predicted f for the entire period September 6, 1974, to August 31, 1975.[24] By permission of Pergamon Press Ltd.

Randall and Whitson defined t as the ratio of observed ground level direct normal intensity to extraterrestrial intensity at normal incidence, and p as the ratio of observed global insolation to the extraterrestrial radiation on a plane parallel to a horizontal plane. They examined data at four locations to infer the dependence of t on p. They found the average value of t corresponding to any range of values of p, $\bar{t}$, as well as the distribution function for variations in t about $\bar{t}$. The average values $\bar{t}(p)$ are illustrated in Figure 3.39. A computer program was written to estimate direct normal intensities from global insolation values. At any hour for which the global insolation is known, the value of p is calculated, and $\bar{t}(p)$ is inferred from the correlation. The distribution function of t values about $\bar{t}(p)$ is used, together with a random number generator to select a deviation of t from $\bar{t}(p)$. Thus, a value of t is selected. This procedure was applied to the standard year corrected global insolation data from the NOAA network. A summary of their findings for average daily insolation on a horizontal surface and average daily direct normal intensity as a function of month for 26 locations is shown in Tables 3.3 and 3.4. Unfortunately, the errors in the standard year irradiance data for total insolation at certain sites such as Seattle have been propagated into this work for direct normal intensities.

In testing the performance of various conceptual designs of solar energy systems by computer simulation, it is desirable to have a model year of hourly values of solar intensities and climatic variables. If long-term hourly data are available, either from actual instruments or by estimation, a reasonable

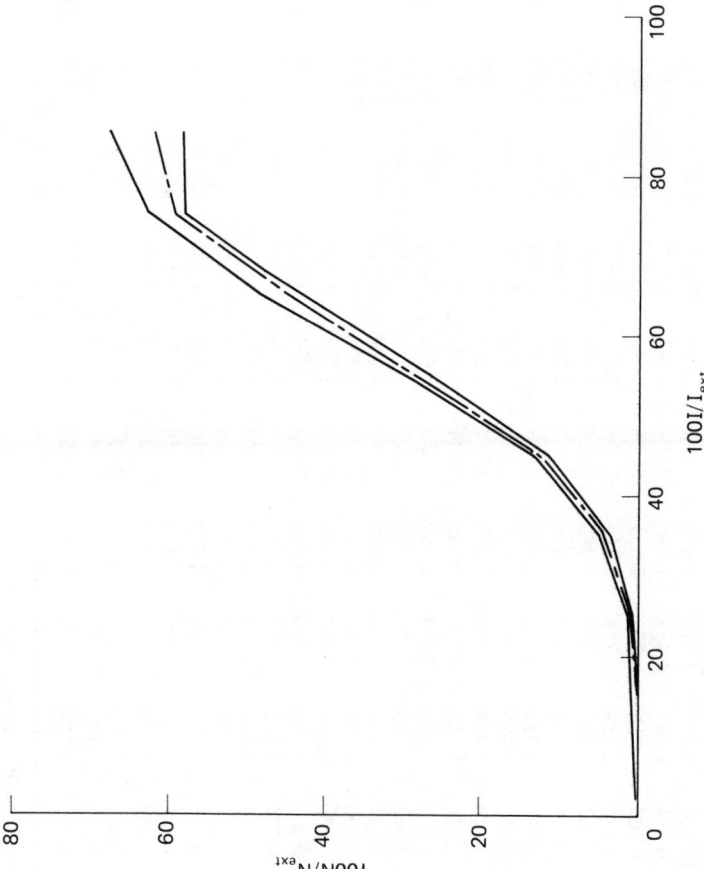

Figure 3.39 Correlation of N/N_{ext} with I/I_{ext} for several sites. The lines show the range of average values. By permission, C.M. Randall and M.E. Whitson, Jr., *Hourly Isolation and Meteorological Data Bases*, Aerospace Corp. Rept ATR-78 (7592)-1, 1 Dec. 77.

TABLE 3.3 Average levels of insolation on a horizontal surface[25] (10^3 Btu/ft^2-day)*

	Jan	Feb	Mar	Apr	May	Jun	Jul	Aug	Sep	Oct	Nov	Dec	Avg
Albuquerque	1.018	1.339	1.767	2.227	2.538	2.677	2.487	2.290	1.970	1.545	1.132	0.926	1.827
Apalachicola	0.853	1.123	1.475	1.878	2.090	1.998	1.814	1.688	1.535	1.370	1.037	0.818	1.472
Bismarck	0.463	0.774	1.167	1.459	1.846	2.059	2.182	1.875	1.351	0.907	0.504	0.371	1.250
Boston	0.476	0.707	1.015	1.326	1.668	1.818	1.748	1.488	1.259	0.888	0.501	0.403	1.104
Brownsville	0.910	1.132	1.459	1.735	1.929	2.116	2.211	2.024	1.694	1.440	1.050	0.860	1.548
Cape Hatteras	0.685	0.948	1.326	1.773	1.960	2.036	1.919	1.703	1.469	1.136	0.872	0.660	1.377
Caribou	0.419	0.723	1.132	1.415	1.576	1.757	1.760	1.500	1.101	0.688	0.365	0.311	1.063
Charleston	0.745	0.993	1.339	1.732	1.859	1.843	1.799	1.586	1.393	1.193	0.933	0.720	1.345
Columbia	0.612	0.872	1.180	1.526	1.878	2.087	2.116	1.878	1.450	1.101	0.704	0.523	1.329
Dodge City	0.828	1.120	1.475	1.884	2.090	2.357	2.293	2.055	1.684	1.301	0.895	0.733	1.561
El Paso	1.126	1.478	1.910	2.360	2.601	2.684	2.449	2.284	1.986	1.637	1.243	1.031	1.897
Ely	0.818	1.139	1.605	2.008	2.309	2.512	2.446	2.230	1.935	1.405	0.926	0.723	1.675
Ft. Worth	0.803	1.063	1.415	1.615	1.900	2.157	2.160	1.986	1.618	1.285	0.933	0.761	1.472
Fresno	0.657	1.012	1.564	2.090	2.484	2.731	2.684	2.423	1.982	1.427	0.888	0.574	1.713
Great Falls	0.419	0.720	1.170	1.488	1.846	2.100	2.328	1.932	1.377	0.895	0.498	0.339	1.259
Lake Charles	0.726	1.009	1.313	1.570	1.849	1.970	1.786	1.656	1.484	1.304	0.914	0.704	1.358
Madison	0.514	0.803	1.136	1.396	1.741	1.948	1.935	1.710	1.297	0.910	0.504	0.387	1.193
Medford	0.406	0.736	1.132	1.637	2.033	2.277	2.474	2.119	1.589	0.980	0.504	0.336	1.354
Miami	1.053	1.313	1.602	1.856	1.843	1.707	1.764	1.630	1.456	1.304	1.113	1.018	1.475
Nashville	0.580	0.822	1.129	1.542	1.824	1.963	1.891	1.735	1.396	1.113	0.711	0.520	1.269
New York	0.501	0.720	1.037	1.364	1.637	1.710	1.688	1.481	1.212	0.895	0.533	0.403	1.094
Omaha	0.634	0.888	1.221	1.557	1.862	2.119	2.103	1.859	1.370	1.047	0.644	0.511	1.323
Phoenix	1.021	1.370	1.814	2.354	2.677	2.737	2.487	2.290	2.014	1.570	1.148	0.933	1.871
Santa Maria	0.853	1.139	1.583	1.919	2.141	2.347	2.341	2.106	1.729	1.351	0.974	0.803	1.611
Seattle	0.260	0.495	0.853	1.294	1.719	1.799	1.979	1.605	1.155	0.650	0.339	0.213	1.034
Washington	0.574	0.812	1.126	1.459	1.094	1.900	1.789	1.621	1.332	0.993	0.650	0.482	1.205

*Note: multiply table values by 11,360 to convert to J/m^2.

TABLE 3.4 Average levels of estimated direct normal intensity[25] (10^3 Btu/ft²-day)*

	Jan	Feb	Mar	Apr	May	Jun	Jul	Aug	Sep	Oct	Nov	Dec	Avg
Albuquerque	1.792	1.976	2.147	2.446	2.674	2.810	2.534	2.455	2.385	2.236	1.944	1.751	2.262
Apalachicola	0.983	1.158	1.342	1.662	1.745	1.561	1.354	1.342	1.342	1.494	1.307	1.021	1.358
Bismarck	0.774	1.098	1.342	1.421	1.726	1.916	2.220	2.078	1.589	1.323	0.837	0.679	1.418
Boston	0.663	0.809	0.952	1.082	1.348	1.450	1.370	1.215	1.240	1.078	0.641	0.590	1.034
Brownsville	0.961	1.085	1.209	1.316	1.453	1.735	1.976	1.852	1.516	1.472	1.164	0.964	1.393
Cape Hatteras	0.888	1.085	1.269	1.602	1.656	1.688	1.529	1.408	1.345	1.266	1.218	0.942	1.326
Caribou	0.625	0.952	1.215	1.266	1.281	1.405	1.465	1.361	1.088	0.796	0.447	0.463	1.028
Charleston	0.904	1.056	1.212	1.434	1.402	1.301	1.288	1.145	1.098	1.240	1.218	0.958	1.186
Columbia	0.866	1.091	1.155	1.339	1.589	1.805	1.941	1.802	1.462	1.396	1.015	0.774	1.351
Dodge City	1.408	1.596	1.396	1.919	2.002	2.309	2.281	2.135	1.954	1.811	1.497	1.332	1.830
El Paso	1.849	2.084	2.287	2.620	2.741	2.795	2.490	2.411	2.303	2.258	2.014	1.795	2.303
Ely	1.478	1.713	2.005	2.186	2.360	2.582	2.572	2.550	2.547	2.154	1.662	1.396	2.103
Ft. Worth	1.091	1.259	1.405	1.377	1.519	1.849	1.941	1.843	1.589	1.519	1.310	1.139	1.481
Fresno	0.796	1.183	1.710	2.189	2.541	2.899	2.896	2.706	2.414	1.970	1.297	0.733	1.944
Great Falls	0.695	1.025	1.389	1.412	1.691	1.976	2.509	2.151	1.688	1.313	0.844	0.609	1.443
Lake Charles	0.752	1.006	1.101	1.129	1.364	1.494	1.301	1.272	1.237	1.358	1.063	0.844	1.155
Madison	0.764	1.050	1.180	1.218	1.459	1.580	1.656	1.570	1.310	1.120	0.669	0.574	1.177
Medford	0.374	0.771	1.040	1.450	1.824	2.122	2.655	2.319	1.824	1.183	0.533	0.314	1.370
Miami	1.186	1.275	1.364	1.456	1.304	1.085	1.155	1.098	1.006	1.148	1.205	1.231	1.212
Nashville	0.701	0.895	0.996	1.256	1.421	1.523	1.446	1.443	1.234	1.259	0.898	0.682	1.139
New York	0.634	0.768	0.936	1.094	1.247	1.224	1.218	1.155	1.059	0.961	0.625	0.511	0.952
Omaha	1.012	1.177	1.272	1.434	1.659	1.913	1.963	1.852	1.472	1.402	0.983	0.837	1.418
Phoenix	1.599	1.894	2.084	2.547	2.820	2.788	2.468	2.363	2.316	2.132	1.792	1.548	2.195
Santa Maria	1.215	1.421	1.691	1.840	1.929	2.195	2.265	2.081	1.824	1.684	1.405	1.285	1.741
Seattle	0.222	0.441	0.682	0.967	1.364	1.377	1.821	1.516	1.139	0.641	0.333	0.190	0.895
Washington	0.745	0.904	1.040	1.189	1.339	1.484	1.342	1.310	1.215	1.123	0.844	0.631	1.094

*Note: Multiply table values by 11,360 to convert to J/m².

procedure for selecting a model year is as follows. The monthly totals of solar intensity and heating and cooling degree-days (see Sec. 7.1) are tabulated for as many years as data are available. For each month of the year, an average of each of these quantities is computed. Then a scan is made of all the Januarys to see which January has average values that are closest to the long-term average. This is repeated for all the months of the year. Finally, a model year is produced by compounding together the 12 most average months.

A different approach for estimating solar intensities is presented in a series of papers by Liu and Jordan.[26] These methods are widely used.

3.11 Circumsolar Radiation

The radiation received at the earth's surface is usually divided into the categories direct and diffuse. Strictly speaking, direct rays should be those received within a half-angle of 0.27° to correspond to the angle subtended by the solar disc. Diffuse radiation would then be radiation that has been scattered and reaches ground level at an angle greater than 0.27° from the center of the solar disc. In practice, this definition is not very useful, because pyrheliometers used for measuring direct radiation have an acceptance half-angle of about 2.9°. Therefore, pyrheliometers record direct radiation plus a small amount of diffuse radiation from the vicinity of the sky near the solar disc. In practical terms, this measured value is referred to as the *direct radiation*. The diffuse radiation, in practical terms, is then the radiation outside of a half-angle of 2.9°. Another term that has come into use is *circumsolar radiation*. This term refers to radiation received within a range of angles just outside the solar disc. In practical terms, the designer of solar collectors is concerned with what fraction of the measured direct normal intensity (half-angles up to 2.9°) can be imaged by a solar concentrator. This depends on the angle of acceptance of the concentrator as shown in Figure 3.40. For a linear array, the maximum acceptance half-angle is given by[27]

$$\theta_m = \arc\sin \frac{1}{c}$$

where c is the concentration. For concentrations of 10 to 30, θ_m varies from 5.7° to 1.9°. For real collectors, the acceptance angles are perhaps about half of these values. Thus, one is concerned with light rays within about 1° to 3° of the center of the solar disc for concentrations ranging from 30 to 10. For two-dimensional collectors with dish geometry the appropriate

[26]B. Y. H. Liu and R. Jordan, "Availability of Solar Energy for Flat Plate Solar Heat Collectors," in *Low Temperature Engineering Application of Solar Energy*, American Society of Heating, Refrigerating and Air Conditioning Engineers, New York, 1977.

[27]See Chapter 10.

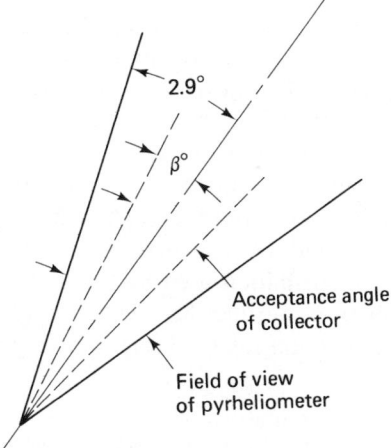

Figure 3.40 Comparison of acceptance angle of concentrating collector with field of view of normal incidence pyrheliometer.

formula is

$$\theta_m = \text{arc} \sin \frac{1}{\sqrt{c}}$$

For concentrations ranging from 100 to 900, the acceptance angles are about the same as for linear arrays with concentrations between 10 and 30. A reasonable average acceptance half-angle for many concentrating collectors is perhaps 1.5°.

We may therefore define the following pragmatic definitions:

N = direct normal intensity
= intensity from 0 to 2.9° half-angle from the center of the solar disc

I = global insolation
= intensity over a hemisphere of acceptance

C = circumsolar radiation
= intensity from 1.5° to 2.9° half-angle from the center of the solar disc

With these definitions, the effective intensity for many solar concentrators is approximately $N - C$.

The circumsolar telescope is a very expensive, sophisticated device which scans the differential solar intensity in each 1.5 min of arc across the solar disc. This allows determination of the intensity vs. angle from the center of the solar disc. Results from use of this device have been reported by Grether, Hunt and Wahlig.[28] These authors defined the circumsolar

[28]D. F. Grether, A. Hunt, and M. Wahlig, "Results from Circumsolar Radiation Measurements," *Proc. 1976 International Solar Energy Society Meeting*, Winnipeg, Canada, August, 1976, Vol. 1, p. 363.

radiation to be the radiation outside the solar disc (approximately from half-angles 0.27° to 2.9°). This is rather different than the definition adopted in this book. The definition used by Grether *et al.* is scientific, but it is not very useful in determining the effect on solar collector performance. In most cases, there is a rapid decrease in differential solar intensity between half-angles of about 0.27° and about 0.5°, and there is little intensity left outside of 0.5°. Thus, even though a moderately large fraction of radiation might come from outside the solar disc, most of this can be concentrated by solar collectors. The definition used by Grether *et al.* emphasizes the occurrence of radiation from outside the solar disc but gives a false impression of the defocussing effect on solar collectors. In Figure 3.41 are shown some typical

Figure 3.41 Brightness as a function of position in a sweep across the solar disc.[28] By permission of the author and the American Section of the International Solar Energy Society, Inc.

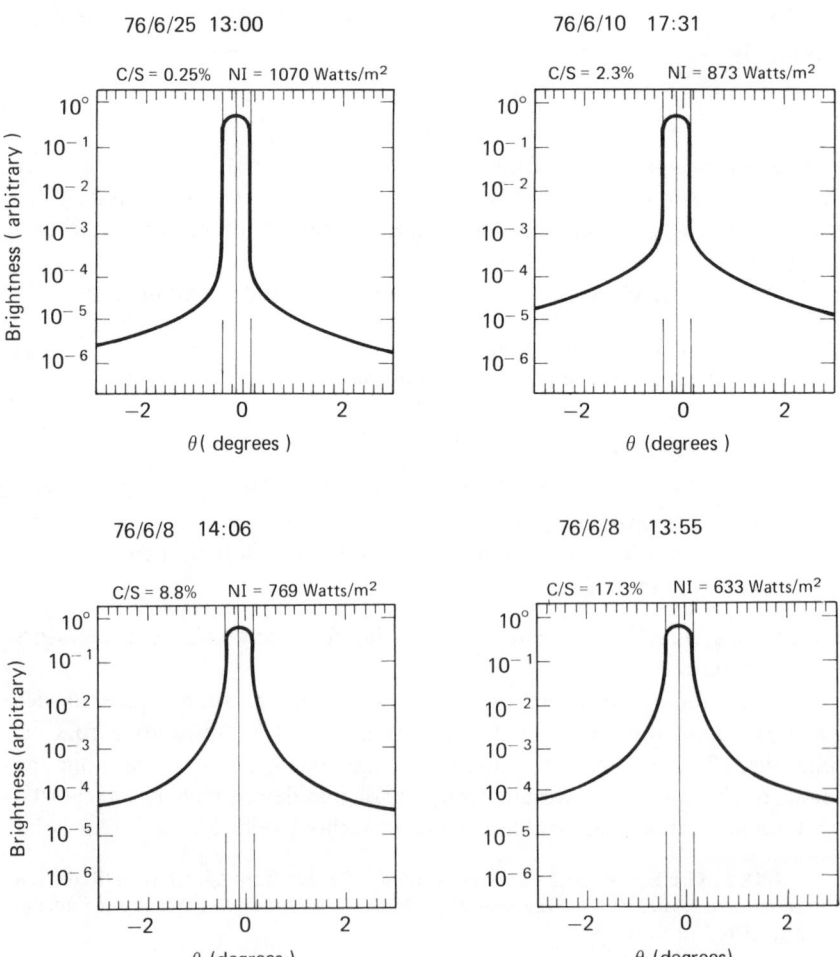

results reported by Grether *et al.* Four nearly instantaneous sky conditions are illustrated, ranging from very clear to moderately cloudy. In the clearest condition, the integrated intensity outside the solar disc to 2.9° is estimated as 0.25% of the total from the center of the solar disc to 2.9°. As the solar intensity is reduced, the *percent circumsolar* (outside the solar disc) rises, reaching 17.3% in the worst case illustrated. However, using the pragmatic definition of circumsolar radiation given here, the percent circumsolar is quite small under all conditions illustrated in Figure 3.41. Circumsolar radiation, as defined by Grether *et al.*, only becomes a factor in solar collection for concentrators with acceptance half-angles less than about 0.5°. For collectors with half-angles of acceptance greater than about 1.5°, circumsolar radiation is generally of no importance.

Worked Examples

1. It is found that at a site at 40°N latitude at 1100 solar time on February 3 the total insolation on a horizontal surface is 670 W/m², and the direct normal intensity is 800 W/m². What are the fractions of direct and diffuse intensity on the horizontal surface?

   Direct component on horizontal surface $= D = N \cos \theta_h$

   $$\sin d = -\sin(23.45°) \cos\left[\frac{(D_s - 1)}{182} 180°\right]$$

   $$\sin d = -0.3979 \times 0.737 = -0.293$$

   $$d = -17.1°$$

   $$\cos \theta_h = \cos L \cos d \cos 15° + \sin L \sin d$$

   $$\cos \theta_h = 0.707 - 0.189 = 0.518$$

   $$\theta_h = 58.8°$$

   $$D = 800 \times 0.518 = 414 \text{ W/m}^2$$

   Diffuse fraction $= (670 - 414)/670 = 0.381$

   Direct fraction $= 0.619$

2. Estimate the solar intensity on a surface tilted at 40° to the south for the conditions given in Example 1 above.

   $$\cos \theta_t = \cos d \cos 15° = 0.923$$

   $$\theta_t = 22.6°$$

   Direct component on tilted surface

   $$= \frac{\cos \theta_t}{\cos \theta_h} \times 414 = \frac{0.923}{0.518} \times 414 = 738 \text{ W/m}^2$$

Diffuse component on tilted surface
$\approx$ diffuse component on horizontal surface $= 670 - 414 = 256 \text{ W/m}^2$

Total intensity on tilted surface $= 738 + 256 = 994 \text{ W/m}^2$

3. What are the extraterrestrial intensities at normal incidence, and on a horizontal plane, at 40°N latitude at 1100 solar time on February 3?

According to Eqs. (3.4) and (3.5),

$$\phi = 32 \times \frac{360}{365.2} = 31.5°$$

$$N_{ext} = \frac{1377(1 + 0.0167 \cos 31.5°)^2}{(0.9997)} = 1417 \text{ W/m}^2$$

$$I_{ext} = N_{ext} \cos \theta_h = 1417 \times 0.518 = 734 \text{ W/m}^2$$

Problems

3.1. A flat plate collector is tilted toward the south at angle T. A horizontal field of snow with reflectivity R lies south of the collector. The direct normal intensity and total insolation on a horizontal surface at solar noon are N and I_0, respectively. What is the solar intensity on the collector?

3.2. Assume that the variables in Problem 3.1 are $T = 45°$, latitude $= 33°N$, day $=$ December 21, $R = 0.7$, $N = 280 \text{ Btu/hr-ft}^2$ (882 W/m²), $I_0 = 175 \text{ Btu/hr-ft}^2$ (551 W/m²), time $=$ solar noon. Calculate the direct, diffuse, and reflected components on the tilted surface.

3.3. How could you use a pyranometer to make a direct measurement of the diffuse component of solar intensity on a surface?

3.4. On March 21 at 1400 solar time at a site with latitude $= 35°N$ in clear weather, it is found that the ground level solar intensities are 800 W/m² on a horizontal surface and 830 W/m² for the direct normal intensity. What are the percentages of extraterrestrial intensity in each case?

3.5. What is the average extinction coefficient of the atmosphere for the conditions given in Problem 3.4?

3.6. If the weather remains clear, what would you expect N and I to be 2 hr later at 1600 solar time for the case given in Problem 3.4?

3.7. The solar intensity on a horizontal surface is measured to be 180 Btu/hr-ft² at 1300 on January 21 at a latitude of 35°N. Estimate the solar intensity on surfaces tilted at 20°, 35°, and 50° toward the south.

3.8. The hourly insolation on a horizontal surface in clear weather at Fort Hood, Texas, has been found to satisfy the equation (Btu/hr-ft²)

$$I_h = \left[192 + 129 \sin \frac{\pi D_s}{365} \right] \cos \left[(H - 12)(0.34 - 0.00056 d_s) \right]$$

where D_s is the day of the solar year measured from December 21 as day 1, d_s is the number of days from December 21 in the shortest direction, and H is the solar time. Integrate this equation from sunrise to sunset to obtain an analytic formula for daily total insolation on a horizontal surface at that location. Plot the daily total vs. day of the year.

3.9. Roughly estimate the clear daily total insolation on a tilted surface at Fort Hood by multiplying the values from Problem 3.8 by tilt correction factors for each day corresponding to 2 hr from solar noon. Plot the intensities on a tilted surface (use 21°, 31°, and 46° tilt) vs. day of the year.

3.10. Using the expression for I_h given in Problem 3.8, write an expression for I_t from tilt correction factors for a surface tilted at any angle. Suppose you had a mountain that was a perfect cone with a 45° half-angle. What would be the daily total solar irradiance on the south face and the north face as a function of day of the year? Plot graphically. Use a latitude = 31°N.

3.11. A house is located at 40°N latitude with its central axis along the east-west direction. The roof is peaked with north- and south-facing slopes at 30° to the horizontal. A snow storm drops 12 in. of snow on the roof during the night (12 in. snow is equivalent to 1 in. of water). The next day is clear and sunny, and the total insolation on a horizontal surface is 1300 Btu/ft² (14,760 kJ/m²). If the air is at 32°F (273°K) and saturated with moisture, what fraction of the snow is melted off on the north and south faces at the end of the day? Assume the reflectivity of the snow is 50%, the roof insulation is R15, the interior temperature is 70°F (294°K), and the average tilt correction factor is that corresponding to 2 hr from noon on January 1.

3.12. The actual data on measured solar intensities at three diverse sites on clear November days are given in the table below. From these data, at each hour:
(a) Calculate the percentage of diffuse radiation on horizontal (and, where appropriate) tilted surfaces.
(b) At the sites where tilted pyranometer measurements are available, compare the observed ratios I_t/I_h with the calculated factors $\cos \theta_t/\cos \theta_h$.
(All intensities in Langleys/hr, temperatures in °F.)

Hour Ending at (local time)	Approx Solar Time at Midpoint of Hour	Fort Hood, TX, Lat. = 31.08° Day = 11/26/76 I_h $I_{30°}$ N T_{amb}				Maynard, MA, Lat. = 42.42° Day = 11/14/77 I_h N T_{amb}			Yuma, AZ, Lat. = 32.83° Day = 11/14/77 I_h $I_{45°}$ N T_{amb}			
8	7:30	6	16	31	56	3	9	35	7	22	32	59
9	8:30	23	40	60	58	16	55	37	25	50	65	68
10	9:30	37	60	71	66	27	69	39	39	70	74	72
11	10:30	48	72	76	69	36	77	41	51	84	78	76
12	11:30	54	81	79	72	42	80	42	55	91	79	79
13	12:30	54	82	78	76	42	81	43	55	90	79	80
14	13:30	49	76	77	77	37	80	42	49	82	77	82
15	14:30	38	63	73	76	28	75	41	38	67	72	83
16	15:30	24	44	62	76	15	64	39	24	45	63	82
17	16:30	6	12	25	72	2	16	37	8	19	41	80

3.13. From the following data:

Hour = solar noon,   Day = December 1

Latitude = 33°N, N_{ext} = 142 mW/cm²

I = 63 mW/cm² N = 93 mW/cm²

(a) Calculate I/I_{ext}, where I = insolation on a horizontal surface.
(b) Calculate the percentage of diffuse radiation on a horizontal surface at ground level.
(c) Calculate the average extinction coefficient for passage through the atmosphere.
(d) Estimate N at 3 hr from solar noon from the extinction coefficient.

4
Flat Plate Collectors

This chapter provides an elementary description of the principles and performance characteristics of flat plate solar collectors.[1] From this description, semiquantitative evaluations result of the effects of additional glazings or selective surfaces on performance. A more sophisticated treatment is given at the end of the chapter on the effect of tube spacing on the collector plate, and the collector efficiency as a function of inlet temperature.

4.1 Static Temperatures Attainable

4.1.1 Introduction

Flat plate solar collectors operate on the principle that when a black sheet of metal is placed in the sun the sheet absorbs sunlight which appears as heat. The plate will gradually heat up until its temperature is high enough above ambient that the rate of heat loss from the plate to the ambient air just balances the rate of heat gain from absorption of solar rays. However, it is a matter of common experience that if there is a wind blowing, the heat transfer from the plate to the air will be so effective that the plate will not be more than a few degrees warmer than ambient. Nevertheless, if the collector

---

[1]For other information, the reader is referred to *Proc. Workshop on Solar Collectors for Heating and Cooling of Buildings, NSF Rept. NSF-RA-N-75-019*, New York, November 21, 1974.

plate can be thermally insulated from the ambient air, the solar heat gain can produce a substantial temperature rise. The simplest mechanism for doing this is to cover the black plate with a transparent cover separated by an air gap. This will allow most of the incident light to reach the plate but will severely cut down the rate of heat loss to the ambient. In bright sunlight near solar noon, solar intensities of ~ 300 Btu/hr-ft² (945 W/m²) are the highest commonly available. A metal sheet (say, $\frac{1}{16}$ in. thick) has a heat capacity of about 0.5 Btu/°F per sq ft. Thus, if the absorbed heat is not carried off, the plate will rise at the rate of 600°F per hr in bright sunlight! In actuality, the plate will begin to lose heat to its surroundings at a faster rate as it becomes hotter and will eventually reach an equilibrium temperature where the rate of heat loss exactly balances the rate of heat gain from solar absorption. In practice, a hot metal sheet is not of any value by itself. A solar collector involves movement of a fluid, either as air blown over the plate or water flowing through tubes attached to the plate, to carry off the heat that is collected.

A well-designed solar collector can convey more than half of the incident solar energy into the circulating fluid as heat, the remainder being lost as heat or reflected light rays to the ambient.

4.1.2 A Static Evacuated Collector

First, consider the idealized flat plate collector shown in Figure 4.1. This collector consists of a shallow box which is very well insulated on the sides and back and which has a glass cover on top. A sheet metal plate is

Figure 4.1 Black metal sheet in evacuated box.

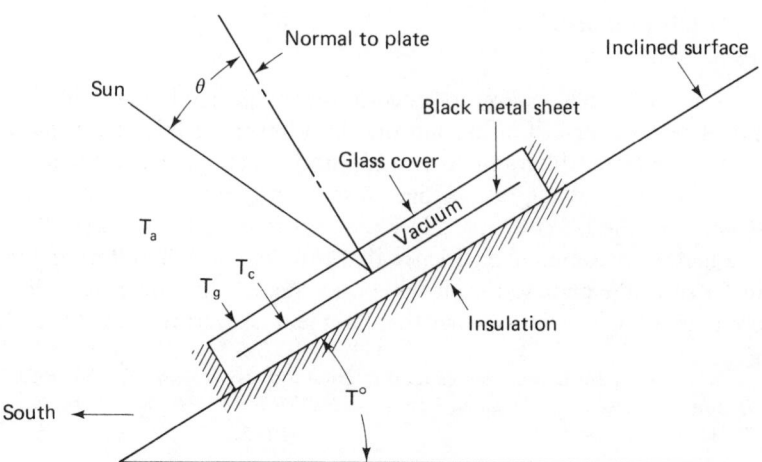

suspended in the box parallel to the glass, and its top surface is painted black. The box is placed on an inclined plane tilted at angle T from the horizontal, facing south. The metal plate becomes heated when exposed to the sun. As it becomes warmer than its surroundings, it begins to lose heat by conduction, convection, and radiation if air is present in the box. We shall consider a simplified model in which the box is evacuated in order to eliminate conduction and convection by air surrounding the plate. Thus, the only mechanism for heat loss by the plate is radiation.

To estimate the temperature that the plate will reach under bright sunny conditions, consider the following simple approximate analysis. The tilt angle will be set equal to the latitude, and this will be taken as 33°N. The ambient temperature will be taken as 30°F (272°K) in winter and 90°F (305°K) in summer. The peak solar intensity on a horizontal surface will be assumed to be 200 Btu/hr-ft² (630 W/m²) in winter and 300 Btu/hr-ft² (945 W/m²) in summer. With these assumptions, rough estimates of the equilibrium temperatures reached will be made on bright sunny winter and summer days near noon.

As the plate becomes heated, it will radiate heat to the inside of the insulation on the sides and bottom of the box until the temperatures of these surfaces rise to nearly the temperature of the plate. If the insulation is very effective, almost no heat will flow out the sides and bottom of the box compared to the heat flow through the glass on top. Thus, heat losses through the sides and back will be neglected. At equilibrium, the collector plate reaches temperature T_c and the glass reaches T_g. The glass becomes heated by absorption of incident sunlight, reflected visible light from the plate, and by infrared radiation from the collector plate. We shall consider only the radiative heating from the hot collector. The net rate of radiative heat transfer from the hot plate to the glass (assuming emissivities of unity) is

$$R_{cg} = \sigma(T_c^4 - T_g^4) \tag{4.1}$$

where R_{cg} is in units of W/m², $\sigma = 5.67 \times 10^{-8}$ W/m²-°K⁴, and the temperatures are in °K. The temperature of the glass will rise until it loses heat to the ambient air at the same rate that it gains heat by radiation from the collector. The glass loses heat to the ambient by radiation and by conduction and convection. The radiative loss term is similar to Eq. (4.1), except that T_c and T_g are replaced by T_g and T_s, where T_s is the effective temperature of the radiation heat sink represented by the sky and the clouds over the solid angle above the collector. When the sky is clear, this can be a rather low temperature, because the radiation emitted by the glass requires a considerable path length of atmosphere for absorption. In cloudy weather, T_s may be roughly approximated by the ambient temperature T_a at the height of the clouds. In the following treatment, the approximation of replacing T_s by T_a will be made even in clear weather. This will tend to underestimate radiative heat losses and overestimate collector temperatures. This approximation is made because it is difficult to estimate effective sky temperatures, and because only a rough estimate of collector temperatures is desired. The conductive/convective loss term can be represented by a linear term in $(T_g - T_a)$. Thus, the total rate of heat loss to the ambient is approximated as

$$R_{ga} = \sigma(T_g^4 - T_a^4) + U(T_g - T_a) \tag{4.2}$$

where U is a heat transfer coefficient for conduction and convection from the glass to the outside air. A rough reasonable value for U in 10 mph wind is ~ 13.6 W/m²-°K. The rate of heat gain by the collector plate by solar absorption is the product of the solar flux impinging on the cover glass, the transmissivity of the cover glass, and the absorptivity of the black surface. The solar flux impinging on the cover glass is calculated from the flux on a horizontal surface, using the tilt correction factor described in Sec. 2.6. At 12 noon on December 21, this factor is 1.59, while at noon on June 21 it is 0.92. The transmissivity of the glass is taken as 0.90, and the absorptivity of the black surface is assumed to be 0.92. Thus, the rate of heat gain by the collector plate is estimated at

$$R_c = I_h f_T (0.92)(0.90) \tag{4.3}$$

where I_h is the insolation on a horizontal surface, and f_T is the tilt correction factor. When thermal equilibrium is reached, the flow of power from the sun to the collector plate, from the collector plate to the glass, and from the glass to the ambient are all equal. We may, therefore, generate two equations by setting

$$R_c = R_{cg} = R_{ga} \tag{4.4}$$

Only two unknown quantities remain, namely T_g and T_c. These may then be evaluated.

In the winter,

$$R_c = 630 \text{ W/m}^2 \times 0.92 \times 0.90 \times 1.59$$

$$R_c = 829 \text{ W/m}^2 \tag{4.5}$$

$$R_{ga} = 5.67 \times 10^{-8} \frac{\text{W}}{\text{m}^2 \text{ °K}^4}[T_g^4 - (272)^4] + 13.6(T_g - 272) \tag{4.6}$$

where the temperatures are in °K.

By setting $R_c = R_{ga}$, one may solve for T_g by trial and error. First, a value of T_g is guessed, and R_{ga} is calculated. Further guesses for T_g are made until R_{ga} closely approximates 829 W/m². The results are as shown below:

| T_g (guess) | | R_{ga} (calculated) |
(°F)	(°K)	(W/m²)
150	339	1345
130	328	1098
120	322	980
110	317	861
106	314	815
107	315	828

The correct value of T_g is, therefore, about 107°F (315°K). This value may be used in the equation

$$R_{cg} = \sigma(T_c^4 - T_g^4) = 829 \text{ W/m}^2$$

to solve for T_c. It is found that

$$T_c = \left(\frac{829}{\sigma} + T_g^4\right)^{1/4} = 396°\text{K} = 252°\text{F}$$

If the same calculation is repeated for the conditions in the summer, it is found that

$$R_c = 945 \times 0.92 \times 0.90 \times 0.92 = 720 \text{ W/m}^2$$

Then, using $T_a = 90°$F (305°K) it is found that $T_g = 166°$F (347°K) and $T_c = 279°$F (410°K). Note that there is a large difference between summer and winter in the glass temperature, but there is a smaller difference in the plate temperature. It is clear that quite high static temperatures can be achieved on a metal plate in an evacuated container in the sun.

Although a solar collector surface absorbs visible and infrared light from the sun, its radiative losses are mainly in the far infrared. Special collector surfaces have been developed that absorb very well in the solar spectrum but are poor radiators in the far infrared. Such surfaces generally may tend to have slightly lower overall absorptance of incident solar light than black paint but have much lower radiative heat loss rates. The achievable static temperature can be increased considerably by using a selective surface on the collector plate, which absorbs well in the solar spectrum but which radiates poorly in the far infrared. Such surfaces have been produced and used in solar collectors. We shall assume a selective surface with values of absorptivity for the solar spectrum of 0.85 and emissivity in the far infrared of 0.3. Based on these values, R_c is reduced by the fraction 0.85/0.92, but the radiative loss R_{cg} is reduced by 0.30/1.00.

With these assumptions, the equations for winter conditions are

$$R_c = 630 \times 0.85 \times 0.90 \times 1.59 = 766 \text{ W/m}^2$$
$$R_{ga} = 5.67 \times 10^{-8} [T_g^4 - (272)^4] + 13.6(T_g - 272)$$
$$R_{cg} = 0.3\sigma(T_c^4 - T_g^4)$$

The solution is $T_g = 102°F$ (312°K) and $T_c = 410°F$ (484°K). In the summer, a similar calculation yields $T_g = 153°F$ (340°K) and $T_c = 415°F$ (486°K). When conduction and convention from the collector plate have been eliminated by surrounding it with a vacuum, a reduction in emissivity of the collector plate in the far infrared produces a much higher static temperature.

In a flat plate collector filled with air, heat loss from the collector plate occurs by both convection/conduction and radiation. Evacuation eliminates convection/conduction, and selective surfaces reduce radiation. Actual collectors filled with air can still reach static temperatures of 200–400°F (360–480°K), depending on the selectivity of the collector surface. Such collectors must be designed so they do not fail from thermal stresses due to overheating when no fluid is circulating.

4.2 Performance of Liquid-Cooled Flat Plate Collectors[2]

4.2.1 Heat Balance

In this section, a brief approximate analysis is made of a flat plate collector which utilizes metal tubes containing flowing water attached to the collector plate to carry off the absorbed heat. The collector is illustrated in Figure 4.2. A single cover glass is suspended above a collector plate with water tubes attached. The incident light flux upon the tilted glass plate is I_0 (power/area). Part of this is reflected and absorbed by the glass, and a fraction t_g is transmitted through the glass. Thus, the light flux incident upon the collector plate is $t_g I_0$. The collector plate absorbs the fraction ϵ_c (sol) of the light falling upon it, which appears as heat in the plate. The notation ϵ_c (sol) denotes the average emissivity of the plate over the solar spectrum. The remainder, $[1 - \epsilon_c \text{ (sol)}]t_g I_0$ is reflected back up toward the

[2]Other useful references on flat plate collector performance include H. Tabor, "Solar Energy Collector Design," Chapter 1 in *Trans. Intl. Conf. on the Use of Solar Energy—The Scientific Basis*, Tuscon, AZ, October 31, 1955; H. C. Hottel and A. Whillier, "Evaluation of Flat Plate Collector Performance," Chapter 6, *loc. cit.*; H. C. Hottel and B. B. Woertz, "Performance of Flat Plate Solar Heat Collectors," *Trans. Am. Soc. Mech. Eng.*, *64*, 91, 1942.

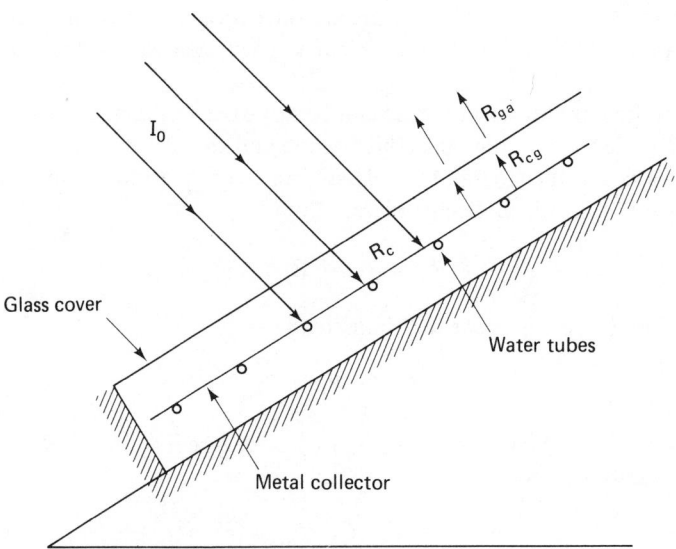

Figure 4.2 Model for flat plate collector.

glass. For our approximate treatment, it is sufficient to assume that this reflected light is entirely lost through the glass, even though some is actually absorbed and reflected by the glass.

The rate at which energy is absorbed by the collector plate (per unit area) is

$$R_c = I_0 t_g \epsilon_c \text{ (sol)} \qquad (4.7)$$

The rate of energy gain by the circulating water per unit area of the collector plate is the principal unknown and is designated as R_w. The rate at which energy is lost by the collector plate to the glass (per unit area) is

$$R_{cg} = \epsilon_c \text{ (ir)} \sigma(T_c^4 - T_g^4) + U_{cg}(T_c - T_g) \qquad (4.8)$$

In this equation, ϵ_c (ir) is the emissivity of the collector surface in the spectral region where it emits (in the far infrared), and U_{cg} is the heat transfer coefficient for free convection between the collector and the glass. The rate of energy loss per unit area from the glass to the ambient is

$$R_{ga} = \epsilon_g \text{ (ir)} \sigma(T_g^4 - T_a^4) + U_{ga}(T_g - T_a) \qquad (4.9)$$

where ϵ_g (ir) is the emissivity of the glass where it emits in the far infrared, and U_{ga} is the heat transfer coefficient for convection between the glass and the ambient air. If there is a wind blowing, U_{ga} is best approximated by a coefficient for forced convection. Forced convection is more effective than free convection.[3] Therefore, the coefficient U_{ga} will generally be larger than U_{cg}.

In practice, the water flow rate can be adjusted to maintain reasonable values of T_c. We shall assume that this is done and that T_c and T_a are known. The unknowns are then R_w and T_g. There are two equations. The first is a heat balance around the collector plate:

$$R_c = R_{cg} + R_w \qquad (4.10)$$

The second is the requirement that in a steady state

$$R_{cg} = R_{ga} \qquad (4.11)$$

These equations may be solved for R_w and T_g, and the collector efficiency may then be calculated to be

$$\eta = \frac{R_w}{I_0} \qquad (4.12)$$

which is the fraction of incident light that ends up as heat in the water.

4.2.2 Linearized Heat Transfer Coefficients

The heat flow from collector plate to glass and from glass to ambient consists of two terms, one being linear in a temperature difference and the other depending on the difference of fourth powers of temperatures. If the total rate of heat flow is plotted vs. temperature difference, as in Figure 4.3, it is found that the heat flow rate is the sum of linear and curved functions. Over limited ranges of temperature, the radiative term does not curve excessively, and a linear function can be made to approximate the total heat flow curve. Thus, one can approximate the function

$$\epsilon\sigma(T_1^4 - T_2^4) + U(T_1 - T_2)$$

by the linear term

$$U'(T_1 - T_2)$$

where U' is a linearized heat transfer coefficient for convection and radiation.

---

[3]For more accurate work, the reader is referred to references on natural convection in inclined air layers (K. G. T. Hollands, L. Konicek, T. E. Unny, and G. D. Raithby, "Free Convection Heat Transfer Across Inclined Air Layers," *J. Heat Transfer*, *98*, 189, 1976) and wind related heat losses (E. M. Sparrow and K. K. Tien, "Forced Convection Heat Transfer at an Inclined and Yawed Square Plate," *J. Heat Transfer*, 99, 507, 1977).

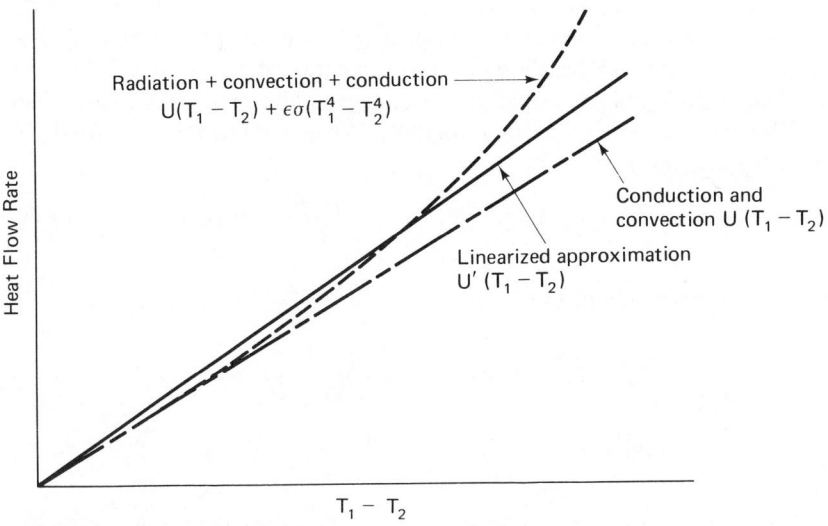

Figure 4.3   Linearized heat transfer coefficient.

The value of U' is given by

$$U' = U + \frac{\epsilon\sigma(T_1^4 - T_2^4)}{T_1 - T_2} \tag{4.13}$$

where T_1 and T_2 are evaluated at the midpoint of the expected range of variation.

4.2.3 Collector Efficiency

Using linearized heat transfer coefficients, the heat flow rates are

$$R_{cg} = U'_{cg}(T_c - T_g)$$
$$R_{ga} = U'_{ga}(T_g - T_a)$$

and Eq. (4.11) takes the form

$$U'_{cg}(T_c - T_g) = U'_{ga}(T_g - T_a)$$

This may be solved for T_g in terms of T_c and T_a:

$$T_g = \frac{U'_{cg}T_c + U'_{ga}T_a}{U'_{ga} + U'_{cg}} \tag{4.14}$$

It can be seen that T_g is a mean temperature part-way between T_c and T_a, and the weighting factors are the linearized heat transfer coefficients. If $U'_{ga} \gg U'_{cg}$, T_g will be nearly equal to T_a, whereas if $U_{ga} \ll U'_{cg}$, T_g will be nearly equal to T_c. Since, in practice, U'_{ga} is generally somewhat greater than U'_{cg}, T_g will tend to lie closer to T_a than T_c. When Eq. (4.14) is used for T_g in Eq. (4.10), the result is

$$R_w = I_0 t_g \epsilon_c \text{ (sol)} - U'_{cg} T_c - \frac{U'_{cg} T_c + U'_{ga} T_a}{U'_{ga} + U'_{cg}} \tag{4.15}$$

This may be rewritten in the form

$$R_w = I_0 t_g \epsilon_c \text{ (sol)} - U'_{cg} \frac{T_c U'_{ga} - T_a U'_{ga}}{U'_{cg} + U'_{ga}} \tag{4.16}$$

$$R_w = I_0 t_g \epsilon_c \text{ (sol)} - \frac{U'_{cg} U'_{ga}}{U'_{cg} + U'_{ga}} (T_c - T_a) \tag{4.17}$$

An overall heat transfer coefficient for transfer from the collector to the glass and then on to the ambient may now be defined as the geometric mean of U'_{cg} and U'_{ga}:

$$U_{ov} = \frac{U'_{cg} U'_{ga}}{U'_{cg} + U'_{ga}} \tag{4.18}$$

Then the efficiency is

$$\eta = \frac{R_w}{I_0} = t_g \epsilon_c \text{ (sol)} - \frac{U_{ov}(T_c - T_a)}{I_0} \tag{4.19}$$

The form of Eq. (4.19) is very instructive. The first term on the right side represents the *optical efficiency* of the collector, namely, the product of the transmissivity of the glass and the absorptivity of the collector surface. This would be the collector efficiency if there were no heat losses. The heat losses are calculated making the following assumptions:

1. All heat losses are through the cover glass. Losses through back and sides may be neglected.
2. Linearized coefficients may be used to represent convection plus radiation between surfaces.

The result is that the second term on the right side of Eq. (4.19) represents the reduction in efficiency due to heat losses. This term represents the ratio of overall rate of heat loss to incident solar intensity. Equation (4.19) is of the general form

$$\eta = A - \frac{B(T_c - T_a)}{I_0} \tag{4.20}$$

where A is the optical efficiency, and B is the overall heat loss coefficient. It has been found that actual collectors tend to follow this law in that a plot of η vs. $(T_c - T_a)/I_0$ is linear over moderate ranges of $(T_c - T_a)/I_0$. This interpretation is illustrated in Figure 4.4. Over very wide ranges of temperature, some curvature does show up in the plot of η vs. $(T_c - T_a)/I_0$ due to the nonlinear radiative terms in the heat loss.

Figure 4.4 Collector efficiency composed of optical efficiency and heat loss terms.

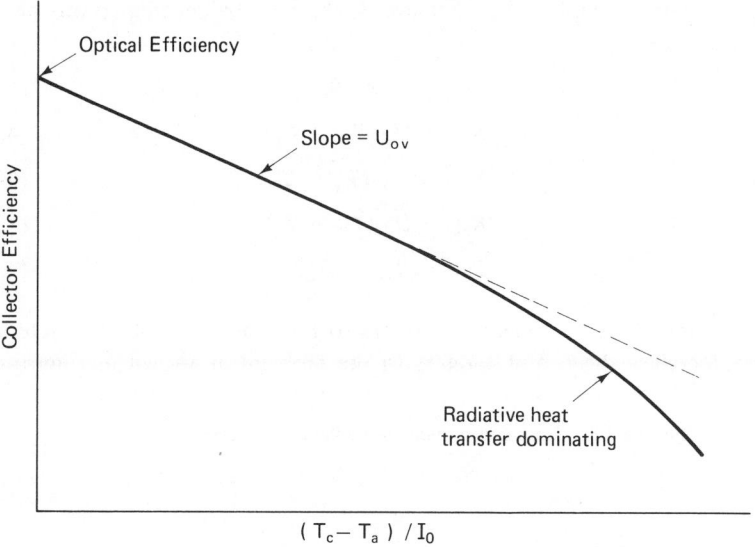

4.2.4 Effects of Multiple Glazing

One method for increasing the efficiency of flat plate collectors when operating at elevated temperatures relative to the ambient is to add additional layers of glass above the collector plate as shown in Figure 4.5. The effect of additional glass cover plates is to reduce the optical efficiency due to reflection and absorption of incident sunlight, and to decrease the overall heat transfer coefficient from the collector plate to the ambient. At moderate operating temperatures, the optical losses can actually reduce the overall performance of multiglazed collectors compared to single-glazed collectors. However, at elevated temperatures, the reduction in heat losses tends to outweigh the loss in optical efficiency, and a multiglazed collector will tend to outperform a single-glazed collector. The choice of the optimum number of glazings in any application depends on the operating and ambient temperatures as well as cost tradeoffs. In many applications for domestic hot water

heating in mild climates, single glazing is optimum. For domestic hot water
heating in cold climates and for space heating, double glazing may be advantageous. Only rarely, under extreme conditions, would triple glazing even be
considered.

An analysis of a double-glazed collector leads to an equation of the
general form of Eq. (4.20), except that the optical efficiency term is

$$A = t_g^2 \epsilon_c \text{ (sol)} \tag{4.21}$$

and B is more complex. To calculate B, the heat balance equations are set
up as follows:

$$R_w = R_c - R_{c1} \tag{4.22}$$

$$R_{c1} = U'_{c1}(T_c - T_1) \tag{4.23}$$

$$R_{12} = U'_{12}(T_1 - T_2) \tag{4.24}$$

$$R_{2a} = U'_{2a}(T_2 - T_a) \tag{4.25}$$

$$R_{c1} = R_{12} = R_{2a} \tag{4.26}$$

where c refers to the collector, 1 refers to the inner glass sheet, 2 refers to
the outer glass sheet, and a refers to the ambient as shown in Figure 4.5.

Figure 4.5   Heat flows in multiglazed flat plate collector.

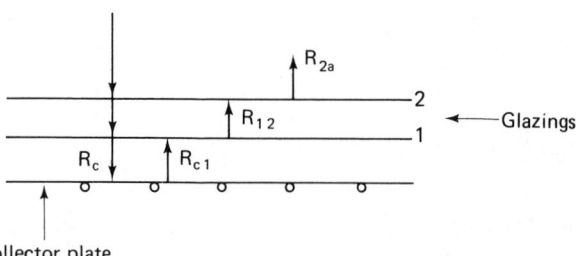

The U' factors are linearized heat transfer coefficients, including both convection and radiation. By setting $R_{c1} = R_{12}$ and solving for T_1, the result is

$$T_1 = \frac{U'_{c1}T_c + U'_{12}T_2}{U'_{c1} + U'_{12}} \tag{4.27}$$

Similarly, by setting $R_{2a} = R_{12}$, it is found that

$$T_2 = \frac{U'_{12}T_1 + U'_{2a}T_a}{U'_{12} + U'_{2a}} \tag{4.28}$$

Equations (4.27) and (4.28) may be combined to solve for either T_1 or T_2 in terms of only T_c and T_a. The result for T_1 is

$$T_1 = \frac{U'_{12}U'_{2a}T_a + U'_{c1}(U'_{2a} + U'_{12})T_c}{U'_{2a}U'_{12} + U'_{2a}U'_{c1} + U'_{12}U'_{c1}} \qquad (4.29)$$

Then it follows that

$$U'_{c1}(T_c - T_1) = \frac{U'_{2a}U'_{12}U'_{c1}}{U'_{2a}U'_{12} + U'_{2a}U'_{c1} + U'_{12}U'_{c1}}(T_c - T_a) \qquad (4.30)$$

The first term on the right side in Eq. (4.30) may be defined as a mean linearized overall heat transfer coefficient U_{ov} for heat flow from the collector plate through the glass panes to the ambient. Then the efficiency of the two-pane collector is

$$\eta = t_g^2 \epsilon_c \,(\text{sol}) - \frac{U_{ov}(T_c - T_a)}{I_0} \qquad (4.31)$$

which has the same form as Eq. (4.20) with $B = U_{ov}$. In comparing the one-pane and two-pane cases, it is found that the overall heat transfer coefficients are

1 *pane:*

$$U_{ov} = \frac{U'_{cg}U'_{ga}}{U'_{cg} + U'_{ga}} \qquad (4.32)$$

2 *panes:*

$$U_{ov} = \frac{U'_{c1}U'_{12}U'_{2a}}{U'_{c1}U'_{12} + U'_{c1}U'_{2a} + U'_{12}U'_{2a}} \qquad (4.33)$$

This pattern may readily be generalized to larger numbers of glass panes. For three panes, the result is

3 *panes:*

$$U_{ov} = \frac{U'_{c1}U'_{12}U'_{23}U'_{3a}}{U'_{c1}U'_{12}U'_{23} + U'_{c1}U'_{12}U'_{3a} + U'_{c1}U'_{23}U'_{3a} + U'_{23}U'_{12}U'_{3a}} \qquad (4.34)$$

If all the coefficients U'_{c1}, U'_{12}, ... were approximately equal (which is only a very crude approximation) the formulas would reduce to

$$U_{ov} = \frac{(U')^{N+1}}{(N + 1)(U')^N} = \frac{U'}{N + 1} \qquad (4.35)$$

where N is the number of glass panes, and U' is any one of the coefficients (since they are presumed equal). This would imply that in going from 1 to 2 to 3 glass panes the overall heat loss coefficient would go from $U'/2$ to $U'/3$

to $U'/4$. In actual practice, it is found that the free convection that takes place between the collector plate and the first glass pane, or between a pair of glass panes, is considerably less effective than the forced convection produced by wind blowing over the outer glass pane. For example, reasonable values of $U'_{c1} \sim U'_{12}$ are ~ 7.8 W/m²-°K, whereas in a 10 mph wind, U_{2a} might be about 13.6 W/m²-°K. With these approximate values, it is found that

1 *pane:*

$$U_{ov} \cong 4.94 \text{ W/m}^2\text{-°K} = 0.87 \text{ Btu/hr-ft}^2\text{-°F}$$

2 *panes:*

$$U_{ov} \cong 3.00 \text{ W/m}^2\text{-°K} = 0.53 \text{ Btu/hr-ft}^2\text{-°F}$$

3 *panes:*

$$U_{ov} \cong 2.17 \text{ W/m}^2\text{-°K} = 0.38 \text{ Btu/hr-ft}^2\text{-°F}$$

Thus, realistically, one may expect a 39 % reduction in U_{ov} by adding a second pane to a one-pane collector, but only a 28 % further reduction in U_{ov} by adding a third pane of glass. In Figure 4.6, η vs. $(T_c - T_a)/I_0$ is plotted for one-, two-, and three-pane collectors based on the above values of U_{ov}, and assuming $t_g \approx 0.90$ and ϵ_c (sol) ~ 0.92. It can be seen that at low values of $(T_c - T_a)/I_0$, the single-pane collector is superior, but that successively larger numbers of panes perform better as $(T_c - T_a)/I_0$ increases. The increase in performance in going from two to three panes is almost never worth the investment.

4.2.5 Effects of a Selective Surface

The heat loss from a collector plate to the inner glass (which, of course, is also the outer glass in single-pane collectors) takes place by convection and radiation. The radiative losses can be reduced by using a special absorbing surface on the collector plate.

It can be shown that all bodies continually emit and absorb radiant energy, and that a body reaches thermal equilibrium when its rate of heat gain balances its rate of heat loss. The spectrum of radiant energy vs. wavelength for a hypothetical perfect black body at various temperatures is shown in Figure 4.7. As a body becomes heated, it emits at shorter wavelengths, whereas at room temperature the emitted radiation is mostly in the (invisible) far infrared. This is why a body appears successively dull red, bright red, yellow, white, and then blue hot as it is brought to very high temperatures. The total amount of radiant energy emitted by a hypothetical black body varies with the fourth power of absolute temperature. Thus, a hot body not only radiates at shorter wavelengths but also emits much more power than a cool body. A white hot body will therefore appear much brighter than a

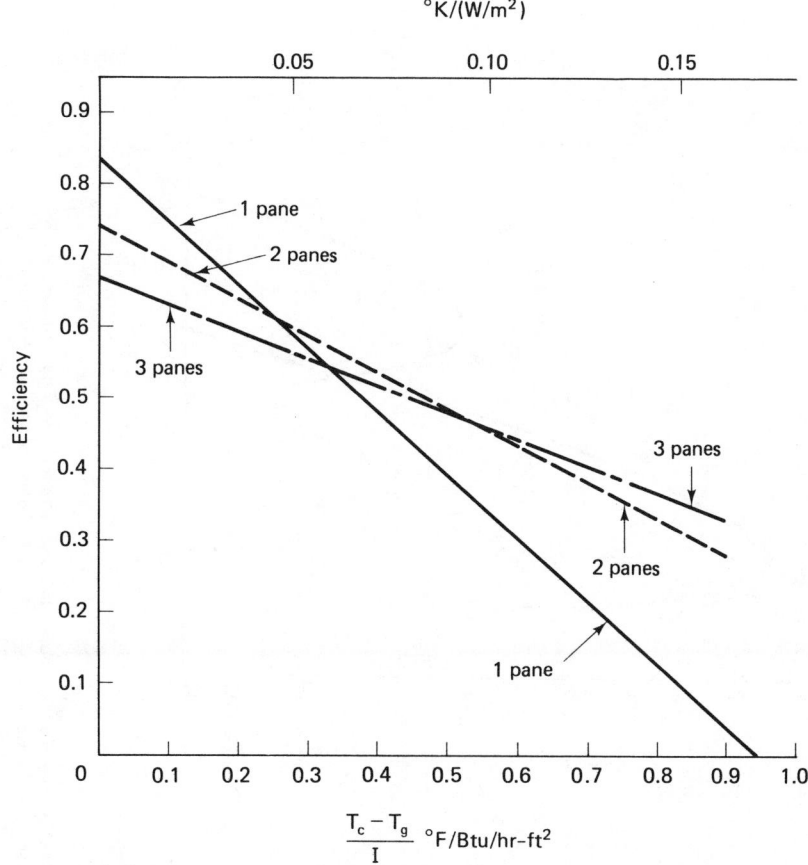

Figure 4.6 Estimated efficiencies of typical flat plate collectors as a function of number of cover glazings.

red hot body at a lower temperature. No real bodies radiate exactly like a "black body," but some are close approximations. The radiative properties of a real body are described by its emissivity $\epsilon(\lambda)$ at any wavelength. The emissivity $\epsilon(\lambda)$ is the ratio of actual power radiated at wavelength λ to the power that would be emitted by a perfect black body at that wavelength. It can be shown that the absorptivity (fraction of incident radiation that is absorbed by a surface) at any wavelength is equal to the emissivity of a surface at that wavelength. Thus, if a body is a poor emitter at any wavelength, it is also a poor absorber at the same wavelength. Some special surfaces can be prepared that have large emissivities (~ 0.8 to 0.9) in the solar spectrum and low emissivities (~ 0.1 to 0.4) in the far infrared. These are called *selective surfaces* and are useful for solar collector plates.

The primary purpose of a solar collector plate is to absorb incident

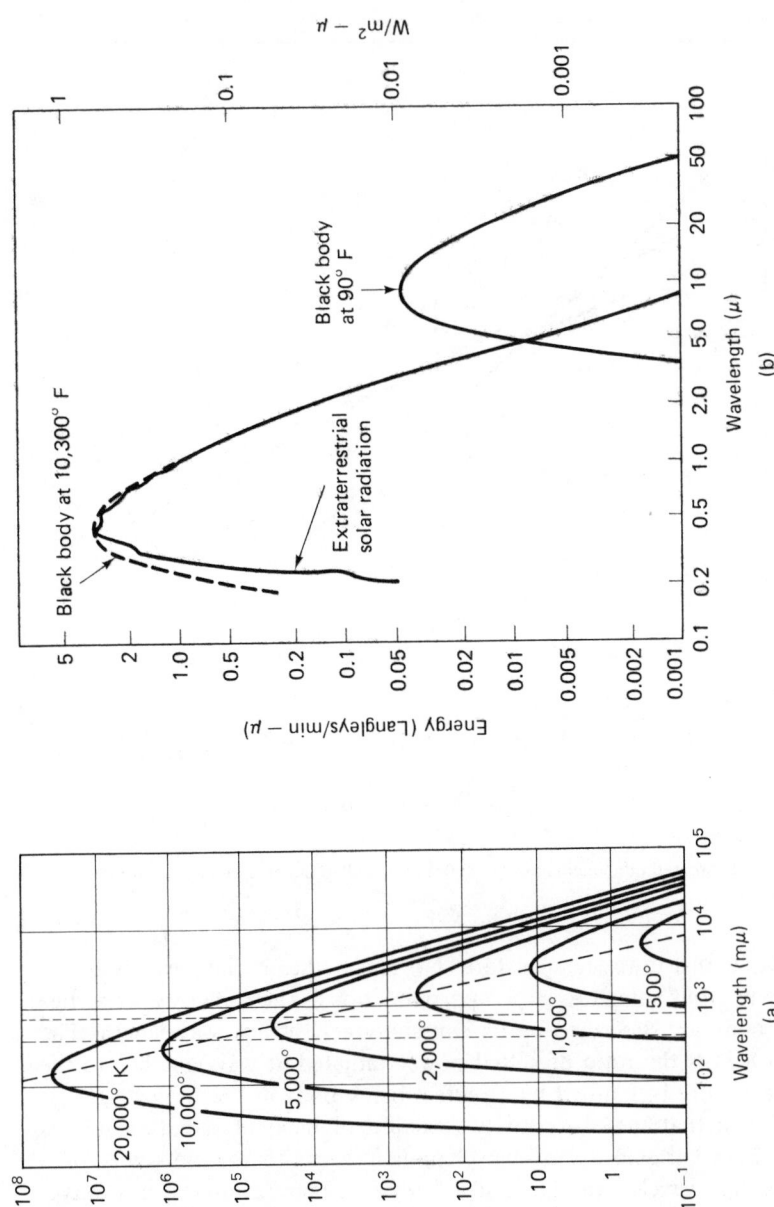

Figure 4.7 (a) Spectral emittance of a black body at various temperatures. The vertical dashed lines indicate the boundaries of the visible spectrum. (b) Radiation of the sun and terrestrial bodies. Adapted from F.W. Sears, *Optics*, Fig. 12-5 Addison-Wesley Publishing Co., 1949, by permission of the publisher.

radiation from the sun. It should, therefore, have as high an emissivity (absorptivity) as possible in the visible and near infrared parts of the spectrum, where most of the solar energy is concentrated. When the collector plate rises in temperature, it radiates in the far infrared, and it would be desirable to have a low emissivity in this spectral range. The emissivities of an ideal selective surface and of actual achievable surfaces are illustrated in Figure 4.8. Although a selective surface greatly reduces radiative losses, it

Figure 4.8 Ideal and achievable selective surfaces.

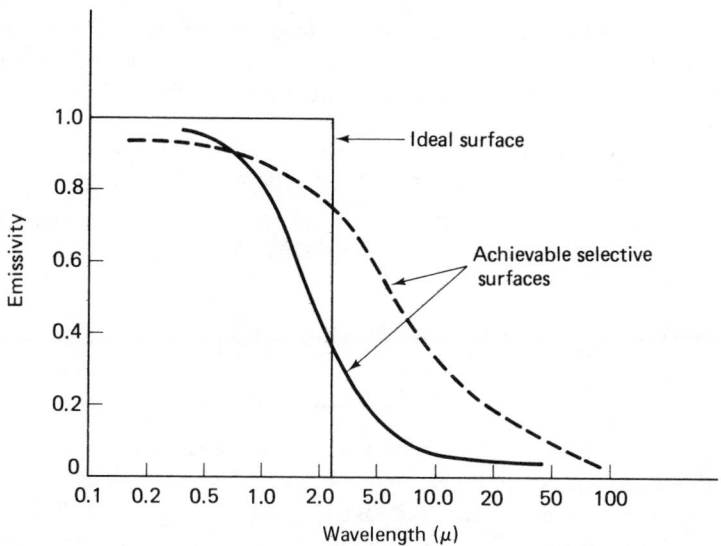

may have a lower overall absorption coefficient for solar radiation than black paint because some of the solar spectrum reaches into the infrared where the absorptivity of the selective surface is falling off with increasing wavelength. Meinel and Meinel[4] present the spectral characteristics of a number of selective surfaces. The group working at the Optical Sciences Center of The University of Arizona has done a great deal of pioneering research on selective surfaces.[5] These surfaces may be of extreme importance at high temperatures, where radiative losses tend to dominate.

The effect of a selective surface on collector performance is manifested in terms of lower heat loss due to radiation from the collector plate. The

[4]A. B. Meinel and M. P. Meinel, *Applied Solar Energy*, Addison-Wesley, Reading, MA, 1976, pp. 263–315.

[5]A review article describing much of this work is B. O. Seraphin, "Spectrally Selective Surfaces and Their Impact on Photothermal Solar Energy Conversion," *Topics in Applied Physics*, Springer-Verlag, Berlin, 1979.

effectiveness of the radiant heat loss reduction depends on the fraction of heat loss due to radiation. For systems where heat losses are dominated by radiation, use of a selective surface can be very effective. However, when convection losses are important, a selective surface may not produce very much improvement. The expected emissivities over the range of the far infrared where the collector plate emits are ~ 0.9 for flat black paint and ~ 0.3 for a selective surface. A reasonable value of U for convection and conduction between the collector and the first cover glass is 7.8 W/m²-°K. Then it follows that the radiative heat loss term is $\epsilon(5.67 \times 10^{-8})(T_c^4 - T_g^4)$ W/m², while the convective loss term is $7.8(T_c - T_g)$ W/m²: the flat black surface is compared with the selective surface for various arbitrary combinations of T_c and T_g in Table 4.1. It can be seen that when flat black paint ($\epsilon = 0.9$) is used on the collector plate, heat loss from the plate is roughly

TABLE 4.1 Comparison of convective and radiative losses from collector plate

	T_c		T_g		Convective Loss (W/m²)	Radiative Loss (W/m²)	
Case	(°F)	(°K)	(°F)	(°K)		$\epsilon = 0.9$	$\epsilon = 0.3$
1	120	322	80	300	172	137	45
2	160	344	100	311	344	240	80
3	200	366	120	322	517	372	124

equally balanced between convection and radiation, with radiation becoming increasingly important at higher temperatures. Use of a selective surface ($\epsilon = 0.3$) reduces the overall heat loss by 30%, 32% and, 35%, in Cases 1, 2, and 3, respectively. These reductions in heat loss should be compared with possible reductions in absorption of solar radiation of perhaps 10% caused by use of a selective surface. If a selective surface reduces light absorption at the collector plate by perhaps 10% and reduces heat loss by perhaps 30%, the selective surface on the collector plate acts very much like an extra pane of cover glass. Thus, a collector with one pane of glass and a selective surface acts in some respects like a collector with a flat black-paint collector surface and two cover glazings.

4.2.6 Effect of Transparent Honeycombs

Transparent honeycombs can be installed between the collector plate and the first glass cover pane to reduce large scale convection.[6] An illustration of this approach is shown in Figure 4.9. The plastic honeycomb breaks up

[6]"Optimization of Thin-Film Transparent Plastic Honeycomb-Covered Flat Plate Solar Collectors," Lockheed Missiles and Space Co., Palo Alto, CA, *Rept. LMSC/D623838* or *SAN/1256-78/1*, May 25, 1978.

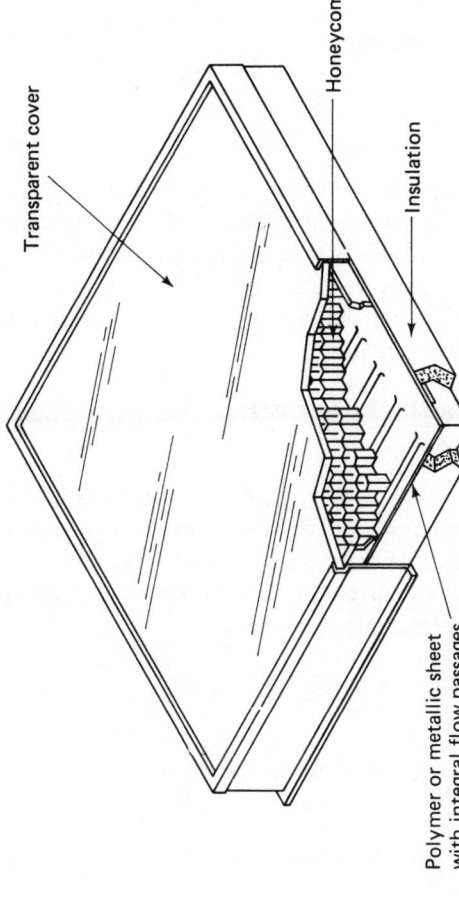

Figure 4.9 Honeycomb-covered flat plate collector design concept. Adapted from 'Optimization of Thin-Film Transparent Plastic Honeycomb Covered Flat Plate Collectors,'' LMSC/D623838, May 25, 1978.

121

the space into channels which restrict large-scale movements of air due to thermally induced density changes. The honeycomb material does not substantially affect the passage of light energy to the collector plate because transmitted and reflected light are both directed down toward the collector plate, while absorbed light heats the honeycomb and contributes to the flow of heat to the collector plate via heat transfer to the air.

4.3 Effect of Tube Spacing

4.3.1 Equations of Heat Flow

The spacing of the water tubes is an important design consideration for a flat plate collector. Assuming that good thermal contact is made between the plate and the tubes, it is desired to estimate the temperature rise on the plate in the gaps between the tubes. In any three-dimensional body where there is a heat source, the temperature variation across the body will satisfy the differential equation

$$\frac{\partial^2 T}{\partial x^2} + \frac{\partial^2 T}{\partial y^2} + \frac{\partial^2 T}{\partial z^2} + \frac{Q}{\kappa} = 0 \qquad (4.36)$$

where Q is the heat source at any point (x, y, z) in units of energy/volume-time, and κ is the thermal conductivity in energy/length-time-temperature. In a solar collector plate, the temperature variations across the thickness of the plate and in the direction parallel to the tubes may be approximately neglected. Thus, an approximate equation is

$$\frac{d^2 T}{dx^2} + \frac{Q}{\kappa} = 0 \qquad (4.37)$$

where x is the distance perpendicular to the tubes.

4.3.2 Heat Flow in a Flat Plate

The heat source term for any small slab of collector plate (as shown in Figure 4.10) is the *net heat gain* by that slab, taking into account the solar input and the rate of heat loss to the ambient. The solar input to a slab of area Wdx is $\alpha I W dx$, where I is the solar intensity, and α is the optical efficiency. The heat loss from this slab can be approximated by a linear term $U(T - T_a)Wdx$, where U is a linearized overall heat loss coefficient (energy/time-area-temperature), T is the plate temperature at that point, and T_a is

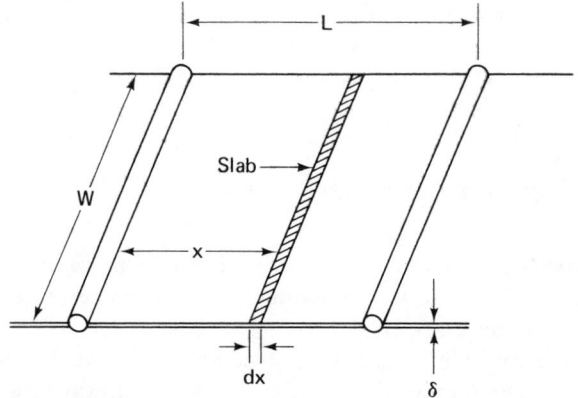

Figure 4.10 Illustration of a slab of collector of dimensions (W by dx by δ).

the ambient temperature. The rate of heat gain per unit area is then $[\alpha I - U(T - T_a)]$. The rate of heat gain per unit volume is this expression divided by the plate thickness δ:

$$Q = \left[\frac{\alpha I - U(T - T_a)}{\delta} \right] \tag{4.38}$$

Thus, the differential equation for variation of T with x is

$$\frac{d^2T}{dx^2} + A - B^2T = 0 \tag{4.39}$$

where

$$A = \frac{\alpha I + UT_a}{\kappa\delta}, \qquad B^2 = \frac{U}{\kappa\delta} \tag{4.40}$$

If a new variable S is defined such that

$$S = T - \frac{A}{B^2} \tag{4.41}$$

Eq. (4.39) is transformed to

$$\frac{d^2S}{dx^2} - B^2S = 0 \tag{4.42}$$

The solution of this equation is

$$S = c_1 e^{Bx} + c_2 e^{-Bx} \tag{4.43}$$

where c_1 and c_2 are constants determined by the boundary conditions. Thus,

$$T = c_1 e^{Bx} + c_2 e^{-Bx} + \frac{A}{B^2} \qquad (4.44)$$

4.3.3 Single Tube with Infinite Sheet

Equation (4.44) can be applied to several specific cases. Consider first a single tube of water at $x = 0$ maintained at a temperature T_0, with an infinite collector plate attached to it as in Figure 4.11. The temperature variation $T(x)$ along the plate may now be estimated. At very large x there is essentially no conduction through the metal sheet toward the tube, and a heat balance on a small slab of collector at large x requires that

$$\alpha I = U(T - T_a) \qquad (4.45)$$

Figure 4.11 Illustration of a single tube with an infinite sheet of metal.

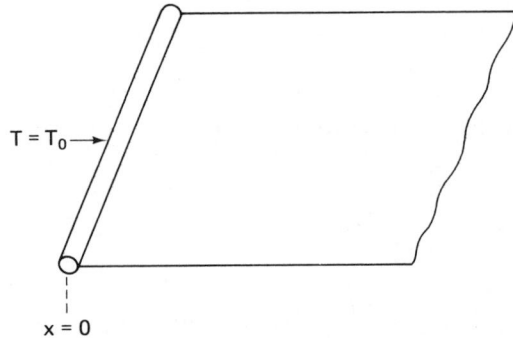

Thus, as $x \longrightarrow \infty$,

$$T(x) \longrightarrow T_a + \frac{\alpha I}{U} \qquad (4.46)$$

In bright sunlight with a single cover glass, typical values are $\alpha = 0.75$, $I = 300$ Btu/hr-ft^2 (946 W/m^2), and $U = 1.0$ Btu/hr-ft^2-°F (5.68 W/m^2-°K) so that

$$T(\infty) \cong T_a + 225°F \qquad (4.47)$$

According to Eq. (4.44) the temperature would become infinite at large x if c_1 is not chosen equal to zero. Therefore, we must choose $c_1 = 0$. The boundary condition at $x = 0$ is $T(0) = T_0$. Thus, the constant c_2 is evaluated from

$$T_0 = c_2 + \frac{\alpha I}{U} + T_a \tag{4.48}$$

so that

$$c_2 = T_0 - \left(\frac{\alpha I}{U} + T_a\right) \tag{4.49}$$

In general then for all x,

$$T(x) = \left[T_0 - \left(\frac{\alpha I}{U} + T_a\right)\right]\exp\left[-\left(\frac{U}{\kappa\delta}\right)^{1/2}x\right] + \left(\frac{\alpha I}{U} + T_a\right) \tag{4.50}$$

It can be seen from this expression that when

$$x \longrightarrow 0,\, T \longrightarrow T_0, \quad \text{and when} \quad x \longrightarrow \infty,\, T \longrightarrow \frac{\alpha I}{U} + T_a$$

For small distances from the tube, a useful approximate expression can be found by expanding the exponential in a power series and retaining only the leading terms. The result is

$$T(x) = T_0 + \left(\frac{\alpha I}{U} + T_a - T_0\right)\left(\frac{U}{\kappa\delta}\right)^{1/2}x \tag{4.51}$$

showing that at first the temperature rises linearly for small distances from the tube. Assuming $U \simeq 1.0$ Btu/hr-ft^2-°F, the quantity $(U/\kappa\delta)^{1/2}$ is equal to 0.10 in.$^{-1}$ for 1/32 in. copper, 0.14 in.$^{-1}$ for 1/32 in. aluminum, and 0.22 in.$^{-1}$ for 1/16 in. steel. An illustration of the exact equation, and this approximation, are given in Figure 4.12 for $(\alpha I/U) + T_a = 300$°F and $T_0 = 150$°F. Curves are shown for $(U/\kappa\delta)^{1/2} = 0.1$ in.$^{-1}$ and 0.2 in.$^{-1}$. It can be seen that the temperature rises quite sharply in just a few inches of sheet metal.

4.3.4 Multiple Parallel Tubes

A more interesting case is where there is a series of parallel water tubes as shown in Figure 4.10. In this case, the full expression, Eq. (4.44), must be used for $T(x)$. The boundary conditions at $x = 0$ and $x = L$ are $T = T_0$. Thus,

$$T_0 = c_1 + c_2 + \frac{A}{B^2} \tag{4.52}$$

$$T_0 = c_1 e^{BL} + c_2 e^{-BL} + \frac{A}{B^2}$$

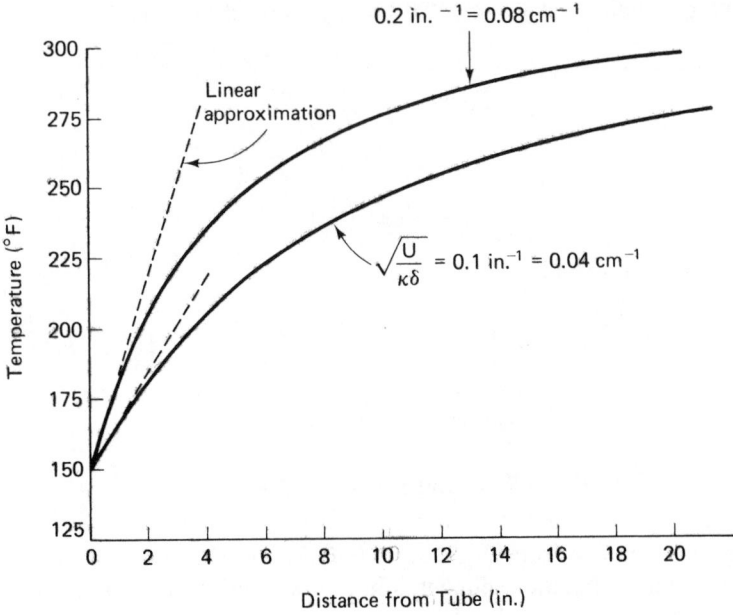

Figure 4.12 Variation of temperature with distance from a water tube in an infinite sheet of metal.

The solution of these equations for c_1 and c_2 is

$$c_{\frac{1}{2}} = \pm \frac{(T_0 - A/B^2)(1 - e^{\mp BL})}{(e^{BL} - e^{-BL})} \tag{4.53}$$

Consider an application where $T_0 = 150°F$, $B^2 = 0.0128$ in.$^{-2}$, $A = 3.84°F/in.^2$, corresponding to bright sunlight (300 Btu/hr-ft^2), an optical efficiency of 0.75, an ambient temperature of 75°F, $U = 1.0$ Btu/hr-ft^2-°F, and a copper sheet 1/32 in. thick. The constants c_1 and c_2 can be calculated for any value of L. The dependence is shown in Table 4.2. With these values,

TABLE 4.2   Dependence of c_1 and c_2 on the spacing between tubes (multiply by 0.556 to convert to °K)

L (in.)	$-c_1$ (°F)	$-c_2$ (°F)
1	70.70	79.30
2	66.45	83.55
4	58.33	91.67
6	50.51	99.49
8	43.20	106.80
∞	0	150

the temperature profiles can be calculated for any arbitrary spacing. The results for 4 in. (0.1 m) spacing and 8 in. (0.2 m) spacing are shown in Figures 4.13 and 4.14. It can be seen that with 1/32 in. (0.08 cm) thick copper in bright sunlight there is a 3.8°F (2.1°K) rise at the midpoint between 4 in. tube spacing and a 14.1°F (7.8°K) rise at the midpoint between 8 in. spacing.

Figure 4.13 Temperature profile for 4 in. (10 cm) tube spacing.

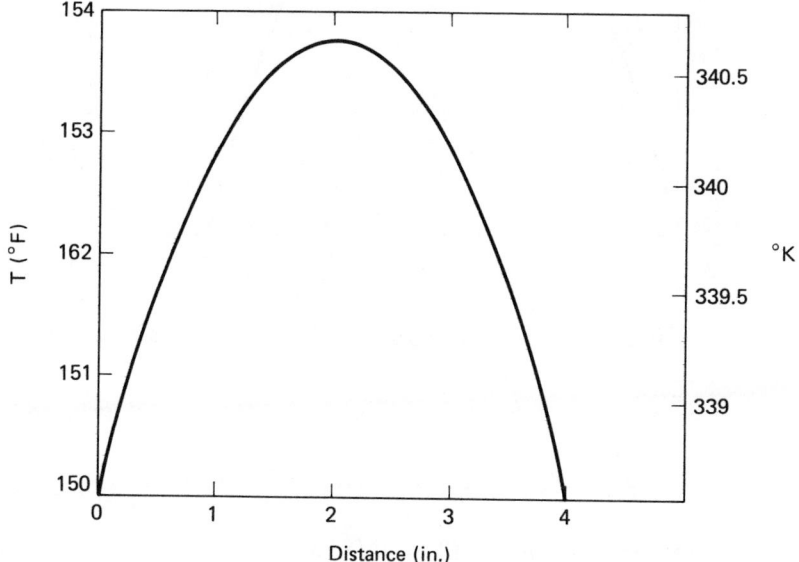

For aluminum or steel plates, the temperature rises are greater. The approximate thermal conductivities of these metals are as follows:

	$Btu\text{-}in./hr\text{-}ft^2\text{-}°F$	$W/m\text{-}°K$
Copper	2500	300
Aluminum	1200	144
Steel	310	37.2

It appears that a suitable spacing for most applications using copper or aluminum is about 4 in. (10 cm) and may be as small as 2 in. (5 cm) with steel collector plates.

The maximum temperature rise from a tube to the midpoint between tubes can be evaluated in the following way:

$$\Delta T_{\text{max}} = T\left(\frac{L}{2}\right) - T_0 = c_1(e^{BL/2} - 1) + c_2(e^{-BL/2} - 1) \qquad (4.54)$$

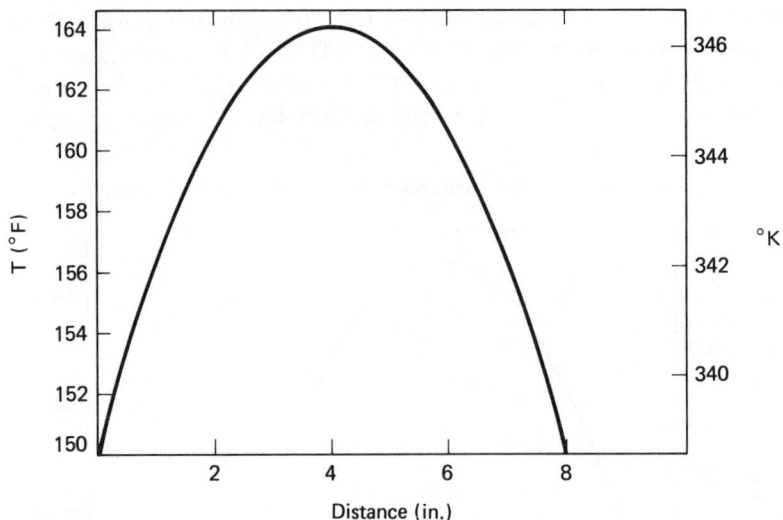

Figure 4.14 Temperature profile for 8 in. (20 cm) tube spacing.

When Eq. (4.53) is used for c_1 and c_2, the result is

$$\Delta T_{\max} = \frac{(T_0 - A/B^2)}{2 \sinh (BL)} \left[4 \sinh \left(\frac{BL}{2} \right) - 2 \sinh (BL) \right] \qquad (4.55)$$

$$\Delta T_{\max} = \left(T_0 - \frac{A}{B^2} \right) \left[\operatorname{sech} \left(\frac{BL}{2} \right) - 1 \right] \qquad (4.56)$$

The hyperbolic secant can be expanded in a power series, and if $BL/2$ is somewhat smaller than unity, only the leading terms are retained. Since

$$\operatorname{sech} \chi = 1 - \frac{\chi^2}{2} + \cdots$$

it follows that

$$\Delta T_{\max} \cong - \left(T_0 - \frac{A}{B^2} \right) \frac{B^2 L^2}{8} \qquad (4.57)$$

showing that the maximum temperature rise is proportional to L^2. For the example given previously and plotted in Figures 4.13 and 4.14, the results of using Eq. (4.56) for $L = 4$ in. (10 cm) and $L = 8$ in. (20 cm) are $\Delta T_{\max} = 3.76°F$ (2.09°K) and 14.14°F (7.86°K), respectively. The results of using Eq. (4.57) are 3.84°F (2.13°K) and 15.36°F (8.53°K), respectively.

4.3.5 Temperature Rise along the Tubes

The preceding treatment is based upon the assumption that the fluid flow rate is high enough that the fluid temperature rise along the tubes is small. In actual operation, there is a temperature rise both along the tubes and perpendicular to the tubes. The rise perpendicular to the tubes can be taken as given by Eqs. (4.44) and (4.54), with T_0 taken as the local value of the fluid temperature at any distance y from the fluid inlet $T_f(y)$. To calculate $T_f(y)$, we may set up the heat balance

$$(F)(C)\frac{dT_f(y)}{dy} = \{\alpha I - U[T_f(y) - T_a]\}L \tag{4.58}$$

where F is the fluid flow rate per tube (mass/time), C is the specific heat, and L is the spacing between tubes. This equation is not quite correct, because the surface losing heat to the ambient is warmer than $T_f(y)$. Nevertheless, with this approximation it follows that

$$\frac{dT_f}{\left(\frac{\alpha I}{U} + T_a - T_f\right)} = \frac{UL}{FC}\,dy \tag{4.59}$$

and after integration from $y = 0$ to y,

$$\frac{\frac{\alpha I}{U} + T_a - T_f(y)}{\frac{\alpha I}{U} + T_a - T_f(0)} = e^{-(UL/FC)y} \tag{4.60}$$

The fluid temperature at any point y along a tube is determined by this expression.

The expression for collector efficiency given in Eq. (4.19) is expressed in terms of T_c, the average collector plate temperature. In some applications, the fluid inlet temperature $T_f(0)$ is known but T_c is not. We may, therefore, estimate T_c from $T_f(0)$ by utilizing Eqs. (4.57) and (4.60). If the tube length is w and we assume the average collector plate temperature is determined at the midpoint between the tubes at a height $y = w/2$ along the tubes, the result is

$$T_c \cong T_f\left(\frac{w}{2}\right) + \frac{[A - B^2 T_f(w/2)]L^2}{8} \tag{4.61}$$

and $T_f(w/2)$ is determined by using $y = w/2$ in Eq. (4.60). This expression can be used to convert Eq. (4.16) into a function of $T_f(0)$ instead of a function of T_c. A more instructive procedure is as follows. If Eq. (4.60) is used at $y = w$, it is found that

$$T_f(w) = \left(\frac{\alpha I}{U} + T_a\right) - \left[\frac{\alpha I}{U} + T_a - T_f(0)\right][e^{ULw/FC}] \qquad (4.62)$$

By subtracting $T_f(0)$ from each side of this equation and multiplying through by $FC/(ILw)$, where A_c is the collector area, the result is obtained:

$$\frac{FC}{ILw}[T_f(w) - T_f(0)] = \frac{FC}{LwU}(1 - e^{-ULw/FC})\left[\alpha - \frac{T_f(0) - T_a}{I}\right] \qquad (4.63)$$

The left side of this equation is the ratio of heat gained by the fluid divided by the solar input and is therefore the collector efficiency, by definition. The right side has the same general form as Eq. (4.19), except for the factor out front. Thus, we may put

$$\eta = \frac{FC}{LwU}(1 - e^{-ULw/FC})\left[\alpha - U\frac{T_f(0) - T_a}{I}\right] \qquad (4.64)$$

in which all the variables on the right side are known even if T_c is not. Equation (4.64) gives the collector efficiency in terms of $T_f(0)$ instead of T_c. The term $(FC/LwU)(1 - e^{-ULw/FC})$ is sometimes called the *collector efficiency factor*.

4.4 Air-Cooled Collectors

Flat plate collectors can be operated with air as the heat transfer fluid. When air-cooled collectors are employed, they are invariably used in conjunction with a rock-bed storage unit as described in Sec. 7.6. Instead of having water tubes embedded in the absorber plate, fins are attached to the absorber plate to improve the heat transfer from the plate to the air. A typical schematic design is shown in Figure 4.15. Such air-cooled collectors tend to be slightly less efficient than liquid-cooled collectors because:

1. The air does not cool the absorber as effectively as a liquid, and the resulting higher plate temperature (under the same conditions) produces a higher heat loss than when the absorber is liquid-cooled.
2. The rapidly flowing air between the absorber and the inner glazing transfers heat readily from the plate to the glazing, greatly reducing the insulating effect of the air gap between the absorber and the glazing. By contrast, in a liquid-cooled collector that air gap is stagnant.

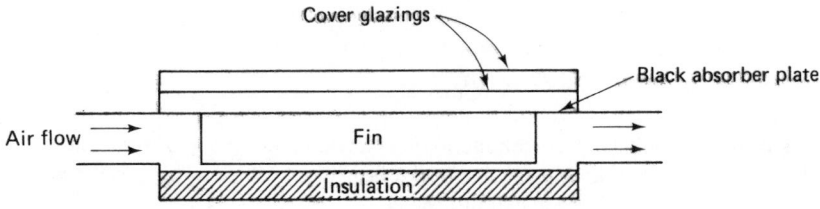

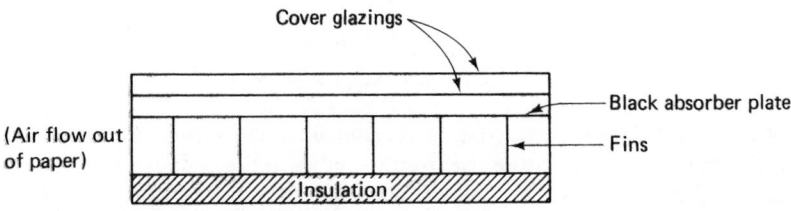

Figure 4.15 Cross section of a typical air-cooled collector.

However, Oonk, Löf, Shaw, and Cole-Appel[7] have argued that one should not compare liquid-cooled and air-cooled flat plate collectors under the *same* conditions because of constraints due to other system components. They argue that air- and liquid-cooled collectors perform equally well under typical conditions required by their respective systems. This will be discussed further in Sec. 7.6.

Worked Examples

1. A flat plate collector is given a static test by exposing it to the sun on a clear bright day. The temperature reached at midday is 225°F (380°K) when the solar intensity on the collector plane is 1000 W/m². Assuming that the average absorptivity of the plate for the solar spectrum is 0.9, the emissivity of the plate is 0.3 in the far infrared, the transmissivity of the glazing is 0.85, and the ambient temperature is 50°F (283°K), what is the overall conductive/convective heat loss coefficient for the collector?

   Heat gain by collector plate $= 0.85 \times 0.9 \times 1000$

   $= 765 \text{ W/m}^2$

   Heat loss rate by radiation $= 0.3 \times 5.67 \times 10^{-8}(380^4 - 283^4)$

   $= 246 \text{ W/m}^2$

[7]R. L. Oonk, G. O. G. Löf, L. E. Shaw, and B. E. Cole-Appel, *Proc. 1976 Amer. Sect. International Solar Energy Society*, Winnipeg, Canada, 1976.

Overall heat loss rate by conduction/convection $= 765 - 246$

$$= 519 \text{ W/m}^2 = U(380 - 283)$$

$$U = 5.35 \text{ W/m}^2\text{-}°\text{K}$$

2. If the heat loss coefficient by conduction/convection from the glazing to the air is 13.6 W/m²-°K in Example 1, what is the heat loss coefficient by conduction/convection from the collector plate to the glass?

Overall resistance = resistance (plate-to-glass) + resistance (glass-to-air)

$$\frac{1}{5.35} = \frac{1}{U} + \frac{1}{13.6}$$

$$U = 8.82 \text{ W/m}^2\text{-}°\text{K}$$

3. A flat plate collector is operated in bright sunlight at 900 W/m² on the collector plate. What is the temperature rise from the tubes to the midpoint between the tubes if the parameters are as follows: plate is 0.08 cm thick copper, tube spacing $= 10$ cm, inlet water temperature $= 320°$K, ambient temperature $= 280°$K, overall linearized heat loss coefficient $= 5$ W/m²-°K, optical efficiency $= 0.72$.

From Eq. (4.56),

$$\Delta T_{\max} = \left(T_0 - \frac{A}{B^2}\right)\left[\text{sech}\left(\frac{BL}{2}\right) - 1\right]$$

$$A = \frac{\alpha I + U T_a}{\kappa \delta} = \frac{0.72 \times 900 + 5 \times 280}{300 \times 0.0008} = 8533°\text{K/m}^2$$

$$B^2 = \frac{U}{\kappa \delta} = \frac{5}{300 \times 0.0008} = 20.8 \text{ m}^{-2}$$

$$\Delta T_{\max} = \left(320 - \frac{8533}{20.8}\right)[\text{sech}\,(0.228) - 1] = 2.30 \;°\text{K}$$

4. Estimate the efficiency of the collector described in Example 3 if the tubes are 2 m long and the flow rate per tube is 10 kg/hr.

According to Eq. (4.64),

$$\eta = \frac{FC}{LwU}(1 - e^{-ULw/FC})\left[\alpha - \frac{U(T_f(0) - T_a)}{I}\right]$$

$$\frac{FC}{LwU} = \frac{(10.0/3600) \text{ kg/sec} \times 4186.5 \text{ J/kg-}°\text{K}}{0.1 \text{ m} \times 2 \text{ m} \times 5 \text{ W/m}^2\text{-}°\text{K}} = 11.63$$

$$\eta = 11.63(1 - 0.9176)\left[0.72 - \frac{5(320 - 280)}{900}\right]$$

$$\eta = 0.958(0.72 - 0.22) = 0.479$$

5. A black plate is mounted in a glass-covered box with one water tube along the line $x = 0$, x being the distance perpendicular to the tube and y the distance along the tube. Use the following data:

$$U = 1.0 \text{ Btu/hr-ft}^2\text{-}°F \text{ (overall heat loss coefficient)}$$
$$I = 300 \text{ Btu/hr-ft}^2 \text{ (intensity on glass)}$$
$$T_0 = 100°F \text{ (entering water temperature)}$$
$$\alpha = 0.8 \text{ (optical efficiency)}$$
$$T_{\text{amb}} = 50°F \text{ (ambient temperature)}$$
$$T(x, y) = \text{plate temperature at } (x, y)$$

(a) Without using any differential equations, what is the plate temperature at large x?

(b) What differential equation governs the behavior of $T(x, 0)$? What are the boundary conditions on this equation?

(c) The general solution of the equation

$$\frac{d^2 T}{dx^2} + a - bT = 0$$

is

$$T = c_1 e^{bx} + c_2 e^{-bx} + \frac{a}{b}$$

Use this result to evaluate c_1 and c_2 for the function $T(x, 0)$. Demonstrate that the equation for T goes to the result from (a) as $x \longrightarrow \infty$.

(a) At large x:

$$\text{Solar heat gain} = \text{heat loss per unit area}$$
$$\alpha I = U(T - T_{\text{amb}})$$
$$T = \frac{\alpha I}{U} + T_{\text{amb}} = \frac{0.8 \times 300}{1.0} + 50 = 290°F$$

(b) In any small volume of metal, the rate of heat gain per unit volume from external sources is $Q(x, y)$:

$$Q(x, y) = \frac{\alpha I - U[T(x, y) - T_{\text{amb}}]}{\delta}$$

$$Q(x, y) = \kappa \left(\frac{\partial^2 T}{\partial x^2} + \frac{\partial^2 T}{\partial y^2} \right)$$

Setting $y = 0$,

$$Q(x, 0) = \kappa \frac{\partial^2 T}{\partial x^2}$$

$$\frac{d^2 T}{dx^2} = \left(\frac{\alpha I + U T_{\text{amb}}}{\kappa \delta} \right) - \frac{UT}{\kappa \delta}$$

$$\frac{d^2 T}{dx^2} = a - bT$$

$$a = \frac{\alpha I + U T_{\text{amb}}}{\kappa \delta}, \qquad b = \frac{U}{\kappa \delta}$$

Boundary conditions:

$$\text{As } x \longrightarrow 0, \qquad T \longrightarrow T_0$$

$$\text{As } x \longrightarrow \infty, \qquad T \longrightarrow \frac{\alpha I}{U} + T_{amb}$$

(c) As $x \longrightarrow \infty$, T cannot go to ∞. Thus, $c_1 = 0$.
As $x \longrightarrow 0$, $T \longrightarrow T_0$. Thus,

$$T_0 = c_2 + \frac{a}{b}$$

$$c_2 = T_0 - \frac{a}{b} = 100 - \frac{(240 + 50)}{1} = -190°F$$

As $x \longrightarrow \infty$,

$$T \longrightarrow c_2 e^{-\infty} + \frac{a}{b} = \frac{\alpha I}{U} + T_{amb}$$

6. A flat plate collector has an optical efficiency of 0.7 and a linearized overall heat
loss coefficient of 0.8 Btu/hr-ft²-°F. If the ambient temperature is 40°F and the
solar intensity on the collector is 300 Btu/hr-ft²,
   (a) What static temperature can be reached on the plate when there is no water
   flow?
   (b) When water flows through the collector at a rate fast enough to produce
   40% collection efficiency, what will the collector plate temperature be
   reduced to?

(a) Static temperature:

$$\alpha I = U(T - T_{amb})$$

$$T = \frac{\alpha I}{U} + T_{amb} = \frac{0.7 \times 300}{0.8} + 40 = 302.5°F$$

(b)

$$\eta = \eta_{opt} - \frac{U(T - T_{amb})}{I}$$

$$0.4 = 0.7 - \frac{0.8(T - 40)}{300}$$

$$T = 152.5°F$$

Problems

4.1. A flat plate collector tilted at the latitude with one pane of glass and a selective surface with absorptivity ≈ 0.9 across the solar spectrum, and emissivity
in the far infrared of ≈ 0.3, is in bright sunlight on a winter day with no fluid
in the tubes. The convective/conductive heat transfer coefficient between col-

lector plate and glass is 9.68 W/m²-°K, and the conductive/convective heat transfer coefficient between the glass and the air is 13.6 W/m²-°K. The total insolation on a horizontal surface at 1300 on January 21 at a latitude of 35°N is 180 Btu/hr-ft² (567 W/m²). The ambient temperature is 30°F (272°K). What static temperature will be reached on the collector plate? Assume the transmissivity of glass is 0.9, and assume 20% diffuse radiation as in Problem 2.7.

4.2. If an extra pane of glass were put on the collector in Problem 4.1, what static temperature could be reached? Assume that the two glass panes can be treated as a single pane, with conductive/convective heat transfer coefficients of 5.81 W/m²-°K to the collector plate and 13.6 W/m²-°K to the air.

4.3. A flat plate collector system with total area $= 48$ ft² (4.46 m²) is used to measure the heat loss coefficient by running the collector at night. Water at 120°F (322°K) is pumped through the collector at the rate of 0.3 gal/min (63 cm³/sec), and it is found that when the ambient temperature is 40°F (277°K) the water exiting the collector is at 108°F (315°K). If the collector has one pane of glass and a plate absorptivity of ≈ 0.85 what is the expression for the efficiency of the collector?

4.4. Show by differentiating Eq. (4.44) and setting $dT/dx = 0$ that the maximum temperature is reached at $x = L/2$.

4.5. Evaluate the maximum temperature rise from a tube to the midpoint between tubes. Use Eqs. (4.56) and (4.57). Assume $T_0 = 130°F$ (328°K), $T_a = 50°F$ (283.2°K), $\alpha = 0.75$, $U = 0.8$ Btu/hr-ft²-°F (4.54 W/m²-°K) $I = 300$ Btu/hr-ft² (946 W/m²), and carry out the calculation for the following cases:
(a) Aluminum (1/32 in.):

$$L = 2 \text{ in. and } 4 \text{ in. } [0.08 \text{ cm}, L = 5 \text{ cm and } 10 \text{ cm}]$$

(b) steel (1/16 in.):

$$L = 2 \text{ in. and } 4 \text{ in. } [0.16 \text{ cm}, L = 5 \text{ cm and } 10 \text{ cm}]$$

(c) copper (1/32 in.):

$$L = 2 \text{ in. and } 4 \text{ in. } [0.08 \text{ cm}, L = 5 \text{ cm and } 10 \text{ cm}]$$

4.6. A solar collector is cooled down by low night temperatures to 25°F (269°K) at sunrise. How much heat per unit area does it take to bring it up to operating temperature where the plate [1/32 in. (0.08 cm) copper] is at 120°F (322°K) and the cover glass [3/16 in. (0.24 cm)] is at 65°F (292°K)? The solar intensity on the collector can be approximated by the function $I = 75(H)$ Btu/hr-ft² $= 236H$ W/m², where H is the number of hours past sunrise during the early morning hours. If the collection efficiency is 20%, how long does it take to raise the collector up to temperature? (Density: Cu $= 555$ lb/ft² (8890 kg/m²), glass $= 156$ lb/ft^3 (2499 kg/m^3); specific heat: Cu $= 0.09$ Btu/lb°F (377 J/kg-°K), glass $= 0.16$ Btu/lb-°F (670 J/kg-°K)).

4.7. Compare the performance of a flat plate collector (a) at fixed tilt equal to the latitude, with a flat plate collector having (b) seasonally adjusted tilt (winter = latitude $+ 25°$, spring/fall = latitude, summer = latitude $- 25°$). Let the latitude be 35°N and the collector efficiency be

$$\eta = 0.7 - \frac{0.8(120 - T_a)}{I_t}$$

where $T_a = 30°F$, 60°F, and 85°F in winter, spring/fall, and summer, respectively, and I_t is the solar intensity in (Btu/hr-ft^2) on the tilted collectors. Use the following clear day intensities on a horizontal surface (Btu/hr-ft^2):

Hours from Noon	Dec. 21	Mar./Sept. 21	June 21
0	183	283	320
1	171	264	303
2	137	225	279
3	90	163	240
4	35	84	185
5	0	37	114

Assume the yearly energy collected is proportional to the energy collected on these 4 clear days. How big a difference does periodic adjustment make in yearly energy collected? What additional costs or other problems are introduced by periodic adjustment?

4.8. Using the data from Problem 4.7, estimate the improvement in yearly performance of a flat plate collector with fixed tilt equal to the latitude produced by mounting it on a turntable (like a merry-go-round) and rotating it at 15° per hr.

4.9. Repeat Problem 4.8, using seasonally adjusted tilt and rotation on a turntable.

4.10. A flat plate collector with a 1/32 in. (0.08 cm) thick aluminum collector plate has an optical efficiency of 0.7 and a heat loss coefficient of 0.8 Btu/hr-ft^2-°F (4.54 W/m^2-°K). If the solar intensity on the collector is 260 Btu/hr-ft^2 (819 W/m^2), the liquid flow rate is 2.0 gal/min (0.45 m^3/hr), the 60 collector tubes are each 6 ft (1.83 m) long, and the spacing between tubes is 4 in. (10 cm), what will the collector efficiency be if the fluid inlet temperature is 120°F and the ambient temperature is 40°F? What will be the temperature rise of the fluid? (Use Eqs. 4.57 and 4.64.)

4.11. The transmission of light by a sheet of glass can be roughly approximated by a function tabulated in terms of the angle of incidence:

Angle of Incidence	Transmission
0	0.87
10	0.87
20	0.87
30	0.87
40	0.86
50	0.82
60	0.78
70	0.67
80	0.39
90	0.00

A flat plate collector is tilted at 35° at a latitude of 35°. Calculate the optical efficiency (assuming 90% absorption by the receiver) of a two-pane collector as a function of the hour of the day on December 21, March 21, and June 21, taking into account (a) variation in transmission of the glass and (b) shading of the collector plate by the sides of the collector boxes. Assume the collector panels are 3 ft wide by 6 ft high and that the collector plate is set back 2 in. behind the outer glass.

4.12. The temperature rise on the collector plate from tube to midpoint between tubes is 4°F, under certain operating conditions. If a new collector plate is used with double the thermal conductivity, double the plate thickness, and double the tube spacing, what will the temperature rise be? Give your reasoning.

5 | Economic Analysis

In this chapter we explain a method for economic appraisal of the desirability of making an initial investment in a solar energy system that will displace a certain amount of conventional fuel each year of its life. The discussion centers on two basic variables—the net present value and the payback period. The net present value is the net value above the investment and interest payments on the loan, of all fuel saved during the lifetime of the system, expressed in current dollars. The pay-back period is the number of years required for the yearly energy savings to pay back the investment and all interest payments. These two parameters generally determine whether a particular system is a good investment. Part of the difficulty in this work is that one must guess a "scenario" for future fuel prices and inflation. The effect of such guesses on the overall economy is discussed. Even though the economics of a solar energy investment might pay for an individual, this could create hardship for a utility. The impact of large-scale usage of solar energy on utilities is briefly considered. Typical examples are worked out.

5.1 Net Present Value Concept

Solar energy installations are capital intensive, which means that a large initial investment is made in order to produce annual savings over the lifetime of the equipment. To determine whether a solar energy system is economically desirable, one must add up the fuel savings over the lifetime of the installation and compare the savings with the value of the investment. The value of the investment includes more than the lump sum for the initial purchase because that money, when invested for the same lifetime, produces considerable income. There are alternative (but similar) descriptions. One is that the user invests a lump sum of cash and therefore foregoes the income from that cash, had it been invested in income-bearing securities. The other is that the user has no cash, but instead borrows the sum required for the solar energy system. The cost is then the sum invested plus the interest payments on the loan over the lifetime of the investment. The two descriptions are very similar mathematically, and both will be considered here.

5.1.1 Investment from Cash on Hand

First consider the situation where an investment is made from cash on hand. When an initial investment I is made for a new energy technology which will save (on average) E units of energy per year during a lifetime of L years, the question is whether the cumulative value of energy saved over L years will exceed the cumulative balance achievable by investing the money in high-quality investments for L years. To answer this question, the rate of inflation, the rate of investment return, and the value of energy must be known for all L years. Since these variables must be guessed, only an approximate numerical treatment is possible.

The net present value of the investment is the cumulative value of energy saved over L years, corrected for inflation to be in terms of current year-zero dollars, less the cumulative value of the money (in year-zero dollars) if it was invested for L years in high-grade investments. We next proceed to calculate each of these quantities and subtract them.

The following nomenclature is adopted:

I = initial investment (year-zero dollars)
L = lifetime of solar installation (years)
E = average energy saved per year (Btu)
t_N = after-tax rate of return on high-quality bonds, certificates of deposit, or bank deposits invested in year N to mature in year L.
r_N = inflation rate in year N
e_N = rate of escalation of fuel prices above the inflation rate in year N
W_L = value of the original investment after L years of earning interest (in year-L dollars)
V_L = the value W_L corrected for inflation into year-zero dollars
F_N = value of fuel in year N (dollars per Btu)
G_N = F_N corrected for inflation into year-zero dollars per Btu
X = sum of values of G_N for all years from 1 to L
P = net present value of the solar energy system

p_N = rate of productivity increase in year N (nonenergy area)

q_N = rate of increase in usage of energy in year N

f_N = fraction of GNP in year N used for gathering, producing, and distributing energy

The net present value is then $P = X - V_L$.

First, the ultimate value of the investment I is calculated if it is invested in high-grade bonds that mature in year L. The interest rate on the bonds is t_1, and the interest earned in each year is $t_1 I$ for L years. The first year's interest, $t_1 I$, is invested at the beginning of year 2, and it earns interest at the rate t_2 for $L - 1$ years. The second year's interest, consisting of the sum of terms $t_1 T$ and $t_2 t_1 I$, is invested at the beginning of year 3 to earn $t_3[t_1 I + t_2 t_1 I]$ for $L - 2$ future years. Consider the following table:

Year 0	Year 1	Year 2	Year 3	Year 4		Year L
I	I	I	I	I	$\cdots$	I
	$t_1 I$	$t_1 I$	$t_1 I$	$t_1 I$	$\cdots$	$t_1 I$
		$t_2 t_1 I$	$t_2 t_1 I$	$t_2 t_1 I$	$\cdots$	$t_2 t_1 I$
			$t_3 t_1 I$	$t_3 t_1 I$		$\cdot$
			$t_3 t_2 t_1 I$	$t_3 t_2 t_1 I$		$\cdot$
				$t_1 t_4 I$		$\cdot$
				$t_1 t_2 t_4 I$		
				$t_1 t_3 t_4 I$		
				$t_1 t_2 t_3 t_4 I$		

The total value after L years may be calculated from

$$W_L = I[1 + Lt_1 + (L - 1)t_2 t_1 + (L - 2)(t_3 t_1 + t_3 t_2 t_1)$$
$$+ (L - 3)(t_1 t_4 + t_1 t_2 t_4 + t_1 t_3 t_4 + t_1 t_2 t_3 t_4) + \ldots] \qquad (5.1)$$

Since one does not generally know in advance what the values of $t_2, t_3, \ldots,$ t_L are, it may be sufficient to assume $t_1 \cong t_2 \cong t_3 \cong \ldots t_1 = $ "t". In this case, W_L reduces to

$$W_L \cong I\{1 + Lt + t^2[(L - 1) + (L - 2) + (L - 3) + \ldots]$$
$$+ t^3((L - 2) + 2(L - 3) + \ldots)\} \qquad (5.2)$$

This can be shown to be reducible to

$$W_L \cong I(1 + t)^L \qquad (5.3)$$

To calculate the value in terms of year-zero dollars, a correction must be made for L years of inflation. Thus,

$$V_L = \frac{W_L}{(1 + r_1)(1 + r_2) \dots (1 + r_L)} \tag{5.4}$$

If $r_1 \cong r_2 \cong r_3 \dots = $ "r," this reduces to

$$V_L \cong \frac{W_L}{(1 + r)^L} \tag{5.5}$$

Thus, if the interest and inflation rates are assumed constant in time,

$$V_L \cong \frac{I(1 + t)^L}{(1 + r)^L} \tag{5.6}$$

This is the approximation that is usually employed, because the variation of t_J and r_J over future years is unknown. Note that if $t > r$, the potential value of I dollars in year 1 is greater than I, whereas when $t < r$, I dollars invested over time are in a sense worth less than the current value of I, because inflation erodes the value faster than interest can be earned.

Some economists prefer to work in terms of the "real rate of interest," which is the excess of t above r. The real rate of interest is the gain per year in purchasing power from interest on fixed-income securities. According to some economists,[1] ". . . use of the real interest rate . . . eliminates the need to forecast inflationary influences and associated price adjustments." This is, of course, a specious argument, since one must have a scenario for inflation in order to estimate the real rate of interest. In 1978, the real rate of interest was negative, since inflation was greater than interest returns. In terms of the real rate of interest, $(t - r)$, Eq. (5.6) is approximately

$$V_L \cong I[1 + (t - r)]^L \tag{5.7}$$

The value of the energy saved during the lifetime of the investment is the sum of the values of energy saved in each year from 1 to L. It will be assumed that there is no salvage value after L years and that there is no depreciation allowance. In any arbitrary year N, the amount of energy saved is E Btu. The value of fuel in dollars per Btu in year N is denoted as F_N. This may be calculated by considering the cumulative effect of escalation of fuel costs. For example, if in any year the inflation rate is 7% and the escalation

[1]S. Ben-David, W. D. Schulze, J. D. Balcomb, R. Katson, S. Noll, F. Roach, and M. Thayer, "Near-Term Prospects for Solar Energy: An Economic Analysis," *Natural Resources Journal,* **17,** 169, 1977.

rate is 10%, the actual rate of increase of fuel prices is 17%. The price of fuel in year-N dollars in year N is

$$F_N = F_1 \prod_{j=1}^{N} (1 + e_j + r_j) \tag{5.8}$$

The value of the energy saved in year N is simply EF_N. In terms of year-zero dollars, the value is

$$G_N = EF_1 \prod_{j=1}^{N} \frac{(1 + e_j + r_j)}{(1 + r_j)} \tag{5.9}$$

The total value (expressed in year-zero dollars) of energy saved over the lifetime of the solar energy system is

$$X = EF_1 \sum_{N=1}^{L} \prod_{j=1}^{N} \frac{(1 + e_j + r_j)}{(1 + r_j)} \tag{5.10}$$

For the restricted special case where e_j and r_j may be treated as approximately constant, this reduces to

$$X = EF_1 \sum_{N=1}^{L} \left(\frac{1 + e + r}{1 + r}\right)^N \tag{5.11}$$

Since $P = X - V_L$, if Eq. (5.10) is used for X, and Eqs. (5.3) and (5.4) are used for V_L, we find that

$$P = EF_1 \sum_{N=1}^{L} \prod_{j=1}^{N} \frac{(1 + e_j + r_j)}{(1 + r_j)} - \frac{I(1 + t)^L}{(1 + r_1)(1 + r_2)\ldots(1 + r_L)} \tag{5.12}$$

It is useful to rewrite this equation in the form

$$P = EF_1 R_L - IS_L \tag{5.13}$$

where

$$R_L = \sum_{N=1}^{L} \prod_{j=1}^{N} \frac{(1 + e_j + r_j)}{(1 + r_j)} \tag{5.14}$$

$$S_L = \prod_{j=1}^{L} \frac{(1 + t)}{(1 + r_j)} \tag{5.15}$$

The factor R_L is the quantity required to multiply the value of fuel saved in year 1 to obtain the value of fuel saved over the L years of the system (expressed in year-zero dollars). The factor S_L multiplies the initial investment to yield the value of that investment including interest earned for L years, corrected back to year-zero dollars. If there were no inflation

$(r_J = 0)$, S_L would reduce to $(1 + t)^L$. If there were no inflation and no fuel price escalation $(e_J = r_J = 0)$, R_L would reduce to simply L.

The point of these calculations is to provide expressions for R_L, which is the factor that multiplies the annual fuel saving in year 1 to give the cumulative fuel saving over L years. When the initial investment (multiplied by S_L) is subtracted off, and the result corrected for the difference between interest and inflation rates over L years, the net present value in year-zero dollars is obtained.

5.1.2 Investment with Borrowed Money

The treatment of the case where the user borrows the initial investment proceeds in a similar way. Because the interest payments are tax deductible, the borrower receives part of the payments back in the form of income tax credits. Therefore, the interest rate used in computations should be the after-tax rate. For a borrower with income in the 30% bracket, for example, the after-tax interest rate is 70% of the actual interest rate. The user pays off an amortized loan in L years at rate h, so that principal and loan are paid back at the end of L years. The level monthly payments M can readily be computed and are shown for typical cases in Table 5.1. These values are arranged so

TABLE 5.1 Monthly payment to amortize a loan of $1000

Period (yr) → Interest Rate ↓	5	10	15	20	25
0.06	19.34	11.11	8.44	7.17	6.45
0.07	19.81	11.62	8.99	7.76	7.07
0.08	20.28	12.14	9.56	8.37	7.72
0.09	20.76	12.67	10.15	9.00	8.40
0.10	21.25	13.22	10.75	9.66	9.09

that every month one pays interest of $1/12$ of h times the principal remaining at the end of the previous month, plus a variable contribution toward the principal. In the beginning payments are mostly interest, but in later years they become mainly principal. To evaluate the breakdown of each payment in terms of principal and interest, the following procedure is used: (1) Start by computing the interest on the previous month's balance. (2) Deduct that interest from the month's payment to obtain the contribution to paying off the principal during that month. (3) Deduct the payment toward principal from the balance remaining at the end of the month. For example, with a 6% loan on $1000 for 5 years, the monthly payment is $19.34. The interest is $5.00 for the first payment, and therefore the first payment consists of

$5.00 interest and $14.34 principal. For the second month, the balance remaining is $1000 − $14.34 = $987.66. A month's interest on this amount is $4.94. Thus, for the second month's payment there is $4.94 interest and $14.40 principal. This can be continued for all payments.

For the case where the initial investment is borrowed, V_L is defined as the sum of payments over the lifetime of the loan, corrected for inflation to year-zero dollars. The sum of payments in any year is $12M$. The payments in year J, corrected for inflation back to year zero, amount to $(12M)/\prod_{I=1}^{J} (1 + r_I)$. The value of V_L is therefore

$$V = 12M \sum_{J=1}^{L} \prod_{I=1}^{J} (1 + r_I)^{-1} \qquad (5.16)$$

This expression can be used in the formula for net present value $P = X - V_L$. Thus, using Eq. (5.10) for X, it is found that

$$P = \sum_{N=1}^{L} \left[EF_1 \prod_{J=1}^{N} \frac{(1 + e_J + r_J)}{(1 + r_J)} - 12M \prod_{J=1}^{N} (1 + r_J)^{-1} \right] \qquad (5.17)$$

This can be re-expressed as

$$P = EF_1 R_L - 12M T_L \qquad (5.18)$$

where

$$T_L = \sum_{N=1}^{L} \prod_{J=1}^{N} (1 + r_J)^{-1} \qquad (5.19)$$

Equations (5.18) and (5.19) are analogous to Eqs. (5.13) and (5.15). If there is no inflation and $r_J = 0$, T_L reduces to L.

5.2 Pay-back Periods

In order to apply the net present value theory, one must adopt a scenario for interest, inflation, and fuel escalation rates over the entire lifetime of the project. Usually these values are unknown and difficult to estimate. Furthermore, the effect of yearly compounding of fuel escalation rates becomes very large toward the end of the lifetime of a solar installation, and in some cases the bulk of the savings occurs in later years. In addition, the lifetime L is often unknown. For all these reasons it is often desirable to ask how many years it will take to recover the cost of the investment. Instead of calculating the net present value for L years, one estimates how many years it will take until the net present value will become zero. The net present value starts off negative and gradually moves up toward zero as there are more and more yearly fuel savings. The "pay-back period" of "break-even time" is the number of years required for the sum of fuel savings to cancel out the cost of the investment. It is an important quantity because it usually requires a shorter period (than net present value) over which e_J and r_J need to be estimated and can therefore be expected to be more accurate. Furthermore, most residents are not certain how long they expect to live in a house and cannot count on savings to be produced 20 years hence. Therefore, the pay-back period may be a more useful quantity than net present value in some applications.

To calculate the pay-back period, one must find the number of years N that it takes before the accumulated gain $EF_1 R_N$ just balances the total cost of the investment over the life time of L years. When the investment is made from cash on hand, the requirement is to find a value N such that

$$EF_1 R_N - IS_L = 0 \qquad (5.20)$$

When the investment is borrowed, the requirement is

$$EF_1 R_N - 12MT_L = 0 \qquad (5.21)$$

This must usually be accomplished by trial and error, using tables of R_L, S_L, and T_L for any scenario of inflation and escalation. In this latter case, it follows that the pay-back period N is the value that results in

$$R_N = \frac{12MT_L}{EF_1} \qquad (5.22)$$

5.3 Illustrative Example

An example of specific net present values and pay-back periods is given next. Consider the scenarios listed in Table 5.2.

A solar energy installation costs $10,000 and saves $EF_1 = \$400$ of fuel the first year. If the lifetime is $L = 15$ years and scenario I from Table 5.2 is assumed, what is the net present value and the pay-back period if (a) the cash is on hand and could have been used to purchase bonds yielding $t = 0.07$, (b) the investment is borrowed at an interest rate $h = 0.08$?

(a) According to Table 5.2, $R_{15} = 46.25$, and therefore $X = 46.25 \times \$400 = \$18,500$. Since $S_{15} = 1.33$,

$$P = 18500 - 1.33 \times 10000 = \$5200$$

TABLE 5.2   Arbitrary scenarios for future inflation and escalation rates

Year (L)	Scenarios for R_L			Scenario for S_L and T_L	
	Scenario I	Scenario II	Scenario III		
	$r = 0.05$ e_1 to $e_5 = 0.2$ e_6 to $e_{10} = 0.1$ e_{11} to $e_{20} = 0.05$	$r = 0.05$ e_1 to $e_{10} = 0.1$ e_{11} to $e_{20} = 0.05$	$r = 0.05$ $e_i = 0.25, 0.20, 0.20,$ $0.15, 0.10, 0.10, 0.05,$ $0.05, 0.02, 0.02, 0.00,$ and 0.00 thereafter	$t = 0.07$ $r = 0.05$	
Year (L)	R_L	R_L	R_L	S_L	T_L
1	1.19	1.10	1.24	1.02	0.95
2	2.61	2.29	2.71	1.04	1.86
3	4.30	3.60	4.46	1.06	2.72
4	6.30	5.04	6.46	1.08	3.55
5	8.69	6.62	8.65	1.10	4.33
6	11.31	8.35	11.05	1.12	5.08
7	14.18	10.24	13.56	1.14	5.79
8	17.32	12.31	16.19	1.16	6.46
9	20.76	14.58	18.87	1.19	7.11
10	24.53	17.06	21.60	1.21	7.72
11	28.48	19.66	24.33	1.23	8.31
12	32.61	22.39	27.06	1.25	8.86
13	36.95	25.25	29.79	1.28	9.39
14	41.49	28.24	32.52	1.30	9.90
15	46.25	31.37	35.25	1.33	10.38
16	51.24	34.65	37.98	1.35	10.84
17	56.46	38.05	40.71	1.38	11.27
18	61.93	41.61	43.44	1.40	11.69
19	67.66	45.34	46.17	1.43	12.09
20	73.67	49.25	48.90	1.46	12.46

To find the pay-back period, look in Table 5.2 for the row for which $(R_N \times 400 - S_L \times 10{,}000) = 400 R_N - 13{,}300 = 0$.

It is found that in years 11 through 13,

$N = Year$	$EF_1 R_N$	IS_L	$EF_1 R_N - IS_L$
11	11,392	13,300	$-8{,}092$
12	13,044	13,300	-256
13	14,780	13,300	$+1{,}480$

Thus, the pay-back period is about $12\frac{1}{4}$ years.

(b) Use Eq. (5.19) and Tables 5.1 and 5.2.

$$P = 400 \times 46.25 - 12 \times 95.60 \times 10.38$$
$$= 18,500 - 11,508 = \$6992$$

$N = Year$	$EF_1 R_N$	$12MT_L$
11	11,392	11,908
12	13,044	11,908

Thus, with borrowed capital under the assumed conditions the net present value is higher, and the pay-back period is shorter (about $11\frac{1}{2}$ years) than with invested cash. This is due to the erosion of the value of future money produced by the assumed inflation in the scenario.

5.4 Trends in Fuel Prices

5.4.1 Effect on the GNP

In 1977, approximately 12% of the gross national product (GNP) was involved in gathering, producing, and distributing energy. During the years 1973–1977, fuel prices escalated much faster than the overall inflation, which indicates that each year energy is making successively larger contributions to the GNP. A necessary consequence of long-term fuel price escalation is that energy costs become an increasingly larger part of the GNP.

It will be assumed that the GNP increases each year in terms of two parts, the energy part E and the nonenergy part N. The nonenergy part increases each year due to increases in productivity and due to general price inflation. The energy part increases due to increases in energy usage and to increases in energy prices. We may then write

$$G_1 = E_1 + N_1 \tag{5.23}$$

where G_1 is the GNP for year 1, and E_1 and N_1 are the energy and nonenergy contributions in year 1. In any subsequent year N,

$$G_N = E_1 \prod_{J=1}^{N} (1 + r_J + e_J + q_J) + N_1 \prod_{J=1}^{N} (1 + r_J + p_J) \tag{5.24}$$

where q_J is the increase in energy usage in year J, and p_J is the increase in productivity in year J. After the passage of N years, the fraction of the GNP that is related to energy is

$$f_N = \frac{E_1 \prod_{J=1}^{N} (1 + r_J + e_J + q_J)}{E_1 \prod_{J=1}^{N} (1 + r_J + e_J + q_J) + N_1 \prod_{J=1}^{N} (1 + r_J + p_J)} \qquad (5.25)$$

$$f_N = \frac{f_1}{f_1 + (1 - f_1) \prod_{J=1}^{N} \dfrac{(1 + r_J + p_J)}{(1 + r_J + e_J + q_J)}} \qquad (5.26)$$

In 1977, f_1 was approximately 12%. For any arbitrary scenario of the future, f_N can therefore be calculated. There is a historical tendency for p_J and q_J to be of the same general magnitude, and therefore $r_J + p_J$ in the numerator of this expression tends to cancel $r_J + q_J$ in the denominator. Thus, as a rough approximation, only the escalation rate changes f_N:

$$f_N \cong \frac{f_1}{f_1 + (1 - f_1) \prod_{J=1}^{N} \dfrac{1}{(1 + e_J)}} \qquad (5.27)$$

For the scenarios listed in Table 5.2, f_N may therefore be calculated. At the end of 20 years, f_N is equal to 43%, 32%, and 31% in scenarios I, II, and III, respectively, from that table. It is clear that continued escalation of fuel prices over long periods leads to energy costs playing a dominant role in the GNP. Whether this is possible without great economic dislocation is uncertain. Some authors have used assumed escalation rates as high as 0.20 for as long as 20 years. This assumption leads to a GNP that is 86% energy costs in 20 years!

It does not seem possible that fuel prices can continue to escalate faster than inflation for very long. The sudden and dramatic round of fuel price increases that took place in the years 1973–1979 was the result of the release of pressures that had built up from many years of artificial price controls and temporary excess of supply. In a sense, fuel prices "caught up" with other prices that had increased at 4% per year for 20 years. As fuel prices continue to increase faster than inflation, they will become a larger part of the cost of commerce and manufacturing and will drag overall prices up with increases in fuel. Indeed, as pointed out previously, continued escalation of fuel prices will tend to make energy the largest contributor to the GNP. It appears that there might be room left for one doubling of fuel prices relative to overall prices (enough to bring energy up to approximately 25% of the GNP) before great economic dislocation occurs. This author's best guess for the future is scenario III in Table 5.2.

5.4.2 Relative Cost of Fossil Fuel vs. Electricity

Another factor that needs to be considered is the relative cost of various fuels. When a solar energy system displaces oil, gas, coal, or electricity the value can be quite different. Part of this difference lies in the basic difference between electricity and the fossil fuels. Since fossil fuels can only be converted to electricity at an efficiency of about 33%, the price of electricity tends to be at least three times higher per Btu than fossil fuels. Government regulation of fossil fuel prices creates an artifical hodgepodge of fossil fuel costs.

The relation between fossil fuel costs and electricity costs is illustrated by considering a resident of Dallas, Texas, in 1978, when the cost of natural gas was about $2.20 per 1000 ft^3. Each ft^3, when burned, yields about 1000 Btu. However, in most home applications, the burners are only perhaps 60% efficient, in the sense that 40% of the heat released goes up the flue or is lost in other ways. Thus, the cost of delivered Btu's from natural gas was $2/0.6 = $3.67 per million Btu delivered. In the same year (1978), the cost of electricity was about $0.04 per kWh. Since 3413 Btu is equivalent to 1 kWh, the cost of 1 million Btu of electricity was $11.72. When electricity is used for resistance heating it is nearly 100% efficient, so no efficiency factor is used. Thus, in Dallas in 1978 electricity was slightly more than three times as expensive as natural gas per Btu delivered. From the point of view of the consumer, when solar energy displaces electricity the value of the solar energy installation is therefore about three times as high when it displaces electricity as when it displaces natural gas. For example, a solar energy installation that displaced 100 million Btu per year (on average) in a Dallas residence that used gas in 1978 would have saved $367 per year. Had that same installation been placed on an all-electric house, the saving would have been $1172 per year. These are the values of EF_1 from previous sections. Clearly, there is a great advantage to the consumer in installing solar energy equipment in all-electric houses.

5.5 The Viewpoint of the Utilities

5.5.1 Baseload and Peaking Loads

Electrical power companies tend to divide their costs into several broad categories. One of these is the amortized cost of the fixed investment in plant facilities over the lifetime of the plant. When this is coupled with costs of maintenance and operations, the result is an annual fixed cost that is roughly independent of how much electricity is actually sold. The annual cost of fuel may vary from a modest fraction to a large fraction of the fixed costs, depending on the type of fuel used and the duty cycle of the plant. Generally, the fixed costs tend to exceed the fuel costs, even with recent fuel price escalation.

Power companies divide their plants into baseload plants and peaking plants. A baseload plant is designed to operate continuously for 24 hr a day, whereas peaking plants are designed to come on and off stream rapidly to provide electricity in response to varying demand. Baseload plants are usually very capital intensive. They are also very expensive to build but utilize the cheaper fuels (hydro, nuclear, or coal). The main costs of such plants are construction, not expended fuel. A peaking plant, on the other hand, is designed for rapid start-up and shut-down and is usually less efficient. Because of the shorter duty cycle, it is not profitable to invest very large amounts of money into peaking plant design and construction. Peaking plants require fuels such as natural gas or oil to give them rapid start-up performance. The fuel costs are therefore a much higher fraction of total costs than in the case of baseload plants.

From the point of view of the utilities, when a consumer installs a solar energy system there is a great difference between whether baseload or peaking power is displaced. Since the costs of having baseload electricity available are mainly fixed, displacement of baseload electricity on an interruptable basis does not save a power company much money. Indeed, since customers are mainly paying for fixed plant investment in baseload capacity, the main consequence of solar displacement of baseload by some customers is that rates per kWh for the remaining customers will have to go up! If a large enough number of customers displace baseload electricity, the power company could conceivably reduce its baseload capacity. The savings to the power company would then be at the baseload electricity rate, which is usually considerably lower than peak power rates. However, if the utility must provide backup to these solar customers, it may not be able to reduce its baseload capacity.

It has been argued that the installation of a solar energy system with storage automatically implies that the savings in electricity be computed at baseload rates.[2] The reason for this is that one could simply install the storage system without the solar collection system, and use that as a means of shifting peak loads to non-peak hours. Thus, one should compare the cost of the solar system (including storage) with the cost of a storage system to be used in conjunction with an ordinary electrical system. The electrical energy displaced by the solar system would be at the baseload rate, and the net cost would be for the entire system less the storage subsystem. On this basis, the value of a solar energy system to a electric utility is considerably

[2]J. G. Asbury and R. O. Mueller, *Science*, **195**, 445, 1977.

reduced compared to the case if it were displacing peak power. Some solar systems that look good economically to the individual residential customer may not help the electric utility very much.

5.5.2 Seasonal Factors

Another aspect of the economics of solar displacement of electric consumption is the seasonal factor. For example, in Dallas, Texas, a summer-peaking area, the installation of a solar space heating system for winter use is almost automatically a displacement of baseload electricity. Space heating systems in Texas do not reduce the required electrical plant capacity of utilities in that summer-peaking area. As space heating systems are installed in all-electric homes in Dallas, the net effect will be to exacerbate the difference between winter and summer electrical usage rates and therefore reduce the load factor (ratio of average instantaneous load to peak instantaneous load). This can only lead to higher charges for residents in the area. In a winter peaking area, solar space heating systems can reduce both peak and average loads. However, periods of greatest energy demand can be associated with low solar availability, in which case the ultimate peak electrical usage may be almost the same whether solar energy space heating systems are installed or not. Unless the storage systems and auxiliary energy input are made to function with a timer or utility controller, there may be almost as great a need for peak power with widespread use of solar energy heating systems as without. In areas where utilities charge a declining rate per kWh with increasing amounts of kilowatt-hours, the residence with solar energy pays a much higher net rate per kWh than a residence without solar energy. Or, in other words, the electrical energy displaced by solar energy is at the lowest rate paid by the residence.

When solar heating/cooling systems are employed, there are additional problems. One is that there is only a limited range of localities where the heating/cooling ratio is properly balanced so that the collector field is not oversized in one season or undersized in the other season. As a result, performance will be poor in one season or the other. In Dallas, Texas, where summer cooling loads are about as high as winter heating loads, the low efficiency of single-stage absorption chillers and the low efficiencies of flat plate collectors at collector temperatures near 200°F (367°K) lead to poor summer performance compared to winter performance. Therefore, a collector sized to provide adequate winter performance will be undersized in summer, and a collector sized to provide adequate summer performance will be oversized in winter. The method for utilizing auxiliary make-up energy for air conditioning is of great importance. The most efficient method is to use electricity generated at a central power station to drive a compression-type chiller. However, large-scale use of this system may not reduce peak summer utility requirements if most of the residential solar systems go onto auxiliary energy during the same peak periods. When auxiliary energy is used to fire an absorption chiller that operates on solar energy when available, the situation is worse. Not only is this procedure wasteful of energy, but the low coefficient of performance (COP) of the chiller requires a much higher instantaneous power input than a compression chiller for the same cooling effect. As a result, widespread installation of such systems with electrical make-up energy can actually increase the peak utility requirements when many of the residential solar units go onto make-up energy during the same peak period.

It may be concluded that the interaction between individual users of solar energy and utilities is a complicated one. When an individual saves energy or money, the utility might not gain much, or could even lose. Utilities are fundamentally concerned with peak usage rates and load factors. Individuals who install solar systems that do not reduce peak rates, and which lower load factors, are not helping the utility very much. If utility charges to customers reflected the true cost of supplying them power (installation, hook-up, and peak usage being dominant) the economics of solar energy might look much worse than when average rates are used.

5.6 Commercial vs. Residential Installations

Commercial installations of solar energy have greater economic advantages over residential installations because the investment can be depreciated and maintenance costs are deductible by commercial owners. Furthermore, many businesses are in higher income tax brackets, and therefore even the interest payment deductions (which are deductible by businesses and individuals alike) lead to greater tax savings.

5.6.1 Investment Tax Credits

Both residential and commercial investors are entitled to income tax credits at the time of investment in solar energy installations. For investors with net positive income, this constitutes a direct effective reduction in the cost of investment. As of mid-1979, the income tax credit provisions were:

(a) residences—30% of the first $2000 and 20% of the next $8000, for a maximum credit of $2200;

(b) commercial—20% of the investment (no upper limit).

5.6.2 Depreciation

Commercial investors in solar energy are entitled to deduct the annual depreciation of the installation from their income before computing their income tax each year. Thus, if an installation costs $50,000 and straight-line depreciation is used over a 10 year lifetime, the company can deduct $5000 from its income each year for ten years. If the company is in the 50% tax bracket, this produces a $2500 saving in taxes each year for 10 years. Gains due to depreciation in terms of year-zero dollars in future years should be reduced by the cumulative amount of inflation during intervening years.

5.6.3 Maintenance Costs

Solar energy systems are so comparatively new that insufficient data exist for maintenance requirements. Initial reports of operational results on actual heating and cooling systems have recently been presented.[3] The costs of maintenance are a legitimate business expense and can be deducted from income by a business. Maintenance costs for individuals in residences are not deductible.

5.6.4 Illustrative Example

Consider a comparison between the economics of residential and commercial installations defined as follows:

	Residential	Commercial
Basic cost	$8000	$80,000
Interest rate	10%	10%
Lifetime of loan	15 years	15 years
Lifetime of equipment	15 years	15 years
Value of energy saved in first year	$600	$6,000
Annual maintenance cost	$200	$2,000
Income tax bracket	30%	50%

The amount of money that must be borrowed to finance the installations is the basic cost less the first year's tax credit. Thus, the loan value can be taken as

[3]Conference Proceedings–Solar Heating and Cooling Systems Operational Results, Coord. by Solar Energy Research Institute, Colorado Springs, CO, Nov. 28–Dec. 1 1978, SERI Rept. No. TP-49-063.

Residential: $8000 − [0.3 × $2000 + 0.2 × $6000] = $6200

Commercial: $80,000 × 0.8 = $64,000

The annual payments on a 10% loan are found from Table 5.1 to be

Residential: $ 800

Commercial: $8256

The tax savings in each year due to interest payments should be subtracted from these figures. A simpler method is to treat the effective interest rate as 5% (after taxes) for commercial and 7% (after taxes) for residential applications. Thus, the annual payments (after tax basis) may be treated as:

Residential: $ 669

Commercial: $6067

The depreciation allowance each year for the commercial installation is $\frac{1}{15}$ × $80,000 = $5333. The annual tax savings due to depreciation is then 50% × $5333 = $2667.

The annual tax savings on maintenance for the business are estimated at 50% × $2000 = $1000.

Thus, in summary, the annual carrying costs including maintenance are

Residential: $869

Commercial: $6067 − $2667 + $1000 = $4400

If scenario III from Table 5.2 is used, the net present values are found to be

Residential:

$$\text{NPV} = 35.25 \times \$600 - \$869 \times 10.38$$
$$= \$21,150 - \$9020 = \$12,130$$

Commercial:

$$\text{NPV} = 35.25 \times \$6000 - \$4400 \times 10.38 = \$165,830$$

The pay-back periods are found as follows:

Residential:

$$\$600 R_N = \$9020$$
$$R_N = 15.0$$
$$N = 7\tfrac{1}{2} \text{ years}$$

Commercial:

$$\$6000 R_N = \$45,670$$
$$R_N = 7.6$$
$$N = 4\tfrac{1}{2} \text{ years}$$

Thus, the commercial system, which is nominally 10 times larger than the residential system, leads to a much shorter pay-back period and a net present value that is more than 10 times higher than for a residential system.

Worked Examples

1. A homeowner borrows $8000 to install a solar heating/hot water system. The loan is at 10% for 20 years. The solar energy system has a lifetime of 15 years, and the first year savings in fuel saved is $600. The homeowner is in the 40% income tax bracket. Using scenario III in Table 5.2, what is the net present value and the pay-back period?

   As a rough approximation, we may take the after-tax interest rate as 6%. The monthly payments, from Table 5.1, are $8 \times 7.17 = \$57.36$. The net present value is

$$\text{NPV} = 35.25 \times \$600 - 12 \times \$57.36 \times 12.46$$
$$= \$21,150 - \$8576 = \$12,574$$

   The pay-back period is found from

$$0 = R_N \times \$600 - \$8576$$
$$R_N = 14.3$$
$$N = 7\tfrac{1}{2} \text{ years, from Table 5.2.}$$

2. What effect would the Energy Tax Act of 1978 have on the economics of the system in Example 1?

   The tax credit in the first year is 30% of the first $2000 plus 20% of the remainder, making a total of $1800 as a tax credit. Thus, the effective amount borrowed is reduced to $8000 − $1800 = $6200. The results are

$$\text{NPV} = 35.25 \times \$600 - 12 \times \$44.45 \times 12.46$$
$$= \$21,150 - \$6646 = \$14,504$$
$$R_N = \frac{\$6646}{\$600} = 11.08$$
$$N = 6 \text{ years}$$

3. The same system described in Examples 1 and 2 is installed by a commercial business. Assuming the business is also in the 40% tax bracket, what is the effect of depreciating the solar installation over its lifetime?

Using straight-line depreciation of $6200, the business deducts $6200/15 = $413 per year from its reportable income. This produces an annual tax saving of $0.4 \times \$413 = \$165$. This should be subtracted from the annual payments, leading to

$$\text{NPV} = \$21{,}150 - \$6646 + 12.46 \times \$165 = \$16{,}560$$

$$R_N = \frac{\$4590}{\$600} = 7.65$$

$$N = 4\tfrac{1}{2} \text{ years}$$

Problems

5.1. A new investment is made in a solar energy system by borrowing $5000 at an interest rate such that the annual payments are $1000 for 10 years (principal plus interest). The solar energy system has a lifetime of 15 years, and the first year's saving in fuel is valued at $500 in current dollars. Assume there is 5% annual inflation and that fuel prices escalate at 10% annually over the 15 year period. Find the net percent value and the pay-back period. What percentage of the GNP will be fuel-related after 15 years of escalation?

5.2. A corporation paying income tax at the rate of 50% on earnings can borrow money at 12% for 10 years and expects inflation to average 9% per year over that period. What is the value in year-zero dollars paid out over the course of the loan?

5.3. The designer of the system described in Problem 5.1 can add extra solar panels at a cost increase of 30% to collect 20% more energy. Show that this additional investment increases the net present value but slightly lengthens the pay-back period.

5.4. The Energy Tax Act of 1978 provides for a 30% tax credit on the first $2000 invested in solar energy and a 20% credit on the next $8000 invested in solar energy. How does this act affect the net present values and pay-back periods for the systems of Problems 5.1 and 5.3?

Domestic Hot Water Systems

6

This chapter describes typical solar domestic hot water systems and estimates their performance as a function of collector size, storage volume, use pattern, collector tilt, and other factors. The analysis is performed by computer simulation, and by "running" conceptual systems against 8760 hourly solar intensities and ambient temperatures for a model year stored in a computer. System performance is analyzed on an hourly basis and on monthly and yearly bases. The economics of various systems is evaluated, and it is shown how an optimized design can be selected for any locality for which solar data are available and collector performance is known. It is shown that storage volumes of 1–2 gal per ft² of solar collector lead to the best designs.

6.1 Conventional Systems

A conventional hot water heater for a residence is illustrated in Figure 6.1. Tap water is introduced through a tube to the bottom of the tank. The entire tank is under the cold water line pressure (usually 40–80 psi). Hot water is drawn from the top of the tank when a hot water tap is opened anywhere in the house. A sensor monitors the temperature on the wall at the middle of the tank, and when the temperature drops below a preset value [usually manually adjustable in the range 120°F–150°F (322°K–339°K)] the gas jets or electrical resistance rods are turned on automatically by a thermostat. When the desired temperature is reached, the heaters are turned off by the

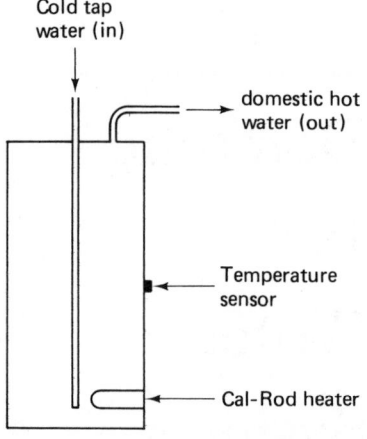

Cold tap
water (in)

domestic hot
water (out)

Temperature
sensor

Cal-Rod heater   **Figure 6.1**   Conventional hot
water heater.

thermostat. The volumetric capacity of the heater is usually sized to be about
25–35 % of daily usage. For a family of four that uses roughly 120 gal/day
(0.455 m³/day) of hot water, a 30 or 40 gal (0.114 to 0.152 m³) tank is typically
used. If the tap water temperature is say 60°F and the hot water temperature is 130°F, then 40 gal $\times$ 8.3 lb/gal $\times$ 1 Btu/lb-°F $\times$ 70°F = 23,240 Btu
(24,520 kJ) are required to bring a full tank of tap water up to hot temperature. For a gas heated unit with, say, 80,000 Btu/hr (23.45 kW) input, if the
burner efficiency is 60 %, 48,000 Btu/hr (14.1 kW) can be added to the water.
Thus, the recovery time after large usage has depleted the tank of hot water
is typically of the order of 23,240/48,000 = $\frac{1}{2}$ hr.

Commercial hot water units can either be essentially scaled-up resi-
dential-type units or can be "rapid recovery" units. A rapid recovery unit
uses a very small volume compared to a day's usage (perhaps 5 %) but has
very high power input for rapid heating as hot water is used. This kind of
unit reduces the space required for the hot water tank in an apartment
complex.

6.2 Solar Water Heating

6.2.1 Heating without "Back-up"

Solar water heating can be accomplished in several ways. In Israel and
some other localities, solar water heaters are installed without "back-up"
from conventional energy sources. The hot water tank is heavily insulated and
is usually sized to be roughly one day's usage and is mounted on the roof of a
building where the solar collectors are also located. Water is automatically

circulated through the collectors when a sensor determines that the collectors are several degrees hotter than the water in the tank. On cloudy days, or periods of high usage, there may not be any hot water available. The people who install these systems are prepared to do without hot water on occasion, with perhaps 70–80% of the hot water needs being satisfied in a hot sunny climate. If freezing temperatures are encountered, care must be taken to ensure the the collectors are higher than the tank and that when the pump is off the water in the collectors will drain down into the tank.

6.2.2 Heating with "Back-up"

In most installations in the United States, it is expected that the solar hot water system will not "stand alone" but will be "backed-up" by conventional energy sourced. The most convenient way to do this is to use solar collectors to heat water in a "preheat tank" which feeds preheated water to a conventional hot water heater running on gas, oil, or electricity. When the solar preheater brings tap water up to the desired temperature (usually around 130°F), no auxiliary back-up energy is used. As long as the solar preheater raises the temperature of the tap water, the amount of conventional back-up energy is reduced. For the remainder of this chapter, only hot water systems *with* back-up will be discussed.

6.2.3 Antifreeze vs. Drain-down Systems

In many solar energy applications, the working fluid in the solar collectors is an antifreeze/water mixture that contains non-potable chemicals. This fluid is used to transfer heat to the potable hot water, but it must be kept chemically isolated from the potable water. A method for doing this that has grown in acceptance is a double-wall jacketed preheat storage tank, as illustrated in Figure 6.2. The heat exchange jacket has its own inner wall outside the wall of the inner storage tank. Thus, there are two metal walls separating the transfer fluid from the potable water. The transfer fluid is circulated to the solar collectors and then through the heat exchange jacket. Cold tap water is introduced into the bottom of the preheat tank, and preheated potable water is drawn off at the top. Some systems utilize a "drain-down" system in which potable water is circulated directly to the collectors and no heat exchange jacket is required. This approach has the advantage that the temperature drop across the heat exchange jacket is avoided, and higher collector efficiencies will result. The disadvantage is that it is not always convenient to have drain-down plumbing, and damage to the collectors could result if the

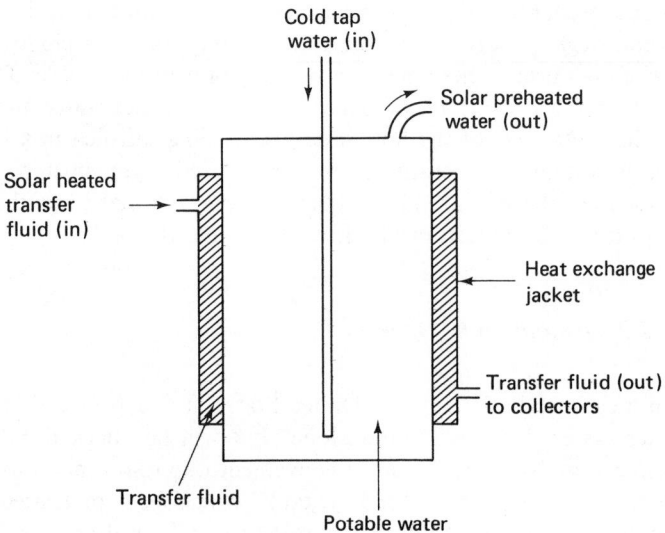

Figure 6.2   Solar preheat tank.

drain-down system malfunctions in freezing weather. It is also not very good practice to tie the potable water to extended plumbing lines to the roof. Corrosion can be more of a problem in drain-down systems because there is more oxygen in the water and rust inhibitors cannot be added.

6.2.4 Flow Diagrams

To use solar energy for heating domestic hot water, a very effective approach is to use a tank such as shown in Figure 6.2 as a preheater for a conventional hot water heater. Consider the system illustrated in Figure 6.3. An antifreeze/water mixture is used in the circuit consisting of the heat exchange jacket on the solar storage tank, a pump, an expansion tank, and the collector modules. Cold tap water is admitted to the bottom of the inner part of the solar preheat tank, and clean solar preheated water is drawn off the top. A tempering valve is included so that solar preheated water will automatically be mixed with cold tap water if the solar heated water exceeds a preset temperature on the tempering valve. Thus, if the tempering valve is set at 130°F and the solar preheated water in the tank is at 150°F, the tempering valve will automatically mix enough cold tap water with the solar heated water so that the water entering the conventional hot water tank is about 130°F.

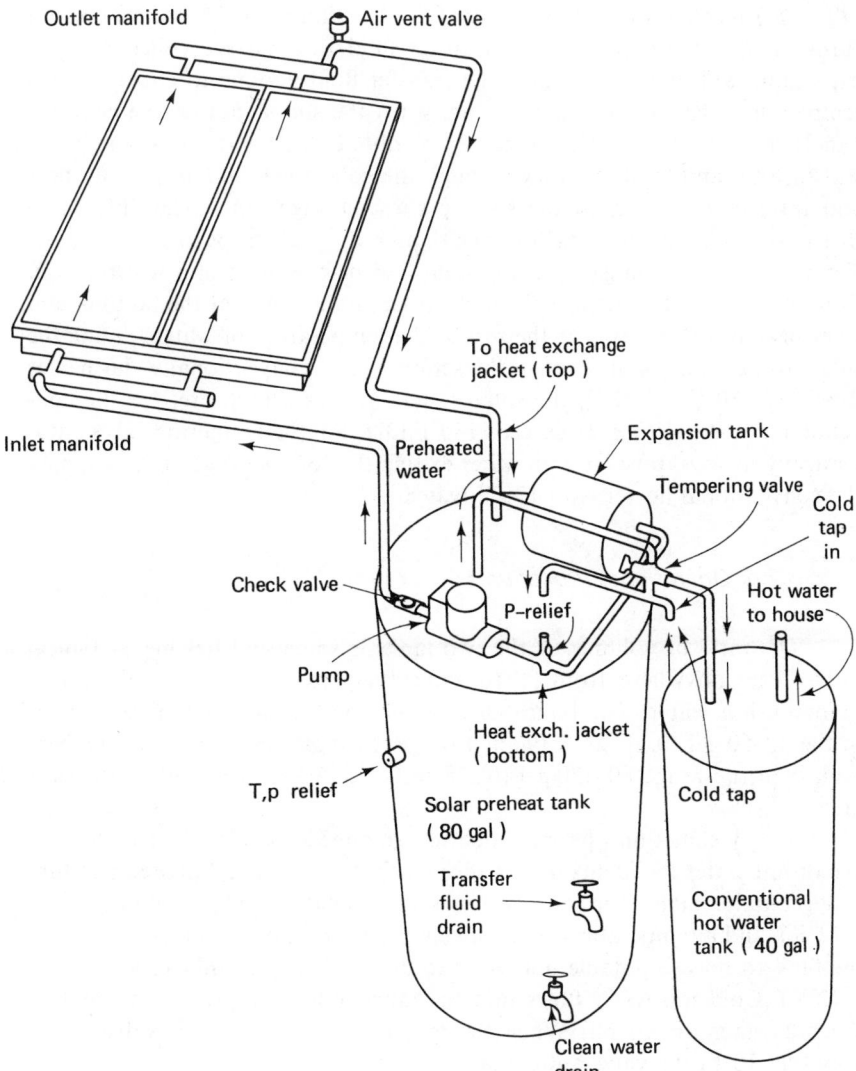

Figure 6.3 Solar domestic hot water system. Adapted from drawings supplied by Lennox Industries, Inc., by permission.

6.2.5 Control Strategy

Electronic temperature sensors are placed on the collector plate, T_c, and the wall of the preheat storage tank, T_s. An electronic controller monitors the difference in temperatures $T_c - T_s$. When the temperature difference

$(T_c - T_s)$ reaches some preset value (usually around 10°F), the thermostat turns on the pump to circulate transfer fluid through the collectors. The expansion tank is used to allow the transfer fluid volume to change as the temperature changes. In the morning, when the sun shines on the collector panels, T_c will rise until it exceeds T_s by about 10°F, whereupon the pump will turn on and fluid will flow through the collector loop and pick up heat and transfer that heat to the solar preheat storage tank. The differential thermostat is designed to stay on even though $T_c - T_s$ drops to a lower value. Typically, $T_c - T_s$ must drop to about 2°F before the pump will turn off. Tap water enters the main body of the solar storage tank at the bottom, and solar heated water leaves at the top. If the temperature of water leaving the solar storage tank is lower than the setting of the conventional water heater [typically 130°F (328°K)], the solar system acts as a preheater for the conventional heater. When T_s exceeds 130°F (328°K), the tempering valve automatically mixes tap water with water exiting the solar storage tank to supply the conventional heater with 130°F water.

6.2.6 Operational Procedures

The ratio of volumes of the two tanks is somewhat flexible. A typical family uses anywhere from 80 to 140 gal/day (0.303 to 0.531 m³/day) of domestic hot water. The typical size of the conventional hot water tank is in the 30–50 gal (0.113 to 0.190 m³) range. In most cases, the solar preheat tank is in the range 60–120 gal (0.228 to 0.455 m³), depending on collector area.

It may sometimes happen, due to prolonged good weather or lack of use of hot water (residents on vacation?), that the storage tank temperature can get quite high. The storage tanks are usually rated for about 210°F (372°K), and a temperature relief valve is placed on the potable water side of the tank to release potable hot water to drain when the tank reaches 210°F (372°K). Cold tap water flows into the tank under line pressure to replace the hot water drawn off. When the temperature at the relief valve drops about 5°–10°F, the valve automatically closes.

In some cases, the lines to the collector become blocked for some reason; the transfer fluid line is provided with a pressure relief valve that will open when the fluid pressure reaches about 100 psi (690 kP). The transfer fluid will be vented to drain.

An air vent valve is placed at the highest point in the system to bleed air out of the lines. Valves are placed at low points on the storage tank for draining clean water and transfer fluid. A check valve is placed between the pump and the collectors to prevent gravity-induced backward flow through the pump.

To have even distribution of the flow between the flat plate collector modules, the flow resistance in each module should be balanced by having parallel flow in the exit manifold in the same direction as in the entrance manifold. Consider Figure 6.4(a). In the reverse flow arrangement, the flow resistance will be lower for the left module, and it will get more than 50% of the total flow. However, in Figure 6.4(b) the parallel flow arrangement gives the same total flow resistance for both modules. In the parallel flow arrangement, the plumbing lines are longer, and one should arrange to have the outlet close to the down pipe to the storage tank to reduce heat losses. All plumbing lines should be insulated.

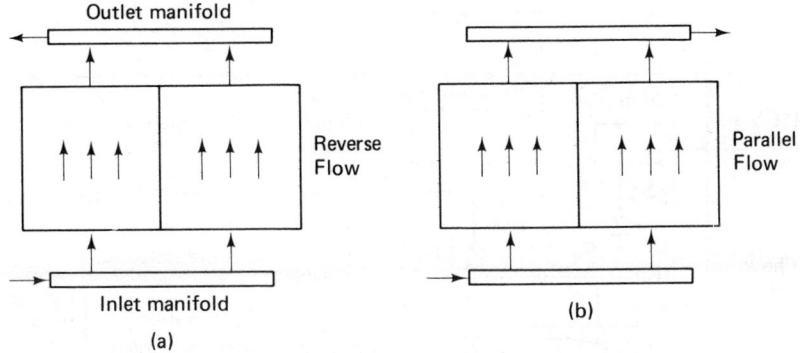

Figure 6.4   Flow schemes for flat plate collector modules.

The full flow diagram for the solar preheat domestic hot water system is shown in Figure 6.5.

Before filling the system, the lines should be flushed. This is accomplished for the potable water system by simply running clean water in and releasing hot water to drain. The tank should also be drained at the bottom to remove sediment. The transfer fluid loop should be flushed with tap water and then filled with a mixture of antifreeze and distilled water.

6.3 Component Performance

One of the key elements in the solar hot water system is the solar collector module. The properties of flat plate solar collectors were discussed in Chapter 4. The type of collector to be selected depends on the climate. In the southern part of the United States, collectors with one glazing are most appropriate, whereas double glazing may be more desirable in the north. If an average solar intensity under operation is, say, 250 Btu/hr-ft^2 (789 W/m^2) and an

Typical piping and wiring

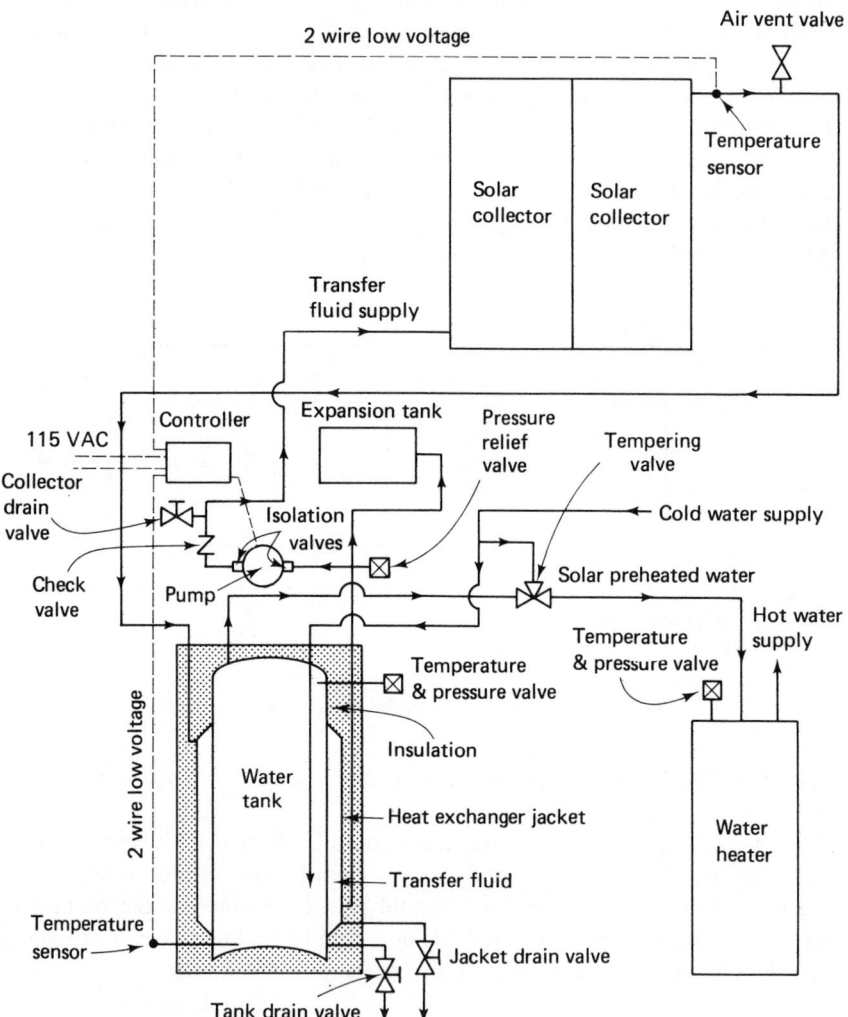

Figure 6.5 Flow diagram for solar preheat domestic hot water system. Adapted from drawings supplied by Lennox Industries, Inc., by permission.

average winter ambient temperature is, say, 30°F, $\Delta T/I = 0.4$°F-hr-ft^2/Btu. In the summer, the operating range for $\Delta T/I$ probably is about two-thirds of this. Over these ranges of operating conditions, double glazing has a small advantage over single glazing. Whether that advantage is worth the additional cost depends on the climatological conditions at the point of application.

The pump is another key component in a domestic hot water system. It should be able to supply the head to pump fluid through the piping, collectors, and heat exchanger. In an open drain-down system, the required head is the sum of frictional losses (usually a few feet) and the difference in height between the collector and the storage tank. In a closed pressurized system, only the frictional losses need to be supplied. The flow rate required for operation can be calculated in terms of the temperature rise across the collectors. In a parallel flow arrangement, in bright sunlight near solar noon [~ 300 Btu/hr-ft^2 (947 W/m^2)], a flow rate of f gal/min will produce a temperature rise ΔT °F if the collector efficiency is 0.5, where

$$f = \frac{300 \times 0.5}{60 \times 8.3\Delta T} \tag{6.1}$$

Any reasonable collector should easily achieve at least 50% efficiency under these conditions. Thus,

$$f = \frac{0.3}{\Delta T} \text{ gal/min} \tag{6.2}$$

per sq ft of collector. A reasonable design point would allow for $\Delta T = 30°$F in bright sunlight. Thus, a sensible flow rate is $f = 0.01$ gal/min per sq ft of solar collector (or 1336 cm^3/min per m^2). In typical hot water installations, collector sizes of 40 to 80 sq ft (3.72 m^2 to 7.43 m^2) are used. Thus, flow rates in the range 0.4 to 0.8 gal/min (1512 to 3024 cm^3/min) are appropriate. The pump required for circulating water at these flow rates is typically in the range 1/20 to 1/10 hp and draws about 50 to 100 watts of power. If the pump runs for say 6 hr/day on 300 days/year, this implies total electrical usage of 90 to 180 kWh per year. At \$0.04 per kWh, the cost of running the pump is about \$3.60 to \$7.20 per year.

The expansion tank could be a tank exposed to the air, but then it would have to be elevated to the highest point in the system (above the collectors). In most actual applications, the expansion tank is of the diaphragm type, in which the liquid expands against a diaphragm backed by air under pressure. Thus, the entire transfer fluid system is under pressure and air can be bled from the lines at the highest point.

6.4 Control System

The control system requires a differential thermostat that will sense the collector temperature T_c and the storage temperature T_s. When $T_c - T_s$ rises to some preset value ΔT_{on}, the thermostat actuates the main flow pump. The pump remains on until $T_c - T_s$ drops to some smaller value ΔT_{off}. When the sun heats up the collectors sufficiently, water will automatically be circulated through the collectors.

In practice, the temperature sensors are thermistors, which are solid-state devices that decrease rapidly in resistance as the temperature increases. By wiring the thermistors into a bridge circuit, a change in temperature will unbalance the bridge and cause a current to flow. This current can be amplified and used to actuate a relay to switch devices like pumps on and off. A typical curve of resistance vs. temperature for a thermistor is shown in Figure 6.6.

Figure 6.6 Temperature dependence of the resistance of a typical thermistor.

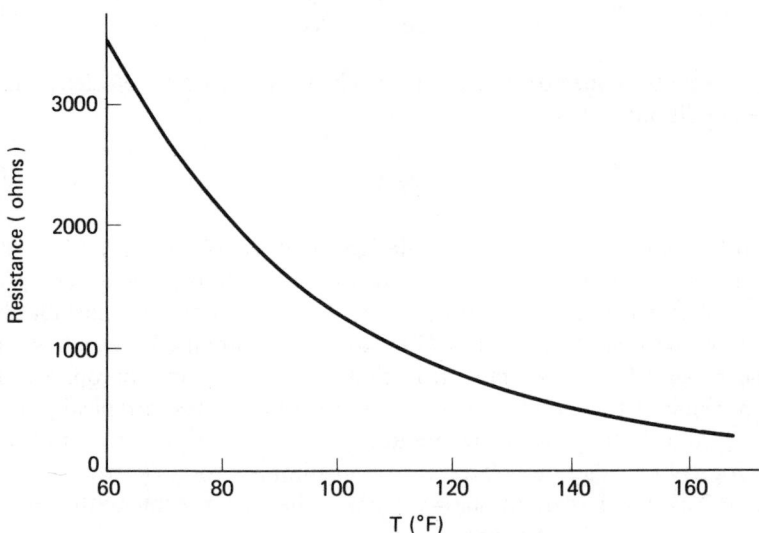

Consider the electronic circuit shown in Figure 6.7. A bridge is constructed from two ordinary fixed resistors R_1 and $(R_2 + R_3)$ and two thermistors R_s and R_c. A dc power supply is used to energize the bridge. First, consider the case where the trim potentiometer R_3 is adjusted so that $R_1 = R_2 + R_3$. When $T_c = T_s$, it follows that $R_c = R_s$, and the bridge is in balance. Under this condition there is no potential difference between points A and B in Figure 6.7. As the collector temperature rises relative to storage, R_c decreases relative to R_s, and the voltage of point B rises relative to point A. The voltage difference between B and A is fed into the input of an operational amplifier (such as a 741) which is wired to have feedback on both channels. The output voltage from the operational amplifier (at pin 6) is small regardless of input voltage until the input voltage rises to a critical value, whereupon the output voltage rises rapidly and drives transistor T_1

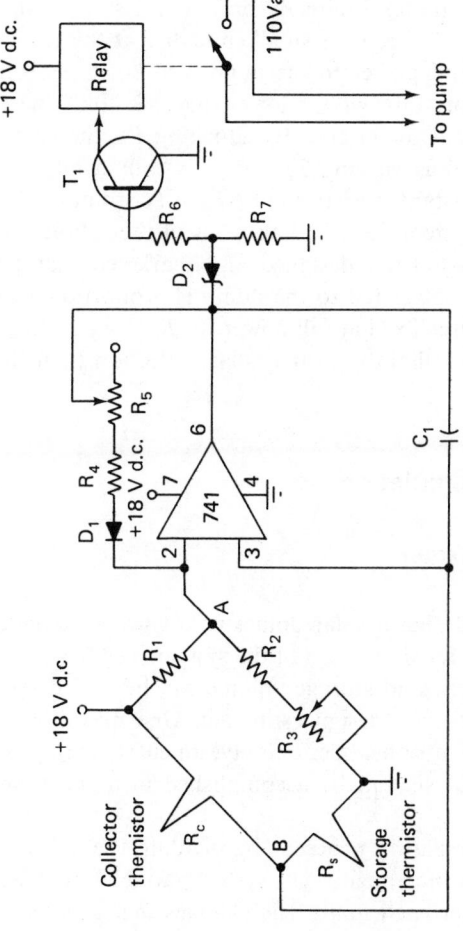

Figure 6.7 Circuit for control of domestic hot water system.

to actuate the relay, which turns on the switch feeding line voltage to the pump. The ratio of resistances R_c/R_s required to fire the relay is typically about 0.8. This implies about a 10°–15°F excess of T_c over T_s to actuate the pump. By changing the position of the wiper arm on trimpot R_3 so that $R_2 + R_3$ is either greater (or less) than R_1, actuation of the pump will occur at smaller (or greater) ratios R_c/R_s and therefore at higher (lower) values of $T_c - T_s$. When the pump is already on and T_c is reduced, a point will be reached where $T_c - T_s$ becomes small enough that the relay will go back to the open position and power to the pump will be turned off. The value of $T_c - T_s$ required for "off" will be lower than for "on" due to the hysteresis generated by the feedback circuit. By adjusting R_5, the amount of feedback can be controlled, thus varying $T_c - T_s$ for "off" while $T_c - T_s$ for "on" remains roughly constant. Variation of ΔT_{on} is accomplished by adjusting R_3.

The control system described above is of the simple on/off type. Some control systems have been designed and marketed that have interruptor circuits so that the power fed to the pump is proportional to $T_c - T_s$ from 0 to ΔT_{on}, and remains fixed at full power for $T_c - T_s > \Delta T_{on}$. These circuits should be capable of slightly greater solar collection than the simple on/off type.

6.5 Computer Simulation

6.5.1 Introduction

To properly design a solar domestic hot water system, it is necessary to perform a computer simulation of the system. The basic design parameters such as collector area and storage volume can be varied systematically and the performance of each design estimated. One may then establish which design produces the most heat per unit investment. Once a program is written, a great deal of analysis can be accomplished in a short time on a digital computer.

Computer simulation proceeds by dividing the day into increments— usually 24 one-hour increments. The system variables are treated as approximately constant over each hour. The changes that accrue in each hour are calculated and used to obtain system parameters for the next hour. The key to the analysis is to perform a heat flow balance around the preheat storage tank for each hour. From the net heat flow to or from the tank, the temperature change for the hour can be calculated, and then the temperature for the next hour can be found. This can be repeated from hour to hour. During this hour, it is necessary to calculate:

1. The solar energy falling on the collectors.

2. The fraction of solar energy converted by the collectors into heat in the transfer fluid.
3. The rate of hot water usage.
4. The rate of heat loss.

Models must be constructed for each of these energy flows.

6.5.2 Solar and Climatological Data

Data on solar intensities are commonly given as the total insolation falling on a horizontal surface. In any area where the National Climatic Center makes such measurements, hourly data are available on computer tape. Unfortunately, 15 or 20 years of hourly data are very unwieldy to use. It is much more convenient to statistically analyze the long-term data to estimate the average intensity per month for each month. One may then select a set of 12 months to create a model year, in which the actual hourly data are used for each month which is chosen to have the nearest approximation to average levels of solar intensity and ambient temperature. Thus, an artificial year is created from 12 months of actual hourly data, where the averages for each month approximate the long-term averages for that month. These data, comprising 12 months of hourly ambient temperatures and solar intensity on a horizontal surface, can be put on computer data cards or a computer tape. At each hour, the temperature and solar intensity are read from the data deck or tape comprising 8760 hourly data values of each variable.

6.5.3 Calculation Procedure

The fraction of incident solar energy converted to heat in the transfer fluid can be estimated at any hour. The procedure is carried out in steps. The solar intensity on a horizontal surface (read from data cards) is converted to intensity on the tilted collector surface by means of Eq. (2.20). At this intensity, the collector efficiency is estimated. When the efficiency is multiplied by the solar intensity on a tilted surface and by the collector area, the result is the heat gain by the collectors. After subtracting thermal losses from the transfer fluid tubes, the result is the solar heat input to the preheat storage tank. In actual practice, the pump goes on whenever the collector temperature exceeds the storage tank temperature by some preset value ΔT_{on}. The computer program is not equipped to do this. A reasonable approximation for the computer program is to assume the pump runs continuously for any hour when the average solar intensity exceeds some minimum value [in this case 80 Btu/hr-ft^2 (252 W/m^2)], and does not run at all for hours when the average solar intensity is less than this value. This turns out to be a reasonable approximation for calculating the solar heat input to the tank.

6.5.4 Hot Water Use Patterns

In any actual household, the rate of hot water usage will vary from day to day and hour to hour. It is not possible to formulate simple algorithms to characterize the actual hot water usage because of the variability in the use patterns. Fortunately, the performance of a solar hot water system is not closely tied to the specific hourly rate of usage, but depends mainly on the daily usage regardless of the hourly pattern. Thus, it is feasible to estimate the daily total of hot water usage and arbitrarily set up a reasonable hourly pattern which sums up to the daily total. This is sufficient for computer estimation of system performance. Computer simulation can test the effect of varying the hourly pattern at constant daily total, and some results on this aspect will be presented in a later section. This depends on the personal habits of the inhabitants—number of showers, baths, dishwashings, clothes washings, etc. Many of the old handbooks recommended 20–25 gal/day (0.0756–0.0907 m^3/day) per person. The United States Army recommends 40 gal/day (0.145 m^3/day) for soldiers in barracks. It is the personal experience of this author that in his household, including two adults and two children, a good estimate of usage is 140 gal/day (0.508 m^3/day), or 35 gal/day (0.127 m^3/day) per person. Some hot water is wasted each day waiting for the hot water lines from the heater to the user to heat up. There is probably a vast difference in total usage and use pattern between weekdays and weekend days, which is difficult to characterize here. In this work, it will be assumed that about 30 gal/day (0.109 m^2/day) are used per person and that the hourly pattern for a household is as follows:

1. No usage during the first 6 hours of the day.
2. Peak usage (16 gal/hr) during the 7th and 8th hours.
3. Constant usage (4 gal/hr) from the 9th through 16th hours.
4. Peak usage (16 gal/hr) during the 17th and 18th hours.
5. Constant usage (4 gal/hr) during the 19th through 24th hours.

This is rather artificial but is reasonable enough to give meaningful results. The effects of varying this use pattern are given in a later section.

6.5.5 Tap Water Temperature

The tap water temperature will vary throughout the year. In any given locality, the variation of tap water temperature can be obtained from the local water company. In this work, it will be assumed that the tap water temperature varies according to the formula (in °F):

$$TT = 59 + 14 \cos \left[\frac{(D_s - 220)\pi}{183} \right] \qquad (6.3)$$

where D_s is the day of the solar year measured from December 21 as day 1. Thus, on $D_s = 37$ (January 26) the tap water temperature is a minimum at 45°F (280°K), and on day $D_s = 220$ (July 29) the maximum of 73°F (296°K) is reached. This is a rough approximation appropriate to Dallas, Texas.

6.5.6 Tempering Valve

The heat removed from storage in any hour is the product of hot water usage and the temperature rise from tap water to exit temperature. The exit temperature is regulated by a tempering value not to exceed the desired final hot water temperature. Thus, if 130°F (328°K) domestic hot water is desired, the exit temperature is equal to the storage temperature if storage is not in excess of 130°F. If, on the other hand, storage is above 130°F, the tempering valve keeps the exit temperature at 130°F.

6.6 Computer Program

6.6.1 General Description

The purpose of the computer program is to model the performance of the solar energy system illustrated in Figure 6.5. The program listing is given in Figure 6.8. A list of variable names and definitions is given in Figure 6.9.

The design parameters such as collector area, storage volume, latitude, collector tilt angle, and print control index are read in Statement (200). The print control index (LBJ) determines whether tabular data or graphical data will be printed and also how frequently the tabular data will be printed. After initializing some constants and sums, 24 hourly solar intensities, 24 hourly ambient temperatures, and the month and day are read for the first day at Statement (1). Then calculations are performed for all quantities that depend on the day but are independent of hour. These include the day of the solar year, the tap water temperature, and the solar declination.

A DO LOOP over the 24 hours of the day is originated next. At each hour, the solar heat input to the tank is calculated. If the measured solar intensity (on a horizontal surface) is less than 40 Btu/hr-ft² (126 W/m²), the solar heat input to the tank SHITT is set equal to zero. When the storage temperature exceeds 190°F (361°K) SHITT is also set equal to zero. The tilt

Figure 6.8 Program to simulate domestic hot water preheat system.

```
C     PROGRAM TO SIMULATE DOMESTIC HOT WATER PRE-HEAT SYSTEM
C
      REAL L,LR,LMT,MSH,MAUX,MTOT,MSHU
      INTEGER SH,TA,HWD,ZSH,ZM
      DIMENSION ZSH(31,24),ZST(31,24),ZTO(31,24),ZTS(31,24),ZAUX(31,24)
      DIMENSION TA(24),SH(24),MOBGN(12),HWD(24),MD(12)
      DATA MOBGN/12,43,71,102,132,163,193,224,255,285,316,346/
      DATA HWD/0,0,0,0,0,0,16,16,4,4,4,4,4,4,4,4,16,16,4,4,4,4,4,4/
      DATA MD /31,28,31,30,31,30,31,31,30,31,30,31/
      PRINT 100
  100 FORMAT(1H1,5X,'PROGRAM TO EVALUATE PERFORMANCE OF SOLAR DOMESTIC H
     1OT WATER SYSTEM')
      ISTART=0
      PI=3.141592
      MS=1
  200 READ 101,L,T,AC,VS,LBJ
  101 FORMAT(4F10.2,I5)
      PRINT 102,L,T,AC,VS
  102 FORMAT(1H0,10X,'LATITUDE = ',F5.1,1H ,14X,'TILT = ',F5.1,/1H ,
     1  14X,'COLLECTOR AREA = ',F6.1,' SQ.FT.',/1H ,4X,'STORAGE VOLUME = ',
     2 F6.1,' GALLONS'//)
C     ****
C         SCHEDULE FOR LBJ - PRINTING INDEX
C             LBJ.GT.0 =  PRINT ALL DATA INCLUDING HOURLY VALUES
C             LBJ.EQ.0 =  PRINT ONLY DAILY,MONTHLY,YEARLY TOTALS
C             LBJ.LT.0 BUT GT. -10 = PRINT ONLY MONTHLY,YEARLY TOTALS
C             LBJ.LE.-10 = PLOT HOURLY DATA
C     ****
      LR=L*PI/180.
      TR=T*PI/180.
      LMT= LR-TR
      SINL=SIN(LR)
      COSL= COS(LR)
      CLMT=COS(LMT)
      SLMT=SIN(LMT)
      DSHW= 0.
      DAUX= 0.
      DTOT= 0.
      MSHW= 0.
      MAUX= 0.
      MTOT= 0.
      YSHW= 0.
      YAUX= 0.
      YTOT= 0.
      MONTH=0
      SINDMX= SIN(23.45*PI/180.)
C     ****
C     DATA ARE STORED ON CARDS. 24 SOLAR INTENSITIES ON 1ST CARD, 24
C     AMBIENT TEMPS ON 2ND CARD, FOR 1ST DAY. ETC.
C     ****
    1 READ 103, M,ID,(SH(J),J=1,24)
      READ 104, (TA(J),J=1,24)

  103 FORMAT(26I3)
  104 FORMAT(6X,24I3)
      IF(M.NE.MS) GO TO 60

C  CALCULATE DAY OF SOLAR YEAR
C
    2 IDS=MOBGN(M)+ID-1
      IF(IDS.GT.365) IDS= IDS-365
      JDS= IDS
      IF(JDS.GT.183) JDS= 366-IDS
```

172

Figure 6.8   (Continued)

```
C
C   TAPWATER TEMPERATURE
C
      TT= 59. + 14.*COS(( IDS-220)*PI/182.5)
      IF(ISTART.EQ.0) TS=TT
      ISTART=ISTART + 1
C
C   SOLAR DECLINATION
C
      SIND   = -SINDMX*COS(PI*(JDS)/182.5)
      DEC= ARSIN(SIND)
      COSD   = COS(DEC)
      DEC= DEC*180./PI
C
C   DAILY HEADING
C
      IF(LBJ.LE.0) GO TO 4
      PRINT 105,M,ID,IDS,JDS,DEC,TT
  105 FORMAT(1H0,4I5,F8.2,F8.1 /)
      PRINT 106
  106 FORMAT(1H0,*     HOUR  SH   TA    TS   TSNEW    ST    EFF SHITT     HWD
     1HHW     AUX  TOTHT  FSOLAR*/)
    4 CONTINUE
C
C   BEGIN HOUR LOOP
C
      DO 50 IHR=1,24
      ST=0.0
      EFF=0.0
      SHITT =0.0
      IF(SH(IHR).LT.40) GO TO 20
      IF(TS.GT.190.) GO TO 20
C
C   TILT CORRECTION FACTOR
C
      HRANG= (IHR-12.5)*PI/12.
      COSH= COS(HRANG)
      SINH= SIN(HRANG)
      COSTH= COSD*COSL*COSH+SIND*SINL
      COSTT= COSD*CLMT*COSH+SIND*SLMT
      IF(COSTH.LE.0.0) GO TO 20
      IF(COSTT.LE.0.0) GO TO 20
      TFCTR= COSTT/COSTH

      IF(TFCTR.GT.3.) TFCTR=3.
C
C   INTENSITY ON TILTED SURFACE
C
      ST=SH(IHR)*TFCTR
      IF(ST.LT.60.) GO TO 20
C
C   COLLECTOR EFFICIENCY
C
      TCOLL= TS + ST/10.
      DELT=TCOLL-TA(IHR)
      EFF = 0.7 - 0.9*DELT/ST
      IF(EFF.LT.0.0) EFF = 0.0
C
   20 CONTINUE
C
C   HOT WATER DEMAND = HWD(IHR) GAL/HR
C   HEAT FLOW TO HOT WATER = HHW   BTU/HR
C
      TFINAL = TS
      IF(TS.GT.130.) TFINAL =130.
      HHW=HWD(IHR)*8.3*(TFINAL-TT)
```

Figure 6.8 (Continued)

```
C
C     SOLAR HEAT INPUT TO TANK (SHITT
C
      SHITT = EFF*ST*AC
C
C
C     CHANGE IN STORAGE TEMPERATURE
C
      TSNEW = TS + (SHITT-HHW)/(VS*8.3)
      AUXHT = 0.
      IF(TS.LT.130.) AUXHT = HWD(IHR)*8.3*(130.-TS)
      TOTHT = HWD(IHR)*8.3*(130.-TT)
      FSOL = 0.0
      IF(TOTHT.EQ.0.0) GO TO 25
      FSOL = HHW/TOTHT
   25 CONTINUE
C
C     PRINT HOURLY DATA
C
      IF(LPJ.LE.0) GO TO 30
      PRINT 107,IHR,SH(IHR),TA(IHR),TS,TSNEW,ST,EFF,SHITT,HWD(IHR),HHW
     1 AUXHT,TOTHT,FSOL
  107 FORMAT(1H ,I7,I5,I4,3F6.1,F6.3,F7.0,I4,3F7.0,F7.3)
   30 CONTINUE
C
C     ADD HOURLY VALUES TO DAILY SUMS
C
      DSHW= DSHW+ HHW
      DAUX= DAUX+ AUXHT
      DTOT= DTOT+ TOTHT
C     IF PLOTTING PROGRAM, STORE HOURLY VALUES FOR PLOT
C
      ZSH(ID,IHR) = SH(IHR)
      ZST(ID,IHR) = ST
      ZTS(ID,IHR) = TS
      ZTO(ID,IHR) = TOTHT
      ZAUX(ID,IHR) = AUXHT
      ZM = MS
C
   50 TS= TSNEW
C
C     DAY IS OVER
C
      IF(LBJ.LT.0) GO TO 55
      PRINT 108
  108 FORMAT(1H0,'   MON   DAY   DSHW     DSAUX     DSTOT    FSOLAR'/)
      FSOL = DSHW/DTOT
      PRINT 109, M,ID,DSHW,DAUX,DTOT,FSOL
  109 FORMAT(1H ,I7,I6,F9.0,F8.0,F9.0,F8.3 /)
   55 CONTINUE
C
C     MONTHLY SUMS
C
      MSHW = MSHW + DSHW
      MAUX = MAUX + DAUX
      MTOT = MTOT + DTOT
      DSHW=0.
      DAUX=0.
      DTOT= 0.
      GO TO 1
   60 CONTINUE
C
C     NEW MONTH
C
```

Figure 6.8 (Continued)

```
      YSHW=YSHW+ MSHW
      YAUX=YAUX+ MAUX
      YTOT=YTOT+ MTOT
      MONTH= MONTH+1
      IF(LBJ.LE.-10) GO TO 65
      PRINT 110
 110  FORMAT(1H0,'     MONTH    MSHW        MSAUX       MSTOT       FSOLAR')
      FSOL = MSHW/MTOT
      PRINT 113,MS,MSHW,MAUX,MTOT,FSOL
 113  FORMAT(1H0,I7, F11.0,F9.0,F11.0,F9.3)
  65  CONTINUE
      MSHW=0.0
      MAUX=0.0
      MTOT=0.0
      IF(LBJ.GT.-10) GO TO 75
      K=MD(MS)
      CALL PLOT(ZSH,ZST,ZTS,ZTO,ZAUX,ZM,K)
  75  MS=M
      IF(MONTH.EQ.12) GO TO 80
      GO TO 2

  80  IF(LBJ.LE.-10) GO TO 90
      FSOL = YSHW/YTOT
      COST=400. + (4.*VS +60.) + 10.*AC
      VALUE = FSOL*YTOT/COST
      PRINT 111
 111  FORMAT(1H0,' MONTHS       YSSH        YSAUX       YSTOT       FSOLA
     1R      VALUE'/)
      PRINT 112, MONTH,YSHW,YAUX,YTOT,FSOL,VALUE
 112  FORMAT(1H ,I5,3F14.0,F8.3,F13.2)
  90  STOP
      END



      SUBROUTINE PLOT(ZSH,ZST,ZT2,ZTO,ZAUX,ZM,KDAYS)
      INTEGER ZSH,ZM
      DIMENSION ZSH(31,24),ZST(31,24),ZT2(31,24),ZTO(31,24),ZAUX(31,24)
      DIMENSION KK(5,24),IFIELD(130,24)
      DATA IBK  /' '/
      DATA IDOT /'.'/
      DATA ISTR /'*'/
      DATA IHOR /'I'/
      DATA IV   /'-'/
      DATA IPL  /'+'/
      DO 200 KL=1,KDAYS
C
C  BLANK OUT THE FIELD FOR ONE DAY
C
      DO 1 I=1,130
      DO 1 J=1,24
    1 IFIELD(I,J) = IBK
C
C  VERTICAL AXIS
C
      DO 2 I=1,130
    2 IFIELD(I,1) = IV
C
C  TICK MARKS
C
      DO 3 I=1,12
      K = 10 + 10*(I-1)
    3 IFIELD(K,1) = IHOR
C
C  HORIZONTAL AXES
C
```

```
      DO 4 I=1,24
      IFIELD(1,I) = IHOR
    4 IFIELD(130,I)=IHOR
C
C     THE DATA FOR EACH DAY WILL BE SCALED AND CONVERTED TO INTEGERS
C     AND STORED IN ARRAYS AS FOLLOWS...
C       KK(1,J) = SH AT HOUR J
C       KK(2,J) = ST
C       KK(3,J) = T2
C       KK(4,J) = TOTAL HEAT
C       KK(5,J) = AUX HEAT
C
      DO 10 I=1,24
      KK(1,I) = ZSH(KL,I)/10+1
      KK(2,I) = ZST(KL,I)/10+1
      KK(3,I)=  ZT2(KL,I)*4/10+20
      KK(4,I) = ZTO(KL,I)/250 + 80
      KK(5,I) = ZAUX(KL,I)/250+ 80
      IFIELD(KK(1,I),I)=IPL
      IFIELD(KK(2,I),I)=ISTR
      IFIELD(KK(3,I),I)=IDOT
      IFIELD(KK(4,I),I)=IPL
   10 IFIELD(KK(5,I),I)=ISTR
      DO 20 I=1,10
   20 PRINT 100,(IFIELD(J,I),J=1,130)
  100 FORMAT(1H ,130A1)
      PRINT 101,(IFIELD(J,11),J=1,130),ZM
      PRINT 101,(IFIELD(J,12),J=1,130),KL
  101 FORMAT(1H ,130A1,I2)
      DO 30 I=13,24
   30 PRINT 100,(IFIELD(J,I),J=1,130)
  200 CONTINUE
      RETURN
      END
```

Figure 6.8 (Continued)

correction factor is calculated from standard geometrical formulas (Eq. (2.20) and the solar intensity on a tilted surface ST is evaluated. If ST is less than 80 Btu/hr-ft^2 (252 W/m^2), SHITT is set equal to zero.

The hot water demand at any hour HWD (IHR) in gallons is given in a data statement. The heat flow from storage to the hot water is simply calculated from the hot water flow and the inlet and outlet temperatures. The outlet temperature is set equal to 130°F (328°K) if the storage temperature is greater than 130°F; otherwise the outlet temperature is set equal to the storage temperature. The change in storage temperature for each hourly increment is computed by calculating the net heat input to the tank (solar input minus heat required for hot water used) and dividing by the mass of water in storage. The storage temperature is then changed in preparation for the next hour, and the procedure is repeated.

The storage temperature at the beginning of the hour, TS, is assumed to stay constant for the whole hour. The storage temperature at the end of

Figure 6.9 List of variable names for the program in Figure 6.8.

```
        LIST OF   VARIABLE NAMES FOR MAIN
AC = AREA OF COLLECTOR (SQ FT)
AUXHT = AUXILIARY HEAT REQUIRED AT ANY HR (BTU)
COSD = COS(DECLINATION)
COSH = COS(HOUR ANGLE)
COST = COST OF SOLAR INSTALLATION
COSTH= COS(ANGLE BETWEEN SUN AND VERT LINE)
COSTT= COS(ANGLE BETW SUN AND PERPENDICULAR TO TILTED SURFACE)
DAUX = DAILY TOTAL OF AUX HEAT
DEC  = DECLINATION ANGLE
DELT = TEMP OF COLLECTOR - AMBIENT TEMP.
DSHW = DAILY SUM OF SOLAR HEAT TO HOT WATER
DTOT = DAILY SUM OF TOTAL HEAT TO PRODUCE HOT WATER
EFF  = COLLECTOR EFFICIENCY
FSOL = FRACTION OF TOTAL HEAT SUPPLIED BY SOLAR
HHW  = HEAT REMOVED FROM STORAGE BY HOT WATER
HRANG= HOUR ANGLE
HWD  = HOT WATER DEMAND (GAL/HR)
ID   = DAY OF MONTH
IDS  = DAY OF SOLAR YEAR (DEC 21 = DAY 1, DEC 20 = DAY 365)
IHR  = HOUR (LOCAL TIME)
JDS  = DAYS FROM DEC 21 IN SHORTEST DIRECTION
K    = NUMBER OF DAYS IN A MONTH
L    = LATITUDE
LBJ  = INDEX TO DETERMINE HOW MUCH PRINTING TO DO
LMT  = LATITUDE - TILT
LR   = LATITUDE(IN RADIANS)
M    = MONTH
MAUX = MONTHLY TOTAL OF AUX HEAT
MD   = NUMBER OF DAYS IN A MONTH
MOBGN= DAY OF SOLAR YR CORRESPONDING TO 1ST DAY OF A MONTH
MONTH= INDEX GIVING HOW MANY MONTHS COMPLETED
MS   = MONTHSAVE = PREVIOUS MONTH
MSHW = MONTHLY SUM OF SOLAR INPUT TO HOT WATER
MTOT = MONTHLY SUM OF TOTAL HEAD REQUIRED
PI   = 3.141592
SH   = SOLAR INTENSIY ON HORIZONTAL SURFACE
SHIT = SOLAR HEAT INPUT TO TANK
SIND = SIN(DECLINATION)
SINH = SIN(HOUR ANGLE)
SINDMX= SIND(23.45 DEG)
ST   = SOLAR INTENSITY ON TILTED SURFACE
T    = TILT ANGLE
TA   = AMBIENT TEMPERATURE
TFCTR = TILT CORRECTION FACTOR (ST= TFCTR*SH)
TFINAL= TEMPERATURE OF WATER ENTERING CONVENTIONAL HEATER
TR   = TILT ANGLE (RADIANS)
TS   = TEMPERATURE OF STORAGE AT START OF AN HOUR
TSNEW = TS AT END OF AN HOUR
TT   = TAPWATER TEMPERATURE
TOTHT = TOTAL HEAT REQUIRED TO BRING TAPWATER TO 130 DEG F
VS   = VOLUME OF STORAGE (GALLONS)
VALUE =(BTU'S SAVED PER YEAR)/(COST)
YAUX = YEARLY SUM OF AUX HEAT
YSHW = YEARLY SUM OF SOLAR HEAT TO HOT WATER
YTOT = YEARLY SUM TOTAL HEAT
ZAUX = AUX HEAT
ZM   = MONTH
ZSH  = SOLAR INTENSITY(HORIZ.SURF.)
ZST  = SOLAR INTENSITY(TILTED SURF.)
ZTO  = TOTAL HEAT
ZTS  = STORAGE TEMP
```

the hour, TSNEW, is calculated from a heat balance, and is used as TS for the next hour. At each hour, provision is made for summing hourly values of various heat flows into daily sums. Provision is also made for printing hourly values if LBJ is greater than zero. At the end of 24 hours, the daily sums are printed, the daily sums are added to the monthly sums, and then the daily sums are reset equal to zero in preparation for a new day. Then, a new set of hourly data are read in for the next day. Each time a new day's data are read in, a comparison is made with the current month, M, and the previous month ("month save" = MS). If M is not equal to MS, this implies that the new day is the first day of a new month, and the program goes to Statement (60). Here, the monthly sums are added to the yearly sums, the monthly sums are printed, and then the monthly sums are reset to zero. Then MS is set equal to M, and the program goes to Statement (2) where the first day of the new month is processed as before. When 12 months have been processed in this way, the variable MONTH goes to 12, and the program goes to Statement (80). Here, the cost is calculated and yearly sums are printed. The cost is estimated as $400 for fixed costs, $(4.*VS+60) for the storage tank, pump, and expansion tank, and $10.*AC for the collectors. The "value" is the number of Btu displaced per year per dollar invested. The program stops after yearly data are printed out.

6.6.2 Plotting on the Line Printer

If the print parameter LBJ is less than or equal to -10, the program does not print tabular data but plots results using the line printer as a plotting machine. This is accomplished in SUBROUTINE PLOT. The key to plotting with a line printer on a computer is to define variables in "A1" format. Such variables contain one alpha-numeric symbol which can be printed on the line printer. The line printer prints symbols on pages which have 68 lines and 132 columns, as shown in Figure 6.10. Usually it is convenient to define movement from left to right on a page as "vertical" and from top to bottom on a page as "horizontal." Consider the following program:

```
      DIMENSION B (132)
      DATA IBK/' '/
      DATA ISTR /'*'/
      DATA IPL /'+'/
      DO 1 I = 1,132
1     B(1) = IBK
      B(2) = ISTR
      B(4) = IPL
      PRINT 2, [B(J), J = 1,132]
2     FORMAT (132A1)
      STOP
      END
```

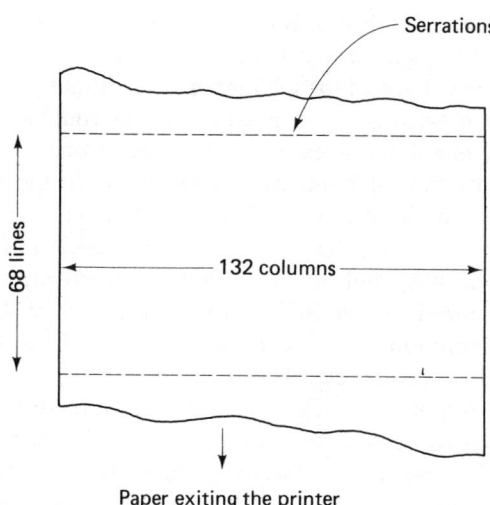

Figure 6.10 Paper on the line printer.

Paper exiting the printer

Figure 6.11 Typical line of print.

This program will store a "blank" in all 132 places in the B-array. Then it will store a (*) in B(2), and a (+) in B(4). Finally, it will print the array on a single line, showing a (*) in column 2, a (+) in column 4, and the rest of the line blank as shown in Figure 6.11. Based on this principle, the PLOT subroutine can plot data from the computer simulation. In the main program, a number of important parameters are stored at each hour into the arrays beginning with the letter Z, so that they can be plotted if the parameter LBJ is less than −9. The solar intensities on horizontal and tilted surfaces, the storage temperature, the total heat required, and the auxiliary heat required are all stored at each hour IHR of each day ID. The month is stored as ZM. At the conclusion of each month, if LBJ is less than −9, SUBROUTINE PLOT is called for plotting. The variable K is simply the number of days in the month. SUBROUTINE PLOT prints hourly data for all the days in the month.

The outline of SUBROUTINE PLOT is that KDAYS is equivalent to K in MAIN and is therefore the number of days in month ZM. The DO 200 loop is taken over all KDAYS days in the month. For each one of these days, the program is the same. First, IFIELD (130,24) is filled with blanks. Then a vertical axis is constructed by filling in dashes horizontally across the page. Tick marks (capital I) are placed on this axis at every tenth position. Horizontal axes are placed at the top and bottom of the plot (left and right of the page). The borders of the plot are now complete. Before plotting, the data must be scaled and converted to integer form. The five KK-arrays are defined for this purpose. It is desired to plot solar intensities at the bottom of the page, using roughly the first thirty divisions. Since the maximum values expected for solar intensity are slightly over 300 Btu/hr-ft², these values are divided by 10 to put them in the rough range 0 to 30. In order to keep the minimum index unity, the number one is added to the expression. Thus, when the solar intensity is less than 30, the index will be one. When the intensity is 30 or greater, and less than 60, the index will be two, etc. Arrays KK(1,1) and KK(2,1) contain such scaled values for the solar intensities on horizontal and tilted surfaces, respectively. The storage temperature is arranged so that 50°F corresponds to tick #30 on the vertical axis, and each 10 additional ticks correspond to 25°F. Thus, the 40th tick is 75°F, the 50th tick is 100°F, etc. The total and auxiliary heats are plotted so that zero corresponds to tick #80. The values in Btu/hr are divided by 200 so that each 10 ticks correspond to 2000 Btu/hr (586 W). Thus, ticks #80, 90, and 100 correspond, respectively, to heats of 0, 2000, and 4000 Btu/hr.

At the completion of the scaling, the values are printed for the 24 hours of each day. At the conclusion of the days of the month, the system returns from PLOT to MAIN.

6.6.3 Running the Program

To run the program, one must select certain key parameters on the first data card. These include the latitude, the collector tilt angle, the collector area (ft²), the storage volume (gal), and the print parameter LBJ. The print parameter determines whether tabular data will be printed on either hourly, daily, or monthly bases, or whether plotted data will be printed in lieu of tables. The storage temperature can be started at any arbitrary value, but we have started it at tap water temperature.

The main purpose of the computer simulation program is to systematically vary system parameters, and compare performance as a function of the parameters. The aim is to optimize the selection of parameters to obtain the highest ratio of performance to invested capital. Performance is judged by the fraction of total energy required that is supplied by solar energy during a year of hourly operation. From this point of view, only the yearly printouts are actually required. However, the monthly printouts show how the performance varies from month to month and give indications of how further variation of parameters will alter system performance. The hourly and daily prints are mainly for displaying the operational details of the system from hour to hour and do not really contribute to the overall evaluation of the worth or value of a candidate system. The plotting routine is merely a more descriptive way of visualizing hourly performance.

6.7 Design Results

6.7.1 Hourly Results

The purpose of this section is to illustrate the kind of results to be expected from computer simulation of a domestic hot water system. The illustration is for the North Central Texas region using solar collectors with efficiency $\eta = 0.7 - 0.9\Delta T/I$, with ΔT in °F and I in Btu/hr-ft^2. It is presumed that collectors are available in 18 ft^2 (1.67 m^2) modules, and that storage tanks are available in volumes of 66, 82, and 120 gal (0.25, 0.31, and 0.45 m^3). Heat losses are neglected, assuming that there are very short well-insulated exterior piping runs to the collectors. It is assumed that 120 gal/day (0.45 m^3/day) of hot water are used, and two different hourly patterns have been analyzed. The hot water temperature is selected as 130°F (328°K).

A typical hourly printout of results is shown in Figure 6.12. The

Figure 6.12 Hourly printout after start-up of solar hot water system.

```
    LATITUDE  =     33.3            TILT =    33.3
COLLECTOR AREA =     36.0 SQ.FT.
STORAGE VOLUME =     82.0 GALLONS


 1     1    12    12    -22.92      46.3
```

HOUR	SH	TA	TS	TSNEW	ST	EFF	SHITT	HWD	HHW	AUX	TOTHT	FSOLAR
1	0	40	46.3	46.3	0.0	0.0	0.	0	0.	0.	0.	0.0
2	0	39	46.3	45.3	0.0	0.0	0.	0	0.	0.	0.	0.0
3	0	38	46.3	46.3	0.0	0.0	0.	0	0.	0.	0.	0.0
4	0	37	46.3	46.3	0.0	0.0	0.	0	0.	0.	0.	0.0
5	0	36	46.3	46.3	0.0	0.0	0.	0	0.	0.	0.	0.0
6	0	35	46.3	46.3	0.0	0.0	0.	0	0.	0.	0.	0.0
7	0	35	46.3	46.3	0.0	0.0	0.	16	0.	11112.	11112.	0.0
8	0	33	46.3	46.3	0.0	0.0	0.	16	0.	11112.	11112.	0.0
9	3	34	46.3	46.3	0.0	0.0	0.	4	0.	2778.	2778.	0.0
10	40	38	46.3	46.3	73.6	0.0	0.	4	0.	2778.	2778.	0.0
11	96	42	46.3	51.4	164.2	0.586	3467.	4	0.	2778.	2778.	0.0
12	140	46	51.4	58.4	232.7	0.589	4934.	4	169.	2609.	2778.	0.061
13	155	48	58.4	65.6	257.6	0.574	5320.	4	402.	2376.	2778.	0.145
14	176	50	65.6	73.7	301.1	0.563	6105.	4	641.	2136.	2778.	0.231
15	117	53	73.7	78.3	215.4	0.524	4060.	4	908.	1870.	2778.	0.327
16	100	52	78.3	82.6	220.1	0.502	3980.	4	1062.	1716.	2778.	0.382
17	63	51	82.6	80.1	189.0	0.460	3127.	16	4817.	6295.	11112.	0.433
18	4	50	80.1	73.5	0.0	0.0	0.	16	4487.	6625.	11112.	0.404
19	0	46	73.5	72.2	0.0	0.0	0.	4	903.	1875.	2778.	0.325
20	0	44	72.2	70.9	0.0	0.0	0.	4	859.	1919.	2778.	0.309
21	0	42	70.9	69.7	0.0	0.0	0.	4	817.	1961.	2778.	0.294
22	0	41	69.7	68.6	0.0	0.0	0.	4	777.	2001.	2778.	0.280
23	0	40	68.6	67.5	0.0	0.0	0.	4	739.	2039.	2778.	0.266
24	0	39	67.5	66.5	0.0	0.0	0.	4	703.	2075.	2778.	0.253

MON	DAY	DSHW	DSAUX	DSTOT	FSOLAR
1	1	17283.	66055.	83338.	0.207

Figure 6.12 (Continued)

1	2	13	13	-22.83	46.2

HOUR	SH	TA	TS	TSNEW	ST	EFF	SHITT	HWD	HHW	AUX	TOTHT	FSOLAR
1	0	40	66.5	66.5	0.0	0.0	0.	0	0.	0.	0.	0.0
2	0	40	66.5	66.5	0.0	0.0	0.	0	0.	0.	0.	0.0
3	0	39	66.5	66.5	0.0	0.0	0.	0	0.	0.	0.	0.0
4	0	39	66.5	66.5	0.0	0.0	0.	0	0.	0.	0.	0.0
5	0	38	66.5	66.5	0.0	0.0	0.	0	0.	0.	0.	0.0
6	0	37	66.5	66.5	0.0	0.0	0.	0	0.	0.	0.	0.0
7	0	37	66.5	62.5	0.0	0.0	0.	16	2689.	8437.	11125.	0.242
8	0	36	62.5	59.3	0.0	0.0	0.	16	2164.	8961.	11125.	0.195
9	22	35	59.3	58.7	0.0	0.0	0.	4	435.	2346.	2781.	0.157
10	86	37	58.7	62.2	158.0	0.486	2766.	4	414.	2367.	2781.	0.149
11	141	39	62.2	68.0	240.8	0.523	4537.	4	529.	2252.	2781.	0.190
12	180	41	68.0	75.3	298.7	0.529	5683.	4	724.	2057.	2781.	0.260
13	199	43	75.3	83.0	330.2	0.522	6204.	4	966.	1815.	2781.	0.347
14	193	46	83.0	90.1	329.6	0.509	6038.	4	1222.	1560.	2781.	0.439
15	172	47	90.1	96.1	315.9	0.487	5541.	4	1457.	1325.	2781.	0.524
16	132	48	96.1	100.7	289.4	0.460	4797.	4	1656.	1125.	2781.	0.595
17	73	48	100.7	94.6	219.0	0.393	3101.	16	7236.	3889.	11125.	0.650
18	6	48	94.6	85.2	0.0	0.0	0.	16	6430.	4696.	11125.	0.578
19	0	45	85.2	83.3	0.0	0.0	0.	4	1294.	1488.	2781.	0.465
20	0	43	83.3	81.5	0.0	0.0	0.	4	1231.	1551.	2781.	0.442
21	0	42	81.5	79.8	0.0	0.0	0.	4	1171.	1611.	2781.	0.421
22	0	40	79.8	78.1	0.0	0.0	0.	4	1114.	1668.	2781.	0.400
23	0	38	78.1	76.6	0.0	0.0	0.	4	1059.	1722.	2781.	0.381
24	0	36	76.6	75.1	0.0	0.0	0.	4	1008.	1774.	2781.	0.362

MON	DAY	DSHW	DSAUX	DSTOT	FSOLAR
1	2	32797.	50641.	83438.	0.393

1	3	14	14	-22.73	46.1

HOUR	SH	TA	TS	TSNFW	ST	EFF	SHITT	HWD	HHW	AUX	TOTHT	FSOLAR
1	0	34	75.1	75.1	0.0	0.0	0.	0	0.	0.	0.	0.0
2	0	32	75.1	75.1	0.0	0.0	0.	0	0.	0.	0.	0.0
3	0	31	75.1	75.1	0.0	0.0	0.	0	0.	0.	0.	0.0
4	0	30	75.1	75.1	0.0	0.0	0.	0	0.	0.	0.	0.0
5	0	30	75.1	75.1	0.0	0.0	0.	0	0.	0.	0.	0.0
6	0	30	75.1	75.1	0.0	0.0	0.	0	0.	0.	0.	0.0
7	0	30	75.1	69.4	0.0	0.0	0.	16	3846.	7292.	11138.	0.345
8	0	32	69.4	64.9	0.0	0.0	0.	16	3096.	8042.	11138.	0.278
9	24	31	64.9	64.0	0.0	0.0	0.	4	623.	2162.	2784.	0.224
10	90	33	64.0	67.0	164.9	0.441	2617.	4	593.	2192.	2784.	0.213
11	146	38	67.0	72.6	248.8	0.505	4526.	4	691.	2093.	2784.	0.248
12	186	44	72.6	79.9	308.1	0.526	5839.	4	878.	1906.	2784.	0.315
13	205	46	79.9	87.6	339.5	0.520	6358.	4	1120.	1664.	2784.	0.402
14	202	47	87.6	94.7	344.2	0.504	6245.	4	1376.	1409.	2784.	0.494
15	177	49	94.7	100.6	324.3	0.483	5639.	4	1613.	1171.	2784.	0.579
16	133	50	100.6	104.9	290.4	0.453	4737.	4	1810.	975.	2784.	0.650
17	72	51	104.9	97.9	216.0	0.385	2996.	16	7810.	3328.	11138.	0.701
18	6	50	97.9	87.8	0.0	0.0	0.	16	6871.	4267.	11138.	0.617
19	0	48	87.8	85.7	0.0	0.0	0.	4	1383.	1402.	2784.	0.497
20	0	45	85.7	83.8	0.0	0.0	0.	4	1315.	1469.	2784.	0.472
21	0	43	83.8	82.0	0.0	0.0	0.	4	1251.	1534.	2784.	0.449
22	0	42	82.0	80.2	0.0	0.0	0.	4	1190.	1595.	2784.	0.427
23	0	41	80.2	78.6	0.0	0.0	0.	4	1132.	1653.	2784.	0.406
24	0	40	78.6	77.0	0.0	0.0	0.	4	1077.	1708.	2784.	0.387

MON	DAY	DSHW	DSAUX	DSTOT	FSOLAR
1	3	37675.	45860.	83534.	0.451

preheat storage tank is filled with tap water at 46.3°F (281°K) on January 1, and hourly prints of system variables are given. At the end of each day, daily totals are given. It can be seen that no changes occur in the system from midnight through the end of the sixth hour of each day because no hot water is used and there is no solar heat input. In Figure 6.13 the hourly output is shown for the last day of January, and the daily and monthly sum prints are also displayed. A plot of the results for the first few days of January is given in Figure 6.14. The (+) and (*) symbols at the bottom of the page represent solar intensities on horizontal and tilted surfaces, respectively. The vertical scale is such that each division represents 10 Btu/hr-ft² (31.6 W/m²). The dots represent storage temperature. The vertical scale begins at 10 divisions above the bottom of the page at 0°F, and each 10 divisions represents 25°F. At the top of the page, the (+) symbols represent the total heat required at any hour, and the (*) symbols represent the auxiliary back-up heat required. When the two are equal, the (*) overrides the (+). The vertical gap between the (*) and (+) represents the solar heat input to the hot water. The scale begins at the 80th vertical division, and each 10 divisions represent 2500 Btu (2638 kJ). The plots are given both directly as they come off the line printer (Figure 6.14), and after hand-drawn lines are used to connect the symbols (Figure 6.15).

6.7.2 Monthly and Yearly Results

These hourly printouts are useful for understanding the way the system reacts to different hourly conditions of hot water demand and solar availability. However, for most design purposes, only the monthly and yearly totals are required. By setting the print parameter LBJ to a value such as −5, only monthly and yearly sums are printed. A typical printout is shown in Figure 6.16. When a series of such runs are compared and analyzed, some interesting conclusions can be drawn. In Figure 6.17 the fraction of total energy required that is supplied by solar energy over the course of a year is plotted vs. collector area for various fixed volumes of the preheat storage tank. The figures are for collectors tilted at the latitude, which is taken as 33°N. For small collector areas, the increase in performance as the collector area is increased is roughly linear. However, for larger collector areas, a region of diminishing returns is approached. The dependence on storage volume is rather small, with the larger tanks outperforming the smaller ones mainly for large collector areas. Figure 6.17 is based on an assumed hot water cycle usage of 0, 0, 0, 0, 0, 0, 16, 16, 4, 4, 4, 4, 4, 4, 4, 4, 16, 16, 4, 4, 4, 4, 4, 4 gal/hr for each of the 24 hours of a day. To test whether this specific form of the usage cycle has any effect on overall performance, calculations were repeated for a different cycle, namely, 0, 0, 0, 0, 0, 0, 24, 20, 8, 4, 0, 8, 4, 2, 2,

Figure 6.13 Hourly printout of last day of month showing monthly sums.

HOUR	SH	TA	TS	TSNEW	ST	EFF	SHITT	HWD	HHW	AUX	TOTHT	FSOLAR
13	234	52	81.9	90.1	354.6	0.534	6819.	4	1224.	1597.	2821.	0.434
14	229	55	90.1	97.7	354.0	0.521	6635.	4	1497.	1324.	2821.	0.531
15	196	57	97.7	103.4	318.2	0.495	5671.	4	1748.	1073.	2821.	0.620
16	152	58	103.4	107.4	276.8	0.462	4606.	4	1939.	882.	2821.	0.687
17	92	59	107.4	100.7	242.2	0.430	3752.	16	8277.	3007.	11285.	0.734
18	22	58	100.7	89.8	0.0	0.0	0.	16	7394.	3890.	11285.	0.655
19	0	54	89.8	87.7	0.0	0.0	0.	4	1488.	1333.	2821.	0.527
20	0	51	87.7	85.6	0.0	0.0	0.	4	1415.	1406.	2821.	0.502
21	0	50	85.6	83.6	0.0	0.0	0.	4	1346.	1475.	2821.	0.477
22	0	47	83.6	81.7	0.0	0.0	0.	4	1281.	1541.	2821.	0.454
23	0	47	81.7	79.9	0.0	0.0	0.	4	1218.	1603.	2821.	0.432
24	0	46	79.9	78.2	0.0	0.0	0.	4	1153.	1662.	2821.	0.411

MON	DAY	DSHW	DSAUX	DSTOT	FSOLAR
1	30	40209.	44426.	84635.	0.475

```
1   31   42   42   -17.36   45.0
```

HOUR	SH	TA	TS	TSNEW	ST	EFF	SHITT	HWD	HHW	AUX	TOTHT	FSOLAR
1	0	44	78.2	78.2	0.0	0.0	0.	0	0.	0.	0.	0.0
2	0	44	78.2	78.2	0.0	0.0	0.	0	0.	0.	0.	0.0
3	0	43	78.2	78.2	0.0	0.0	0.	0	0.	0.	0.	0.0
4	0	42	78.2	78.2	0.0	0.0	0.	0	0.	0.	0.	0.0
5	0	43	78.2	78.2	0.0	0.0	0.	0	0.	0.	0.	0.0
6	0	43	78.2	78.2	0.0	0.0	0.	0	0.	0.	0.	0.0
7	0	43	78.2	71.7	0.0	0.0	0.	16	4407.	6876.	11282.	0.391
8	0	45	71.7	66.5	0.0	0.0	0.	16	3547.	7736.	11282.	0.314
9	12	47	66.5	65.5	0.0	0.0	0.	4	714.	2107.	2821.	0.253
10	61	50	65.5	66.9	98.5	0.468	1661.	4	679.	2142.	2821.	0.241
11	92	53	66.9	69.8	141.5	0.521	2657.	4	727.	2094.	2821.	0.258
12	114	57	69.8	73.5	172.0	0.543	3364.	4	821.	2000.	2821.	0.291
13	125	58	73.5	77.5	188.6	0.536	3640.	4	945.	1876.	2821.	0.335
14	124	59	77.5	81.2	190.8	0.523	3591.	4	1076.	1744.	2821.	0.382
15	110	59	81.2	84.1	177.6	0.498	3182.	4	1199.	1622.	2821.	0.425
16	85	60	84.1	86.0	153.5	0.469	2590.	4	1296.	1525.	2821.	0.459
17	51	59	86.0	81.0	131.7	0.426	2018.	16	5436.	5847.	11282.	0.482
18	13	59	81.0	73.9	0.0	0.0	0.	16	4769.	6514.	11282.	0.423
19	0	58	73.9	72.5	0.0	0.0	0.	4	960.	1861.	2821.	0.340
20	0	56	72.5	71.2	0.0	0.0	0.	4	913.	1908.	2821.	0.324
21	0	55	71.2	69.9	0.0	0.0	0.	4	868.	1952.	2821.	0.308
22	0	56	69.9	68.7	0.0	0.0	0.	4	826.	1995.	2821.	0.293
23	0	58	68.7	67.6	0.0	0.0	0.	4	786.	2035.	2821.	0.279
24	0	62	67.6	66.5	0.0	0.0	0.	4	747.	2073.	2821.	0.265

MON	DAY	DSHW	DSAUX	DSTOT	FSOLAR
1	31	30713.	53905.	84618.	0.363

MONTH	MSHW	MSAUX	MSTOT	FSOLAR
	839755.	1772630.	2612381.	0.321

```
2   1   43   43   -17.09   45.1
```

HOUR	SH	TA	TS	TSNEW	ST	EFF	SHIT	HWD	HHW	AUX	TOTHT	FSOLAR

Figure 6.14 Computer plot of hourly results.

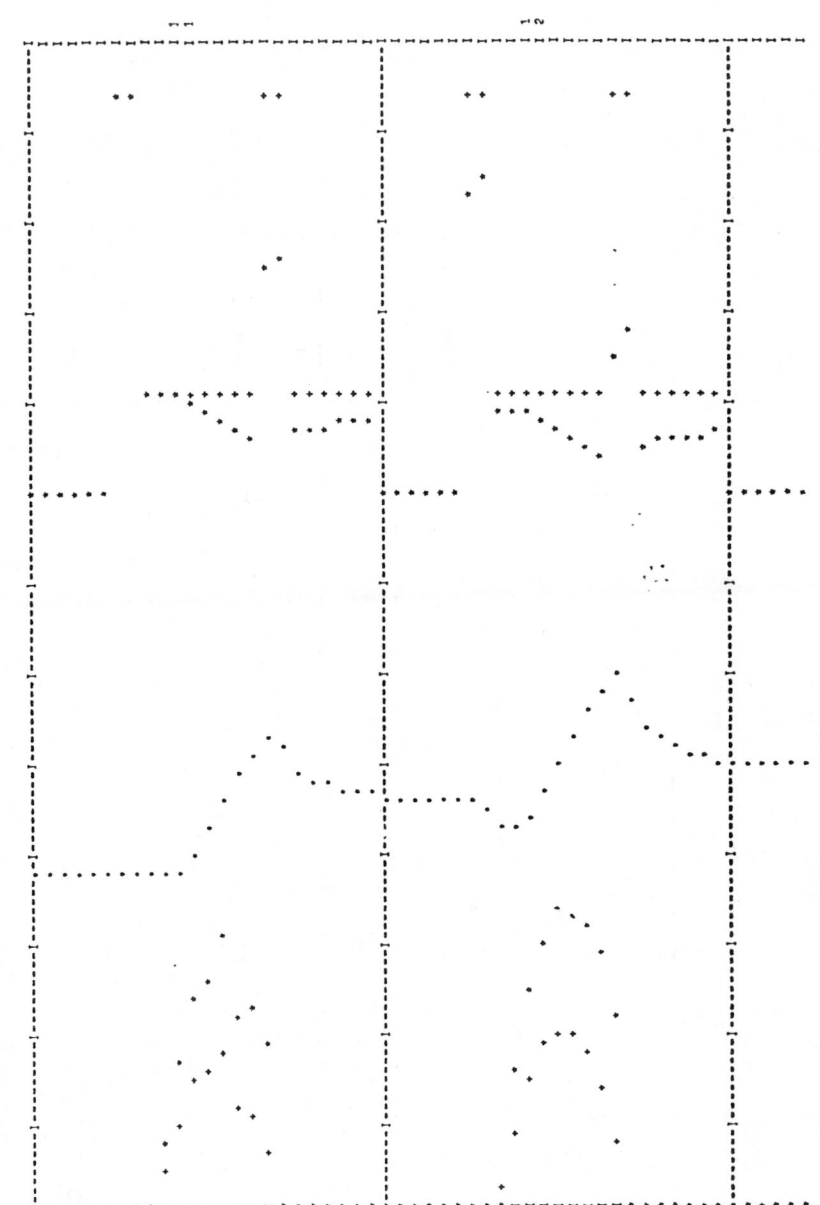

PROGRAM TO EVALUATE PERFORMANCE OF SOLAR DOMESTIC HOT WATER SYSTEM

LATITUDE = 33.3   TILT = 33.3
COLLECTOR AREA = 36.0 SQ.FT.
STORAGE VOLUME = 82.0 GALLONS

Figure 6.15 Computer plots of hourly results with lines drawn between data.

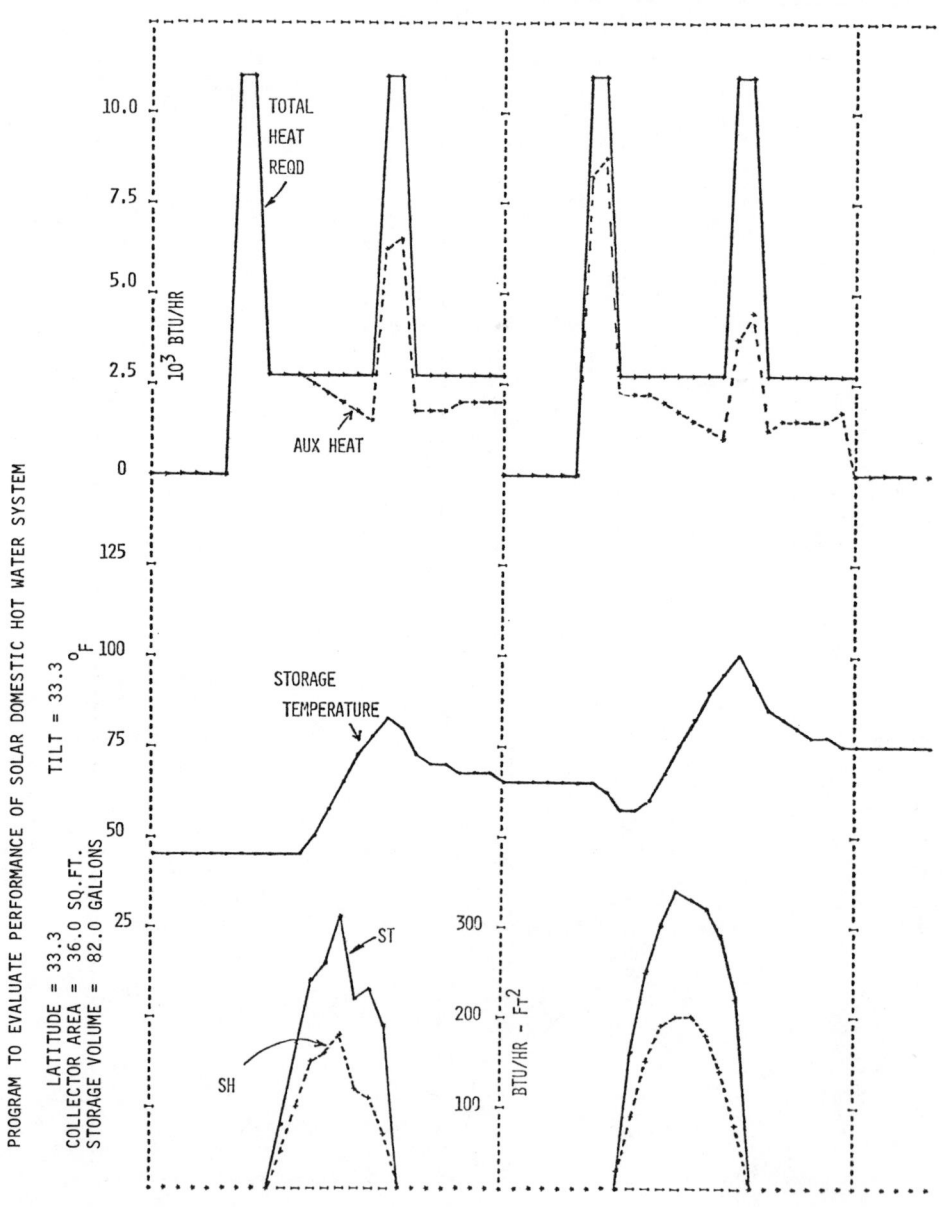

Figure 6.16 Monthly sums.

```
PROGRAM TO EVALUATE PERFORMANCE OF SOLAR DOMESTIC HOT WATER SYSTEM

          LATITUDE =    33.3                 TILT =   33.3
    COLLECTOR AREA =    36.0 SQ.FT.
    STORAGE VOLUME =    82.0 GALLONS
```

MONTH	MSHW	MSAUX	MSTOT	FSOLAR
1	839755.	1772630.	2612381.	0.321
MONTH	MSHW	MSAUX	MSTOT	FSOLAR
2	884499.	1461712.	2346205.	0.377
MONTH	MSHW	MSAUX	MSTOT	FSOLAR
3	963564.	1515354.	2478915.	0.389
MONTH	MSHW	MSAUX	MSTOT	FSOLAR
4	864259.	1343806.	2208056.	0.391
MONTH	MSHW	MSAUX	MSTOT	FSOLAR
5	904928.	1155262.	2060187.	0.439
MONTH	MSHW	MSAUX	MSTOT	FSOLAR
6	966290.	847356.	1813641.	0.533
MONTH	MSHW	MSAUX	MSTOT	FSOLAR
7	1016138.	757977.	1774113.	0.573
MONTH	MSHW	MSAUX	MSTOT	FSOLAR
8	1094670.	693007.	1787674.	0.612
MONTH	MSHW	MSAUX	MSTOT	FSOLAR
9	1026601.	822500.	1849096.	0.555
MONTH	MSHW	MSAUX	MSTOT	FSOLAR
10	1064375.	1045609.	2109977.	0.504
MONTH	MSHW	MSAUX	MSTOT	FSOLAR
11	898470.	1357596.	2256060.	0.398
MONTH	MSHW	MSAUX	MSTOT	FSOLAR
12	855922.	1575747.	2431663.	0.352

MONTHS	YSSH	YSAUX	YSTOT	FSOLAR	VALUE
12	11379467.	14348554.	25727920.	0.442	5912.43

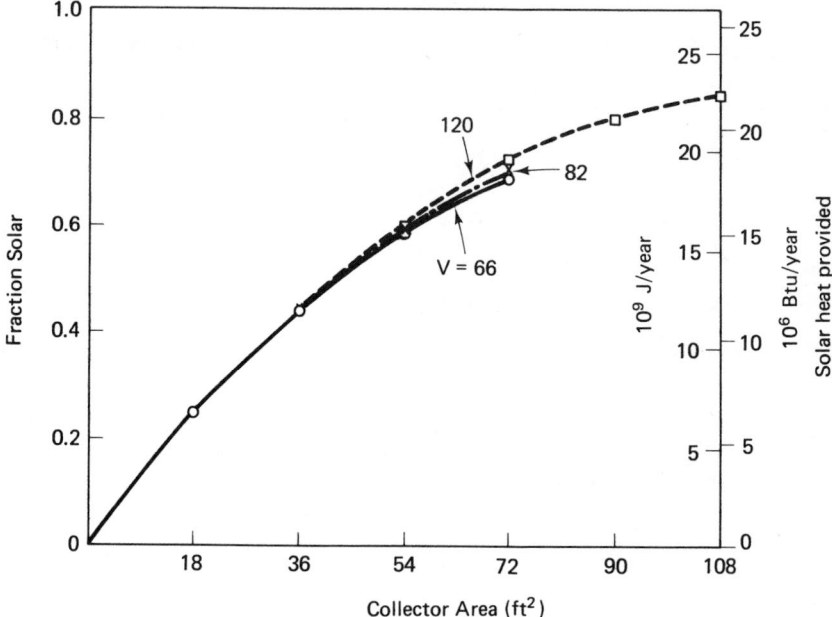

Figure 6.17 Fraction of total energy required by a domestic hot water system supplied by solar energy at latitude = tilt = 33°N.

4, 10, 10, 8, 4, 10, 2, 0, 0, with the same total daily total usage (120 gal/day). The results are compared in Table 6.1. It can be seen that the net difference in performance is small, and that the performance of a solar domestic hot water system depends only on the daily usage and not how the usage is distributed over the hours of a day.

Another way of viewing the results in Figure 6.17 is to plot the monthly performance across a year as shown in Figure 6.18. In this figure, the storage volume is fixed at 120 gal (0.454 m³) and the number of collector modules [18 ft² (1.78 m²) module, tilted at the latitude = 33°] is varied from 1 to 6. As each module is added, the gain in performance diminishes. This is especially true in midsummer, where the fraction solar approaches 100% even though there is still room for improvement in winter performance. For large collector areas, an increase in collector tilt angle would be beneficial to improve winter performance because summer performance is essentially saturated. For small collector areas, an increase in collector tilt angle would hurt performance more in summer than it would help in winter. A specific example is shown in Figure 6.19, corresponding to 54 ft² (5.02 m²) of collector, 82 gal (0.31 m³) storage, and a latitude of 33°N. The monthly performance for a variety of tilt angles is shown. For this particular design, any tilt angle between about 20° and 33° is optimal, with a slight preference for

TABLE 6.1 Monthly performance of solar hot water systems for different demand cycles. Illustration for latitude = tilt = 33°, collector area = 54 ft² (5.35 m²), storage = 82 gal (0.310 m³)

	Fraction Solar	
Month	*Demand Cycle #1**	*Demand Cycle #2†*
January	0.438	0.438
February	0.509	0.509
March	0.524	0.525
April	0.530	0.531
May	0.592	0.594
June	0.711	0.714
July	0.758	0.760
August	0.808	0.814
September	0.736	0.738
October	0.671	0.672
November	0.537	0.540
December	0.476	0.478

*Hourly usage (gal) = 0, 0, 0, 0, 0, 0, 16, 16, 4, 4, 4, 4, 4, 4, 4, 4, 16, 16, 4, 4, 4, 4, 4, 4.
†Hourly usage (gal) = 0, 0, 0, 0, 0, 0, 24, 20, 8, 4, 0, 8, 4, 2, 2, 4, 10, 10, 8, 4, 10, 2, 0, 0.

Figure 6.18 Fraction of total energy required by a domestic hot water system supplied by solar energy at latitude = tilt = 33°N for a 120 gal tank. One module = 18 ft².

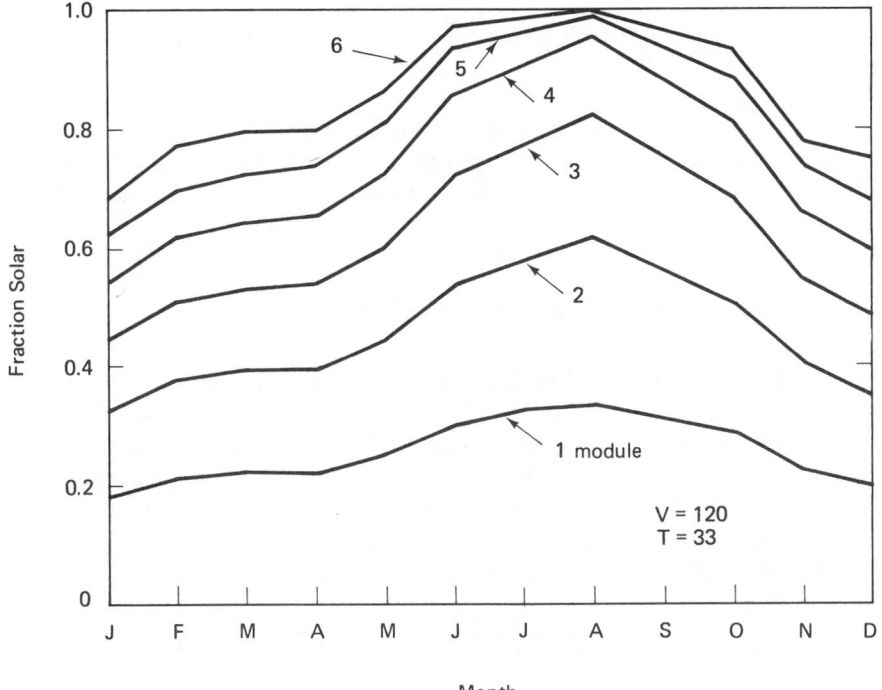

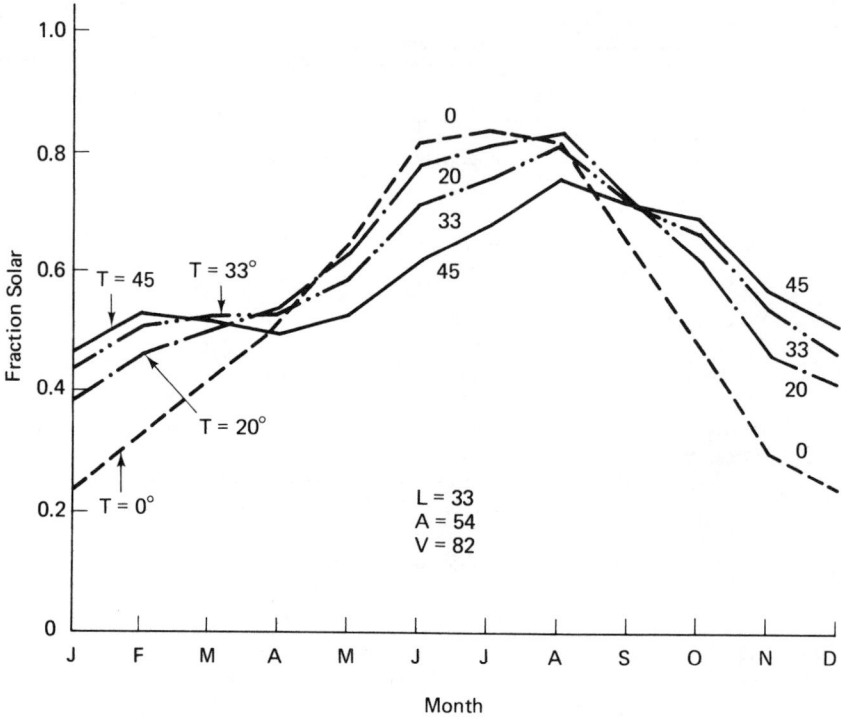

Figure 6.19 Effect of tilt angle of collectors on the performance of domestic hot water system.

the 33° figure. At 0°, the gain in summer is small, but the loss in winter is tremendous. At 45°, the winter gain over 33° is small, but the summer loss is moderately large. Thus, tilting the collectors to about the latitude angle is generally best.

6.7.3 Performance vs. Estimated Cost

The choice of an optimal design can only be made in terms of performance per unit cost. Only a single example will be given here. It will be assumed that the costs can be divided into three parts. The collector modules will be estimated at $10 per ft². The costs for installation are fixed at $400, regardless of collector area or storage volume. The unit consisting of storage tank, pump, control box, and expansion tank is assumed to be given by $(4 \times V + 60)$, where V is the storage volume in gallons. The heat collected (Btu/yr) per dollar invested may then be plotted vs. collector area as in Figure 6.20. It can be seen that even though the larger collector areas produce

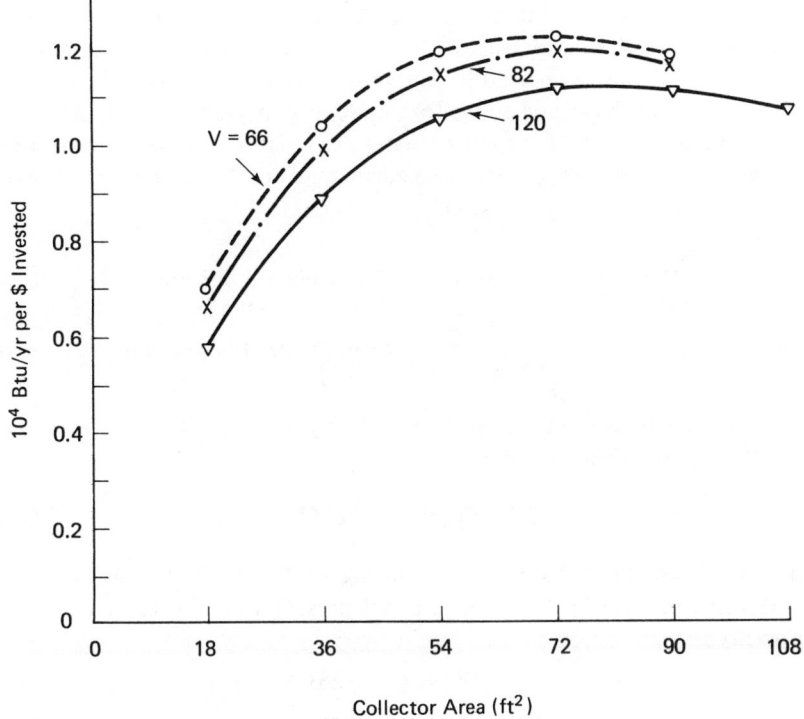

Figure 6.20 Value of solar domestic hot water systems in Btu/yr per dollar invested.

only diminishing returns in performance, the high fixed costs tend to favor the larger systems because they are spread over more collector area. Figure 6.20 indicates that the optimal size is probably in the range 54–90 ft² (5.35–8.36 m²) of collector. The additional cost of going to a 120 gal (0.45 m³) is not justified. The relative costs of 66 and 82 gal (0.249 and 0.310 m³) storage are nearly balanced. These conclusions are based on the crude estimates of tank cost given previously and might be substantially changed if the actual price schedule was considerably different.

6.8 Economics of Solar Domestic Hot Water

The economics of the various designs of solar hot water systems will be discussed in terms of the methodology of Chapter 5. The main intent is to illustrate the type of procedure required. The actual figures are somewhat arbitrary. It will be assumed that scenario III in Table 5.2 can be used for future inflation and escalation rates. The same procedures can be used with any other scenario. For illustrational purposes, it will be assumed that money is borrowed on a 20 year loan at an after-tax interest rate of 8 %, with monthly payments of $M = \$8.37$ per \$1000 borrowed. The costs of gas and electricity will be taken as those prevailing for residences in Dallas, Texas, in 1978, namely, \$2.20 per Mcf[1] for gas and \$0.04 per kWh for electricity. Assuming that gas burners are 60 % efficient, the current cost of delivered energy for the assumed conditions is estimated at

$$F_{1g} = \frac{\$2.20}{0.6} = \$3.67 \text{ per } 10^6 \text{ Btu} = \$3.85 \text{ per } 10^9 \text{ J (from gas)} \qquad (6.4)$$

$$F_{1e} = \$0.04 \text{ kWh} \times \frac{10^3}{3.413} = \$11.72 \text{ per } 10^6 \text{ Btu (from electricity)} \quad (6.5)$$

To calculate the net present value for any installation that replaces gas or electricity, the formula is

$$P = EF_1 R_{20} - 12MT_{20} \qquad (6.6)$$

where a 20 year life is assumed. According to Table 5.2, $R_{20} = 48.90$ and $T_{20} = 12.46$ in scenario III. Thus, the net present value in replacing gas or electricity is

$$P_g = 179.46E - 1251.5C \qquad (6.7)$$

$$P_e = 573.1E - 1251.5C \qquad (6.8)$$

where E is the energy displaced by the solar system (10^6 Btu per yr), and C is the cost of the installation in 1000s of dollars. To calculate the payout period, one must calculate the net present value,

$$P_N = EF_1 R_N - 12MT_{20} \qquad (6.9)$$

for N years of operation as a function of N, and determine the value of N necessary to make $P_N = 0$. The results of a computer program to carry out these calculations are given in Table 6.2 for the domestic hot water systems described in this chapter. It can be seen that for reasonable designs, the net present value of an installation costing about \$1500 is over \$1000 when displacing gas and of the order of \$8000 when displacing electricity. The pay-back periods are about 13 years in displacing gas and about 5 years in displacing electricity. According to Table 6.2, a 72 ft² collector is favored in terms of pay-back period. The net present value continues to rise when an additional collector module is added, but at a much slower rate. A 72 ft² (6.69 m²) collector produces about 70% of the energy required for domestic

[1]One Mcf = 1000 ft³ = 28.3 m³.

TABLE 6.2   Net present value and payout period for solar domestic hot water
systems displacing gas and electricity

Collector Area (ft^2)	Preheat Tank Volume (gal)	Solar Heat Collected (10^6 Btu/yr)	Cost ($)	Gas		Electricity	
				NPV (20 yr)	PBP	NPV (20 yr)	PBP
18	66	6.4	904	17	20	2536	8
36	66	11.3	1084	671	14	5119	6
54	66	15.0	1264	1110	13	7015	5
72	66	17.6	1444	1351	13	8279	5
18	82	6.4	968	−63	20	2456	8
36	82	11.4	1148	609	14	5096	6
54	82	15.3	1328	1084	13	7106	5
72	82	18.1	1508	1361	13	8456	5
18	120	6.4	1120	−253	20	2266	9
36	120	11.5	1300	437	16	4964	6
54	120	15.5	1480	929	14	7031	6
72	120	18.5	1660	1243	13	8525	5
90	120	20.5	1840	1376	13	9446	5
108	120	21.7	2020	1366	14	9908	6

hot water for a family that uses 120 gal/day (0.454 m³/day) of hot water.
The economics look fairly good when solar energy displaces electricity, but
the pay-back period is uncomfortably long when natural gas is displaced.

Worked Examples

1. A solar hot water system is to be installed in a locality where the solar irradiation
on a tilted collector on a clear winter day is 21,000 kJ/m². The hot water demand
is 0.5 m³/day, and the tapwater temperature is 42°F (279°K). What collector
area is required to supply 100% of the hot water demand at 130°F (328°K) if
the collector efficiency averages 43% over a day?

Heat required $= 0.5$ m³ $\times 10^3$ kg/m³ $\times$ 4187 J/kg-°K $\times (328 - 279)$°K

$= 103,000$ kJ

Area required $= \dfrac{103,000 \text{ kJ}}{21,000 \text{ kJ} \times 0.43} = 11.4$ m²

2. Estimate the effect of winter ambient temperature on solar domestic hot water
production by using the efficiency expression

$$\eta = 0.7 - 0.8\frac{(T_s + 20 - T_a)}{230}$$

where T_s is the storage temperature, and T_a is the ambient temperature, both in °F. Assume that in winter T_s averages around 95°F, and consider values of T_a at 10°F, 30°F, and 50°F. What is the effect of T_a on η?

$$\eta = 0.7 - 0.00348(115 - T_a)$$

$T_a\ (°F)$	η
10	0.335
30	0.405
30	0.475

3. A solar domestic hot water system displaces 65% of the annual load of 20×10^9 J. The homeowner is considering adding more collectors at a cost of $600 and estimates this will increase the load displacement to 84%. If the owner's after-tax interest rate is 6%, is it a good investment when replacing electricity at $0.04 per kWh?

Since the Energy Tax Act of 1978 provides only a one-time credit for solar energy systems, an add-on in a later year does not qualify for a tax credit. Using scenario III in Table 5.2, we find that

$$\text{1st year savings} = 0.19 \times 20 \times 10^6 \text{ kJ} \times \$0.04/\text{kWh} \times \tfrac{1}{3600} \text{ hr/sec}$$
$$= \$42.22$$

Based on a 15 year life,

$$\text{NPV} = \$42.22 \times R_{15} - 12 \times \$5.06 \times T_{15}$$
$$\text{NPV} = \$42.22 \times 35.25 - 12 \times \$5.06 \times \$10.38$$
$$\text{NPV} = \$1488 - \$630 = \$858$$
$$0 = \$42.22 R_N - \$630$$
$$R_N = \$14.92$$
$$N = 7\tfrac{1}{2} \text{ year pay-back}$$

Problems

6.1. A solar hot water system has 40 ft (12.2 m) of exposed $\tfrac{1}{2}$ in. (1.27 cm) copper tubing leading from the storage tank to the collectors, 20 ft (6.1 m) of $1\tfrac{1}{8}$ in. (2.86 cm) tubing as collector headers, and the equivalent of 400 linear feet (122 m) of 3/16 in. (0.59 cm) tubing in the collector plates. Calculate the volume of liquid that cools down at night and the amount of heat required to bring this fluid up from 20°F (266.5°K) to 120°F (322°K). If the collectors have

64 ft² (5.95 m²) of receiver and they collect 600 Btu/ft² (6813 kJ/m²) on an average winter day, what fraction of the collected energy is used to bring the collector up to temperature?

6.2. A solar hot water system is found to perform on a clear winter day in Dallas, Texas, as shown on January 3 in Figure 6.12. This 36 ft² (3.34 m²) collector system gains 38,720 Btu (40,850 kJ) in a day. These results indicate that the average values during the 6 hr period centered on noon are

$$ST = 307 \text{ Btu/hr-ft}^2 = 969 \text{ W/m}^2$$

$$TS = 90°F = 305.4°K$$

$$TA = 46°F = 281°K$$

Assuming that the collector plate runs about 30°F (16.7°K) hotter than storage under these conditions, the collector temperature averages about 120°F (322°K). The expression for collector efficiency is

$$\eta = 0.7 - \frac{0.9(TC - TA)}{ST}$$

$$= 0.7 - \frac{0.9(120 - 46)}{307} = 0.483$$

This correlates well with the computer printout, which shows that $38,720/36 = 1076$ Btu/ft² (12,219 kJ/m²) are collected out of a total solar input of 2220 Btu/ft² (25,211 kJ/m²) for the day ($1076/2220 = 0.484$).

Now consider an identical collector in Minnesota on a clear winter day. Reduce the solar intensities by 10%, and use an average ambient temperature of 15°F. What average collection efficiency and total heat collected do you get for Minnesota?

6.3. How low would the heat loss coefficient for the collector in Problem 6.2 have to be in order to collect 1000 Btu/ft² (3155 W/m²) on a clear winter day in Minnesota?

6.4. Call up manufacturers of solar hot water systems and thereby estimate the installed costs of the systems described in Table 5.2. Adjust the NPV and PBP estimates for Dallas, Texas, in that table to your estimated costs.

6.5. Make a crude estimate of the performance of several of the systems in Table 6.2 when installed in your area instead of Dallas, Texas. Do this by multiplying the "solar heat collected" by two factors, one to account for the difference in solar intensity in your area and the other to account for the difference in collector efficiency due to the difference in ambient temperature. Use Table 3.3 for the average levels of solar intensity and, from the Appendix, data on average ambient temperatures. Assume the relative collector efficiency in any area can be taken as

$$\eta = 0.7 - \frac{0.8(120 - T_a)}{230}$$

with T_a in °F.

6.6. Use the computer program in Chapter 6 to estimate the performance of solar hot water systems in your area. Generate a tape of hourly solar intensities and temperatures from a SOLMET tape from the National Climatic Center.

6.7. Determine the effect of the Energy Tax Act on the economics of the 72 ft²/82 gal (6.69 m²/0.310 m³) and 90 ft²/120 gal (8.36 m²/0.454 m³) hot water systems listed in Table 6.2. (Reduce the cost by 30%.)

6.8. Use the data in Problem 3.12 to estimate the performance of a solar hot water system on a clear November day at the three localities: Fort Hood, Texas, Maynard, Massachusetts, and Yuma, Arizona. Assume that the collection efficiency of the collectors is given by $\eta = 0.7 - 0.9(T_c - T_a)/I_t$, where T_c is the collector temperature (°F), T_a is the ambient temperature (°F), and I_t is the solar intensity (Btu/hr-ft²) on the tilted collectors. Assume that $T_c \cong T_s + I_t/10$, where T_s is the temperature of storage (°F). Use the following design parameters.

> Collector area $= 72$ ft² $= 6.69$ m²
> Storage volume $= 82$ gal $= 0.310$ m³
> Collector tilt $=$ latitude
> Storage temperature at dawn $= 85°F = 302.6°K$
> Hot water demand, gal/hr $=$ constant at 10 gal/hr (0.0378 m³/hr)
>   for the 12 hr period ending 8 pm
> Tap water temperature $= 50°F$ (283°K) at Fort Hood, 38°F (277°K)
>   at Maynard, and 70°F (294°K) at Yuma
> Hot water temperature $= 130°F$ (328°K)

Space
Heating
Systems

This chapter concerns the use of solar energy to displace some of the conventional energy ordinarily used to heat a house or building. We show how the heating load required by a house or building depends on the method of construction and on the climate. Conventional systems for providing this load are described briefly. Solar energy systems using water as the working fluid for providing some of this load are described in some detail. Solar energy systems employing air as the working fluid are mentioned briefly. The methods of computer simulation are used to explore performance as a function of design parameters, as well as on operational strategy. We explain why special modes of operation such as "direct heating" and "recirculation" do not offer much advantage over simpler modes. Economic evaluations are made to show that solar energy can be competitive in displacing electricity in buildings heated by electrical resistance heat.

7.1 Demand Patterns

7.1.1 Estimation of Space Heat Load from Monthly Utility Bills

An empirical procedure for estimating demand patterns by month is illustrated for a particular house of 3200 ft² (297.3 m²) located in Dallas, Texas. Over a period of several years, the average demand pattern of monthly gas usage is shown in Figure 7.1. Gas is used for only three purposes in this

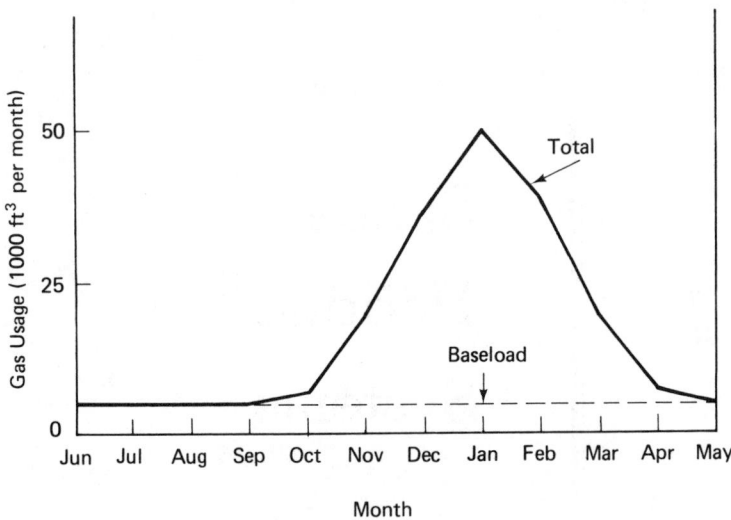

Figure 7.1 Gas usage per month for a 3200 ft² (297 m²) house in Dallas, Texas.

particular house: space heating, domestic hot water, and clothes drying. If it is assumed that the gas usage for hot water and clothes drying does not vary with the months of the year, then the summer monthly use of 5 Mcf (1.42×10^5 m³) can be attributed to these purposes. By subtracting 5 Mcf from the monthly gas usage in the winter, an estimate of the gas usage for space heating is obtained. Each Mcf of gas burned yields approximately 10^6 Btu (37.3×10^6 J). However, the efficiency of a gas heating system (fraction of heat conveyed into residence) is probably of the order of 60%. Thus, the heat input to the residence for space heating (in Btu) is calculated by multiplying the net gas usage (after subtracting 5 Mcf per month) by 6×10^5 Btu/ Mcf. The results are shown in Figure 7.2.

7.1.2 Heating Degree-days

It is obvious that the space heating load is at a maximum during the months of coldest weather. It is common to report ambient temperatures in terms of *heating degree-days* for correlation with heating loads. The heating degree-days are a measure of how far below 65°F the ambient temperature persists. For any arbitrary day, the degree-days are calculated from

$$DD = \frac{1}{24} \sum_{i+1}^{24} (65 - T_i) \tag{7.1}$$

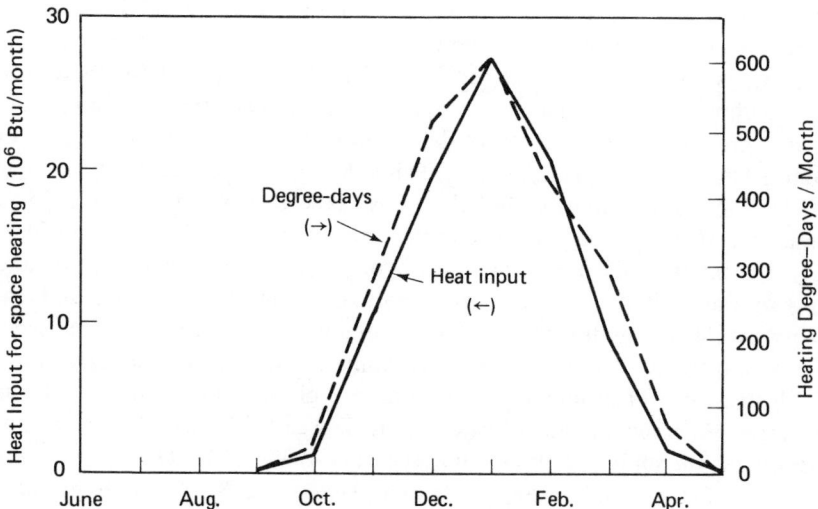

Figure 7.2 Estimated heat input to a 3200 ft² (297 m²) residence in Dallas, Texas, for space heating, compared with heating degree-days.

where T_i is the ambient temperature at hour i, and the sum is taken over 24 hr of a day. Thus, DD is the 24 hr average of the deviation of the ambient temperature below 65°F. In taking the sum, the term $65 - T_i$ is replaced by zero for any hour where $T_i > 65$°F. The heating degree-days for a month is simply the sum of DD over the days of the month. According to a simple theory, the heating load during any day or month should be proportional to the heating degree-days for that period. The rationale is as follows. The rate of heat loss from a building to the ambient should be roughly proportional to the difference in temperatures between the interior and the ambient. This rate of heat loss must be supplied in order to keep the interior temperature stable. Typically, the interior temperature may be kept at about 72°F (295°K). Thus, the rate of heat loss is proportional to $72 - T_i$ at any hour. However, not all of this loss must be supplied by the furnace. Each occupant of the residence gives off heat at the rate of about 200 Btu/hr (59 W). All appliances and machines produce heat. Also, there is direct solar gain from sunlight through windows and walls. It is a matter of experience that for many residences these factors will produce about a 7°F rise in interior temperature above the ambient. Therefore, on average, the furnace will not go on until the ambient temperature drops below 65°F. The rate at which heat must be supplied to the residence from the furnace is then proportional to $65 - T_i$, and this is the rationale for computing degree-days with a 65°F base temperature. The average levels of heating degree-days in Dallas are plotted in Figure

7.2, along with the actual space heat load for a residence. It can be seen that there is close correspondence between these figures. Since the residence requires about 27×10^6 Btu (28.5×10^9 J) in a month with about 600 degree-days, it follows that for this particular residence the heating requirements are 45,000 Btu per degree-day, or 14 Btu per degree-day per ft^2 of house. On an hourly basis, one may define the degree-hour (DH) as the value of $65 - T_i$ at any hour, except that it is taken as 0 when $T_i > 65$. For a residence that requires 14 Btu/DD-ft^2, the average hourly requirement is $14/24 = 0.6$ Btu/DH-ft^2. This is a key figure for any type of construction and provides an estimate of the hourly heat load at any hour for any ambient temperature for any floor area. Values ranging from 0.4 to 0.8 are expected for well-insulated to poorly-insulated homes. The average monthly totals of degree-days for a number of localities are given in the Appendix. These averages are based on data taken by the U.S. Weather Service during various periods in the middle 20th century (typically 1930–1970). There is growing evidence that a climate change is taking place in the Northern Hemisphere that may substantially alter the heat demands of buildings in the United States.[1] The period from 1900–1950 was the warmest period in a thousand years. During the 19th century, much colder winters were experienced in many parts of the United States than during the period that is considered "normal" for weather records. Since 1950, a slight cooling trend has developed, which appears to have accelerated in the late 1970s. It is possible in some cases that heating degree-days for the 1980s and 1990s may be considerably (i.e., more than 30%) higher than the averages compiled by the Weather Service.

7.1.3 Empirical Hot Water Usage

It is interesting to estimate the hot water usage in the residence described in Sec. 7.1.1. The baseload for gas is 5 Mcf per month, and as a guess we assume that 80% of this load is for hot water. Assuming that the gas heater is 60% efficient, the heat delivered for the domestic hot water is estimated as

$$0.6 \times 0.8 \times 5000 \times 1000 \text{ Btu/ft}^3 = 2.4 \times 10^6 \frac{\text{Btu}}{\text{month}}$$

[1] T. Alexander, "Ominous Changes in the World's Weather," *Fortune*, February, 1974, p. 90; O. B. Toon, J. B. Pollack, C. Sagan, A. Summers, B. Baldwin and W. Van Camp, *J. Geophysical Research*, **81**, 1071, 1976; W. W. Kellogg, "Climate Change and the Influence of Man's Activities on the Global Environment," National Center for Atmospheric Research, Boulder, CO, September, 1972; J. Gribbin and H. H. Lamb, "Climatic Change in Historical Times," in *Climatic Change*, ed. J. Gribbin, Cambridge University Press, New York, 1978; *Energy and Climate*, National Academy of Sciences Rept., 1977.

Assuming an average temperature rise of 70°F, the daily usage is estimated as

$$\frac{2.4 \times 10^6 \text{ Btu/month}}{70°\text{F} \times 1 \frac{\text{Btu}}{\text{lb }°\text{F}} \times 8.3 \frac{\text{lb}}{\text{gal}} \times 30 \frac{\text{days}}{\text{month}}} = \begin{matrix} 125 \text{ gal/day} \\ = 0.473 \text{m}^3/\text{day} \end{matrix}$$

This result is not unexpected for a family of four.

7.1.4 Cooling Loads and Cooling Degree-days

The cooling loads can be processed in the same way as heating loads. In Figure 7.3 the monthly electrical usage is plotted for the same residence.

Figure 7.3   Electrical usage per month for a 3200 ft² (297 m²) house in Dallas, Texas.

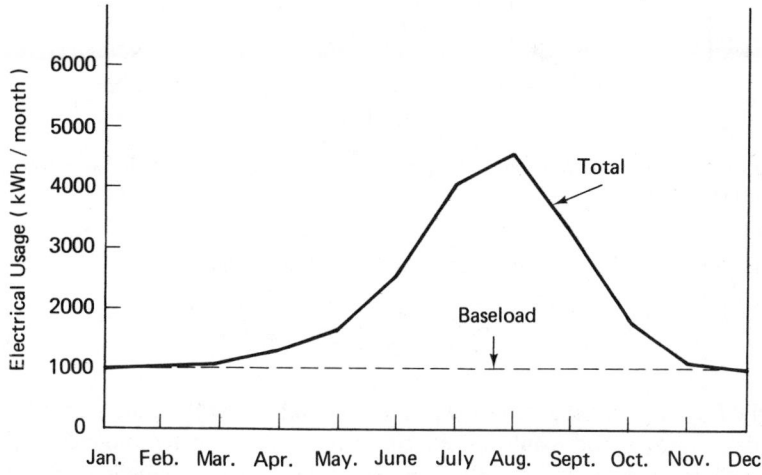

Assuming that the electrical baseload for cooking, lighting, and applicances is independent of month, it may be assumed that 1000 kWh per month are required for these purposes. The excess above this value in any month may be taken as the air conditioning load. The refrigeration machines that operate to cool this residence are rated for a coefficient of performance of about 2.8, which means that 2.8 units of heat energy are pumped out of the house for each unit of electrical energy supplied to the air conditioner. Since there are 3413 Btu in a kilowatt-hour, the heat removed from the residence in Btu by the air conditioners in any month can be estimated by multiplying the net kWh (after subtracting the baseload) by 2.8 × 3414. The results are plotted

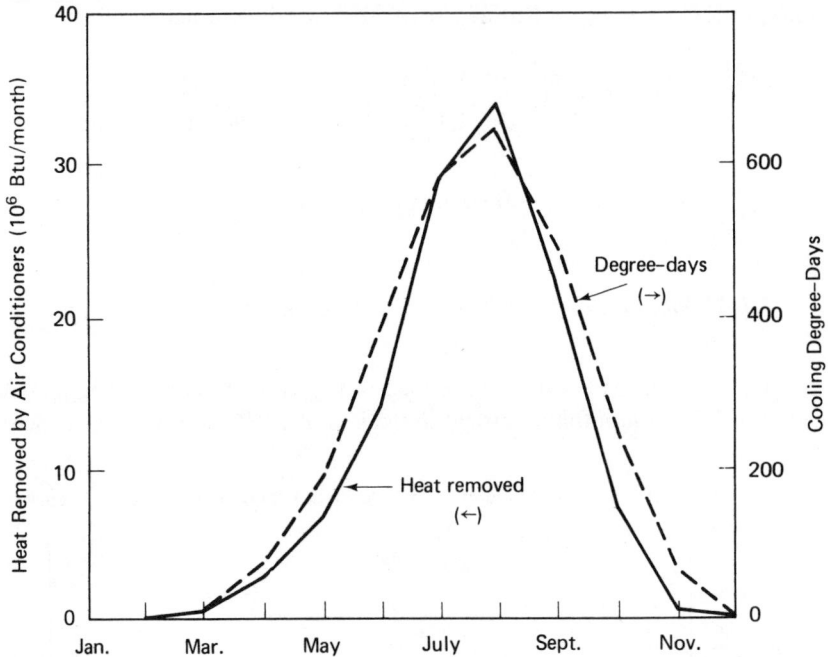

Figure 7.4 Heat removed from building by air conditioners, compared with cooling degree-days per month.

in Figure 7.4. The cooling degree-days are defined as

$$\text{CDD} = \frac{1}{24} \sum_{i=1}^{24} (T_i - 65) \qquad (7.2)$$

with $(T_i - 65)$ taken as zero if $T_i < 65$. It can be seen that, on the strict basis of degree-days, one would expect the summer load to be about 5% higher than the winter load. Actually, it is about 25% higher. This is undoubtedly due to the fact that direct solar gain is higher in the summer and a great deal of energy is required to dehumidify in the summer. Summer loads per degree-day are always expected to be somewhat higher than winter loads. The average midsummer load is roughly

$$3.4 \times 10^6 \text{ Btu}/(31 \times 24) \text{ hr} = 45,700 \text{ Btu/hr (48,200 kJ/hr)}$$

Since the average degree-hours is $645/31 = 20.8$, the summer load is 2200 Btu/hr per degree-hour, or 0.69 Btu/hr per degree-hour per ft² of floor area.

The house for which the above load calculations were made is extensively single glazed, including five sliding glass doors, four 4 ft × 7 ft glass panels, and many large windows. A house with less glass and extra insulation would undoubtedly require less heating and cooling. However, the house is extensively shaded by trees, which helps to keep summer cooling loads from becoming higher.

7.1.5 Estimation of Loads from Structure

In some texts and handbooks, the reader is encouraged to estimate heating and cooling loads by summing the heat losses through the roof, walls, and floor in each room, together with an estimate of losses due to infiltration. Methods are presented for making reasonable estimates of losses through various kinds of walls, floors, windows, and roofs. The main difficulty is in estimating the infiltration, which is nearly impossible to calculate, and which is often makes a large contribution to the total load. The procedure for estimating the heat flow through any section of wall, ceiling, or floor is to assume that this heat flow (per unit surface area) can be expressed as

$$Q = U(T_i - T_0) \qquad (7.3)$$

where U is a heat loss coefficient (usually expressed in Btu/hr-ft^2-°F), and T_i and T_0 are interior and outside temperatures. The heat flow is in Btu/hr per ft^2 of surface. The heat transfer coefficient may be viewed as a *conductance*, which is the reciprocal of a heat flow *resistance*,

$$R = \frac{1}{U} \qquad (7.4)$$

The total resistance to heat flow across a composite surface is the sum of the individual resistances of component parts of the wall. The resistances of standard building components are tabulated in various handbooks.[2] Some of these resistances are tabulated in Table 7.1. An example of the calculation of an overall resistance is now given.

Consider a section of wall as shown in Figure 7.5. The wall consists of an interior gypsum board, a 3 in. insulation batt, a $\frac{1}{2}$ in. insulation board, and $\frac{1}{2}$ in. exterior siding. There are heat flow resistances through each of these materials as well as at the air/surface boundaries on the inside and the outside. On the outside, a low resistance characteristic of a 10–15 mph wind has been used, whereas on the inside a higher resistance characteristic of stagnant air has been used. The insulation batt is partitioned by 2 × 4 wood studs

[2]*ASHRAE Handbook of Fundamentals,* American Society of Heating, Refrigerating and Air-Conditioning Engineers, New York, 1972.

TABLE 7.1   Thermal resistances of materials

Material	$R\ (hr\text{-}ft^2\text{-}°F/Btu)$
Gypsum board ($\frac{1}{2}$ in.)	0.5
Plywood ($\frac{1}{2}$ in.)	0.6
Insulating board ($\frac{1}{2}$ in.)	1.3
Acoustic tile ($\frac{1}{2}$ in.)	1.3
Wood subfloor ($\frac{3}{4}$ in.)	1.0
Concrete (per in. thickness)	0.08
Common brick (per in. thickness)	0.2
Face brick (per in. thickness)	0.1
Carpet and rubber pad	1.2
Tile (asphalt, linoleum)	0.05
Insulation (blanket batt or loose fill)	3.2 per in. thickness
Roofing (per in. thickness)	1.0
Windows (single-pane)	1.0
Windows (double-glazed)	1.7
Inside air/surface film	0.7
Outside air/surface film	0.2
Wood siding ($\frac{1}{2}$ in.)	0.8

Figure 7.5   Section of a 2 in. × 4 in. frame wall.

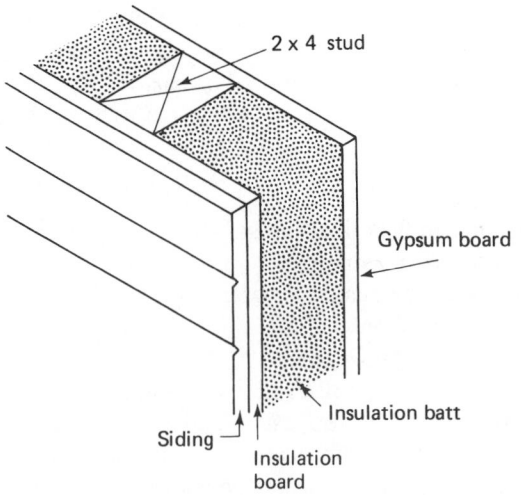

at periodic intervals, but this is neglected. The results are given in Table 7.2. Similar calculations can be performed on other composite walls, floors, and ceilings. Some results are shown in Table 7.3.

The procedure is now to go over the entire shell of a residence, including walls, ceilings, and floors, and evaluate the product of heat transfer coeffi-

TABLE 7.2 Calculation of thermal resistance of a composite wall

	R
1. Inside air/surface film	0.2
2. Gypsum wall board	0.5
3. Insulation batt ($3\frac{1}{2}$ in.)	10.0
4. Insulation board ($\frac{1}{2}$ in.)	1.3
5. Wood siding ($\frac{1}{2}$ in.)	0.8
6. Outside air/surface film	0.7
Total resistance	13.5

$U = 1/R = 0.074$ Btu/hr-ft^2-°F

TABLE 7.3 Overall thermal resistances of composite walls, floors, and ceilings

Construction	Resistance
1. Frame wall, 2 × 4 studs, gypsum, 3 in. insulation batt, insulation board, wood siding	13.5
2. Frame wall, 2 × 6 studs, gypsum, 5 in. insulation batt, insulation board, wood siding	21.0
3. Solid masonry wall, face brick, common brick, gypsum	2.7
4. Insulated ceiling, 6 in. batt, gypsum	20.8
5. Floor, linoleum tile, felt, plywood, wood subfloor	3.0
6. Single-pane window	1.0

cient and area for each surface. For example, one might have a 14 ft × 11 ft room with two exterior walls that are 8 ft high. Depending on the type of construction, the U factors are evaluated for each exterior surface. The windows are treated separately. An example is provided in Table 7.4. The sum of UA factors for surfaces other than the floor is 46.2 Btu/hr-°F, with the windows contributing more than 50%. The value of UA for the uninsulated floor is 51.3 Btu/hr-°F. The total heat loss from this room in winter is the sum of heat losses through the surfaces. If the interior, exterior, and ground temperatures are T_i, T_0, and T_g, °F, respectively, the heat loss is

$$Q = 46.2(T_i - T_0) + 51.3(T_i - T_g)$$

Typical values of $(T_i - T_g)$ are about half of typical values of $(T_i - T_0)$, so this can be approximated by

$$Q \cong 71.9(T_i - T_0) \text{ Btu/hr} \tag{7.5}$$

TABLE 7.4 Example of heat loss calculation for a typical room with two
exterior walls

Surface	Construction	Area (ft²)	R	U	UA
1. Windows	Single-pane	24	1.0	1.0	24.0
2. Wall #1	Frame (2 × 4)	112	13.5	0.074	8.3
3. Wall #2	Frame (2 × 4)	88	13.5	0.074	6.5
4. Ceiling	6 in. insulation batt	154	20.8	0.048	7.4
5. Floor	Linoleum, no insulation	154	3.0	0.333	51.3

On a unit floor area basis, this comes to

$$Q \cong 0.46(T_i - T_i) \text{ Btu/hr-ft}^2 \tag{7.6}$$

The heat loss coefficient of 0.5 Btu/hr-ft²-°F is typical of the average construction of houses in Texas. Houses with better insulation have lower losses. Equation (7.5) gives an estimate of the rate of heat loss through surfaces for fixed interior and exterior temperatures. In addition to such losses, there are losses due to infiltration of exterior air into the room through cracks and other crevices, around doors and windows, and the opening of doors. Such infiltration allows direct entry of exterior air, which must be heated by the furnace. Infiltration losses are difficult to estimate because they depend on the condition of the house (especially doors and windows), the prevailing winds, and the personal habits of the inhabitants. A rough rule of thumb is to use one volume change per hour in rooms with exterior doors and windows. The product of density and specific heat of air is approximately 0.018 Btu/ft³-°F. If this is multiplied by the number of cubic feet infiltrating per hour, the result is an effective coefficient for heat loss which when multiplied by $(T_i - T_0)$ gives the heat loss due to infiltration. For the 11 ft × 14 ft room described above, if one volume change (1232 ft³) infiltrates per hour, the coefficient of $(T_i - T_0)$ is $1232 \times 0.018 = 22$. Thus, to include infiltration in the total heat loss, Eqs. (7.5) and (7.6) should be changed to

$$Q = 94(T_i - T_0) \text{ Btu/hr} = 0.61(T_i - T_0) \text{ Btu/hr-ft}^2 \tag{7.7}$$

If it can be assumed that Eq. (7.7) holds for an entire residence, then the heat loss coefficient for the type of construction described is estimated at about 0.6 Btu/hr-°F per ft² (3.41 W/°K-m²) of the floor area. This is the same value found empirically in the beginning of this chapter for an actual residence in Dallas, Texas. The heat actually required by the residence is the heat loss less the heat gain due to occupants, appliances, and solar gain. These latter heat gains will be roughly approximated as a constant, G, per ft²

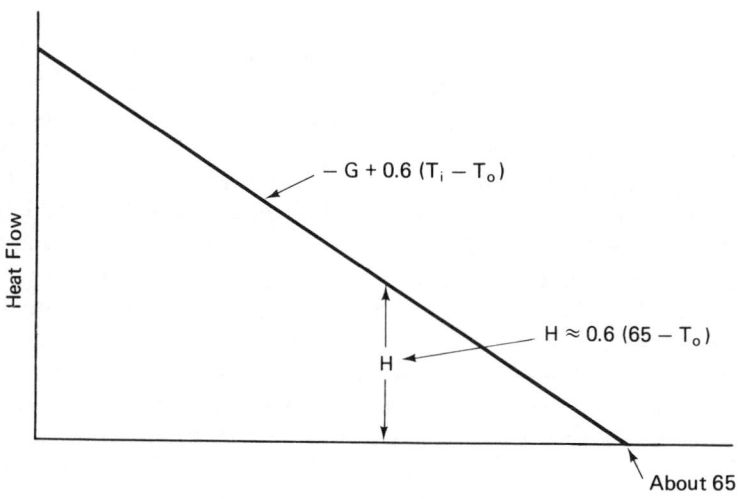

T_o = ambient temperature

Figure 7.6 Schematic plot of heat required by a residence as a function of ambient temperature.

of residence. Then the heat required may be written as

$$H = -G + 0.6(T_i - T_0) \qquad (7.8)$$

per ft² of residence. If T_i is held constant at $\sim 72°F$, a plot of H vs. T_0 is shown schematically in Figure 7.6. The heat required by the residence per ft² may be written as

$$H = 0.6\left(T_i - \frac{G}{0.6} - T_0\right) \qquad (7.9)$$

According to Figure 7.6, $T_i - G/0.6$ is roughly equal to 65°F when $T_i = 72°F$. Thus, the heat required is

$$H = 0.6(65 - T_0) \qquad (7.10)$$

which indicates the basic consistency of the degree-hour and degree-day concepts. Clearly, the heat loss coefficient 0.6 is only appropriate for one kind of construction and can be quite different for other types of construction.

There is some evidence that the above description works well in an average sense when taken over a period of, say, a month. For example, the data in Figures 7.2 and 7.3 support this concept. Another example is the monthly gas consumption data for Fort Hood, a large army base in Central Texas. A comparison of the monthly data against degree-days for Fort Hood is shown in Figure 7.7. The correlation is quite good. However, on a daily basis the correlation is not so good. Although days with high degree-days generally have high gas demand, the proportionality is quite variable. Some results are given in Figure 7.8. Although the correlation between gas load for space heat and degree-days works quite well when averaged over many days, it can be considerably in error on any given day. In general, it is found that variations in energy usage are not as sharp as variations in degree-days. This is probably due to thermal inertia of the structure, which tends to store heat and not immediately react to hourly variations in ambient temperature. There are also variations in heat load with wind velocity and solar intensity. Variations in hourly load tend to roughly follow variations in $65 - T_0$, but the variations in load are usually not as wide as the variations in $65 - T_0$.

Figure 7.7 Monthly gas consumption for heating and heating degree-days. By permission, American Technological University, Killeen, TX.

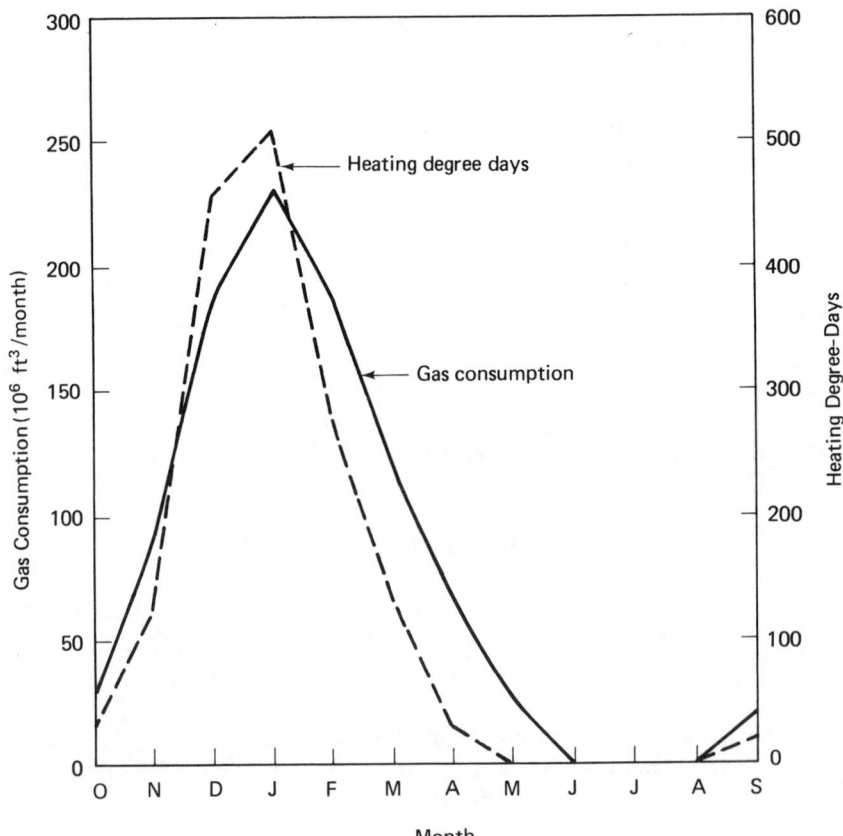

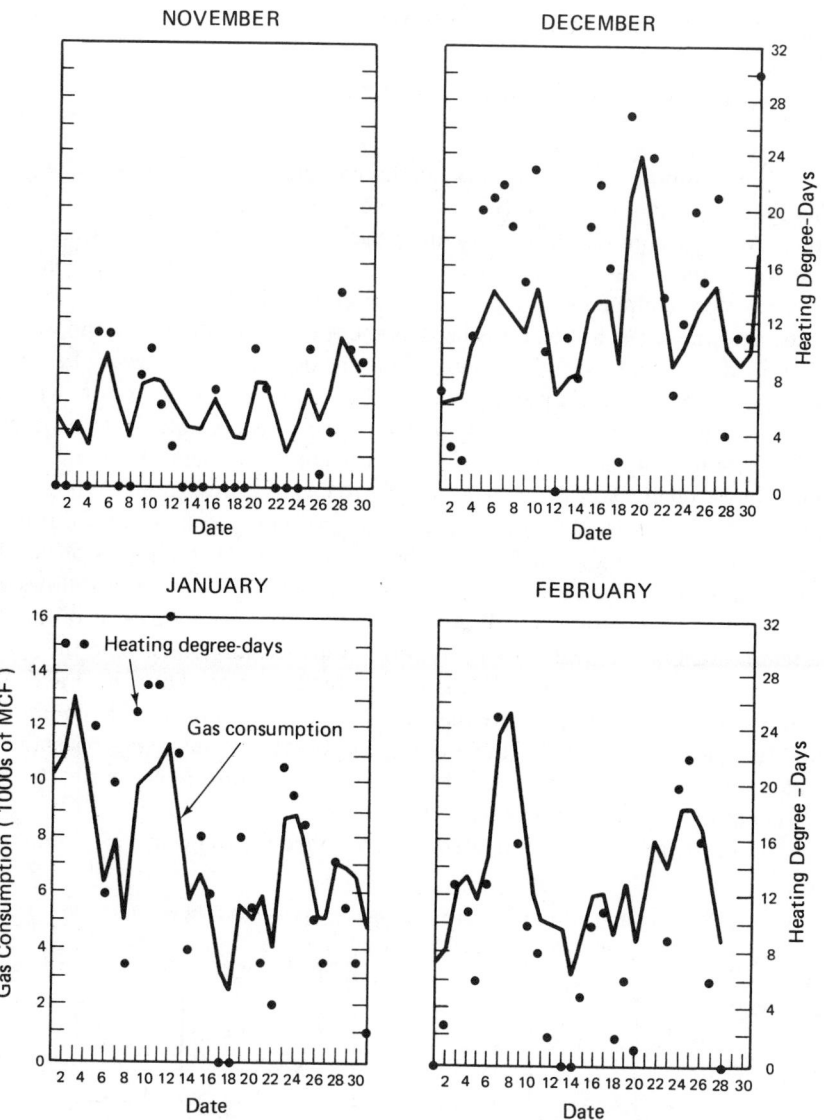

Figure 7.8 Daily gas consumption for heating and heating degree-days at Fort Hood, Texas. By permission, American Technological University, Killeen, TX.

7.2 Conventional Heating Systems

7.2.1 General Description

Conventional heating systems can be characterized in a number of ways. The first consideration is the fuel used, whether electricity, gas, fuel oil, or other. Another consideration is the fluid used to exchange heat between the heater and the house. The most common system is a *forced air* system in which a blower circulates air over heating coils and then to the house via ducts. In some cases, hot water is used as the transfer medium, and the water is pumped to radiators or baseboard units in the rooms. In *radiant* heating systems, the hot water is pumped through pipes in a concrete floor, which becomes warm and then warms the room above. It is possible to have water or air systems coupled to electrical-, gas-, or oil-fired heaters. A heating system is controlled by a thermostat which is located in a room. When the temperature drops to a set value, the heater goes on and stays on until the temperature rises to a few degrees above the set value. It then goes off until the room temperature drops off. Thus, the temperature in a room follows a saw-tooth pattern as shown in Figures 7.9 and 7.10. When a gas- or oil-fired system is used, some heat goes up the flue. Furthermore, each time the system cycles on and off, there are losses and inefficiencies. Thus, in practice, such systems rarely convert more than 50-70% of the heat released in combustion to heat transferred to the house. Electrical systems using resistance heaters are not subject to such losses and can usually supply nearly 100% of the electrical power consumed to the house.

In most cases, the heating plenums are congested to occupy a minimum

Figure 7.9 Cycling of conventional heating system in mild weather.

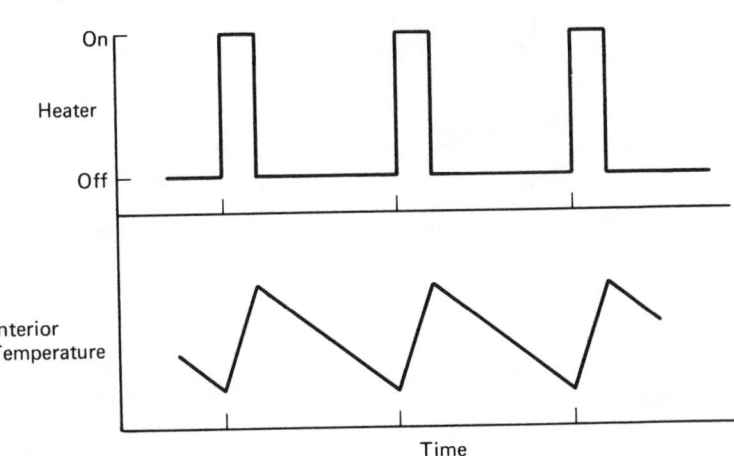

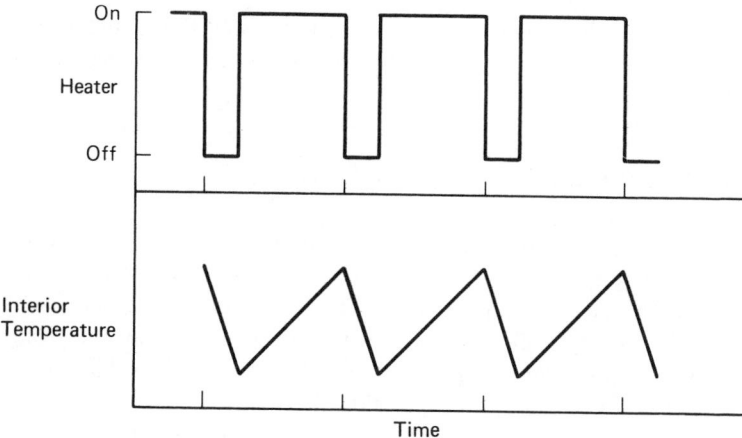

Figure 7.10 Cycling of conventional heating system in cold weather.

of house space. In addition, the rooms in which these systems are placed are often not much larger than the heating systems. Thus, it is often very difficult to physically retrofit a solar energy installation to an existing system.

7.2.2 Combination with Solar Air Systems

It is only feasible to use a solar energy system employing air as the transfer fluid with rock storage on a house that has a forced air heating system. In such applications it is desirable to know the design temperature of heated air leaving the furnace in the conventional system. The solar energy system should be designed to provide air at this temperature as often as possible. Domestic hot water is more difficult to produce in solar air systems. A solar energy system employing water as the transfer/storage fluid can be applied to a house with either hot water or forced air heat. If hot water heat is used, the design temperature of hot water leaving the heater is of critical importance. If forced air heat is used in the house, a hot water coil must be installed in the air plenum so that solar heated water can preheat the air flowing to the conventional heater. The design of this coil is not a simple matter and may require some trial and error.

7.2.3 Fireplaces

It should be noted that many houses have wood-burning fireplaces. Under many conditions, the draft created by the fire causes a rush of gases up the chimney, which creates a reduced air pressure inside the house. As a result, exterior air is drawn in through cracks around doors and windows, etc. This infiltration can actually cool off a house more than the heating effect of the fireplace.

7.2.4 Heat Pumps

Some houses that use electrical heat employ heat pumps. A heat pump is essentially a refrigeration machine that attempts to cool off the outside air as it pumps heat into a residence. The thermodynamic coefficient of performance or energy efficiency ratio is the ratio of heat pumped from the cold side to the electrical energy input. The heat pumped to the hot side is the sum of the heat pumped from the cold side and the electrical energy input. Many manufacturers define the operating energy efficiency ratio [or heating COP (HCOP)] as the heat pumped to the hot side divided by the electrical input. The HCOP is equal to one plus the thermodynamic COP. Existing units are capable of pumping between two and three units of heat into a house for each unit of electrical energy expended. The actual performance depends on the unit and on the temperature range of the ambient air. At low ambient temperatures [below about 35°F (275°K)] heat pumps function poorly and resistance heating begins to take over. In moderate climates, a heat pump should be able to average two units of heat pumped into a house for each unit of electricity expended during a winter season. Heat pumps greatly cut the cost of electricity for heating but involve additional expense for the purchase of the heat pump and considerable wear and tear on the heat pump. In some locations, electricity is about four times as expensive as the available fossil fuel (gas or oil). If electrical resistance heating is used at 100% efficiency and gas or oil is used at 60% efficiency, electrical resistance heating is $4 \times 0.6 = 2.4$ times as expensive as heating with gas or oil. If a heat pump with an average operating coefficient of performance of 2 is used, annual costs for electricity (disregarding heat pump costs) are only 1.2 times as high as for fossil fuels. These figures are only given as a rough example. They vary considerably with location. Heat pump systems will be treated in more detail in another section.

7.3 Space Heating with Liquid Solar System

7.3.1 Basic Flow Arrangements

A retrofit of a solar water system to a heating system employing forced-air distribution is illustrated in Figure 7.11. The main storage tank contains water. The domestic hot water preheat storage tank (60–120 gal) is immersed in the main tank. A differential thermostat turns on pump $P1$ when the

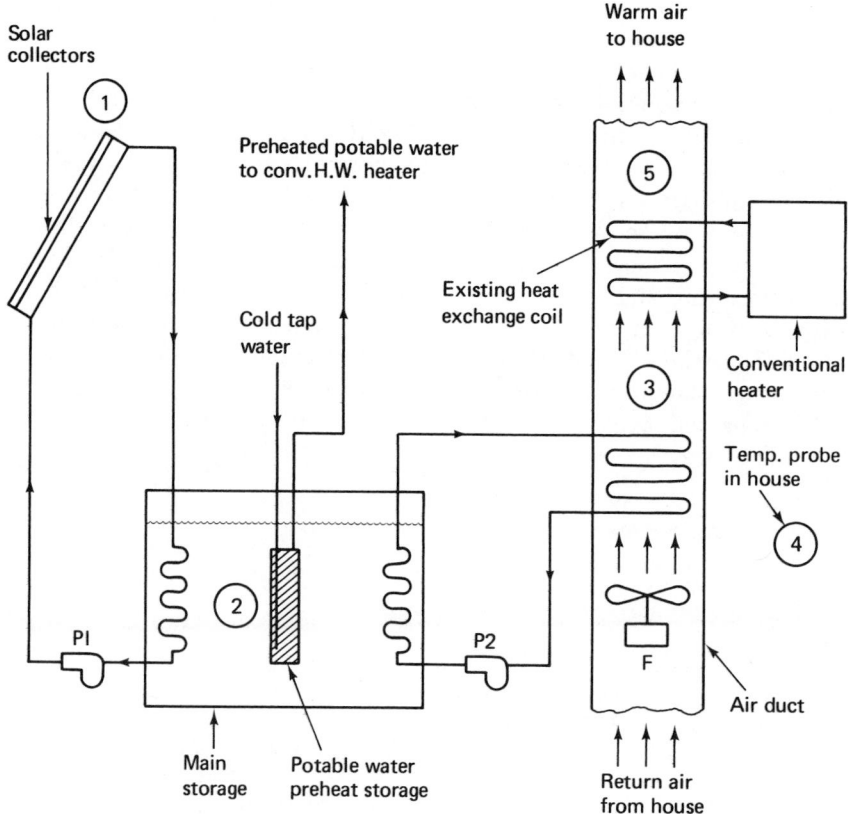

Figure 7.11 Retrofit of solar space heating system, using water as the transfer fluid to conventional forced air system, retaining the existing heater.

difference in temperatures between sensors (1) on the collector and (2) in the storage tank exceeds about 5–10°F. The pump is turned off when the temperature difference $T_1 - T_2$ decreases to about 2°F. The air plenum in the house is modified to include an additional heat exchange coil supplied by pump $P2$. When the regular house thermostat T_4 turns on the heating system, pump $P2$ is turned on if T_2 is greater than some preset value. After a short time delay, the main fan F is turned on. If the air temperature T_3 is lower than the design temperature for heated air entering the house (T_5), the conventional heater goes on. For example, suppose the design temperature for heated air is $T_5 = 110°F$ at the furnace outlet, and return air from the house is about 70°F. If main storage is at $T_2 = 100°F$, the air at T_3 is probably raised to, say, 80°F assuming a 20°F drop across the heat exchanger. The conventional heater then boosts the air temperature from 80°F to ~110°F at T_5. If storage was higher than about 130°F, the air at T_3 would probably be heated to 110°F (or higher), and the conventional heater would not go on.

One major problem with this particular design is the cyling of the conventional heater. The conventional heater is sized to heat return air from 70°F to, say, 110°F. If the solar preheater heats the return air to some intermediate temperature, the conventional heater will cycle on and off in a rapid sawtooth pattern, leading to poor efficiency. The time lags required to heat the solar water coil in the plenum, and nonuniformities in the air flow make control of this system rather sluggish and unsatisfactory. An alternative procedure is to use an in-line water heater as shown in Figure 7.12. In this system, T_3 is the temperature of solar heated water and T_5 is the design hot water temperature. If $T_5 = 130°F$, for example, and T_3 was, say, 110°F, the conventional hot water heater would cycle on and off to keep the water entering the heating coil near 130°F. The control of this system is relatively simple. The cycling of the conventional heater can be reduced by placing a small holding tank between the conventional heater and the coil. Thus, the conventional heater would charge up this holding tank at periodic intervals

Figure 7.12 Solar space heating system with in-line back-up conventional heater.

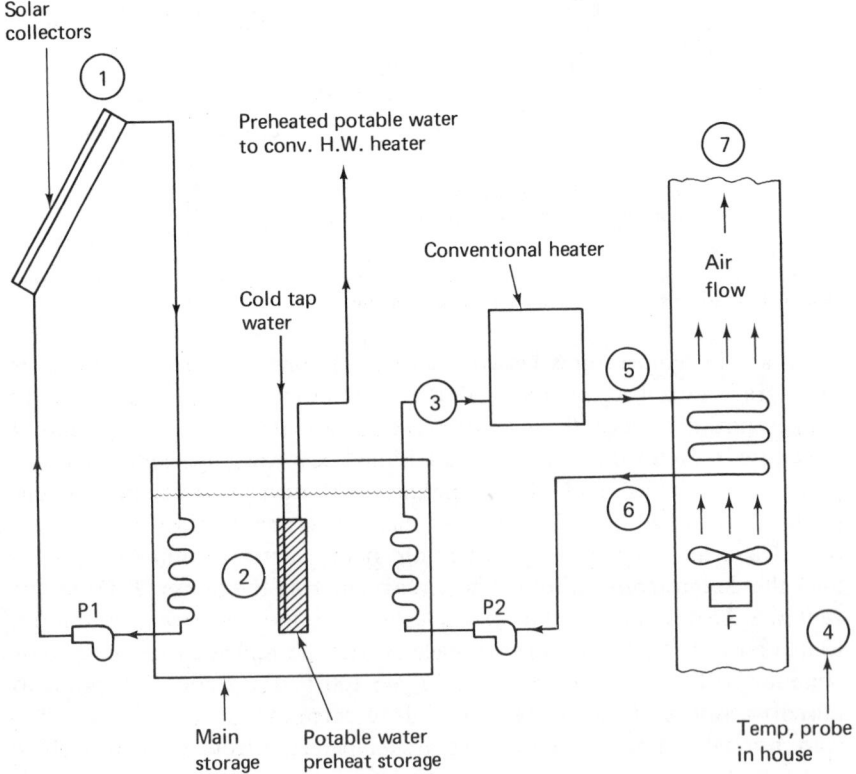

when the tank had cooled several degrees. Another way to reduce cycling is to place two smaller conventional heating units in series so that on many occasions one unit will remain on for considerable periods without cycling.

The same approach can be applied to houses using hot water heat distribution to the rooms as shown in Figure 7.13.

Figure 7.13 Solar space heating system with hot water distribution system.

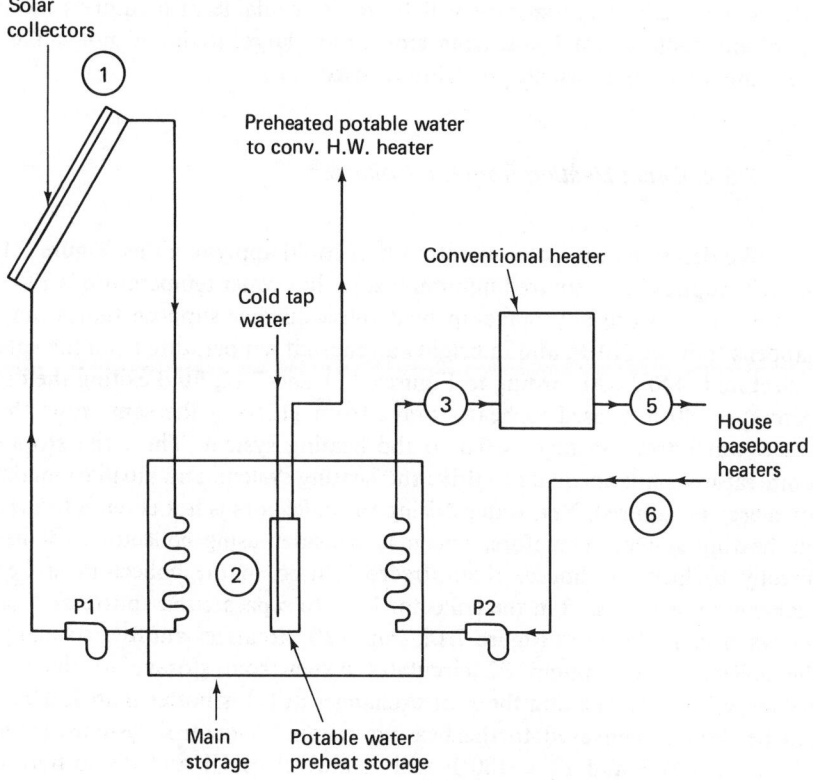

Principal advantages of the in-line conventional heater, as shown in Figures 7.12 and 7.13, are the simplicity and response of the control system and that the heating plenum is kept simple. One possible fault is that the in-line system may tend to require more auxiliary heat under some circumstances. For example, suppose that a heating coil brings circulating air to within 20°F of the temperature of hot water entering the coil, and suppose the design temperature for air leaving the furnace is 110°F. In Figure 7.12, assume a temperature drop of 20°F from T_5 to T_6. Now compare the systems illustrated in Figures 7.11 and 7.12 if storage is at 100°F. In Figure 7.11, the air is heated from 70°F to 80°F at T_3 by solar heat and further heated to 110°F at T_5 by conventional heat. Thus, the fraction solar is $10/40 = 25\%$. In Figure 7.12, $T_3 = 100°F$, $T_5 = 130°F$, and $T_6 = 110°F$. Thus, conventional heat is supplying all the heat required to heat the house plus the additional heat to raise water from storage at 100°F to 110°F before returning to storage. On the other hand, in Figure 7.11 storage is being cooled by operation, whereas in Figure 7.12 storage is being heated by operation. Thus, even though more auxiliary make-up heat is required for that period in Figure 7.12, solar energy from storage will be more available at a future time of operation. Unless heat losses from storage are large, it should not make a great difference as to which procedure is used.

7.3.2 Direct Heating from the Collectors

We describe next a consideration that could apply to either Figure 7.11 or 7.12. Suppose the required minimum solar hot water temperature is 130°F in order not to use any make-up heat. Now further suppose that storage happens to be at 110°F, and in bright sun the exit temperature from the solar collectors is 140°F. According to Figures 7.11 and 7.12, fluid exiting the collectors at 140°F is used to heat storage from 110°F at the same time that 110°F water from storage is fed to the heating system. Thus, the storage temperature is not adequate to drive the heating system, and auxiliary make-up energy is required. Yet, water exiting the collectors is hot enough to drive the heating system. Therefore, one may consider using collector exit fluid directly to heat the house. If antifreeze is used in the collectors, a heat exchanger must be used in the collector loop to separate the antifreeze from the water as is shown in Figure 7.14. Pump $P1$ circulates antifreeze through the collectors, and pump $P2$ circulates water from storage to the heat exchanger. If water exiting the heat exchanger at (7) is hotter than 130°F, it can be directly conveyed to the heating system. A control system sensing that $T_7 > 130°F$ and $T_2 < 130°F$ will rotate valves $V1$ and $V2$ so that no water is allowed up from storage through $V2$, and some of the water from the heat exchanger is passed along the dashed line from $V1$ through $V2$ to the heating system.

Usually, the flow rate in the $P2$ loop is considerably higher than the flow rate in the heating plenum loop, so when valve $V1$ is changed for direct heating, part of the flow is routed to storage and part is carried through to the heating plenum. The return from the heating plenum is conveyed to storage as usual. The control system required for this system is rather complicated. Valves $V1$ and $V2$ must be actuated when $T_7 > 130°F$ and $T_2 <$

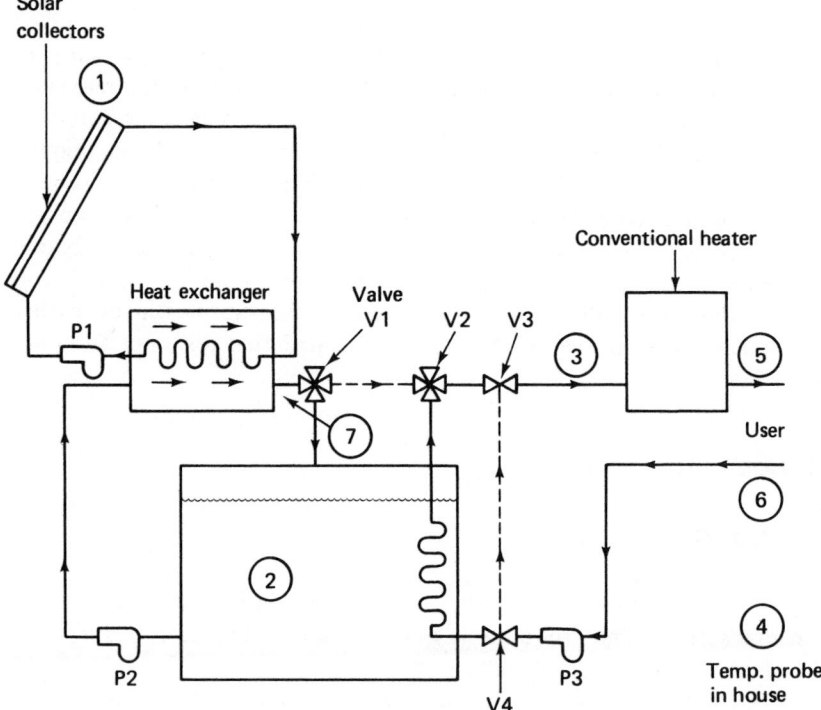

Figure 7.14 Solar space heating system allowing direct heating and recirculation modes. (Domestic hot water preheater not shown for clarity.)

130°F *and* when the household thermostat indicates that heat is required. The gains that can be realized from this direct heating mode are largely illusory. Although under the stated conditions one can reduce the amount of auxiliary make-up energy required for the moment, the price paid is that less heat is dumped into storage and therefore more auxiliary energy will be required at a subsequent time after the sun goes down. If heat losses from storage are severe, this favors use of direct heating since it is desirable to use solar heat while it is available. However, it should be economically desirable to heavily insulate storage.

7.3.3 Recirculation with Auxiliary Heat

Another option that can be considered in the system with an in-line heater is a recirculation mode. For example, in Figure 7.12 if storage is quite cold, say 90°F, and there is a 20°F drop across the heating coil, then $T_3 =$

90°F, $T_5 = 130$°F, and $T_6 = 110$°F. It would therefore appear wasteful to return water at 110°F from the heating coil to the storage tank which is at 90°F. To avoid heating the storage tank with conventional heat, a recirculation mode can be used as illustrated in Figure 7.14. Whenever $T_6 > T_2$, valves $V3$ and $V4$ are automatically actuated, and flow takes place along the dashed line from $V4$ to $V3$. In this case, the heating system is totally isolated from the solar energy system which functions only as a collector/storage system. The net effect of using recirculation over a year is not large because when conventional heat is used to heat storage when storage is cold, collected solar energy will be available at usable temperatures at an earlier time. When averaged over an entire year, recirculation does not add very much efficiency.

7.4 Computer Simulation

7.4.1 Operating Modes

The computer simulation of the system shown in Figure 7.15 is explained in this section. The procedure is quite similar to that described in Secs. 6.4 and 6.5 for domestic hot water systems.

Four modes of operation are defined. In mode 1, direct heating from the solar collectors and recirculation are both allowed when conditions warrant these options. In case both are permitted at the same time, direct heating takes precedence over recirculation. In mode 2, direct heating is allowed under suitable conditions but recirculation is not. In mode 3, recirculation is allowed but direct heating is not. In mode 4, neither of these options is permitted. In any computer run, the mode is fixed. To compare the performance of a system with fixed design parameters in the four modes, four computer runs must be made.

7.4.2 Input Data

Provision is made for multiple runs with a single computer entry. Each run is defined by a data card on which the latitude, tilt, collector area, storage volume, print parameter, house area, and operational mode are indicated. In the first run, as the hourly data on solar intensity and ambient temperature are read in from data cards, the information is stored and retained. When the year is over for the first run, a new data card is read for the latitude, tilt, etc., for the second run. In the second and succeeding runs, solar intensity and ambient temperature data are obtained from storage rather than data cards.

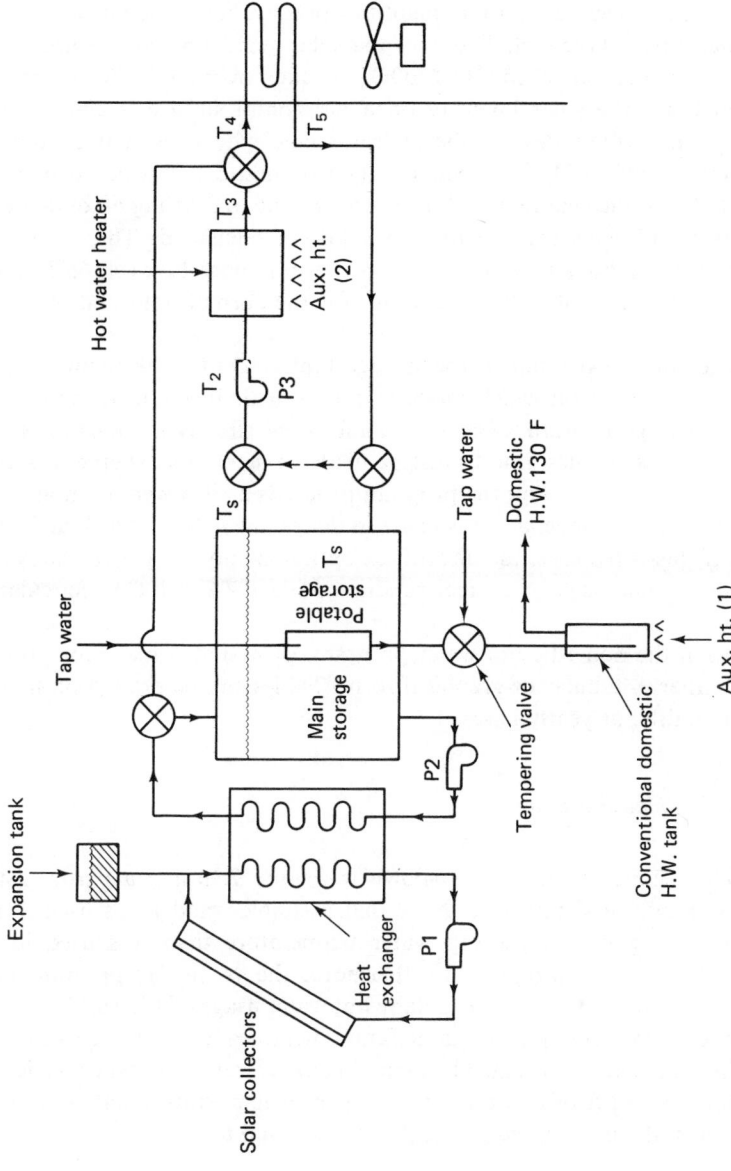

Figure 7.15 Schematic of system used for computer simulation.

7.4.3 Daily Procedure

For each new day, all the quantities that are independent of hour are calculated first. These include day of the solar year, tap water temperature, and solar declination. Then a DO LOOP is taken over the 24 hr of that day. At each hour, the solar intensity on a horizontal surface is converted to intensity on a tilted surface, the collector efficiency is estimated, and the solar heat input (SHI) is calculated. The domestic hot water demand is calculated, and the heat removed from storage due to drawing off of domestic hot water and replacing it with tap water is calculated. The space heat demand for the hour is estimated from the difference between 65°F and a weighted mean of ambient temperatures for the current hour and the previous 2 hr.

The solar heat input to the storage tank and the tank input to space heat are dependent on which mode is operational. In mode 4, the calculations are straightforward. Modes 1, 2, and 3 are slightly more complicated. In each of these modes, a test must be made to determine whether a special option is called for. If not, the program proceeds as if it were in mode 4. In the direct heating mode, a test is made to determine if $TS < 130°F$ and $T1 > 130°F$. If this is the case, the direct heat option is applied. If not, the system proceeds as in mode 4. In the recirculation mode, if $TS < 100°F$ recirculation is used.

As in the domestic hot water program, provision is made for printing results either in tabular or graphical form. Tables can be printed on hourly, daily, monthly, or yearly bases.

7.4.4 Heating Load

The computer program is shown in Figure 7.16 and a partial list of variable names in Figure 7.17. Note that a simpler model was used in this particular program for the tap water temperature than was used in the domestic hot water program. Furthermore, the particular program illustrated in Figure 7.16 is based on a daily hot water usage of 176 gal/day, which is probably representative of a large family with heavy hot water usage. The space heating load is estimated by assuming an overall heat loss coefficient of 0.5 Btu/hr per sq ft of house floor area, per °F temperature difference from 65°F. Thus, the heating requirements are set equal to

$$(0.5)(AH)(65 - TAVG)$$

where AH is the house area and TAVG is the average ambient temperature affecting the heating load. Due to the thermal inertia of the heat capacity of

Figure 7.16 Program to simulate solar space heating/domestic hot water system for a single-family residence.

```
C       SYSTEM FOR A SINGLE FAMILY RESIDENCE
C
C
C
        REAL MSSH
        REAL MSH,MAUX2,MHW
        REAL L,LR,LMT,MSH,MAUX,MTOT,MSHW
        INTEGER TAMB,ZTAM
        INTEGER SH,TA,HWD,ZSH,ZM
        DIMENSION TAMB(3),ZTAM(31,24)
        DIMENSION ZSH(31,24),ZST(31,24),ZTO(31,24),ZTS(31,24),ZAUX(31,24)
        DIMENSION TA(24),SH(24),MORGN(12),HWD(24),MD(12)
        DATA MORGN/12,43,71,102,132,163,193,224,255,285,316,346/
        DATA HWD/0,0,0,0,0,0,16,16,8,8,8,8,8,8,8,8,16,16,8,8,8,8,8,8/
        DATA MD /31,28,31,30,31,30,31,31,30,31,30,31/
        PRINT 100
100 FORMAT(1H1,5X,'PROGRAM TO EVALUATE PERFORMANCE OF A SOLAR SPACE HE
       1ATING AND DOMESTIC HOT WATER SYSTEM',/1H0,'
       2
        PI=3.141592
        MS=1
        TS= 55.
200 READ 101,L,T,AC,VS,LBJ,VMS,AH,MODE
101 FORMAT(4F10.2,I10,2F10.2,I4)
        PRINT 102,L,T,AC,VS,VMS,AH
102 FORMAT(1H0,10X,'LATITUDE = ',F5.1,1H ,14X,'TILT = ',F5.1,/1H ,
       14X,'COLLECTOR AREA = ',F6.1,' SQ.FT.',/1H ,4X,'STORAGE VOLUME = ',
       2 F6.1,' GALLONS',/1H ,3X,'MAIN STORAGE = ',F7.1,' GALLONS',/1H ,
       3 5X,'HOUSE AREA = ',F6.1,' SQ FT'/ )
C
C       MODE = 1    ALLOWS DIRECT HEATING FROM COLLECTORS,AND RECIRCULATION
C       MODE = 2    ALLOWS DIRECT HEATING
C       MODE = 3    ALLOWS RECIRCULATION
C       MODE = 4    IS PLAIN VANILLA
C
        GO TO (5,6,7,8),MODE
      5 PRINT 10
        GO TO 9
      6 PRINT 11
        GO TO 9
      7 PRINT 12
        GO TO 9
      8 PRINT 13
      9 CONTINUE
10 FORMAT(1H ,3X,'ALLOWS DIRECT HEATING OF HOUSE FROM COLLECTORS AND
       1 RECIRCULATION OF PLENUM WATER TO AUX HEATER'//)
11 FORMAT(1H ,3X,'ALLOWS DIRECT HEATING OF HOUSE FROM COLLECTORS'//)
12 FORMAT(1H ,3X,'ALLOWS RECIRCULATION OF PLENUM WATER TO AUX HEATER'
       1 //)
13 FORMAT(1H ,3X,'NO DIRECT HEATING OR RECIRCULATION'//)
C       ****
C       SCHEDULE FOR LBJ - PRINTING INDEX
C          LBJ.GT.0 =  PRINT ALL DATA INCLUDING HOURLY VALUES
C          LBJ.EQ.0 =  PRINT ONLY DAILY,MONTHLY,YEARLY TOTALS
C          LBJ.LT.0 BUT GT. -10 = PRINT ONLY MONTHLY,YEARLY TOTALS
C          LBJ.LE.-10 = PLOT HOURLY DATA
C       ****
        LR=L*PI/180.
        TR=T*PI/180.
        LMT= LR-TR
        SINL=SIN(LR)
        COSL= COS(LR)
        CLMT=COS(LMT)
        SLMT=SIN(LMT)
```

Figure 7.16 (Continued)

```
      DHW  =0.0
      JSH  =0.0
      DAUX2=0.0
      DSHW= 0.
      DAUX= 0.
      DTOT= 0.
      MHW  =0.0
      MSH  =0.0
      MAUX2=0.0
      MSHW= 0.
      MAUX= 0.
      MTOT= 0.
      YHW  =0.0
      YSH  =0.0
      YAUX2=0.0
      YSHW= 0.
      YAUX= 0.
      YTOT= 0.
      MMM  = 0
      MONTH=0
      SINDMX= SIN(23.45*PI/180.)
C   ****
C     DATA ARE STORED ON CARDS. 24 SOLAR INTENSITIES ON 1ST CARD, 24
C     AMBIENT TEMPS ON 2ND CARD, FOR 1ST DAY. ETC.
C   ****
    1 READ 103, M,ID,(SH(J),J=1,24)
      READ 104, (TA(J),J=1,24)
  103 FORMAT(26I3)
  104 FORMAT(6X,24I3)
      IF(M.NE.MS) GO TO 60
C
C   CALCULATE DAY OF SOLAR YEAR
C
    2 IDS=MOHGM(M)+ID-1
      IF(IDS.GT.365) IDS= IDS-365
      JDS= IDS
      IF(JDS.GT.183) JDS= 366-IDS
C
C   TAPWATER TEMPERATURE
C
      TT=55.
      IF(IDS.GT. 85) TT=63.
      IF(IDS.GT.163) TT=70.
      IF(IDS.GT.287) TT=63.
      IF(IDS.GT.334) TT=55.
C
C   SOLAR DECLINATION
C
      SIND    = -SINDMX*COS(PI*(JDS)/182.5)
      DEC= ARSIN(SIND)
      COSD    = COS(DEC)
      DEC= DEC*180./PI
C
C   DAILY HEADING
C
      IF(LBJ.LE.0) GO TO 4
      PRINT 105,M,ID,IDS,JDS,DEC,TT
  105 FORMAT(1H0,4I5,F8.2,F8.1 /)
      PRINT 106
  106 FORMAT(1H ,'HOUR  SH    ST    EFF   TA    TS  TSNEW   SHITT   HHW
     1 TISP   AUX    AUX2   TOTHW   TOTAL   FSOL'/)
    4 CONTINUE
C
C   BEGIN HOUR LOOP
C
```

Figure 7.16　(Continued)

```
      DO 50 IHR=1,24
      ST=0.0
      EFF=0.0
      T1=TT
      SHITT=0.0
      IF(SH(IHR).LT.40) GO TO 20
      IF(TS.GT.190.) GO TO 20
C
C   TILT CORRECTION FACTOR
C
      HRANG= (IHR-12.5)*PI/12.
      COSH= COS(HRANG)
      SINH= SIN(HRANG)
      COSTH= COSD*COSL*COSH+SIND*SINL
      COSTT= COSD*CLMT*COSH+SIND*SLMT
      IF(COSTH.LE.0.0) GO TO 20
      IF(COSTT.LE.0.0) GO TO 20
      TFCTR= COSTT/COSTH
      IF(TFCTR.GT.3.) TFCTR=3.
C
C   INTENSITY ON TILTED SURFACE
C
      ST=SH(IHR)*TFCTR
      IF(ST.LT.80.) GO TO 20
C
C   COLLECTOR EFFICIENCY
C
      TGAIN = 30.+ST/300.

      T3=130.
      T4=100.
      FLOW = SPHT/30.
      AUX2 = FLOW*(130.-TS)
      TISP = -FLOW*(100.-TS)
      GO TO 150
  133 CONTINUE
C
C   AUX HEAT WITH RECIRCULATION
C
      T3=130.
      T4=100.
      T2=100.
      FLOW = SPHT/30.
      AUX2= SPHT
      TISP= 0.
      GO TO 150
  137 CONTINUE
C
C   PARTIAL AUX HEAT
C
      T2=TS
      T3=130.
      T4=100.
      FLOW = SPHT/30.
      AUX2 = FLOW*(130.-TS)
      TISP = FLOW*(TS-100.)
      GO TO 150
  138 CONTINUE
C
C BYPASS STORAGE FROM COLLECTOR
C   DIRECT HEATING MODE - DETERMINE IF COLLECTOR FLOW IS ADEQUATE
C   TO SUPPLY FLOW TO SPACE HEATER
C
      FLOWC= SHI/TGAIN
      FLOW= SPHT/30.
      IF(FLOW.GT.FLOWC) GO TO 139
```

223

Figure 7.16 (Continued)

```
C   FLOW IS ADEGUATE
        T2=T1
        T3=T2
        T4=T3-30.
        FLO1 = FLOWC-FLOW
        SHITT = FLO1*(T1-TS)
        AUX2=0.
        TISP = FLOW*(TS-T4)
        AUX2 = 0.
        GO TO 150
C   FLOW IS INADEQUATE
    139 SHITT=SHI
        IDH=0
        GO TO 137
C
C   NO AUX HEAT
C
        T1 = TS + TGAIN
        DELT = TS + 0.5*TGAIN - TA(IHR)
        EFF = 0.7 - 0.9*DELT/ST
        IF(EFF.LT.0.0) EFF = 0.0
C
    20 CONTINUE
C
C   HOT WATER DEMAND = HWD(IHR) GAL/HR
C   HEAT FLOW TO HOT WATER = HHW   BTU/HR
C
        TFINAL = TS
        IF(TS.GT.130.) TFINAL =130.
        HHW=HWD(IHR)*8.3*(TFINAL-TT)
        AUX=0.
        IF(TS.LT.130.) AUX   = HWD(IHR)*8.3*(130.-TS)
        TOTHW = HWD(IHR)*8.3*(130.-TT)
C
C   SPACE HEAT DEMAND
C
        IF(MMM.EQ.0) GO TO 130
        TAMB(3)= TAMB(2)
        TAMB(2)= TAMB(1)
        TAMB(1)= TA(IHR)
        GO TO 132
    130 DO 131 I=1,3
    131 TAMB(I) = TA(IHR)
    132 TAVG = 0.5*(TAMB(1)+0.7*TAMB(2)+0.3*TAMB(3))
        MMM= MMM + 1
        SPHT = 0.5*AH*(65.-TAVG)
        IF(SPHT.LT.0.0) SPHT=0.0
C
C   SOLAR HEAT INPUT TO TANK (SHITT)
C
        SHI=EFF*ST*AC
        IDH=0
C
C   DETERMINE IF DIRECT HEATING IS POSSIBLE
C   IDH=1 MEANS DIRECT HEATING
C
        IF(MODE.GT.2) GO TO 162
        IF(T1.GT.130..AND.TS.LT.130.) IDH=1
    162 IF(IDH.EQ.0) SHITIT=SHI
        IF(IDH.EQ.1) GO TO 138
        IF(TS.GT.130.) GO TO 140
        IF(TS.GT.100.) GO TO 137
C
C   TS IS LESS THAN 100    AUX HEAT ONLY
C
        IF(MODE.EQ.1.OR.MODE.EQ.3) GO TO 133
```

Figure 7.16 (Continued)

```
C
C   NO RECIRCULATION - AUX HEAT ONLY
C
      T2=TS
C
  140 CONTINUE
      FLOW = SPHT/30.
      T2=TS
      T3=T2
      T4=T3-30.
      AUX2=0.
      TISP=SPHT
  150 CONTINUE
C
C   CHANGE IN STORAGE TEMPERATURE
C
      TSNEW = TS + (SHITT -HHW-TISP)/(VMS*8.3)
      TOTAL = TOTHW + SPHT
      FSOL=0.
      IF(TOTAL.EQ.0.) GO TO 25
      FSOL = 1. - (AUX+AUX2)/TOTAL
      IF(FSOL.LT.0.) FSOL=0.
   25 CONTINUE
C
C   PRINT HOURLY DATA
C
      IF(LBJ.LE.0) GO TO 30
      PRINT 107, IHR,SH(IHR),ST,EFF,TA(IHR),TS,TSNEW,SHITT,HHW,TISP,AUX,
     1 AUX2,TOTHW,TOTAL,FSOL
  107 FORMAT(1H ,I3,I5,F6.1,F5.3,I4,2F5.1,F8.0,F7.0,3F8.0,F7.0,F8.0,
     1 F6.3)
   30 CONTINUE
C
C   ADD HOURLY VALUES TO DAILY SUMS
C
      DAUX2 = DAUX2+ AUX2
      DHW   = DHW  + TOTHW
      DSH   = DSH  + SPHT
      DTOT = DTOT + TOTAL
      DSHW= DSHW+ HHW
      DAUX=DAUX+AUX
C  IF PLOTTING PROGRAM, STORE HOURLY VALUES FOR PLOT
C
      ZSH(ID,IHR) = SH(IHR)
      ZST(ID,IHR) = ST
      ZTS(ID,IHR) = TS
      ZTAM(ID,IHR) = TA(IHR)
      ZTO(ID,IHR) = TOTAL
      ZAUX(ID,IHR) = AUX + AUX2
      ZM = MS
C
   50 TS= TSNEW
C
C   DAY IS OVER
C
      IF(LPJ.LT.0) GO TO 55
      PRINT 108
  108 FORMAT(1H0,'  MONTH  DAY       DTOT       DAUX       DAUX2       DSH
     1     DHW      DSHW     FSOLAR'/)
      FSOL = 1.-(DAUX+DAUX2)/DTOT
      IF(FSOL.LT.0.) FSOL=0.
      PRINT 109, M,ID,DTOT,DAUX,DAUX2,DSH,DHW,DSHW,FSOL
  109 FORMAT(1H ,2I6,6F10.1,F7.3 /)
   55 CONTINUE
```

225

Figure 7.16 (Continued)

```
C
C   MONTHLY SUMS
C
      MAUX2 = MAUX2+ DAUX2
      MHW   = MHW  + DHW
      MSH   = MSH  + DSH
      MSHW  = MSHW + DSHW
      MAUX  = MAUX + DAUX
      MTOT  = MTOT + DTOT
      MSSH= MSH-MAUX2
      DAUX2 = 0.
      DHW   = 0.
      DSH   = 0.
      DSHW=0.
      DAUX=0.
      DTOT= 0.
      GO TO 1
   60 CONTINUE


C
C   NEW MONTH
C
      YAUX2 = YAUX2+ MAUX2
      YHW   = YHW  + MHW
      YSH   = YSH  + MSH
      YSHW=YSHW+ MSHW
      YAUX=YAUX+ MAUX
      YTOT=YTOT+ MTOT
      YSSH = YSH - YAUX2
      YSOL = YSSH + YSHW
      FSH = YSSH/YSH
      FHW = YSHW/YHW
      MONTH= MONTH+1
      IF(LBJ.LE.-10) GO TO 65
      PRINT 110
  110 FORMAT(1H0,9X,'TOTAL      AUX HEAT   AUX HEAT    SPACE HT   SPACE HT
     AHOT W',
     1ATER   HOT WATER  FSOLAR',/1H ,' MONTH   DEMAND    HOT WTR    SPACE
     2HT     DEMAND    (SOLAR)    DEMAND    (SOLAR)')
      FSOL = 1.-(MAUX+MAUX2)/MTOT
      IF(FSOL.LT.0.) FSOL=0.
      PRINT 113, MS,MTOT,MAUX,MAUX2,MSH,MSSH,MHW,MSHW,FSOL
  113 FORMAT(1H0,I5,7F11.1,F8.3)
   65 CONTINUE
      MAUX2 = 0.
      MHW   = 0.
      MSH   = 0.
      MSHW=0.0
      MAUX=0.0
      MTOT=0.0
      IF(LBJ.GT.-10) GO TO 75
      K=MD(MS)
      CALL PLOT(ZST,ZTS,ZTAM,ZTO,ZAUX,ZM,K)
   75 MS=M
      IF(M.EQ.26) GO TO 80
      IF(MONTH.EQ.12) GO TO 80
      GO TO 2
   80 IF(LBJ.LE.-10) GO TO 90
      FSOL = 1.-(YAUX+YAUX2)/YTOT
      CCST = 1500. + 50.*SQRT(VMS) + 10.*AC
      VALUE= YSOL/COST
      PRINT 111
```

Figure 7.16 (Continued)

```
111 FORMAT(1HC,8X,'TOTAL         TOTAL    FSOLAR  SPACE HT    SPACE HT   FSO
   1LAR   HOT WATER   HOT WATER   FSOLAR   HOT WATER SPACE HEAT',/1H ,
   2MONTH    DEMAND        SOLAR      TOTAL     DEMAND         SOLAR      SP HT
   3DEMAND       SOLAR      HOT WTR AUX HEAT     AUX HEAT' /)
      PRINT 112,MONTH,YTOT,YSOL,FSOL,YSH,YSSH,FSH,YHW,YSHW,FHW,YAUX,
   1 YAUX2
112 FORMAT(1H ,I4,2E11.3,F7.3,2E11.3,F7.3,2E11.3,F7.3,2E11.3)
      PRINT 116
116 FORMAT(1H0,' COLLECTOR    MAIN STORAGE              SOLAR BTUS      VAL
   1UE',/1H ,'    AREA            VOLUME        COST     SUPPLIED       (BTU/D
   2OLLAR)'/ )
      PRINT 117, AC,VMS,COST,YSOL,VALUE
117 FORMAT(1H ,F7.0,8X,F6.0,5X,F8.2,E12.3,3X,E11.3)
  90 STOP
      END



      SUBROUTINE PLOT(ZST,ZTS,ZTAM,ZTC,ZAUX,ZM,KDAYS)
      INTEGER ZM
      INTEGER ZTAM
      DIMENSION ZTAM(31,24)
      DIMENSION ZSH(31,24),ZST(31,24),ZTS(31,24),ZTO(31,24),ZAUX(31,24)
      DIMENSION KK(5,24),IFIELD(130,24)
      DATA IBK  /' '/
      DATA IDOT /'.'/
      DATA ISTR /'*'/
      DATA IHOR /'I'/
      DATA IV   /'-'/
      DATA IPL  /'+'/
      DO 200 KL=1,KDAYS
C
C BLANK OUT THE FIELD FOR ONE DAY
C
      DO 1 I=1,130
      DO 1 J=1,24
   1 IFIELD(I,J) = IBK
C
C VERTICAL AXIS
C
      DO 2 I=1,130
   2 IFIELD(I,1) = IV
C
C TICK MARKS
C
      DO 3 I=1,12
      K = 10 + 10*(I-1)
   3 IFIELD(K,1) = IHOR
C
C HORIZONTAL AXES
C
      DO 4 I=1,24
      IFIELD(1,I) = IHOR
   4 IFIELD(130,I)=IHOR
C
C THE DATA FOR EACH DAY WILL BE SCALED AND CONVERTED TO INTEGERS
C AND STORED IN ARRAYS AS FOLLOWS...
C   KK(1,J) = ST AT HOUR J
C   KK(2,J) = TS
C   KK(3,J) = TAMB
C   KK(4,J) = TOTAL HEAT
C   KK(5,J) = AUX HEAT
C
```

Figure 7.16 (Continued)

```
      DO 10 I=1,24
      KK(1,I) = ZST(KL,I)/10 + 1
      KK(2,I) = ZTS(KL,I)*4/10 + 20
      KK(3,I) = ZTAM(KL,I)*4/10+ 20
      KK(4,I) = ZTO(KL,I)/1000 + 70
      KK(5,I) = ZAUX(KL,I)/1000+ 70
      IFIELD(KK(1,I),I)=IPL
      IFIELD(KK(2,I),I)=ISTR
      IFIELD(KK(3,I),I)=IDOT
      IFIELD(KK(4,I),I)=IPL
   10 IFIELD(KK(5,I),I)=ISTR
      DO 20 I=1,10
   20 PRINT 100,(IFIELD(J,I),J=1,130)
  100 FORMAT(1H ,130A1)
      PRINT 101,(IFIELD(J,11),J=1,130),ZM
      PRINT 101,(IFIELD(J,12),J=1,130),KL
  101 FORMAT(1H ,130A1,I2)
      DO 30 I=13,24
   30 PRINT 100,(IFIELD(J,I),J=1,130)
  200 CONTINUE
      RETURN
      END
```

Figure 7.17 Definition of variable names for the program in Figure 7.16.

```
AC = COLLECTOR AREA (FT**2)
AH = HOUSE AREA (FT**2)
AUX = AUXILIARY MAKE-UP HEAT FOR DOMESTIC HOT WATER (BTU/HR)
AUX2= AUXILIARY MAKE-UP HEAT FOR SPACE HEATING (BTU/HR)
COST = ESTIMATED COST FOR INSTALLATION (DOLLARS)
DAUX = DAILY SUM OF AUX HT FOR DOM.HOT WATER (BTU)
DAUX2= DAILY SUM OF AUX HT FOR SPACE HEAT (BTU)
DEC = ANGLE OF DECLINATION
DHW  = DAILY SUM OF TOTAL HEAT REQUIRED FOR DOM.H.W. (BTU)
DSH  = DAILY SUM OF SPACE HEAT REQUIREMENTS (BTU)
DSHW = DAILY SUM OF SOLAR HEAT INPUT TO DOM.H.W.
DTOT = DAILY SUM OF TOTAL HEAT REQD FOR H.W. AND SP.HT.
EFF = EFFICIENCY OF SOLAR COLLECTORS
FHW =  FRACTION OF DOMESTIC H.W. HEAT FROM SOLAR ENERGY
FLOW = FLOW RATE THROUGH SPACE HEATING COIL (LB/HR)
FLOWC = FLOW RATE IN LOOP FROM STORAGE TO HEAT EXCHANGER IN
        COLLECTOR LOOP (LB/HR)
FLO1 = FLOWC - FLOW
FSH =  FRACTION OF 2PACE HEAT FURNISHED BY SOLAR ENERGY
FSOL = FRACTION OF TOTAL HEAT FURNISHED BY SOLAR ENERGY
HWD= DOMESTIC HOT WATER DEMAND (GAL/HR)
IDH = PARAMETER (=0 IF NO DIRECT HTG, =1 WITH DIRECT HTG)
IHR = HOUR
ISS = AN INDEX FOR NUMBERING THE HOURS OF A YEAR FROM 1 TO 8760
KD  = STORED VALUE OF DAY
KM  = STORED VALUE OF MONTH
KSH = STORED VALUE OF SOLAR INTENSITY (HOR.SURF.)
KTA = STORED VALUE OF AMBIENT TEMPERATURE
LBJ= CONTROL PARAMETER FOR PRINTING
M= MONTH
MAUX = MONTHLY SUM OF AUX.ENERGY FOR DOM.H.W. (BTU)
MAUX2= MONTHLY SUM OF AUX.ENERGY FOR SPACE HEAT (BTU)
MD = DAYS IN A MONTH
MHW  = MONTHLY SUM OF TOTAL HEAT REQD.FOR DOM.H.W. (BTU)
MMM = INDEX (=0 FOR 1ST HOUR, AND NON-ZERO THEREAFTER)
```

```
MODE = OPERATIONAL OPTIONS AVAILABLE. MODE=1 ALLOWS DIRECT HEATING
       AND RECIRCULATION, MODE=2 ALLOWS DIRECT HEATING, MODE=3
       ALLOWS RECIRCULATION, MODE=4 ALLOWS NO SPECIAL OPTIONS
MONTH = NUMBER OF COMPLETED MONTHS
MS= MONTHSAVE = PREVIOUS MONTH
MSH  = MONTHLY SUM OF TOTAL HEAT REQD.FOR SP.HT. (BTU)
MSHW = MONTHLY SOLAR HEAT INPUT TO HOT WATER (BTU)
MSSH = MONTHLY SOLAR HEAT INPUT TO SP.HT. (BTU)
MTOT = MONTHLY TOTAL HEAT REQUIRED FOR DOM.H.W. AND SP.HT. (BTU)
SH = SOLAR INTENSITY ON HORIZONTAL SURFACE
SHI  = SOLAT HEAT INPUT
SHITT = SOLAR HEAT INPUT TO TANK
SPHT = SPACE HEAT DEMAND BY HOUSE (BTU/HR)
ST = SOLAR INTENSITY ON TILTED SURFACE
T = TILT ANGLE OF SOLAR COLLECTORS
TA = AMBIENT TEMPERATURE
TAMB(I) = AMBIENT TEMPERATURE - (1 = AT THIS HR, 2 = ONE HR AGO,
          AND 3 = 2 HRS AGO)
TAVG = WEIGHTED MEAN AMBIENT TEMPERATURE FOR LAST 3 HOURS
TFCTR = FACTOR TO CONVERT INTENSITY ON HORIZ.SURF. TO TILTED SURF.
TT = TAPWATER TEMPERATURE
T1 = COLLECTOR EXIT TEMPERATURE
T2 = TEMP.OF WATER ENTERING SP.HT. AUX HTR
T3 = TEMP.OF WATER ENTERING SP.HT. COIL
T4 = TEMP.OF WATER EXITING SP.HT.COIL
TISP = TANK INPUT TO SP.HT. (BTU/HR)
TOTAL= TOTAL HEAT REQD FOR H.W. AND SP.HT. (BTU/HR)
TOTHW= TOTAL HEAT REQD FOR H.W. (BTU/HR)
TS = STORAGE TEMPERATURE AT BEGINNING OF HOUR
TSNEW =STORAGE TEMPERATURE AT END OF HOUR
VALUE= SOLAR BTU'S DELIVERED PER YR PER $ INVESTED
VMS= VOLUME OF MAIN STORAGE (GALLONS)
YAUX = YEARLY SUM OF AUX.HT. FOR DOM.H.W. (BTU)
YAUX2= YEARLY SUM OF AUX.HT. FOR SPACE HEAT (BTU)
YHW  = YEARLY SUM OF TOTAL HEAT REQD FOR H.W. (BTU)
YSH  = YEARLY SUM OF TOTAL HEAT REQD FOR SPACE HEAT (BTU)
YSHW = YEARLY SUM OF SOLAR HEAT INPUT TO DOM.H.W. (BTU)
YSOL = YEARLY SUM OF SOLAR HEAT INPUT TO SP,HT. AND DOM.H.W.
YSSH = YEARLY SUM OF SOLAR HEAT INPUT TO SPACE HEAT (BTU)
YTOT = YEARLY SUM OF TOTAL HEAT REQD FOR SPACE HEAT AND DOM.H.W.(BTU)
```

Figure 7.17   (Continued)

the structure, the heating load does not respond instantaneously to changes in the ambient temperature. Therefore, an arbitrarily weighted mean of ambient temperatures for the current hour TAMB(1), the previous hour TAMB(2), and 2 hours ago TAMB(3) is arbitrarily used for TAVG:

$$TAVG = 0.5[TAMB(1) + (0.7)(TAMB(2)) + (0.3)(TAMB(3))]$$

The heating system is based on a minimum design input temperature of $130°F$ to the space heat coil. It is presumed that there is a $30°F$ temperature drop across the coil, and that the flow cycles on and off as needed. The total flow (in lb) during any hour is the space heat demand divided by $30°F$. When solar heated water is available at $130°F$ or higher, it is fed directly into the space heat coil and no auxiliary make-up heat is required. However, when solar heated water is only available below $130°F$, auxiliary make-up energy is used to bring the water input to the space heat coil up to $130°F$. A holding tank is maintained inside the main storage tank for preheating domestic hot water. It is assumed that such water reaches the storage temperature.

7.5 Design Results

7.5.1 Hourly Results in Various Modes

The design results in this section are given for a house of 2500 ft² in North Central Texas (latitude = 33°N). The data on hourly solar intensities and ambient temperatures are the same as in Sec. 6.6 for domestic hot water systems. Solar collectors with efficiency $\eta = 0.7 - 0.9(\Delta T)/I$ are used with ΔT in °F and I in Btu/hr-ft². As before, printing can be done on hourly, daily, monthly, or yearly tabular bases, or on an hourly plotting basis. Typical hourly printouts starting with tap water in storage on January 1 are shown in Figure 7.18 for simple operation (mode 4) and, in Figure 7.19, allowing direct heating and recirculation (mode 1).

The difference between conventional operation and recirculation is most apparent at start-up when the storage tank is at tap water temperature. In conventional operation, the water from storage at 55°F is heated to 130°F

Figure 7.18 Hourly printout of performance of solar space heating system (no direct heat or recirculation).

```
            LATITUDE  =   33.0          TILT  =   33.0
   COLLECTOR AREA  =    600.0 SQ.FT.
   STORAGE VOLUME  =     80.0 GALLONS
   MAIN STORAGE  =    1200.0 GALLONS
   HOUSE AREA  =   2500.0 SQ FT

NO DIRECT HEATING OR RECIRCULATION


    1      1     12     12   -22.92     55.0
```

HOUR	SH	ST	EFF	TA	TS	TSNEW	SHITT	HHW	TISP	AUX	AUX2	TOTHW	TOTAL	FSOL
1	0	0.0	0.0	40	55.0	59.7	0.	0.	-46875.	0.	78125.	0.	31250.	0.0
2	0	0.0	0.0	39	59.7	64.1	0.	0.	-43652.	0.	76152.	0.	32500.	0.0
3	0	0.0	0.0	38	64.1	68.1	0.	0.	-40409.	0.	74150.	0.	33750.	0.0
4	0	0.0	0.0	37	68.1	71.9	0.	0.	-37164.	0.	72164.	0.	35000.	0.0
5	0	0.0	0.0	36	71.9	75.3	0.	0.	-33983.	0.	70233.	0.	36250.	0.0
6	0	0.0	0.0	35	75.3	78.4	0.	0.	-30890.	0.	68390.	0.	37500.	0.0
7	0	0.0	0.0	35	78.4	80.8	0.	3106.	-27013.	6854.	64513.	9960.	47460.	0.0
8	0	0.0	0.0	33	80.8	83.0	0.	3425.	-25613.	6535.	65613.	9960.	49960.	0.0
9	3	0.0	0.0	34	83.0	85.0	0.	1860.	-21935.	3125.	60685.	4980.	43730.	0.0
10	40	72.9	0.0	38	85.0	86.5	0.	1994.	-16838.	2986.	50588.	4980.	38730.	0.0
11	96	162.9	0.409	42	86.5	91.6	39971.	2093.	-12915.	2887.	41665.	4980.	33730.	0.0
12	140	230.9	0.477	46	91.6	98.7	66101.	2432.	-6632.	2548.	30382.	4980.	28730.	0.0
13	155	255.6	0.477	48	98.7	105.8	73091.	2900.	-934.	2080.	22184.	4980.	26230.	0.075
14	176	298.6	0.487	50	105.8	113.9	87214.	3375.	3639.	1605.	15111.	4980.	23730.	0.296
15	117	213.4	0.398	53	113.9	117.9	50979.	3909.	6937.	1071.	8063.	4980.	19980.	0.543
16	100	217.2	0.382	52	117.9	121.5	49785.	4177.	9698.	803.	6552.	4980.	21230.	0.654
17	63	189.0	0.319	51	121.5	123.0	36202.	8832.	12547.	1128.	4953.	9960.	27460.	0.779
18	4	0.0	0.0	50	123.0	120.6	0.	9030.	14374.	930.	4376.	9960.	28710.	0.815
19	0	0.0	0.0	46	120.6	118.6	0.	4359.	16346.	621.	7404.	4980.	28730.	0.721
20	0	0.0	0.0	44	118.6	116.5	0.	4221.	16248.	759.	10002.	4980.	31230.	0.655
21	0	0.0	0.0	42	116.5	114.5	0.	4085.	15826.	895.	12924.	4980.	33730.	0.590
22	0	0.0	0.0	41	114.5	112.7	0.	3952.	14515.	1028.	15485.	4980.	34980.	0.528
23	0	0.0	0.0	40	112.7	111.0	0.	3829.	13188.	1151.	18062.	4980.	36230.	0.470
24	0	0.0	0.0	39	111.0	109.4	0.	3715.	11865.	1265.	20635.	4980.	37480.	0.416

230

Figure 7.18 (Continued)

MONTH	DAY	DTOT	DAUX	DAUX2	DSH	DHW	DSHW	FSOLAR
1	1	798310.0	38265.4	898407.9	688750.0	109559.6	71294.4	0.0

1	2	13	13	-22.83	55.0

HOUR	SH	ST	EFF	TA	TS	TSNEW	SHITT	HHW	TISP	AUX	AUX2	TOTHW	TOTAL	FSOL
1	0	0.0	0.0	40	109.4	108.4	0.	0.	9779.	0.	21471.	0.	31250.	0.313
2	0	0.0	0.0	40	108.4	107.5	0.	0.	8756.	0.	22494.	0.	31250.	0.280
3	0	0.0	0.9	39	107.5	106.7	0.	0.	8154.	0.	24346.	0.	32500.	0.251
4	0	0.0	0.0	39	106.7	106.0	0.	0.	7267.	0.	25233.	0.	32500.	0.224
5	0	0.0	0.0	38	106.0	105.3	0.	0.	6725.	0.	27024.	0.	33750.	0.199
6	0	0.0	0.0	37	105.3	104.7	0.	0.	6187.	0.	28813.	0.	35000.	0.177
7	0	0.0	0.0	37	104.7	103.5	0.	6598.	5462.	3362.	29538.	9960.	44960.	0.268
8	0	0.0	0.0	36	103.5	102.4	0.	6437.	4194.	3523.	32056.	9960.	46210.	0.230
9	22	0.0	0.0	35	102.4	101.8	0.	3148.	3005.	1832.	34495.	4980.	42480.	0.145
10	86	156.5	0.300	37	98.7	101.2	28166.	2902.	0.	2078.	35000.	4980.	39980.	0.073
11	141	238.8	0.420	39	101.2	106.9	60230.	3071.	1352.	1905.	31148.	4980.	37480.	0.118
12	160	296.3	0.455	41	106.9	111.6	51270.	3443.	365.	1537.	0.	4980.	34980.	0.956
13	199	327.6	0.467	43	111.6	117.7	61673.	3759.	-2533.	1221.	0.	4980.	32480.	0.962
14	193	326.9	0.458	46	117.7	123.9	63867.	4162.	-2126.	818.	0.	4980.	28730.	0.972
15	172	312.9	0.434	47	123.9	129.4	57989.	4575.	-970.	405.	0.	4980.	27480.	0.985
16	132	285.7	0.399	48	129.4	133.6	48116.	4937.	1012.	42.	0.	4980.	26230.	0.998
17	73	219.0	0.303	48	133.6	134.5	39850.	9960.	21250.	0.	0.	9960.	31210.	1.000
18	6	0.0	0.0	48	134.5	131.3	0.	9960.	21250.	0.	0.	9960.	31210.	1.000
19	0	0.0	0.0	45	131.3	128.3	0.	4980.	25000.	0.	0.	4980.	29980.	1.000
20	0	0.0	0.0	43	128.3	125.2	0.	4868.	25952.	112.	1548.	4980.	32480.	0.949
21	0	0.0	0.0	42	125.2	122.3	0.	4662.	24166.	318.	4584.	4980.	33730.	0.855
22	0	0.0	0.0	40	122.3	119.5	0.	4470.	23252.	510.	7998.	4980.	36230.	0.765
23	0	0.0	0.0	38	119.5	116.9	0.	4285.	21981.	695.	11769.	4980.	38730.	0.678
24	0	0.0	0.0	36	116.9	114.4	0.	4110.	20423.	870.	15827.	4980.	41230.	0.595

MONTH	DAY	DTOT	DAUX	DAUX2	DSH	DHW	DSHW	FSOLAR
1	2	832060.0	20562.9	417303.4	722500.0	109559.6	88996.9	0.482

1	3	14	14	-22.73	55.0

HOUR	SH	ST	EFF	TA	TS	TSNEW	SHITT	HHW	TISP	AUX	AUX2	TOTHW	TOTAL	FSOL
1	0	0.0	0.0	34	114.4	112.6	0.	0.	18650.	0.	20100.	0.	38750.	0.481
2	0	0.0	0.0	32	112.6	110.8	0.	0.	17278.	0.	23972.	0.	41250.	0.419
3	0	0.0	0.0	31	110.8	109.3	0.	0.	15344.	0.	27156.	0.	42500.	0.361
4	0	0.0	0.0	30	109.3	107.9	0.	0.	13549.	0.	30201.	0.	43750.	0.310
5	0	0.0	0.0	30	107.9	106.8	0.	0.	11565.	0.	32185.	0.	43750.	0.264
6	0	0.0	0.0	30	106.8	105.8	0.	0.	9872.	0.	33878.	0.	43750.	0.226
7	0	0.0	0.0	30	105.8	104.3	0.	6743.	8426.	3217.	35324.	9960.	53710.	0.282
8	0	0.0	0.0	32	104.3	103.0	0.	6541.	5851.	3419.	35399.	9960.	51210.	0.242
9	24	0.0	0.0	31	103.0	102.3	0.	3188.	4265.	1792.	38235.	4980.	47480.	0.157
10	90	163.3	0.273	33	102.3	104.3	26787.	3138.	3017.	1842.	36983.	4980.	44980.	0.137
11	146	246.8	0.413	38	104.3	109.7	61157.	3276.	4876.	1704.	28874.	4980.	38730.	0.210
12	186	305.7	0.462	44	109.7	115.2	57928.	3629.	-496.	1351.	0.	4980.	31230.	0.957
13	205	336.9	0.470	46	115.2	121.9	68384.	3994.	-2921.	986.	0.	4980.	28730.	0.966
14	202	341.4	0.458	47	121.9	128.6	68114.	4443.	-3106.	537.	0.	4980.	27480.	0.980
15	177	321.2	0.432	49	128.6	134.5	61830.	4886.	-1415.	52.	0.	4980.	24980.	0.996
16	133	286.8	0.390	50	134.5	138.8	67077.	4980.	18750.	0.	0.	4980.	23730.	1.000
17	72	216.0	0.289	51	138.8	139.8	37459.	9960.	17500.	0.	0.	9960.	27460.	1.000
18	6	0.0	0.0	50	139.8	137.0	0.	9960.	18750.	0.	0.	9960.	28710.	1.000
19	0	0.0	0.0	48	137.0	134.3	0.	4980.	21250.	0.	0.	4980.	26230.	1.000
20	0	0.0	0.0	45	134.3	131.3	0.	4980.	25000.	0.	0.	4980.	29980.	1.000
21	0	0.0	0.0	43	131.3	128.0	0.	4980.	27500.	0.	0.	4980.	32480.	1.000
22	0	0.0	0.0	42	128.0	124.9	0.	4850.	26879.	130.	1871.	4980.	33730.	0.941
23	0	0.0	0.0	41	124.9	121.9	0.	4639.	24862.	341.	5138.	4980.	34980.	0.843
24	0	0.0	0.0	40	121.9	119.2	0.	4442.	22813.	538.	8437.	4980.	36230.	0.752

MONTH	DAY	DTOT	DAUX	DAUX2	DSH	DHW	DSHW	FSOLAR
1	3	875810.0	15947.2	357753.3	766250.0	109559.6	93612.5	0.573

1	4	15	15	-22.63	55.0

HOUR	SH	ST	EFF	TA	TS	TSNEW	SHITT	HHW	TISP	AUX	AUX2	TOTHW	TOTAL	FSOL
1	0	0.0	0.0	40	119.2	117.2	0.	0.	19962.	0.	11288.	0.	31250.	0.639
2	0	0.0	0.0	39	117.2	115.3	0.	0.	18589.	0.	13911.	0.	32500.	0.572

231

Figure 7.19 Hourly printout of solar space heating system (with direct heating and recirculation allowed).

PROGRAM TO EVALUATE PERFORMANCE OF A SOLAR SPACE HEATING AND DOMESTIC HOT WATER SYSTEM

```
        LATITUDE  =  33.C              TILT =  33.0
COLLECTOR AREA  =  600.0 SQ.FT.
STORAGE VOLUME  =    80.C GALLONS
MAIN STORAGE  =  1200.0 GALLONS
   HOUSE AREA = 2500.0 SQ FT
```

ALLOWS DIRECT HEATING OF HOUSE FROM COLLECTORS AND RECIRCULATION OF PLENUM WATER TO AUX HEATER

    1   1   12   12   -22.92   55.0

HOUR	SH	ST	FFF	TA	TS	TSNEW	SHITT	HHW	TISP	AUX	AUX2	TOTHW	TOTAL	FSOL
1	0	0.0	0.0	40	55.0	55.0	0.	0.	0.	0.	31250.	0.	31250.	0.0
2	0	0.0	0.0	39	55.0	55.0	0.	0.	0.	0.	32500.	0.	32500.	0.0
3	0	0.0	0.0	38	55.0	55.0	0.	0.	0.	0.	33750.	0.	33750.	0.0
4	0	0.0	0.0	37	55.0	55.0	0.	0.	0.	0.	35000.	0.	35000.	0.0
5	0	0.0	0.0	36	55.0	55.0	0.	0.	0.	0.	36250.	0.	36250.	0.0
6	C	0.0	0.0	35	55.0	55.C	0.	0.	C.	0.	37500.	0.	37500.	0.0
7	0	0.C	0.0	35	55.0	55.0	0.	0.	0.	996C.	37500.	9960.	47460.	0.0
8	0	0.0	0.0	33	55.0	55.0	0.	0.	0.	9960.	40000.	9960.	49960.	0.0
9	3	0.0	0.0	34	55.0	55.0	0.	0.	0.	4980.	38750.	4980.	43730.	0.0
10	40	72.9	0.0	38	55.0	55.0	0.	0.	0.	4980.	33750.	4980.	38730.	0.0
11	96	162.9	0.583	42	55.0	60.7	56993.	0.	0.	4980.	28750.	4980.	33730.	0.0
12	140	230.9	0.598	46	60.7	65.0	82787.	380.	0.	4600.	23750.	4980.	28730.	0.013
13	155	255.6	0.581	48	69.0	77.9	89121.	929.	0.	4051.	21250.	4980.	26230.	0.035
14	176	298.6	0.571	50	77.9	88.0	102319.	1517.	0.	3463.	18750.	4980.	23730.	0.064
15	117	213.4	0.507	53	88.0	94.3	64967.	2189.	0.	2791.	15000.	4980.	19980.	0.110
16	100	217.2	0.480	52	94.3	100.3	62545.	2608.	0.	2372.	16250.	4980.	21230.	0.123
17	63	189.0	0.420	51	100.3	104.5	47659.	6015.	170.	3945.	17330.	9960.	27460.	0.225
18	4	0.0	0.0	50	104.5	103.5	0.	6568.	2785.	3392.	15965.	9960.	28710.	0.326
19	0	0.0	0.0	46	103.5	102.9	0.	3222.	2784.	1758.	20966.	4980.	28730.	0.209
20	0	0.0	0.0	44	102.9	102.3	0.	3181.	2550.	1799.	23700.	4980.	31230.	0.184
21	0	0.0	0.0	42	102.3	101.8	0.	3143.	2241.	1837.	26509.	4980.	33730.	0.160
22	C	0.0	0.0	41	101.8	101.3	0.	3107.	1798.	1873.	28232.	4980.	34980.	0.140
23	0	0.0	0.0	40	101.3	100.9	0.	3075.	1360.	1905.	29890.	4980.	36230.	0.122
24	0	0.0	0.0	39	100.9	100.5	0.	3045.	932.	1935.	31568.	4980.	37480.	0.106

MONTH	DAY	DTOT	DAUX	DAUX2	DSH	DHW	DSHW	FSOLAR
1	1	798310.0	70580.2	674129.4	688750.0	109559.6	38979.7	0.067

    1   2   13   13   -22.83   55.0

HOUR	SH	ST	EFF	TA	TS	TSNEW	SHITT	HHW	TISP	AUX	AUX2	TOTHW	TOTAL	FSOL
1	0	0.0	0.0	40	100.5	100.4	0.	0.	480.	0.	30770.	0.	31250.	0.015
2	0	0.0	0.0	40	100.4	100.4	0.	0.	430.	0.	30820.	0.	31250.	0.014
3	0	0.0	0.0	39	100.4	100.3	0.	0.	400.	0.	32100.	0.	32500.	0.012
4	0	C.0	0.0	39	100.3	100.3	0.	0.	357.	0.	32143.	0.	32500.	0.011
5	C	0.0	0.0	38	100.3	100.3	0.	0.	330.	C.	33420.	0.	33750.	0.010
6	0	0.0	0.0	37	100.3	100.2	0.	0.	304.	0.	34696.	0.	35000.	0.009
7	0	0.0	0.0	37	100.2	99.6	0.	6007.	268.	5953.	34732.	9960.	44960.	0.140
8	0	0.0	0.0	36	99.6	99.0	0.	5923.	0.	4037.	36250.	9960.	46210.	0.128
9	22	0.0	0.0	35	99.0	98.7	0.	2922.	0.	2058.	37500.	4980.	42480.	0.069
10	86	156.5	0.282	37	101.8	103.9	26506.	3107.	2084.	1873.	32916.	4980.	39980.	0.130
11	141	238.8	0.410	39	103.9	109.1	58784.	3249.	4253.	1731.	28247.	4980.	37489.	0.200
12	180	296.3	0.448	41	109.1	115.8	79705.	3591.	9075.	1389.	20925.	4980.	34980.	0.362
13	199	327.6	0.455	43	115.8	122.9	89443.	4038.	14489.	942.	13011.	4980.	32480.	0.570
14	193	326.9	0.443	46	122.9	129.4	86913.	4510.	18150.	470.	5600.	4980.	28730.	0.789
15	172	312.9	0.418	47	129.4	134.6	78498.	4939.	22033.	41.	467.	4980.	27480.	0.981
16	132	285.7	0.382	48	134.6	138.5	65547.	4980.	21250.	0.	0.	4980.	26230.	1.000
17	73	219.0	0.283	48	138.5	139.1	37198.	9960.	21250.	0.	0.	9960.	31210.	1.000
18	6	0.C	0.0	48	139.1	136.0	0.	9960.	21250.	0.	0.	9960.	31210.	1.000
19	0	0.0	0.0	45	136.0	133.0	0.	4980.	25000.	0.	0.	4980.	29980.	1.000
20	0	0.0	0.0	43	133.0	129.7	0.	4980.	27500.	C.	0.	4980.	32480.	1.000
21	0	0.0	0.0	42	129.7	126.3	0.	4960.	28457.	20.	293.	4980.	33730.	0.991
22	C	0.0	0.0	40	126.3	123.1	0.	4737.	27437.	243.	3813.	4980.	36230.	0.888
23	0	0.0	0.0	38	123.1	120.0	0.	4522.	25998.	456.	7752.	4980.	38730.	0.788
24	0	0.0	0.0	36	120.0	117.2	0.	4319.	24221.	661.	12029.	4980.	41230.	0.692

MONTH	DAY	DTOT	DAUX	DAUX2	DSH	DHW	DSHW	FSOLAR
1	2	832060.0	16547.2	370522.2	722500.0	109559.6	93012.5	0.535

```
          1      3     14     14    -22.73     55.0

HOUR   SH     ST     EFF     TA     TS    TSNEW    SHITT    HHW      TISP     AUX      AUX2    TOTHW    TOTAL    FSOL
   1    0    0.0    0.0     34  117.2  115.3      0.       0.    22190.      0.    16560.       0.   38750.   0.573
   2    0    0.0    0.0     32  115.0  112.9      0.       0.    20558.      0.    20692.       0.   41250.   0.498
   3    0    0.0    0.0     31  112.9  111.1      0.       0.    18257.      0.    24243.       0.   42500.   0.430
   4    0    0.0    0.0     30  111.1  109.4      0.       0.    16121.      0.    27629.       0.   43750.   0.368
   5    0    0.0    0.0     30  109.4  108.1      0.       0.    13760.      0.    29990.       0.   43750.   0.315
   6    0    0.0    0.0     30  108.1  106.9      0.       0.    11746.      0.    32004.       0.   43750.   0.268
   7    0    0.0    0.0     30  106.9  105.2      0.    6889.    10026.   3071.    33724.    9960.   53710.   0.315
   8    0    0.0    0.0     32  105.2  103.8      0.    6663.     7118.   3297.    34132.    9960.   51210.   0.269
   9   24    0.0    0.0     31  103.8  102.9      0.    3240.     5373.   1740.    37127.    4980.   47480.   0.181
  10   90  163.3  0.270     33  102.9  104.9   26428.   3182.     3904.   1798.    36096.    4980.   44980.   0.158
  11  146  246.8  0.411     38  104.9  110.1   60867.   3311.     5479.   1669.    28271.    4980.   38730.   0.227
  12  186  305.7  0.460     44  110.1  117.3   84435.   3659.     8836.   1321.    17414.    4980.   31230.   0.400
  13  205  336.9  0.464     46  117.3  125.0   93886.   4138.    13713.    842.    10037.    4980.   28730.   0.621
  14  202  341.4  0.449     47  125.0  131.9   92078.   4645.    18717.    335.     3783.    4980.   27480.   0.850
  15  177  321.2  0.423     49  131.9  137.5   81497.   4980.    20000.      0.        0.    4980.   24980.   1.000
  16  133  286.6  0.380     50  137.5  141.7   65430.   4980.    18750.      0.        0.    4980.   23730.   1.000
  17   72  216.0  0.277     51  141.7  142.6   35901.   9960.    17500.      0.        0.    9960.   27460.   1.000
  18    6    0.0    0.0     50  142.6  139.7      0.    9960.    18750.      0.        0.    9960.   28710.   1.000
  19    0    0.0    0.0     48  139.7  137.0      0.    4980.    21250.      0.        0.    4980.   26230.   1.000
  20    0    0.0    0.0     45  137.0  134.3      0.    4980.    25000.      0.        0.    4980.   29980.   1.000
  21    0    0.0    0.0     43  134.0  130.8      0.    4980.    27500.      0.        0.    4980.   32480.   1.000
  22    0    0.0    0.0     42  130.8  127.4      0.    4980.    28750.      0.        0.    4980.   33730.   1.000
  23    0    0.0    0.0     41  127.4  124.2      0.    4807.    27390.    173.     2610.    4980.   34980.   0.920
  24    0    0.0    0.0     40  124.2  121.2      0.    4592.    25164.    368.     6086.    4980.   36230.   0.821

MONTH   DAY      DTOT         DAUX        DAUX2         DSH          DHW         DSHW     FSOLAR
   1      3   875810.0     14633.4     363398.5     766250.0     109559.6     94926.2     0.572


          1      4     15     15    -22.63     55.0

HOUR   SH     ST     EFF     TA     TS    TSNEW    SHITT    HHW      TISP     AUX      AUX2    TOTHW    TOTAL    FSOL
   1    0    0.0    0.0     40  121.2  119.0      0.       0.    22052.      0.     9198.       0.   31250.   0.706
   2    0    0.0    0.0     39  119.0  116.9      0.       0.    20535.      0.    11965.       0.   32500.   0.632
```

Figure 7.19 (Continued)

by a conventional heater and the water exiting the heating coil is fed back to storage. This condition is illustrated in Figure 7.18, where the storage temperature rises all morning due to large heat additions in (AUX2). In the recirculation mode, illustrated in Figure 7.19, the storage remains at 55°F all morning as water from the space heat coil at 100°F is recirculated to the conventional heater and the water exiting the heating coil is fed back to storage. This condition is illustrated in Figure 7.19, where the storage remains at 55°F (286°K) all morning as water from the space heat coil at 100°F (311°K) is recirculated to the conventional heater to be reheated to 130°F (328°K). No change in storage takes place until the sun comes up and begins to heat storage. At the end of the first day, the auxiliary heat used in conventional operation is about 950,000 Btu (1.00×10^9 J), whereas with recirculation it is about 750,000 Btu (0.789×10^9 J).

The operation of the direct heat mode is illustrated on January 3 in Figure 7.19. At 11 am, collector exit water is still below 130°F, and the system operates conventionally with water drawn from storage at 103°F to feed to the in-line conventional heater before introduction to the space heat coil. At 12 noon, however, collector exit water reaches 138.9°F, which

TABLE 7.5 Results of runs with a house area of 2500 ft², a latitude of 33°, and a collector tilt of 33°

Lat.	Tilt	House Area (ft²)	Coll. Area (ft²)	Stor. Vol. (gal)	Mode	Total Heat Demand (Btu)	Heat Provided by Solar (Btu)	Fraction Solar	Estim. Cost ($)	Btu/$	Fraction Solar Sp. Ht.	Fraction Solar Dom. H.W.
33	33	2500	300	600	1	0.107E9	0.521E8	0.488	5725	0.910E4	0.290	0.882
33	33	2500	300	600	2	0.107E9	0.517E8	0.485	5725	0.904E4	0.282	0.888
33	33	2500	300	600	3	0.107E9	0.516E8	0.484	5725	0.902E4	0.281	0.888
33	33	2500	300	600	4	0.107E9	0.513E8	0.480	5725	0.896E4	0.272	0.894
33	33	2500	300	1200	1	0.107E9	0.536E8	0.502	6232	0.859E4	0.312	0.880
33	33	2500	300	1200	2	0.107E9	0.534E8	0.500	6232	0.856E4	0.307	0.884
33	33	2500	300	1200	3	0.107E9	0.532E8	0.498	6232	0.853E4	0.303	0.887
33	33	2500	300	1200	4	0.107E9	0.530E8	0.496	6232	0.850E4	0.298	0.891
33	33	2500	600	1200	1	0.107E9	0.730E8	0.684	9232	0.790E4	0.556	0.937
33	33	2500	600	1200	2	0.107E9	0.728E8	0.682	9232	0.788E4	0.553	0.939
33	33	2500	600	1200	3	0.107E9	0.723E8	0.678	9232	0.783E4	0.546	0.941
33	33	2500	600	1200	4	0.107E9	0.721E8	0.676	9232	0.781E4	0.542	0.942
33	33	2500	900	1200	1	0.107E9	0.836E8	0.783	12,232	0.683E4	0.694	0.960
33	33	2500	900	1200	2	0.107E9	0.833E8	0.781	12,232	0.681E4	0.690	0.961
33	33	2500	900	1200	3	0.107E9	0.828E8	0.776	12,232	0.677E4	0.683	0.962
33	33	2500	900	1200	4	0.107E9	0.826E8	0.774	12,232	0.675E4	0.679	0.963
33	33	2500	900	1800	1	0.107E9	0.865E8	0.810	12,621	0.685E4	0.733	0.964
33	33	2500	900	1800	2	0.107E9	0.863E8	0.809	12,621	0.684E4	0.730	0.965
33	33	2500	900	1800	3	0.107E9	0.857E8	0.803	12,621	0.679E4	0.721	0.966
33	33	2500	900	1800	4	0.107E9	0.856E8	0.802	12,621	0.678E4	0.719	0.967
33	33	2500	600	1800	1	0.107E9	0.745E8	0.698	9621	0.774E4	0.577	0.938
33	33	2500	600	1800	2	0.107E9	0.743E8	0.696	9621	0.772E4	0.574	0.939
33	33	2500	600	1800	3	0.107E9	0.739E8	0.692	9621	0.768E4	0.566	0.942
33	33	2500	600	1800	4	0.107E9	0.737E8	0.691	9621	0.766E4	0.563	0.944

33	33	2500	300	300	1	0.107E9	0.494E8	0.463	5366	0.921E4	0.257	0.873
33	33	2500	300	300	2	0.107E9	0.487E8	0.457	5366	0.908E4	0.241	0.886
33	33	2500	300	300	3	0.107E9	0.489E8	0.458	5366	0.911E4	0.247	0.877
33	33	2500	300	300	4	0.107E9	0.481E8	0.451	5366	0.897E4	0.230	0.890
33	33	2500	600	600	1	0.107E9	0.687E8	0.644	8724	0.788E4	0.500	0.930
33	33	2500	600	600	2	0.107E9	0.683E8	0.640	8724	0.783E4	0.493	0.934
33	33	2500	600	600	3	0.107E9	0.680E8	0.637	8724	0.779E4	0.488	0.933
33	33	2500	600	600	4	0.107E9	0.676E8	0.633	8724	0.774E4	0.481	0.936
33	33	2500	900	900	1	0.107E9	0.810E8	0.759	12,000	0.675E4	0.660	0.956
33	33	2500	900	900	2	0.107E9	0.807E8	0.756	12,000	0.673E4	0.655	0.958
33	33	2500	900	900	3	0.107E9	0.803E8	0.752	12,000	0.669E4	0.649	0.957
33	33	2500	900	900	4	0.107E9	0.800E8	0.750	12,000	0.667E4	0.644	0.959
33	33	2500	900	2700	1	0.107E9	0.887E8	0.831	13,098	0.677E4	0.763	0.966
33	33	2500	900	2700	2	0.107E9	0.885E8	0.829	13,098	0.676E4	0.760	0.967
33	33	2500	900	2700	3	0.107E9	0.880E8	0.824	13,098	0.672E4	0.752	0.969
33	33	2500	900	2700	4	0.107E9	0.878E8	0.822	13,098	0.670E4	0.748	0.970
33	33	2500	300	900	4	0.107E9	0.524E8	0.491	6000	0.874E4	0.289	0.893
33	33	2500	100	100	4	0.107E9	0.259E8	0.243	3000	0.863E4	0.000	0.740
33	33	2500	100	200	4	0.107E9	0.274E8	0.256	3207	0.853E4	0.006	0.754
33	33	2500	100	300	4	0.107E9	0.278E8	0.261	3366	0.827E4	0.014	0.752
33	33	2500	200	200	4	0.107E9	0.390E8	0.365	4207	0.926E4	0.119	0.855
33	33	2500	200	400	4	0.107E9	0.413E8	0.387	4500	0.917E4	0.149	0.859
33	33	2500	200	600	4	0.107E9	0.421E8	0.394	4725	0.890E4	0.162	0.857
33	33	2500	400	400	4	0.107E9	0.556E8	0.521	6500	0.855E4	0.325	0.911
33	33	2500	400	800	4	0.107E9	0.595E8	0.557	6914	0.860E4	0.377	0.915
33	33	2500	400	1200	4	0.107E9	0.609E8	0.570	7232	0.841E4	0.397	0.915
33	33	2500	500	500	4	0.107E9	0.621E8	0.582	7618	0.815E4	0.409	0.925
33	33	2500	500	1000	4	0.107E9	0.663E8	0.622	8081	0.821E4	0.466	0.930
33	33	2500	500	1500	4	0.107E9	0.678E8	0.635	8436	0.804E4	0.487	0.931
33	33	2500	750	750	4	0.107E9	0.744E8	0.697	10,369	0.718E4	0.571	0.949
33	33	2500	750	1500	4	0.107E9	0.795E8	0.745	10,936	0.727E4	0.638	0.956
33	33	2500	750	2250	4	0.107E9	0.813E8	0.762	11,372	0.715E4	0.664	0.958

signals the system to bypass storage and feed 138.9°F water directly to the space heat coil. As a result, (AUX2) goes to zero at 12 noon even though storage is still below 130°F.

7.5.2 Yearly Results

The yearly results for a series of runs with various collector areas, storage volumes, and collector tilt angles are listed in Tables 7.5 and 7.6. In all cases, the house area is chosen as 2500 ft² (232 m²) and the latitude is 33°. The solar intensities and ambient temperatures are chosen as typical of North Central Texas. The estimated cost of an installation should be interpreted as a very rough guess. The first result to be observed is that, over the course of a year, direct heating from the solar collector heat exchanger does not materially improve the performance. Although direct heating reduces auxiliary make-up heat at any hour, it puts less heat into storage and therefore additional auxiliary make-up heat is required at subsequent hours. The net result at the end of a year is that the improvement due to direct heating is typically less than 1%. The recirculation option is generally even less effective than direct heating in reducing make-up heat over the course of a year. These conclusions are based on the assumption of no heat losses from storage. Heat losses from storage might tend to increase the utility of the direct heat and recirculation options because these operational modes put less heat into storage under certain conditions.

7.5.3 Effect of Heat Losses

In order to test the effect of heat loss, a heat loss term was inserted in the program for a few runs. The heat loss was based on the assumption of a heat loss coefficient from storage of 0.05 Btu/hr per ft² of surface area of storage per °F storage temperature elevation above ground temperature for a tank buried in the earth. The ground temperature is assumed to vary through the year in the same way as the tap water temperature. Note that if the storage tank were located inside the residence then heat losses from storage would leak into the house and would not be "lost" at all. Thus, the runs without heat loss terms would be characteristic of interior storage. For storage buried in the ground, the estimated rate of heat loss is $0.05 \times$ (storage surface area $-$ ft²) $\times$ (TA $-$ TT), where TS is the storage temperature, and TT is the tap water temperature in °F. The storage surface area is estimated by assuming a cylindrical tank with a length LT three times the diameter DT.

TABLE 7.6 Results of runs with house area of 2500 ft², a latitude of 33°, and a variable tilt

Lat.	Tilt	House Area (ft²)	Coll. Area (ft²)	Stor. Vol. (gal)	Mode	Total Heat Demand (Btu)	Heat Provided by Solar (Btu)	Fraction Solar	Estim. Cost ($)	Btu ($)	Fraction Solar Sp. Ht.	Fraction Solar Dom. H.W.
33	45	2500	300	600	1	0.107E9	0.536E8	0.502	5725	0.937E4	0.310	0.884
33	45	2500	300	600	2	0.107E9	0.533E8	0.499	5725	0.931E4	0.303	0.890
33	45	2500	300	600	3	0.107E9	0.531E8	0.498	5725	0.924E4	0.301	0.890
33	45	2500	300	600	4	0.107E9	0.528E8	0.495	5725	0.922E4	0.293	0.896
33	45	2500	600	1200	1	0.107E9	0.755E8	0.707	9232	0.817E4	0.589	0.941
33	45	2500	600	1200	2	0.107E9	0.753E8	0.705	9232	0.815E4	0.586	0.943
33	45	2500	600	1200	3	0.107E9	0.747E8	0.700	9232	0.809E4	0.577	0.945
33	45	2500	600	1200	4	0.107E9	0.745E8	0.698	9232	0.807E4	0.574	0.946
33	45	2500	900	1800	1	0.107E9	0.893E8	0.837	12,621	0.707E4	0.770	0.969
33	45	2500	900	1800	2	0.107E9	0.891E8	0.835	12,621	0.706E4	0.768	0.969
33	45	2500	900	1800	3	0.107E9	0.885E8	0.829	12,621	0.701E4	0.758	0.970
33	45	2500	900	1800	4	0.107E9	0.883E8	0.828	12,621	0.700E4	0.756	0.971
33	55	2500	300	600	1	0.107E9	0.534E8	0.500	5725	0.932E4	0.310	0.879
33	55	2500	300	600	2	0.107E9	0.530E8	0.497	5725	0.927E4	0.302	0.885
33	55	2500	300	600	3	0.107E9	0.529E8	0.496	5725	0.924E4	0.300	0.885
33	55	2500	300	600	4	0.107E9	0.526E8	0.492	5725	0.918E4	0.292	0.890
33	55	2500	600	1200	1	0.107E9	0.758E8	0.710	9232	0.821E4	0.593	0.941
33	55	2500	600	1200	2	0.107E9	0.756E8	0.708	9232	0.819E4	0.590	0.943
33	55	2500	600	1200	3	0.107E9	0.750E8	0.703	9232	0.813E4	0.582	0.945
33	55	2500	600	1200	4	0.107E9	0.748E8	0.701	9232	0.811E4	0.578	0.946
33	55	2500	900	1800	1	0.107E9	0.898E8	0.842	12,621	0.712E4	0.778	0.969
33	55	2500	900	1800	2	0.107E9	0.897E8	0.840	12,621	0.710E4	0.775	0.970
33	55	2500	900	1800	3	0.107E9	0.891E8	0.834	12,621	0.706E4	0.766	0.971
33	55	2500	900	1800	4	0.107E9	0.889E8	0.833	12,621	0.704E4	0.763	0.972

The volume of a tank is $VT = \pi(DT)^2(LT)/4$, and the surface area is

$$AT = \pi\left[(\dot{DT})(LT) + \frac{(DT)^2}{2}\right]$$

If $LT = 3(DT)$, it follows that

$$VT = \left(\frac{3\pi}{4}\right)(DT)^3$$

$$AT = \left(\frac{7\pi}{2}\right)(DT)^2$$

and, therefore

$$AT = \left(\frac{7\pi}{2}\right)\left(\frac{4}{3\pi}\right)^{2/3}(VT)^{2/3}$$

Runs were made with the heat loss term to determine the magnitude of the heat loss and the effect of the heat loss term on differences in performance in modes 1, 2, 3, and 4. A comparison of runs with and without the heat loss term is shown in Table 7.7. It can be seen that roughly 8% of the solar heat collected is lost through R20 insulation around a tank buried in the ground.

TABLE 7.7   Effect of a heat loss term due to $R = 20$ insulation on a tank buried in the ground (mode = 4)

Collector Area (ft^2)	Storage Volume (gal)	Fraction Solar	
		No Heat Loss	With Heat Loss
300	600	0.480	0.439
500	1000	0.622	0.573
600	1200	0.698	0.625
750	1500	0.745	0.691

Clearly, heat losses from such tanks can be significant. When a comparison of modes 1, 2, 3, and 4 is made for systems incorporating the heat loss characteristic of storage buried in the ground, the results are as shown in Table 7.8. Even when heat loss is present, recirculation and direct heating do not improve performance very much.

7.5.4 Plots of Results

The results of the runs shown in Tables 7.5 and 7.6 can be plotted in various ways. The overall fraction of energy required for space heat and hot water that is supplied by solar energy is plotted vs. collector area in Figure

TABLE 7.8   Effect of direct heating and recirculation on
performance when storage is buried in the
ground with R̊20 insulation

Collector Area (ft²)	Storage Volume (gal)	Operating Mode	Fraction Solar
300	600	1	0.450
		2	0.443
		3	0.445
		4	0.439
500	1000	1	0.584
		2	0.579
		3	0.577
		4	0.573
750	1000	1	0.702
		2	0.699
		3	0.694
		4	0.691

7.20. Three curves are shown: for the storage volume (in gal) numerically equal to the collector area (in ft²) and for storage volumes of two times and three times the collector area. It can be seen that the larger storage volumes produce slightly higher performance, with a much smaller improvement in going from $V = 2A$ to $V = 3A$ than in going from $V = A$ to $V = 2A$. Since

Figure 7.20 Fraction of required total energy supplied by solar energy in mode 4 with V in gal and A in ft².

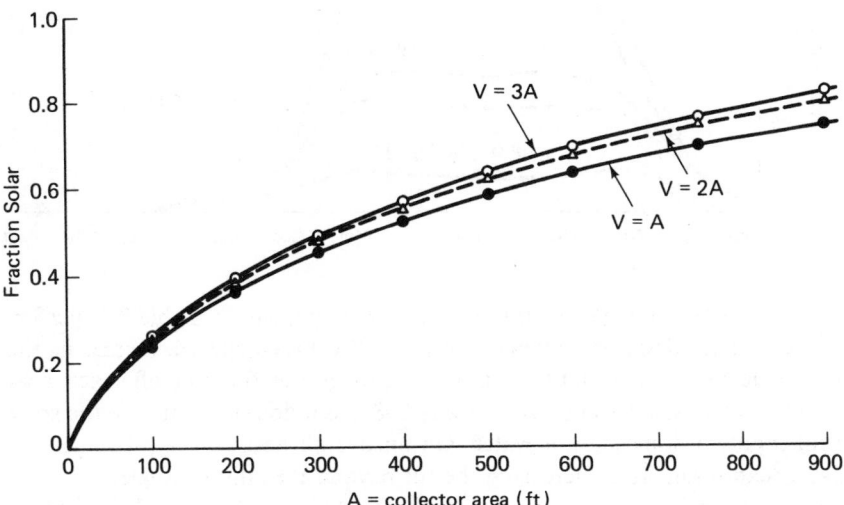

increasing the volume of storage increases the cost, it appears that optimum storage size is probably in the range $V = A$ to $2A$. The shape of the curves in Figure 7.20 are mainly determined by the collector efficiencies at prevailing storage temperatures. For small collector areas, the storage tank tends to remain relatively cool, and the collector efficiency (on average) is very high. As more collector area is added, the typical storage temperatures encountered go up, and the collector efficiencies go down. Thus, the slope of the curve of fraction solar vs. collector area tends to decrease as the collector area is increased. The returns from increasing the collector area eventually diminish until it is not worth adding additional collectors.

Some illustrations of the monthly variation of solar heat supplied are shown in Figures 7.21 and 7.22. In Figure 7.21, the heat load required by the house on a monthly basis is plotted along with the monthly solar heat inputs to space heating for several designs.

Figure 7.21 Monthly space heat demand compared with solar heat inputs of several designs. (House area = 2500 ft², $L = T = 33°$).

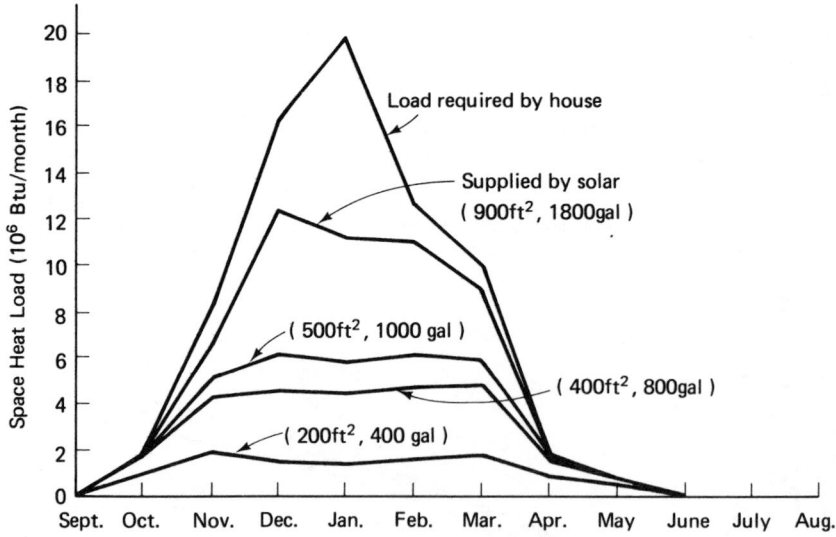

The effect of collector tilt angle is not very great. In Table 7.9, performance at fixed design parameters but variable tilt is compared. Increasing the tilt angle beyond the latitude angle improves the fraction of space heat supplied from solar energy by typically 7–8 % but does not improve the solar heat input to domestic hot water. On an overall basis, an improvement of 2–3 % can result from increasing the tilt beyond the latitude angle.

A quantity given in the tables is the "value," defined as the annual number of Btus provided by solar energy divided by the estimated cost in

Figure 7.22 Monthly results of solar space heating system.

```
PROGRAM TO EVALUATE PERFORMANCE OF A SOLAR SPACE HEATING AND DOMESTIC HOT WATER SYSTEM

        LATITUDE =   33.0              TILT =   33.0
  COLLECTOR AREA =  300.0 SQ.FT.
  STORAGE VOLUME =   80.0 GALLONS
    MAIN STORAGE =  600.0 GALLONS
      HOUSE AREA = 2500.0 SQ FT

ALLOWS DIRECT HEATING OF HOUSE FROM COLLECTORS AND  RECIRCULATION OF PLENUM WATER TO AUX HEATER
```

MONTH	TOTAL DEMAND	AUX HEAT HOT WTR	AUX HEAT SPACE HT	SPACE HT DEMAND	SPACE HT (SOLAR)	HOT WATER DEMAND	HOT WATER (SOLAR)	FSOLAR
1	22987552.0	1120052.0	16249710.0	19591216.0	3341506.0	3396334.0	2276292.0	0.244

MONTH	TOTAL DEMAND	AUX HEAT HOT WTR	AUX HEAT SPACE HT	SPACE HT DEMAND	SPACE HT (SOLAR)	HOT WATER DEMAND	HOT WATER (SOLAR)	FSOLAR
2	15862670.0	724768.7	9419245.0	12795000.0	3375755.0	3067657.0	2342899.0	0.361

MONTH	TOTAL DEMAND	AUX HEAT HOT WTR	AUX HEAT SPACE HT	SPACE HT DEMAND	SPACE HT (SOLAR)	HOT WATER DEMAND	HOT WATER (SOLAR)	FSOLAR
3	13310613.0	621951.9	6490113.0	10101250.0	3611137.0	3209356.0	2587409.0	0.466

MONTH	TOTAL DEMAND	AUX HEAT HOT WTR	AUX HEAT SPACE HT	SPACE HT DEMAND	SPACE HT (SOLAR)	HOT WATER DEMAND	HOT WATER (SOLAR)	FSOLAR
4	4713691.0	171945.7	585254.3	1777501.0	1192246.0	2936152.0	2764246.0	0.839

MONTH	TOTAL DEMAND	AUX HEAT HOT WTR	AUX HEAT SPACE HT	SPACE HT DEMAND	SPACE HT (SOLAR)	HOT WATER DEMAND	HOT WATER (SOLAR)	FSOLAR
5	3690313.0	36471.3	25375.9	656251.4	630875.5	3034065.0	2997594.0	0.983

MONTH	TOTAL DEMAND	AUX HEAT HOT WTR	AUX HEAT SPACE HT	SPACE HT DEMAND	SPACE HT (SOLAR)	HOT WATER DEMAND	HOT WATER (SOLAR)	FSOLAR
6	2639643.0	0.0	0.0	0.0	0.0	2639643.0	2639643.0	1.000

MONTH	TOTAL DEMAND	AUX HEAT HOT WTR	AUX HEAT SPACE HT	SPACE HT DEMAND	SPACE HT (SOLAR)	HOT WATER DEMAND	HOT WATER (SOLAR)	FSOLAR
7	2717064.0	0.0	0.0	0.0	0.0	2717064.0	2717064.0	1.000

MONTH	TOTAL DEMAND	AUX HEAT HOT WTR	AUX HEAT SPACE HT	SPACE HT DEMAND	SPACE HT (SOLAR)	HOT WATER DEMAND	HOT WATER (SOLAR)	FSOLAR
8	2717064.0	0.0	0.0	0.0	0.0	2717064.0	2717064.0	1.000

MONTH	TOTAL DEMAND	AUX HEAT HOT WTR	AUX HEAT SPACE HT	SPACE HT DEMAND	SPACE HT (SOLAR)	HOT WATER DEMAND	HOT WATER (SOLAR)	FSOLAR
9	2766917.0	0.0	0.0	137500.5	137500.5	2629417.0	2629417.0	1.000

MONTH	TOTAL DEMAND	AUX HEAT HOT WTR	AUX HEAT SPACE HT	SPACE HT DEMAND	SPACE HT (SOLAR)	HOT WATER DEMAND	HOT WATER (SOLAR)	FSOLAR
10	4803388.0	58017.2	337736.5	1800001.0	1462264.0	3003389.0	2945371.0	0.918

MONTH	TOTAL DEMAND	AUX HEAT HOT WTR	AUX HEAT SPACE HT	SPACE HT DEMAND	SPACE HT (SOLAR)	HOT WATER DEMAND	HOT WATER (SOLAR)	FSOLAR
11	11093493.0	564838.2	4610819.0	8028750.0	3417931.0	3064736.0	2499903.0	0.533

MONTH	TOTAL DEMAND	AUX HEAT HOT WTR	AUX HEAT SPACE HT	SPACE HT DEMAND	SPACE HT (SOLAR)	HOT WATER DEMAND	HOT WATER (SOLAR)	FSOLAR
12	19423024.0	910689.7	12716547.0	16136250.0	3419403.0	3286775.0	2376094.0	0.298

MONTH	TOTAL DEMAND	TOTAL SOLAR	FSOLAR TOTAL	SPACE HT DEMAND	SPACE HT SOLAR	FSOLAR SP HT	HOT WATER DEMAND	HOT WATER SOLAR	FSOLAR HOT WTR	HOT WATER AUX HEAT	SPACE AUX
12	0.107E+09	0.521E+08	0.488	0.710E+08	0.206E+08	0.290	0.357E+08	0.315E+08	0.882	0.421E+07	0.50

COLLECTOR AREA	MAIN STORAGE VOLUME	COST	SOLAR BTUS SUPPLIED	VALUE (BTU/DOLLAR)
300.	600.	5724.74	0.521E+08	0.910E+04

241

TABLE 7.9   Effect of variation of collector tilt angle on performance of solar
            heating system at $L = 33°$ (mode 4)

Collector Area (ft^2)	Storage Volume (gal)	Tilt Angle (degrees)	Fraction Solar		
			Overall	Sp. Ht.	Dom. H.W.
300	600	33	0.480	0.272	0.894
		45	0.495	0.293	0.896
		55	0.492	0.292	0.890
600	1200	33	0.676	0.542	0.942
		45	0.698	0.574	0.946
		55	0.701	0.578	0.946
900	1800	33	0.802	0.719	0.967
		45	0.828	0.756	0.971
		55	0.833	0.763	0.972

dollars. The design with the highest value involves the invested dollars working to produce the greatest number of Btu/yr per dollar. However, the design
with the highest value is not necessarily the one with the greatest desirability.
As one adds to the system size beyond the point of maximum value, dividends
from the investment continue to climb but at a slower rate. Thus, the optimum design point is usually much larger than the point producing the greatest
number of Btu's per dollar. A discussion of the economics of optimizing the
design is provided in Sec. 7.7. For the system under consideration here, the
plot of value vs. collector area is shown in Figure 7.23. The hourly results of
the space heating/hot water system may also be plotted as in Figure 7.24,
which illustrates the transient behavior of the system after startup.

7.6 Space Heating with Air Systems

7.6.1 General Description

In previous sections, solar space heating with a liquid as the heat
transfer fluid was considered. Solar space heating can also be accomplished
with air as the transfer fluid, using a rock-filled bin as the heat storage
reservoir. The operations of the air system for space heating are illustrated
in Figure 7.25. In Figure 7.25(a) air is blown through the collectors and then
into the hot side of a pebble bed. There is a large temperature drop across
the pebble bed, and one may expect that with 130°F air exiting the collector
the air that has passed through the storage bin will be around 70°F. As more
and more air is circulated in this manner, the demarcation region that separates the 130°F region from the 70°F region in the pebble bed gradually moves

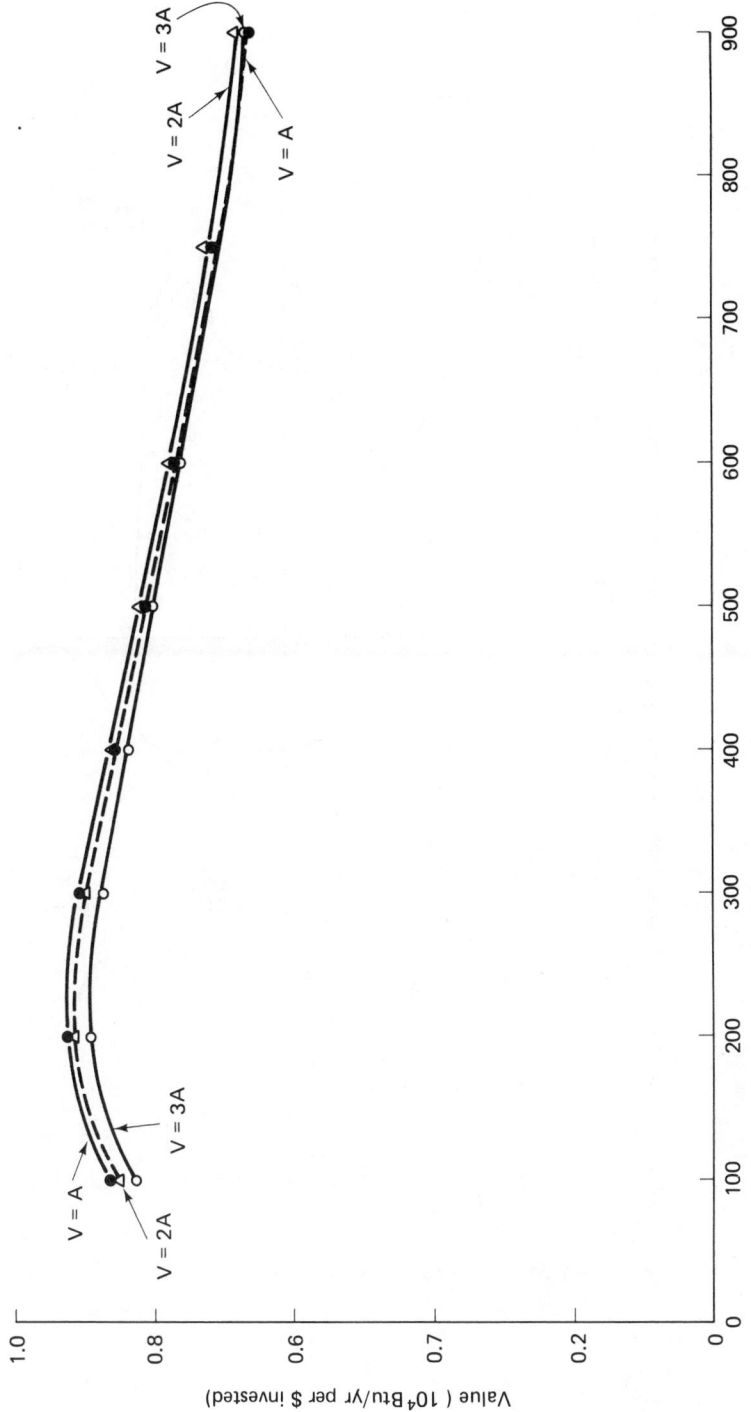

Figure 7.23 Btus per dollar for a solar space heat/domestic hot water system.

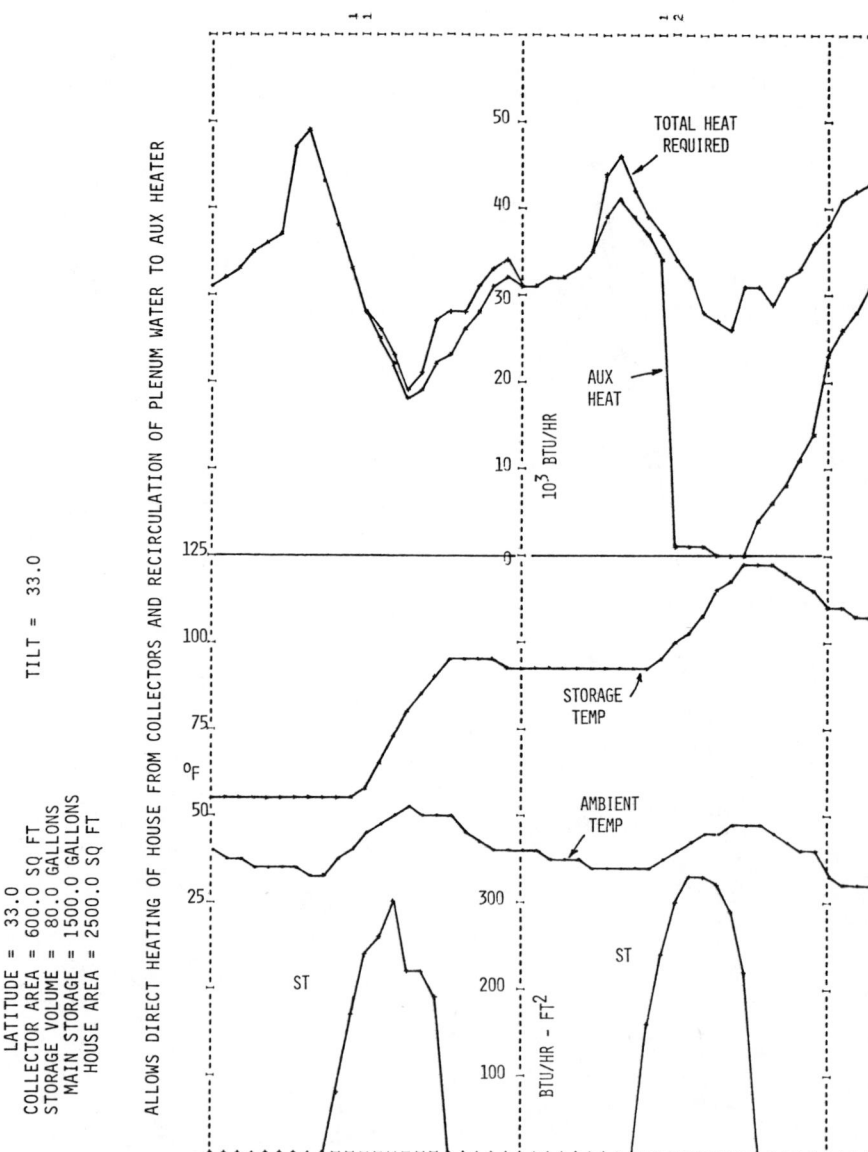

Figure 7.24 Plot of hourly variation of variables.

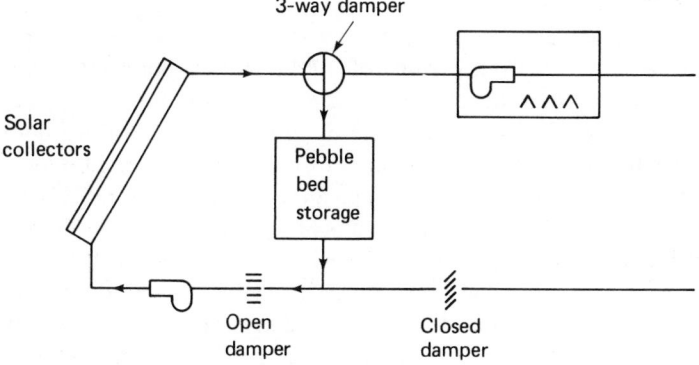

(a) Heating storage from collectors.

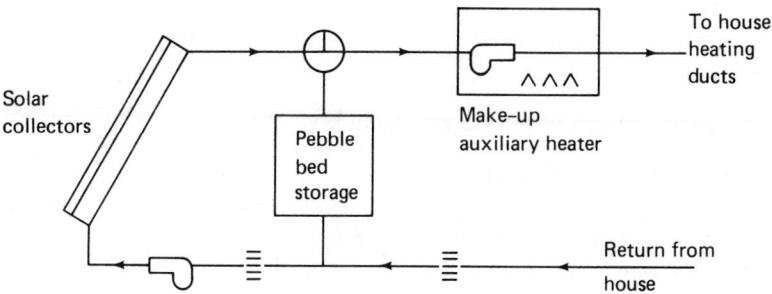

(b) Heating house from collectors

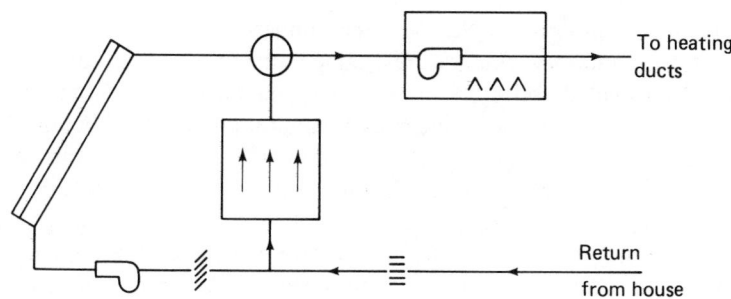

(c) Heating house from storage

Figure 7.25 Operations with a simple solar air heating system.

down. Eventually, when storage is "full of heat" it is all at $\sim 130°F$. In Figure 7.25(b), the house is heated directly from the collectors. The air return from the house is at $\sim 130°F$. During periods of bright sunlight, the system will tend to cycle between modes (a) and (b) because there is more solar energy available than is required to heat the house in an hour. Mode (c) in Figure 7.25 involves heating the house from storage. Return air from the house at $\sim 70°F$ is blown into the cold side of storage and warm air at $\sim 130°F$ exits the storage unit and goes to the house. As this process continues, the demarcation region between the 70°F and 130°F regions move up until the entire pebble bed is at 70°F and auxiliary heat must be used. Actually, there is no demarcation line but only a broad transition zone between the high and low temperatures. The full 130°F exit temperature is not achieved unless storage is nearly "full of heat."

The principal advantages of air systems are:

1. No corrosion problems.
2. No freeze protection required.
3. Integration with forced air heating is simple, and there are no heat exchangers required.

The disadvantages of air systems include:

1. High cost of air circulation.
2. Ductwork and dampers are expensive and occupy considerable space.
3. Collectors may be less efficient.
4. Domestic hot water production is less efficient.

7.6.2 Comparison of Air and Water Systems

It was mentioned in Sec. 4.3 that under identical conditions (same fluid entrance and exit temperatures, same ambient temperature, same solar intensity) air-cooled collectors of the same general method of construction have slightly lower collection efficiencies than liquid-cooled collectors. Oonk, Löf, Shaw, and Cole-Appel[3] point out that, in actual operations, air-and water-cooled collectors are operated under rather different conditions. They claim that the nature of the remaining system components tends to favor the air system, and that when all factors are considered the air and liquid systems perform fairly equally. They indicate that it has been customary in many of the measurements of collector efficiency to plot efficiency vs. (T_{in} −

[3]R. L. Oonk G. O. G. Löf, L. E. Shaw, and B. E. Cole-Appel, *Proc. 1976 Conf. Amer. Sect. International Solar Energy Society*, Winnipeg, Canada, 1976.

$T_{amb})/I$, where T_{in} is the fluid inlet temperature, T_{amb} is the ambient temperature, and I is the intensity falling on the collector.

Some investigators correlate the efficiency with $(T_{in} - T_{amb})/I$ rather than $(T_{coll} - T_{amb})/I$, where T_{coll} is the average temperature of the collector plate. By varying the fluid flow rate, various temperature rises can be produced, and T_{coll} can vary over a considerable range for fixed T_{in}. Even if one is not equipped to measure T_{coll} (and it is doubtful that one should be reporting efficiencies without measuring T_{coll}), T_{coll} may still be estimated as being "slightly" higher than $(T_{in} + T_{out})/2$ for liquid-cooled collectors and "somewhat" higher than T_{out} for air-cooled collectors. The terms "slightly" and "somewhat" can be made more definite with a very short heat transfer analysis. The major point made by Oonk *et al.* is that a house requires forced air in an operating range of temperatures centered around 100°F, and that, in order to produce forced air at such a temperature, liquid and air systems operate quite differently. In a liquid system, taking into account temperature drops across heat exchangers, typical values are $T_{in} = 100°F$, $T_{out} = 120°F$, $T_{coll} = 130°F$. It requires 120°F liquid exiting the collector to deliver 100°F air to the residence due to the temperature drops in exchanging heat with storage and from storage to the forced-air heating system. Oonk *et al.* claim that in an air system the stratification resulting from blowing air through a pebble bed storage implies that unless storage is "full of heat" the air returned from storage to the collector inlet will be close to the room return air temperature, namely 70°F. While heating rooms directly from the collectors this is certainly the case, and when heating storage from the collectors it is at least partly true. The air collector then receives air at $T_{in} = 70°F$, and heats it to about $T_{out} = 100°F$ for circulation back to the house or storage. The average plate temperature will be about $T_{coll} = 130°F$ under these conditions. Note that in air collectors the average collector plate temperature is much higher than $(T_{in} + T_{out})/2$. It may be argued that, under conditions appropriate to actual operations for space heating, T_{coll} will be nearly the same whether the collector is air-cooled or liquid-cooled. If it is assumed that, at identical plate temperatures, air- and liquid-cooled collectors have essentially the same efficiency at the same plate temperature, it would be concluded that the two systems have about the same efficiency under typical operating conditions. Actually, even at the same plate temperatures, the air collector will be a little less efficient due to the greater heat loss rate (from moving air) from the plate to the inner glass.

Integrating the domestic hot water system into a solar air system is usually not very efficient. An air-to-water heat exchanger is used in the collector exit manifold duct, and dampers are provided so that this heat exchanger can be used all summer when the space heating system is turned off. It is usually not possible to do much more than partly preheat tap water before it enters a conventional hot water heater, even under the best of solar conditions. Use of a separate solar hot water heater in conjunction with a solar air space heater is probably the best approach.

Solar air heating systems are useful in conjunction with simple inexpensive cooling systems. For periods of the year when days are hot and nights are cool, one may blow exterior air through the pebble bed with a timer switch during the early hours before dawn to cool off the bed. Then, during the day this reservoir of "cold" can be used for space cooling. Rock beds may also be cooled by evaporative cooling of air.

A particularly interesting way to utilize air systems for space heating is to build the collectors vertically into the outer south-facing wall of a structure. The exterior glazing then becomes the outer wall of the building. Since air systems are manifestly not appropriate for space cooling (except via evaporative or night air cooling), there is no harm in the fact that the collection efficiency is very poor in the summer. Such systems are marketed with a perforated finned-aluminum first absorber plate mounted vertically behind double glazings. A second metal absorber plate is mounted about 2 in. behind the first plate; the sunlight that passes through the perforations heats the second plate. Air is circulated between the two warm plates for heat transfer. The storage consists of thin plastic trays of a eutectic salt mixture that melts around 90°F. This storage system absorbs and receives heat at 90°F and holds a considerable amount of heat per unit volume.

7.7 Economics of Solar Space Heating

The economics of various designs for solar space heating will be discussed in terms of the methodology of Chapter 5. It is assumed that scenario III in Table 5.2 can be used for future inflation and escalation rates. The same procedures could be used with any other assumed scenario. As in Sec. 6.7, it is assumed that money is borrowed for 20 years at 8%, with monthly payments of $8.37 per $1000 borrowed. Equations (6.7) and (6.8) may then be used for the net present value, and Equation (6.9) can be used to evaluate the pay-back period of any design. Calculations are carried out for five designs with storage volume (gal) equal to twice the collector area (ft²). The results are presented in Table 7.10. When the solar space heating system replaces natural gas the economics are very unfavorable, with almost no net present value and long pay-back periods. When electricity is displaced the net present value is high, and the pay-back period is reduced to about 7 years. There is no single optimum design, and any of the designs listed in Table 7.10 could be justified in replacing electricity.

TABLE 7.10 Net present value and pay-back period for selected solar space heating/hot water systems that displace electricity and gas in Dallas, Texas (total yearly heat requirement = 1.07×10^8 Btu)

Collector Area (ft^2)	Storage Volume (gal)	Fraction Solar			Estimated Cost ($)	Replacing Electricity		Replacing Natural Gas	
		Overall	Sp. Ht.	Hot Water		NPV ($)	PBP (years)	NPV ($)	PBP (years)
300	600	0.480	0.272	0.894	5725	22,270	6	2050	16
500	1000	0.622	0.466	0.930	8081	28,030	7	1830	17
600	1200	0.676	0.542	0.942	9238	29,900	7	1430	18
750	1500	0.745	0.638	0.956	10936	31,990	7	620	19
900	1800	0.802	0.719	0.967	12621	33,390	8	−390	21

Worked Examples

1. A house has 1600 ft² of exterior wall and 2500 ft² of roof and floor. Glass windows comprise 20% of the walls, the remainder being uninsulated brick/stud/sheetrock. The roof is R20, and the floor is R3. What percent improvement in heat loss coefficient would be produced by placing 3 in. insulation between the studs of the walls? Neglect infiltration.

   The heat loss coefficient is found by summing UA over all exterior surfaces:

$$
\begin{array}{lll}
\text{roof:} & \text{UA} = 2500 \times 1/20 & = 125 \\
\text{floor:} & \text{UA} = 2500 \times 1/3 \times 1/2 & = 416 \\
\text{walls:} & 1280 \times 1/2.7 & = 474 \\
\text{glass:} & \text{UA} = 320 \times 1/1 & = 320 \\
& & \overline{\text{total} = 1335}
\end{array}
$$

   The factor 1/2 was used in the floor, assuming that ground temperature is midway between interior and exterior temperatures. The heat loss coefficient is then

$$
U = \frac{UA}{2500} = 0.53 \text{ Btu/hr-°F per ft}^2 \text{ of floor area}
$$

   By raising the resistance of the walls to R13, it follows that the UA of the walls is reduced to $1280/13 = 100$, and the sum of the UA is reduced to 961. Thus, the heat loss coefficient is reduced to $961/2500 = 0.38$ Btu/hr-°F per ft² of floor, for a 28% reduction in heat loss.

2. A house has 1600 ft² of wall, 400 ft² of windows, 2000 ft² of roof, and 2000 ft² of floor. The insulation resistances are: ceiling, R20; walls, R10; floor, R5; windows, R1. (a) What is the rate of heat loss when the interior is 70°F and the ambient is 40°F? (b) At what rate should heat be supplied to the house to maintain it at 70°F? Neglect infiltration.

   (a) The sum of UA for all surfaces is

$$
\Sigma\, UA = \frac{1600}{10} + \frac{2000}{20} + \frac{2000}{5} + \frac{400}{1} = 1060
$$

$$
1060 \times (70 - 40) = 31{,}800 \text{ Btu/hr}
$$

   (b) The net rate of heat loss after taking account of heat gain from appliances, people, and other factors is $1060 \times (65 - 40) = 26{,}500$ Btu/hr. This is the required rate of heat supply.

3. For the house of Example 2, the homeowner is considering using double glazing of windows (cost = $10 per ft²; increases windows to R2) or adding extra wall insulation (cost = $2 per ft²; increases walls to R20). Which is more cost effective?

The cost of double-glazed windows is $400 \times 10 = \$4000$. They reduce UA by 200. The cost of wall insulation is $1600 \times 2 = \$3200$. They reduce UA by 80. The investment in double-glazed windows is clearly superior for the case described.

Problems

7.1. Estimate the heat loss coefficient per ft^2 for a house containing:
   (a) 2500 ft^2 of floor area constructed as in item 5 in Table 7.3.
   (b) 2500 ft^2 of ceiling constructed as in item 4 in Table 7.3.
   (c) 1344 ft^2 of wall constructed as in item 1 in Table 7.3.
   (d) 256 ft^2 of single-pane window.
   (e) Infiltration at the rate of one house volume (20,000 ft^3) per 2 hr.

7.2. Obtain data for your home, or the home of someone you know, on monthly energy usage, and estimate the heating, cooling, and baseload rates for each month as in Figures 7.1 and 7.3. Compare the heating and cooling (if appropriate) monthly loads with the monthly degree-days for your area as in Figures 7.2 and 7.4.

7.3. From the results of Problem 7.2 and the area of the house, what is the heat loss coefficient for the house per ft^2 of floor area? Compare with the rest of the class.

7.4. What is the effect of The Energy Tax Act of 1978 on the economics of the systems in Table 7.10 (30% credit on first $2000, 20% credit on next $8000)?

7.5. Compare the effect of tilting the flat plate collectors of a space heating system at Madison, Wisconsin, at angles of 43°, 63°, and 90° (vertical). Use the data in Appendix II, and assume that 30% of the insolation on a horizontal surface is diffuse and that 70% is direct. The house heating load per day is 25,000 (HDD) Btu/day, where HDD is the number of heating degree-days for an average day of a month. There are 800 ft^2 of solar collectors. The efficiency on a daily basis is estimated to be

$$\eta = 0.3 + \frac{T_{max}}{650}$$

where T_{max} is the average afternoon high temperature (°F) for an average day of the month.

8

Thermodynamics,
Engines, and
Heat Pumps

In this chapter we provide a summary of thermodynamic concepts and results as a basis for the following chapters, which deal with heat pumps, air conditioners, and power conversion. It is assumed that the reader has had some thermodynamics background. The Rankine cycle for partial conversion of heat to work is described. Methods for improving cycle efficiency are covered briefly. Refrigeration cycles are discussed. Cooling by compression refrigeration, absorption refrigeration, and desiccant/humidification is discussed. Advanced cycles for solar electrical power generation (Brayton and Stirling cycles) are treated briefly.

8.1 The Simple Rankine Cycle

8.1.1 Physical Description

The Rankine cycle is a closed cycle for conversion of heat to work. Heat is abstracted from a high temperature source to vaporize a liquid in a boiler at high pressure. The high pressure vapor is used to drive an engine (in practice, usually a steam turbine). The exhaust from the engine usually consists of a mixture of low pressure vapor and liquid droplets at a much lower temperature than at the inlet to the engine. This exhaust must be returned to the boiler to complete the cycle. In order to do this, the exhaust vapor must be condensed by heat removal so that the liquid can be pumped

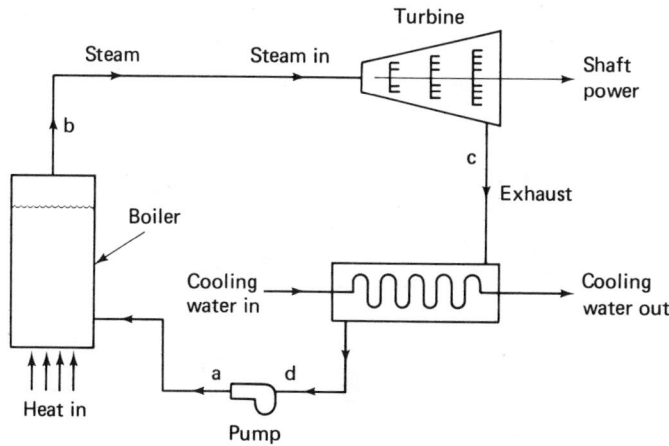

Figure 8.1 The steam Rankine cycle.

up to high pressure before introduction to the boiler. The Rankine cycle is illustrated in Figure 8.1. The condensate return water is introduced into the boiler at a relatively low temperature at point a. This water is heated irreversibly and gradually brought up to the boiling point e, whereupon further heat addition produces saturated steam at b. The steam passes through the turbine, where it expands against turbine wheels, producing mechanical shaft power. The expanded steam from the turbine exhaust enters the condenser at c. The condensate leaves the condenser at d, where it is pumped up to high pressure before reintroduction to the boiler. The absolute pressure in the condenser is determined by the vapor pressure of water at the temperature of the condenser (set by the cooling water). Usually, a vacuum exists in the condenser, since the prevailing condenser temperatures are in the general range of 125°F.

The steam generator may be divided into sections consisting of a preheater and a boiler, even though in practice they may not be physically distinct. This arrangement is illustrated in Figure 8.2. In this case, saturated steam is used at the working pressure of the boiler.

8.1.2 Pressure–Volume Diagram

The Rankine cycle may also be understood by following the path of water around the cycle on a p-V diagram as shown in Figure 8.3. Isotherms are shown at three temperatures. The boiling point of water at the pressure in the boiler is T_3. The temperature of condensate from the condenser is T_1,

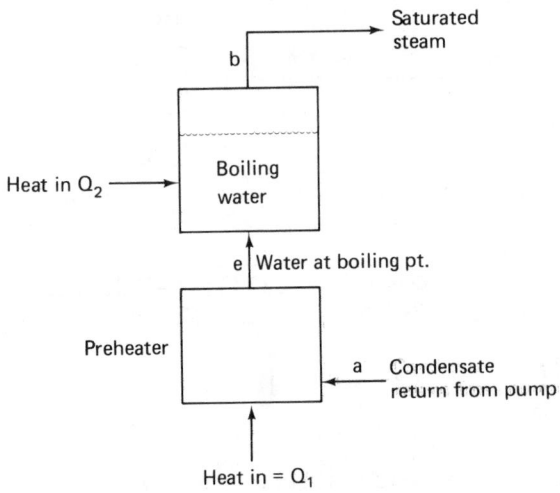

Figure 8.2 Subdivision of boiler.

Figure 8.3 The saturated steam Rankine cycle on a *p-V* plot.

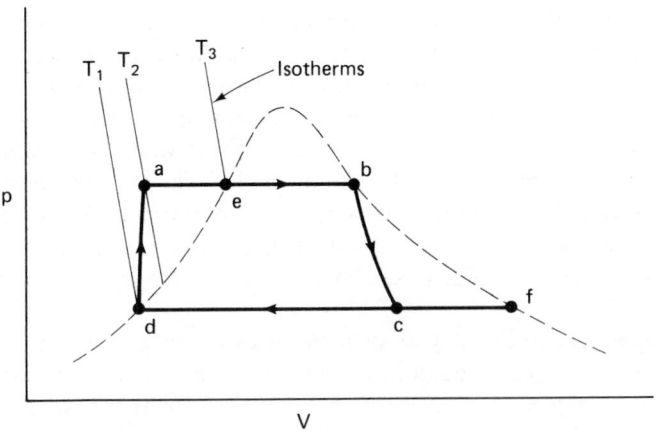

and the temperature of water after being pumped to high pressure is T_2, which is slightly higher than T_1. Water at *a* enters the preheater, where it is heated until it reaches the boiling point at *e*. Further heating boils the water, whereupon saturated steam at *b* is formed. When steam passes through the turbine adiabatically, the steam loses heat along a curve similar to *bc*. Point *c* represents a mixture of liquid and vapor rejected by the turbine. The condenser cools this exhaust to saturated liquid at point *d*. Note that the preheater, boiler, and turbine inlet are all at the same high pressure, and the exhaust and condenser are at the same low pressure. The pump raises the pressure along line da, and the vapor expands through the turbine along line bc.

8.1.3 Cycle Efficiency

The Rankine cycle can be carried out reversibly (in principle) by using a reversible heating mechanism to preheat the condensate return and using a heat source for boiling that is infinitesmally higher in temperature than the boiling point. The turbine would also have to be reversible in the sense of an infinite number of stages, each one with an infinitesimal pressure drop and perfectly operating without nonuniformities or friction. The efficiency of such a reversible Rankine cycle can be calculated. The heat input to the preheater and boiler takes place at constant pressure. Thus, the heat input per pound of water is equal to the enthalpy change in taking 1 lb of water from point a to point b:

$$Q = h_b - h_a$$

The heat removal in the condenser occurs at constant pressure, so it is equal to the enthalpy difference $h_c - h_d$. The work done in the turbine is the heat input in the boiler minus the heat removed in the condenser. Thus,

$$W = h_b - h_a - (h_c - h_d)$$

The efficiency is the ratio of work done divided by the heat input. Thus,

$$\eta = \frac{W}{Q} = \frac{h_b - h_a - h_c + h_d}{h_b - h_a}$$

The enthalpy change across the pump $(h_a - h_d)$ is usually quite small and can often be neglected. Thus, the efficiency can be approximated by

$$\eta \simeq \frac{h_b - h_c}{h_b - h_a}$$

The enthalpies of water and steam are tabulated in steam tables. For example, suppose the working pressure in the boiler is 422 psi (2910 kP) corresponding to a saturated steam temperature of 450°F (505°K), and cooling water is available at 100°F (311°K). According to steam table data,

$$\eta \simeq \frac{1205.6 - h_c}{1205.6 - 68.0}$$

To determine the enthalpy of the turbine exhaust h_c, one must use the fact that for a reversible adiabatic expansion through the turbine, $\Delta S = 0$ because $Q = 0$. Therefore,

$$S_c = S_d$$

where $S =$ entropy. But point c in Figure 8.3 corresponds to a mixture of liquid at d and vapor at f. Let the fraction of vapor be denoted as x. Then

$$xS_f + (1 - x)S_d = S_c = S_b$$

According to steam tables data,

$$x(1.9819) + (1 - x)(0.1296) = 1.4806 \text{ Btu/lb-}^\circ\text{F} = 6.199 \text{ kJ/kg-}^\circ\text{K}$$

This can be solved for x, the result being

$$x = 0.729$$

Then the enthalpy of the exhaust can be evaluated from

$$h_c = xh_f + (1 - x)h_d$$

From the steam tables,

$$h_c = x(1104.7) + (1 - x)(68.0) = 823.8 \text{ Btu/lb} = 1916 \text{ kJ/kg}$$

The efficiency of the reversible Rankine cycle is then calculated to be

$$\eta = \frac{1205.6 - 823.8}{1205.6 - 68.0} = 0.335$$

If a Carnot cycle is operated between a heat source at 450°F (505°K) and a heat sink at 100°F (311°K), the efficiency is

$$\eta_c = \frac{450 - 100}{(450 + 460)} = 0.385$$

This is higher than the Rankine cycle efficiency because in the Carnot cycle all the heat is added at 450°F (505°K). In the Rankine cycle, heat is added over a range of temperatures from 100°F (311°K) to 450°F (505°K). The efficiency of the Rankine cycle is a heat-weighted average of Carnot cycle efficiencies over the entire temperature range. When water is heated at constant pressure from 100°F (311°K) to 450°F (505°K), the enthalpy change is $430.2 - 68.0 = 362.2$ Btu/lb (842 kJ/kg). When 1 lb of water is boiled at 450°F, the enthalpy change is 775.4 Btu/lb (1804 kJ/kg). If it is assumed that the specific heat of water is roughly constant over the range 100°F to 450°F, then equal amounts of heat are supplied over each differential range of temperature. Thus, the heat-weighted average of Carnot cycle efficiencies is

$$\eta = \frac{\dfrac{\displaystyle\int_{100}^{450} \dfrac{T-100}{T+460}\, dT}{\displaystyle\int_{100}^{450} dT}\,(362.2) + 0.385(775.4)}{362.2 + 775.4}$$

This works out to be $\eta = 0.333$, in essential agreement with the value calculated from enthalpies in the Rankine cycle. From this analysis it can be seen that the object of the designer of any power cycle is to inject heat over as high a temperature range as possible.

8.2 Rankine Cycle with Superheat

One method for adding heat at higher temperatures is to use superheat. If a system is limited to some working pressure, then the use of saturated steam (steam in equilibrium with boiling water) limits the temperature to the boiling point at that pressure. If one has a heat source (such as burning of a fossil fuel) with a much higher potential temperature available, then the steam can be superheated at constant pressure above the boiling point at that pressure. This process is illustrated in Figure 8.4. For the case considered here, at a pressure of 422 psi (2910 kP) and a condensate return temperature of 100°F (311°K), $Q_1 = 362.2$ Btu/lb (842 kJ/kg), $Q_2 = 775.4$ Btu/lb (1804 kJ/kg), and Q_3 depends on the temperature to which the steam is superheated. The effect of superheat on the cycle is illustrated in Figure 8.5. Point b is now in the vapor region. If the amount of superheat is great enough, point c can actually occur to the right of point f, and the turbine exhaust will be pure vapor.

The effect of superheat on the efficiency of a Rankine cycle can be evaluated by the usual method of enthalpies. Two cases should be distinguished. One is where there is a fixed working pressure and the temperature of the steam is increased by superheating. In this case, superheat increases the cycle efficiency because it adds heat at higher temperatures. The second case is where the working temperature is fixed by the heat source (such as solar collectors) and the question is what working pressure should be used. For example, if 450°F (505°K) is the working temperature, by using a working pressure of 422 psi (2910 kP), saturated steam is implied. However, at a working pressure of 134.5 psi (927 kP), the boiling point is at 350°F (450°K) and there is 100°F (311°K) of superheat. In this case, superheat reduces the cycle efficiency because the large lump sum of heat required for vaporization

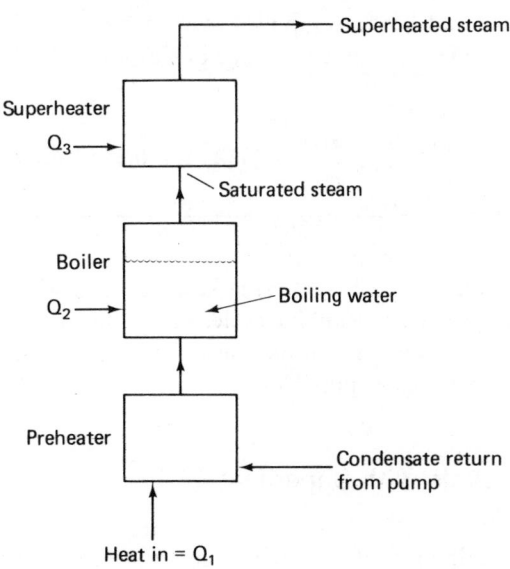

Figure 8.4 Boiler and superheater.

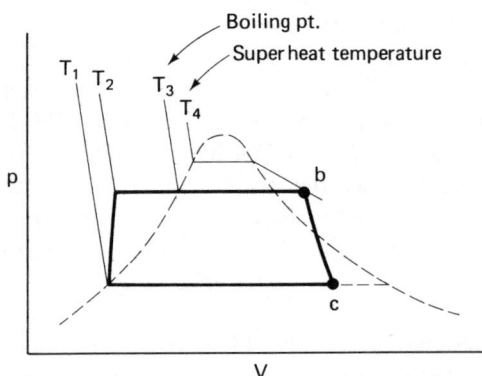

Figure 8.5 Rankine cycle; using superheat.

is transferred at a lower temperature. The results of calculations from steam tables are shown in Tables 8.1 and 8.2. In Table 8.1, the working pressure is held constant at 422 psi (2910 kP), and the amount of superheat is varied from saturation at 450°F (505°K) to 1000°F (811°K). In Table 8.2, the temperature is held constant at 450°F and the pressure is varied to vary the amount of superheat. The results for heat flows and cycle efficiencies are indicated. The conclusion to be drawn is that one should use as high a working pressure and temperature as possible to obtain the highest cycle efficiency. Superheat helps only when a system is pressure-limited. For systems that are limited by the available temperature, saturated steam yields the highest cycle

TABLE 8.1 Heat flows and cycle efficiencies of a Rankine cycle with fixed pressure 422 psi (2910 kP) and variable amounts of superheat

| Steam Temperature (°F) | Preheat (Btu/lb) | Boiling Heat (Btu/lb) | Superheat (Btu/lb) | Turbine Exhaust | | Cycle Efficiency |
				Fraction Vapor	Enthalpy (Btu/lb)	
450	362.2	775.4	0.0	0.729	824.2	0.335
500	362.2	775.4	37.5	0.750	845.5	0.338
600	362.2	775.4	99.5	0.784	880.8	0.343
700	362.2	775.4	156.5	0.812	909.4	0.350
800	362.2	775.4	209.5	0.833	931.6	0.359
1000	362.2	775.4	309.5	0.864	963.7	0.381

TABLE 8.2 Heat flows and cycle efficiencies of a Rankine cycle at fixed temperature 450°F (505°K) and variable superheat (variable pressure)

Steam Pressure (psi)	Boiling Point (°F)	Preheat (Btu/lb)	Boiling Heat (Btu/lb)	Superheat (Btu/lb)	Total Heat Added (Btu/lb)	Cycle Efficiency
522	450	362.2	775.5	0	1137.7	0.335
195.6	380	285.6	845.4	42.6	1173.6	0.301
103	330	232.8	887.5	64.8	1185.1	0.268

efficiency. However, use of saturated steam can cause wear on the turbine blades due to water droplet formation as the steam cools, and it may be desirable, in practice, to sacrifice some cycle efficiency to use a moderate amount of superheat to protect the turbine from this effect. Very high temperature steam cycles are called *supercritical* if no actual boiling occurs. Instead, water is heated above the critical pressure and gradually and continuously acquires the properties of steam as illustrated in Figure 8.6.

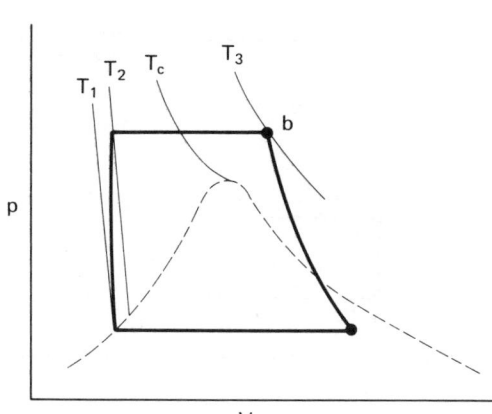

Figure 8.6 The supercritical steam cycle.

8.3 The Reheat Cycle

Another method for improving cycle efficiency is the reheat cycle. The reheat cycle is most appropriate for fossil fuel burning systems. The turbine is divided into two or more sections. The exhaust from the high pressure turbine is steam at reduced pressure, and this is conveyed back to the main burner area to be superheated at constant pressure before injection into the second turbine. Since heat is added to the steam at relatively high temperatures, this increases the overall cycle efficiency. The reheat cycle is illustrated in Figures 8.7 and 8.8. Reheat can be employed several times with multiple turbines.

Figure 8.7 The reheat cycle.

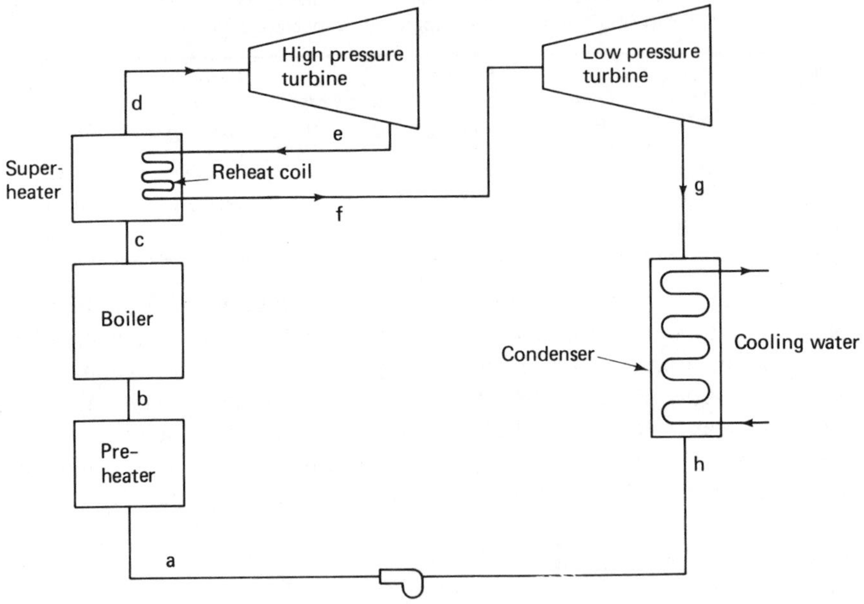

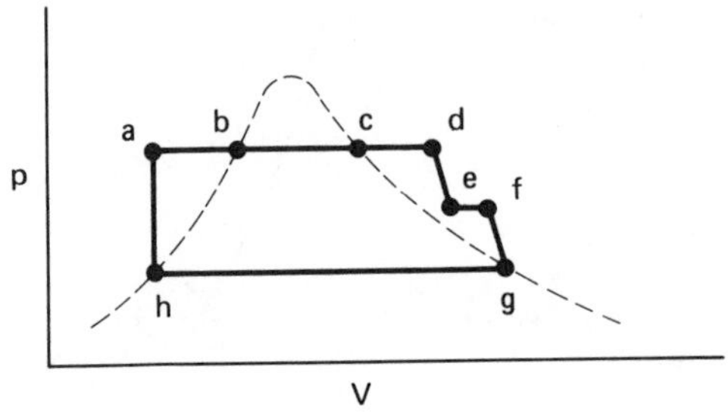

Figure 8-8 *p-V* diagram for reheat cycle.

8.4 The Regenerative Cycle

Another approach for increasing cycle efficiency is the regenerative cycle. In this cycle, the turbine is divided into two or more sections, and some of the exhaust steam from the high pressure turbine is used to preheat condensate return from the low pressure turbine before introduction to the boiler. The net effect is to raise the temperature of the condensate return and increase the average temperature at which heat is added to the boiler. Thus, the efficiency of the cycle is improved.

Let us consider a specific example. Suppose there is a Rankine cycle with a boiler operated at 680 psi (4689 kP), corresponding to a boiling point of 500°F (533°K). The steam is superheated to 700°F (644°K), and the turbine exhaust is at 125°F (325°K). The Carnot efficiency of a cycle operating between these temperatures is $(644 - 325)/644 = 0.496$. The reversible Rankine cycle efficiency is calculated as before. The fraction of vapor is the exhaust x calculated from

$$S_{v,700} = xS_{v,125} + (1 - x)S_{L,125}$$
$$1.569 = x(1.9218) + (1 - x)(0.1732)$$
$$x = 0.798$$

Then the enthalpy of the exhaust is

$$h_{ex} = (0.798)(1115.4) + (0.202)(93.0)$$
$$= 908.9 \text{ Btu/lb} = 2114 \text{ kJ/kg}$$

The reversible Rankine cycle efficiency is then

$$\eta = \frac{1346 - 908.9}{1346 - 93.0} = 0.349$$

The Rankine cycle efficiency is considerably below the Carnot efficiency due mainly to the heat supplied in bringing condensate return up from 125°F (325°K) to 500°F (533°K). Next, consider the regenerative cycle illustrated in Figure 8.9. The high pressure turbine is arbitrarily designed to produce an exhaust pressure of 170 psi (1172 kP), which results in slightly superheated vapor as the exhaust. The turbine exhaust at B has the same entropy as the superheated steam entering the turbine, namely, 1.569 Btu/lb-°F (6.57 kJ/kg-°K). One may then find from steam tables that the exhaust temperature is 380°F (466°K), has an enthalpy of 1204 Btu/lb (2801 kJ/kg), and no liquid water is present in the exhaust. Now, if the exhaust from the first turbine at B is divided as shown in Figure 8.9, a fraction f goes to the regenerative

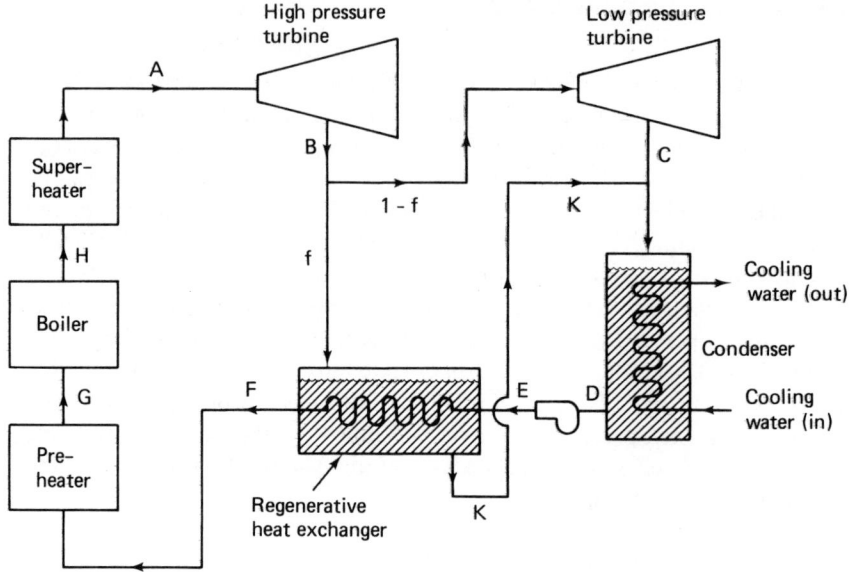

Figure 8.9 The regenerative reheat cycle for preheating feedwater.

heater, and $(1 - f)$ goes to the inlet of the low pressure turbine. A heat balance around the regenerator is carried out for 1 lb of flow through the system. Then

$$h_F - h_E = f(h_B - h_K)$$

The enthalpy change across the pump is neglected, and it is assumed that $h_D \cong h_E$. The state of the exhaust C from the low pressure turbine is found by the usual procedure for steam entering the turbine at 380°F (466°K) at 170 psi (1172 kP) and exhaust at 125°F (325°K). It is found that the exhaust C is 0.798 vapor and has an enthalpy of 908.9 Btu/lb (2114 kJ/kg). The efficiency of the reheat cycle is

$$\eta = \frac{(h_A - h_B) + (1 - f)(h_B - h_C)}{h_A - h_F}$$

The only unknowns are f and h_F. The enthalpy h_F can be estimated in the following way. If one starts at $f = 0$ and f is gradually increased, the flow rate of exhaust from the first turbine is so much smaller than the flow rate of condensate return in the regenerator that the temperature at F is only moderately elevated above the temperature at E. Under such conditions, the steam from the exhaust of the first turbine is fully condensed, and the liquid water at K is cooled to within about 20°F of the water entering at E. This condition is maintained as f is increased, but a point is reached where T_F begins to approach T_B. When T_F reaches perhaps $T_B - 10°F$, increasing f cannot further increase T_F. Instead, increasing f merely increases the enthalpy of fluid at K, thus wasting heat by conveying it from B directly to the condenser. Accordingly, the range of f is restricted to values less than f_{max}, where f_{max} results in $T_F = T_B - 10°F$ and $T_K = T_E + 20°F$. For values of $f < f_{max}$, T_K remains equal to $T_E + 20°F$, while T_F drops below $T_B - 10°F$. For values of $f < f_{max}$, the enthalpies in Btu/lb are related by

$$h_F = h_E + f(h_B - h_K)$$

Assuming that the specific heat of water is roughly 1 Btu/lb-°F (4.187 kJ/kg-°K),

$$h_K \cong h_E + 20 \text{ Btu/lb}$$

Thus,

$$h_F = h_E + f(h_B - h_E - 20)$$
$$h_F = 93 + f(1204 - 93 - 20)$$
$$h_F = 93 + f(1191)$$

The overall cycle efficiency is then

$$\eta = \frac{h_A - h_B + (1 - f)(h_B - h_C)}{h_A - 93 - 1191f}$$
$$= \frac{1346 - 1204 + (1 - f)(1204 - 909)}{1346 - 93 - 1191f}$$
$$= \frac{142 + 295(1 - f)}{1253 - 1191f}$$

The maximum value of f is the value that results in $T_F \cong 370°F$ (461°K) or $h_F = 348$ Btu/lb (798 kJ/kg). This is $f_{max} = 0.21$. For $0.21 > f > 0$, the dependence of η on f is indicated in Table 8.3. As f is increased, the through-

TABLE 8.3 Efficiency of reheat cycle vs. fraction reheat

f	η
0.00	0.349
0.05	0.354
0.10	0.359
0.15	0.366
0.21 (max)	0.374

put of fluid required to produce any fixed amount of power goes up. Thus, the pumping losses are slightly higher than without reheat.

8.5 Organic Rankine Cycles

The variation of the vapor pressure of water as a function of temperature is shown in Figure 8.10. To supply the mechanical force to drive a turbine, the pressure of vapor entering the turbine should be at least ∼100 psi (690 kP) and preferably greater than 300 psi (2070 kP). These pressures are not

Figure 8.10   Vapor pressure of water.

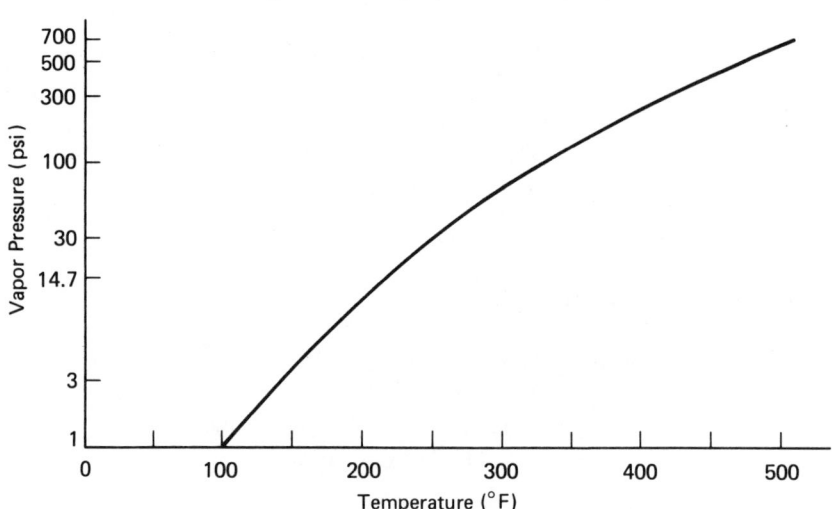

attained in the water system until the temperature reaches the range 300°F to 400°F (422°K to 478°K). There are situations where heat is available at lower temperatures, say 200°F (367°K), and it is desired to run a turbine from this heat source. In such cases, water would be a totally inappropriate working fluid because the pressure would be too low to produce effective action in a turbine. Many organic fluids have much higher vapor pressures than water in the 180°F to 350°F (355°K to 450°K) temperature range, and these organic fluids are more suitable for low temperature turbine cycles. However, the use of organic fluids involves a number of disadvantages. The practical disadvantages in using organics are that they are typically expensive, flammable, and sometimes poisonous, and therefore the seals around pumps and valves must be very carefully designed to reduce leakage. Leakage in water systems is much less of a problem because water can be replenished easily and the environmental effects are nil.

In addition to the practical disadvantages of using organic fluids, there are basic disadvantages in the cycle performance. One reason for this is that organics have much lower heats of vaporization compared to their specific heats, and therefore the preheating of condensate up to the boiling point requires a larger proportional share of heat than boiling, as compared to a water system. Since boiling occurs at a higher temperature then preheating, this makes the reversible cycle efficiency lower with an organic fluid than with water at the same working temperatures. A second problem with the organic Rankine cycles is that the enthalpy changes associated with fixed temperature changes are much lower than for water. Therefore, in order to put a given amount of heat into an organic, many more pounds of material must be moved through the boiler than with water. The internal parasitic losses involved in pumping this large amount of fluid through the system reduces the net amount of mechanical shaft power available to an external user. In water systems, the power required for pumping condensate return may be typically 2-5% of the mechanical shaft power produced, depending on operating conditions. In organic systems, parasitic pumping requirements may run as high as 10-25% of the mechanical shaft power produced by the turbine. Another disadvantage of organic fluids is the shape of the liquid-vapor region in the p-V diagram is different than for water, resulting in superheated vapor produced in the turbine exhaust. A heat exchanger must be provided between the turbine and the condenser to partially recover excess heat from the exhaust to the condensate return fluid. Losses inevitably enter into this step. Consider the organic Rankine cycle shown in Figures 8.11 and 8.12. The preheater, boiler, and superheater are similar to those used for water except that the boiler is much smaller with an organic fluid. A line of constant entropy corresponding to a reversible turbine is ab. The super-heated exhaust at b could be conveyed directly to the condenser, but then the superheat in the exhaust would be lost to the cooling water in the condenser. Therefore, a heat exchanger is placed between the turbine exhaust and the condenser to partially recover the superheat from the exhaust into the con-densate return. The heat gained across ef is equal to the heat loss across bc. Generally, as a rule of thumb, the vapor from b can be cooled at c to within about 10°F of the temperature of condensate at e. The cycle efficiency is, therefore,

$$\eta = \frac{(h_a - h_f) - (h_c - h_d)}{(h_a - h_f)}$$

If the enthalpy change across the pump can be neglected, the approximation $h_d \cong h_e$ can be made. Then

$$\eta = \frac{h_a - h_c - (h_f - h_e)}{h_a - h_f}$$

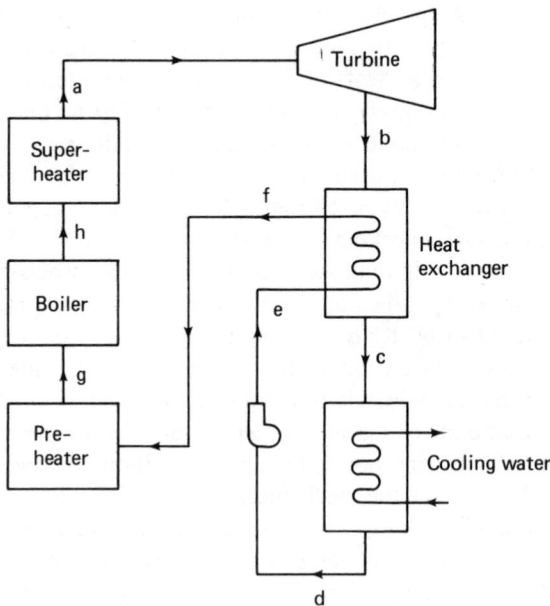

Figure 8.11 Rankine cycle for organic fluids with super-heated exhaust.

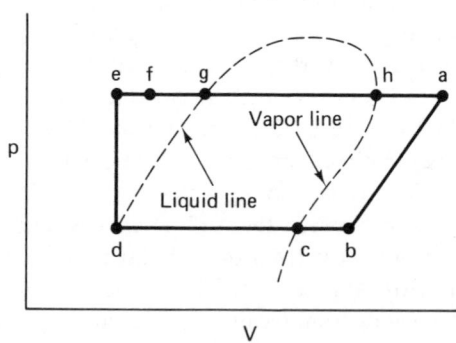

Figure 8.12 p-V diagram for organic Rankine cycle.

But since $(h_f - h_e) = (h_b - h_c)$,

$$\eta = \frac{h_a - h_b}{h_a - h_f} = \frac{(h_a - h_b)}{(h_a - h_e) + (h_e - h_f)}$$

From this formula it can be seen how the heat recovery $(h_f - h_e)$ produces an improved cycle efficiency. In applications, the temperature T_d may be taken as saturated liquid at exhaust pressure. Neglecting any temperature gain across the pump, $T_e \approx T_d$. The temperature at c can be estimated as

$T_c \approx T_e + 10°F$. From this assumption about the effectiveness of the heat exchanger, the enthalpy change $h_b - h_c$ can be evaluated. Then, since $(h_f - h_e) = (h_b - h_c)$, the enthalpy h_f can be calculated. Therefore, the cycle efficiency can be calculated.

8.6 Comparison of Carnot, Steam Rankine, and Organic Rankine Cycles

The above analysis will be applied to a specific example. Suppose there is a heat source at 220°F (378°K) and a heat sink at 100°F (311°K). The efficiencies of the Carnot cycle, the steam Rankine cycle, and a specific organic Rankine cycle will be calculated. The Carnot efficiency is $67/378 = 0.176$. Using saturated steam at 200°F (378°K) and 17.2 psi (119 kP), the reversible steam Rankine cycle efficiency is

$$\eta = \frac{h_v(220°F) - h_{ex}(100°F)}{h_v(220°F) - h_L(100°F)}$$

The enthalpy of the exhaust can be calculated in the usual way by finding the fraction of vapor in the exhaust that produces $S_{ex}(100°F) = S_v(220°F)$. The result is $x = 0.872$, $h_{ex} = 972.0$ Btu/lb, and thus

$$\eta = \frac{1153.5 - 972.0}{1153.5 - 68.0} = 0.167$$

Of course, the low vapor pressure of water (17.2 psi) at 220°F implies that the turbine would, in practice, have very poor performance.

Next, consider the use of the organic fluid known as Refrigerant 114, or simply R114. According to tables in Appendix III, the vapor pressure at 220°F (378°K) is 222 psi (1531 kP). The advantage of the organic fluid is immediately obvious from the fact that the vapor pressure is so high. The pressure of the turbine exhaust is set by the vapor pressure of R114 at the heat sink temperature, 100°F (311°K). This is 45.85 psi (316.1 kP). Following a line of constant entropy through the turbine (*ab* in Figure 8.12), point *b* is located at the intersection of the line $S = 0.170$ Btu/lb-°F, with $p_b = 45.85$ psi (316.1 kP). Thus, $T_b = 130°F$ (328°K) and there is 30°F of superheat in the turbine exhaust. Now, assuming that $T_c = T_e + 10°F = 110°F$, we find that $h_b - h_c \cong 2.8$ Btu/lb. Since $h_f - h_e = h_b - h_c$, it follows that $h_f = 34.6$ Btu/lb and $T_f = 111°F$ (317°K). Therefore,

$$\eta = \frac{100.60 - 90.1}{100.60 - 34.6} = 0.159$$

A comparison of the heat flows in the preheater and boiler using water and R114 is shown in Table 8.4. Using water, 89% of the heat is deposited at 220°F (378°K) in the boiling step. When R114 is used, only 65% is deposited at 220°F (378°K) in the boiling step. The fluid throughput required to produce 1 Btu equivalent of mechanical work is more than 10 times as high when using R114 rather than with water.

TABLE 8.4   Comparison of heat flows using water and R114 between 220°F (378°K) and 100°F (311°K) in Rankine cycles

	Heat Required (*Btu/lb*)	
Step	*Water*	*R114*
Preheat liquid: (Water from 100°F to 220°F) (R114 from 111°F to 220°F)	120.2	29.3
Boil liquid at 220°F	965.3	53.4
Fraction of heat supplied in boiling step	0.89	0.65
Pounds of fluid circulated per Btu equivalent of mechanical work produced	0.0055	0.076

8.7 Effects of Irreversibility

In all the discussions thus far, the cycles were considered to be reversible. Irreversibility in the boiler occurs because the heat source is at a higher temperature than the fluid being heated. However, this merely limits the actual irreversible cycle efficiency to that of a reversible cycle at the *fluid* temperature rather than at the *source* temperature. The calculations given here, based on fluid temperatures, are therefore actually representative of real systems. Irreversibility in the turbine is a more serious problem. In the calculations provided in this book, we have assumed the turbine to be reversible, so that the entropy of the turbine exhaust can be taken equal to the entropy of steam entering the turbine. No actual turbine is even close to being reversible, and the actual entropy of the exhaust is higher than the entropy of the steam. As a result, the enthalpy of the exhaust is higher than if the cycle were reversible, and less heat is removed from the steam and converted to work in the turbine. This effect is illustrated schematically in Figure 8.13. With increasing degrees of irreversibility, the line of expansion through the turbine shifts to the right and less work is produced. Since the excess heat content in the turbine exhaust is removed in the condenser,

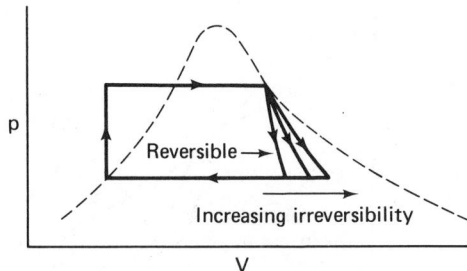

Figure 8.13 Effect of irreversibility on the Rankine cycle.

irreversibility in the turbine acts like a direct heat leak from the boiler to the cooling water. For a case illustrated previously, with superheated steam at 700°F (644°K) and 680 psi (4689 kP) and a 125°F (325°K) exhaust, isentropic expansion leads to a cycle efficiency of 0.345 and 77.5% vapor in the exhaust. Varying degrees of irreversibility can be described in terms of the fraction of vapor in the exhaust. In Table 8.5 the cycle efficiency is shown as a function of fraction of vapor in the exhaust. In practice, typical values of η/η_{rev} in the range 0.6 to 0.8 are obtained.

TABLE 8.5 Effect of irreversibility in the turbine on cycle efficiency for a turbine with inlet steam at 700°F (644°K), 680 psi (4689 kP), and 125°F (325°K) exhaust

Fraction Vapor in Exhaust	h_{ex} (Btu/lb)	η	η/η_{rev}
0.775*	885.4	0.345	1.00
0.85	962.0	0.281	0.82
0.90	1013.2	0.239	0.69
0.95	1064.3	0.197	0.57

*Reversible cycle

8.8 Refrigeration Cycles

8.8.1 Carnot Refrigeration Cycle

The Carnot refrigeration cycle using an ideal gas as the working substance is illustrated in Figure 8.14. It is simply the Carnot heat engine run in reverse. The coefficient of performance is

$$\text{COP} = \frac{T_1}{T_2 - T_1}$$

When a condensible vapor is used as the working substance, the cycle is modified as shown in Figure 8.15. The system pressures and temperatures are

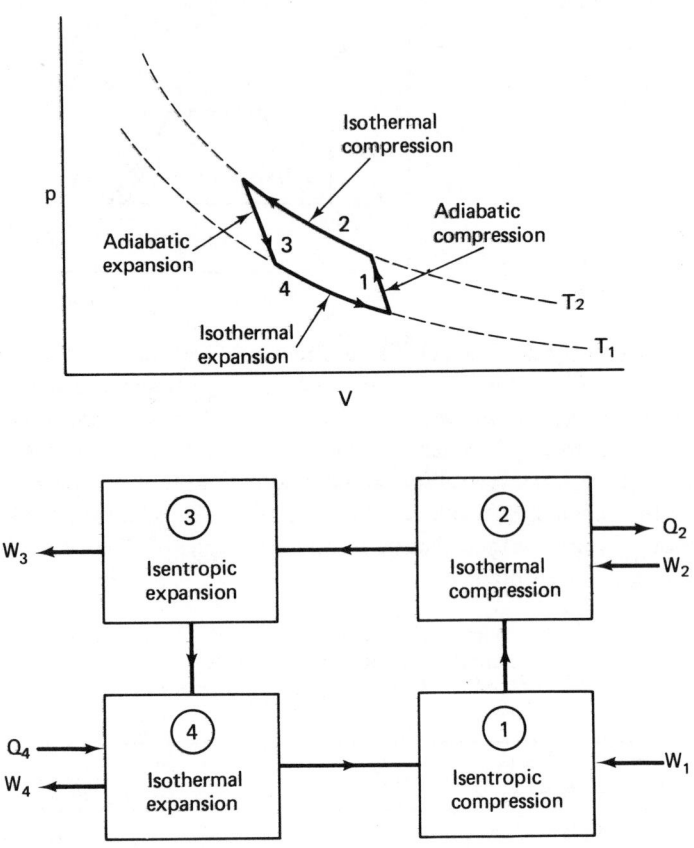

Figure 8.14 Carnot refrigeration cycle, using an ideal gas as a working substance.

adjusted so that the working fluid at step 5 is saturated liquid at T_1. When heat Q_5 is supplied by the external material to be cooled, the working fluid evaporates isothermally and becomes saturated vapor at T_1 at a. Thus, the refrigeration occurs in step 5 with the working fluid absorbing heat from an external source at T_1. The saturated vapor is adiabatically compressed in step 1 to point b, which is superheated vapor at T_2. To further isothermally compress this vapor, the isothermal condenser is used to remove heat at T_2 in step 3, producing saturated liquid at T_2 at d. An adiabatic expansion in step 4 yields a mixture of saturated vapor and saturated liquid at T_1 at e. The vapor merely passes on to a, while the liquid can take up more heat in the evaporator. Since all heat is absorbed at T_1 and all heat rejected is at T_2, this is a Carnot cycle. All steps are reversible. Therefore, the COP is again given by $T_1/(T_2 - T_1)$. To actually build a physical Carnot cycle of this type, one would have to construct compressors for steps 1 and 2, and a turbine for

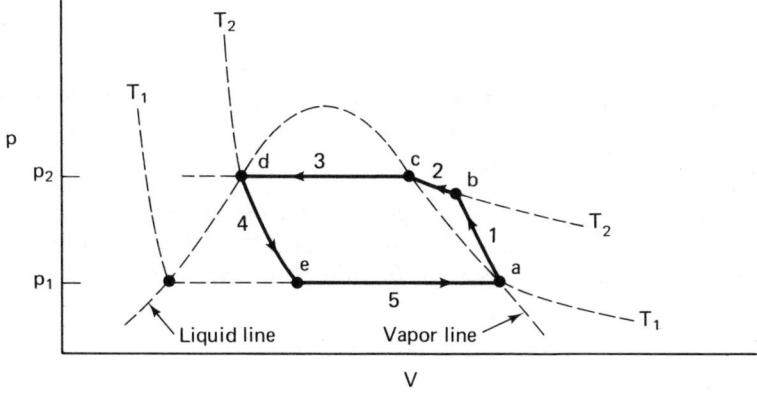

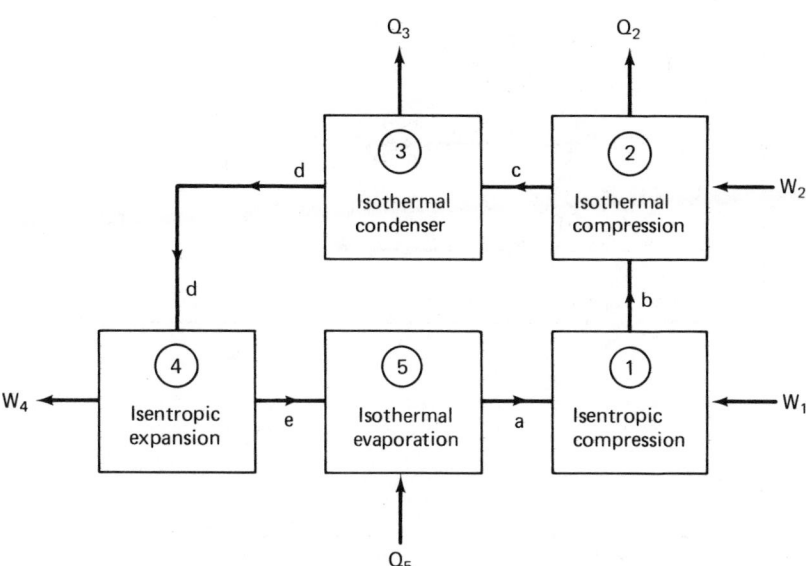

Figure 8.15 Carnot refrigeration cycle, using a condensible vapor as a working substance.

step 4, with connecting rods to a single flywheel. Such a machine would be extremely difficult to construct and operate.

8.8.2 Theoretical Single-stage Cycle

Since the Carnot refrigeration cycle is not feasible in practice, actual refrigeration cycles make compromises that reduce the COP but which allow reasonable construction and operation. The two compromises generally used are:

1. Replace the turbine at step 4 of Figure 8.15 by a simple irreversible expansion valve, and relinquish attempting to utilize the work W_4.
2. Replace steps 1 and 2 of Figure 8.15 by a single isentropic compression to (p_2, T_3), and allow an irreversible cooling from (p_2, T_3) to (p_2, T_2) to occur in the condenser.

These two compromises in design allow the use of a single compressor, which makes the entire cycle feasible to construct. The effect of these design adjustments is shown in Figure 8.16. Step 1 now involves isentropic compres-

Figure 8.16 The single-stage refrigeration cycle.

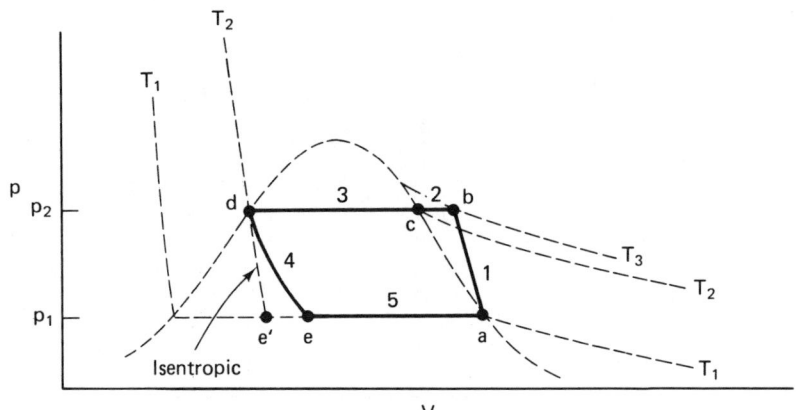

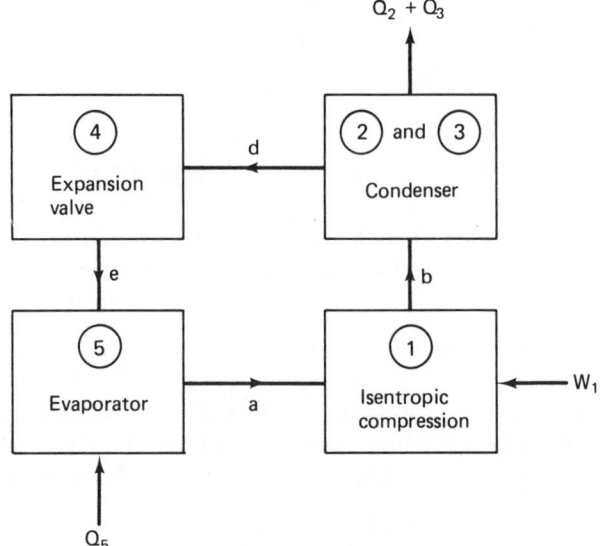

sion to p_2, resulting in superheated vapor at T_3, with $T_3 > T_2$. Steps 2 and 3 occur together in the condenser. Step 2 is a manifestly irreversible cooling of superheated vapor from T_3 to T_2 at constant pressure. Step 3 is a reversible condensation of saturated vapor to saturated liquid at T_2. Step 4 is the irreversible expansion through a throttling valve, resulting in point e being further to the right than if a reversible expansion to e' occurred. Step 5 is the usual reversible evaporation step. The COP of this cycle must be evaluated from the definition

$$\text{COP} = \frac{Q_s}{W_1} = \frac{h_a - h_e}{(h_b - h_d) - (h_a - h_e)}$$

For any specific refrigerant, operating between specific temperatures T_1 and T_2, the enthalpies h_a, h_b, and h_d can quickly be read from charts. The only problem is to locate point e. A useful approximation is to assume that $h_e \cong h_d$, in which case

$$\text{COP} = \frac{h_a - h_d}{h_b - h_a}$$

We now illustrate the use of this equation. Suppose Freon 12 is used as a refrigerant, the evaporator temperature is $T_1 = 40°F$ (278°K), and the condenser temperature is $T_2 = 100°F$ (311°K). Then, from the diagram for Freon 12 in Appendix III,

$$p_2 = 120 \text{ psi (896 kP)}, \quad h_a = 82 \text{ Btu/lb (190.7 kJ/kg)}$$
$$p_1 = 51 \text{ psi (354 kP)}, \quad h_b = 89.5 \text{ Btu/lb (208.1 kJ/kg)}$$
$$T_3 = 113°F \text{ (318°K)}, \quad h_d = 31 \text{ Btu/lb (72.1 kJ/kg)}$$

$$\text{COP} = \frac{82.31}{89.5 - 82} = 6.8$$

The Carnot COP for these conditions is 8.3. Actually, inefficiency in the compressor will reduce the COP considerably below 6.8.

Note that the place where cooling of the external material to be cooled occurs is in the evaporator. Liquid refrigerant at its boiling point at a reduced pressure will spontaneously take up heat and evaporate when it comes into contact with external material that is slightly warmer than the refrigerant. This is a very effective process for pulling heat out of the external material even when that material is very slightly warmer than the liquid refrigerant in the evaporator. For example, in a refrigerator or freezer the evaporator coils are in contact with the air or food in the enclosure. The refrigerant so evaporated is compressed to a higher pressure, cooled in a condenser, and then allowed to expand through an expansion valve. When saturated liquid refrigerant at a relatively high pressure and temperature is allowed to expand through a valve to a region of low pressure, conservation of enthalpy requires that a mixture of saturated vapor and saturated liquid form at the low pressure (and low temperature). Thus, a spontaneous cooling results from this process, which is also known as *flashing*. The low temperature vapor so produced is of no use for refrigeration and merely cycles back to the compressor. However, the low temperature liquid at its boiling point is immediately available for cooling.

8.8.3 Absorption Chillers

It has been noted that conventional refrigeration systems utilize liquid refrigerant at its boiling point at reduced pressure to absorb heat from an external heat source. The problem is to regenerate liquid refrigerant at this temperature. In conventional refrigeration systems this is accomplished by compressing the refrigerant and condensing it at high pressure, followed by flashing through an expansion valve. The major energy input to a conventional refrigeration system is the mechanical energy to drive the compressor. Refrigeration systems can also be operated with the principal energy input in the form of heat. Such systems usually work on the absorption principle. Since heat is much more readily available than mechanical energy, it would seem to be very advantageous to use absorption refrigeration rather than compression refrigeration. Unfortunately, the COP of an absorption chiller is manifestly less than that of a compression chiller. As a result, if one considers the alternatives of (1) using a heat source to drive an absorption chiller or (2) using the heat source to drive a heat engine to produce mechanical power to drive a compression chiller, it is found that for fixed source and sink temperatures and reversible machines the two alternatives have identical COPs. In practice, irreversibilities are different in the two systems, and advantages and disadvantages can be associated with the two approaches.

Absorption refrigeration utilizes two substances, the refrigerant and the absorbent. The two most prevalent combinations in practice are ammonia/water, and water/LiBr as the refrigerant/absorbent. The discussion here will mainly focus on water/LiBr systems. Since water is the refrigerant, the system cannot be used below 32°F (273°K). Therefore, this system is only appropriate for air conditioning where chilled water at 40–45°F (277.6–280.4°K) is a very effective coolant. In this system, liquid water under a vacuum at 40–45°F acts as the refrigerant in the evaporator. Liquid chilled water at atmospheric pressure is used to transfer heat from the circulating air to the liquid water under vacuum in the evaporator as illustrated in Figure 8.17. When the water vapor under vacuum is produced in the evaporator, it would be compressed in a conventional refrigeration system. In an absorption

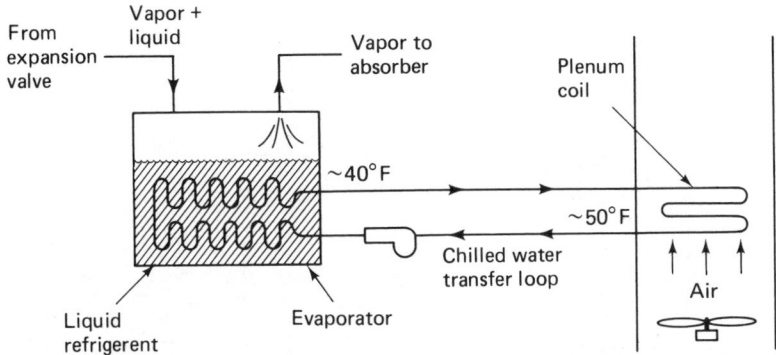

Figure 8.17 The chilled water transfer loop from an absorption chiller to an air plenum.

refrigeration system, the vapor is absorbed by the absorbent, in this case LiBr solution. The LiBr solution is pumped up to a higher pressure (this requires very little energy compared to compressing a vapor), where it is heated to drive off water vapor at the higher pressure. Thus, water vapor at the higher pressure is produced without the mechanical work involved in compression systems. The rest of the system is similar in concept to the compression refrigeration system. The water vapor is condensed and then expanded through a throttling valve to the lower pressure, producing a mixture of low pressure liquid and vapor. The vapor is recirculated to the absorbent, and the liquid is available for evaporative cooling as before. A flow diagram of an absorption chilling system employed for air conditioning is shown in Figure 8.18. The generator contains a strong solution of LiBr in water. Heat is supplied to the generator from hot water or steam at pq, and vapor is produced at b. Cooling water from a cooling tower is introduced at k, and the condensate collects at c. The pressure in the condenser is set by the cooling water temperature. Typically, the condensate temperature is about $100°F$ ($311°K$), corresponding to a pressure of ~ 1 psi (6.9 kP). The condensate expands through shower-head expansion valves to produce liquid and vapor at $\sim 40°F$ ($277.6°K$) in the evaporator. The vapor passes on to e, while the liquid collects in the evaporator at d to absorb heat from the chilled water loop from the house air plenum. The water vapor at e is passed under a shower of strong LiBr solution, which absorbs the vapor and produces a diluted weak solution at f. The absorption process liberates heat, and thus cooling water from l is introduced through the coil mn to cool the weak solution. The weak solution is pumped back up to the generator. To reduce the energy requirements of the system, a heat exchanger is placed between the generator and the absorber. The cool weak solution is thus heated before being introduced to the generator at a, and the hot strong solution is cooled

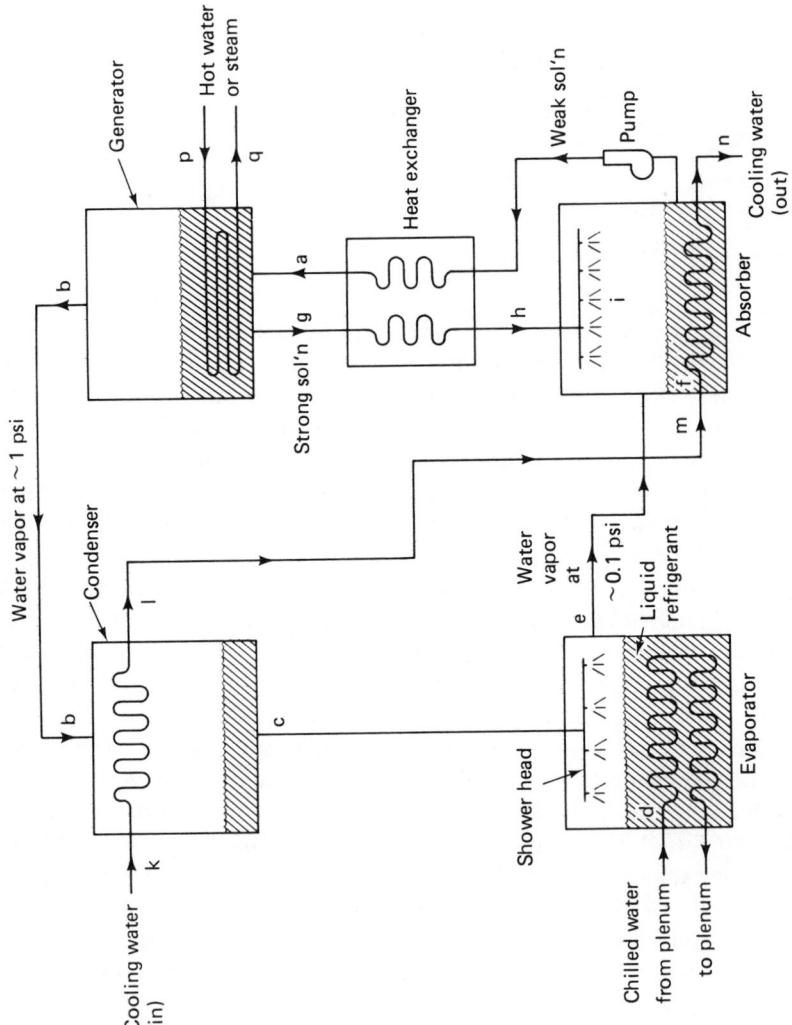

Figure 8.18 Schematic of absorption refrigeration system.

before introduction to the absorber at h. The pressure differential across the shower heads is ~ 1 psi (6.9 kP), which is equivalent to 2 ft (0.61 m) of head of water. Thus, gravity can be used to establish this pressure differential.

The theoretical coefficient of performance for a reversible absorption chiller can be evaluated in terms of the conceptual diagram shown in Figure 8.19. The heat reservoir at T_3 corresponds to the generator ($T_3 > T_2 > T_1$), and the reservoirs at T_2 correspond to the condenser and the absorber (Q_{2a} and Q_{2b}, respectively). The reservoir at T_1 corresponds to the evaporator.

Figure 8.19 Conceptual arrangement for absorption chiller.

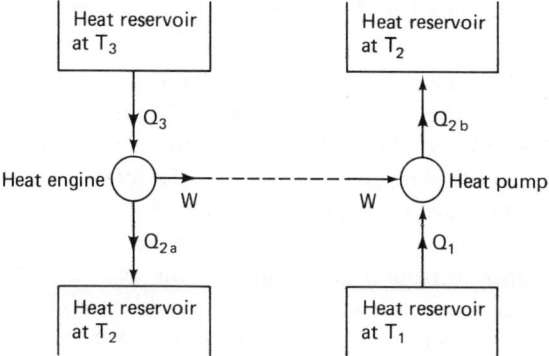

The absorption system neither produces nor requires external work. A conceptual reversible heat engine is operated between T_3 and T_2, and the work produced is used internally to drive a reversible heat pump operating between T_1 and T_2. The net effect on the universe is the equivalent of an absorption chilling system. Heat is absorbed by the chiller at a high temperature (T_3) and a low temperature (T_1), and heat is rejected at a medium temperature (T_2) as illustrated in Figure 8.20. The efficiency of the heat engine is $(T_3 - T_2)/T_3$, and the COP of the heat pump is $T_1/(T_2 - T_1)$. Thus,

$$\frac{W}{Q_3} = \frac{T_3 - T_2}{T_3}$$

$$\frac{Q_1}{W} = \frac{T_1}{T_2 - T_1}$$

Upon eliminating W between these equations, it is found that the COP of the reversible absorption chiller is

$$\text{COP} = \frac{Q_1}{Q_3} = \frac{T_1(T_3 - T_2)}{T_3(T_2 - T_1)}$$

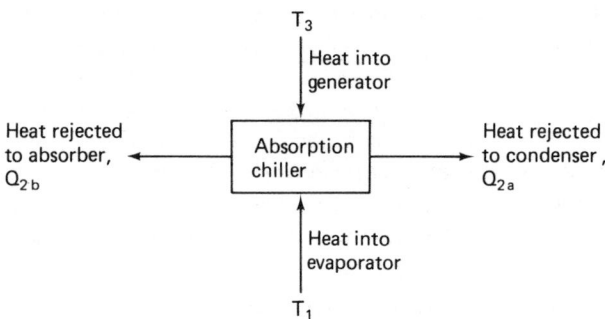

Figure 8.20 Heat flows to an absorption chiller.

For representative values $T_3 = 180°F$ (355°K), $T_2 = 90°F$ (305°K), and $T_1 = 40°F$ (278°K), the reversible COP is 1.41. In actual practice, COPs in the range 0.5 to 0.7 are obtained. The mechanical work required to pump cooling water to a cooling tower is often fairly substantial and should be taken into account when evaluating performance.

The great advantage of the absorption chiller system described above is that it can operate from a heat source in the general range near 200°F (367°K). For certain applications to solar energy this is of crucial importance. However, when absorption chillers are run from fossil energy, or from certain types of concentrating solar collectors, much higher temperatures are potentially available. In such cases, the conventional single-stage absorption chiller is not so desirable because of its low COP. The COP of an absorption chiller can be improved by going to a multiple-effect staging arrangement. This is analogous to multiple-effect evaporation processes used in chemical engineering. The main idea is to use a higher temperature and pressure in the first generator, and use the condenser for the first system as a generator for a second stage at lower pressure as illustrated in Figure 8.21. Because of the higher generator temperature, a higher COP is obtained. In practice, a COP of about 1.0 can be obtained with a heat source in the 300–350°F (422–450°K) range.

The absorption chillers have several disadvantages. They are not easily made in small sizes (under 25 tons)[1] and are very expensive in the smaller size range. They are also bulky and heavy. They do not cycle well and typically require 30 min to reach proper performance after being turned on. Because of the low COPs, a much larger amount of cooling water must be circulated to cooling towers than for equivalent refrigeration from compression refrigeration.

[1]One "ton" of refrigeration is the cooling effect (heat removal) required to produce 1 ton of ice in 24 hr. This amounts to 12,000 Btu/hr, or 3.517 kW.

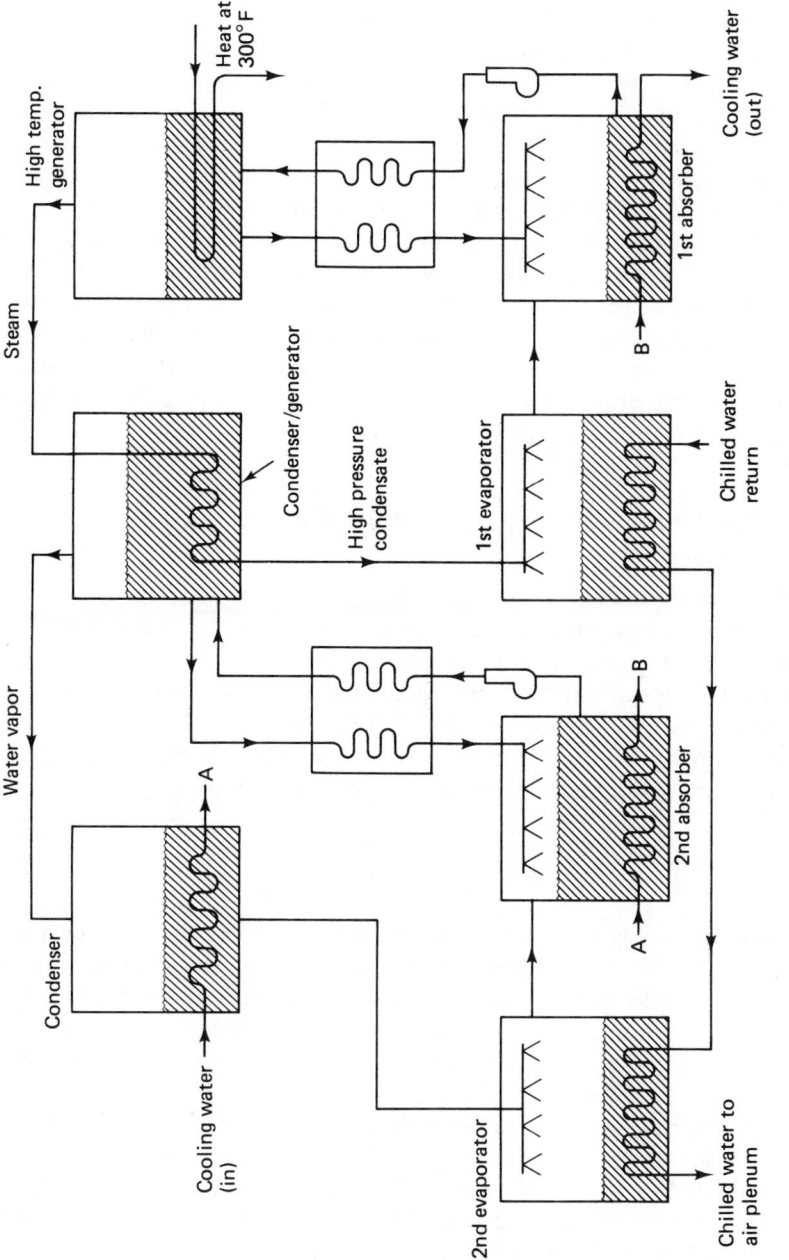

Figure 8.21 Schematic of two-stage absorption chiller.

One interesting aspect of the absorption chilling process is that water (under vacuum) is the refrigerant in the evaporator. In solar air conditioning systems, one typically desires to produce as much chilled water as possible during the day for air conditioning both day and night. Thus, when a solar air conditioning system operates, it cools chilled water in a storage tank. The problem is that it takes a very large volume of chilled water because the chilled water is only useful over a small range of temperature (perhaps 40–55°F (277.6–286.0°K)), and this implies that only 15 Btu of "cold" are stored per pound of water. It may therefore be desirable to store liquid water refrigerant under vacuum that is produced in the evaporator faster than it is needed when the sun is shining. Since each pound of water absorbs over 1000 Btu (1055 kJ) in evaporating, only about 1/100 of the volume is required to store the same amount of "cold." If storage tanks of water could be attached to the evaporator and generator, it might be possible to integrate cold storage with absorption refrigeration.

8.8.4 Refrigeration by Desiccant Drying and Humidification

At any temperature, there is a maximum amount of water that can be evaporated into a fixed amount of air. The dependence of the humidity ratio (lb water/lb dry air) at saturation on temperature is shown in Figure 8.22. Enthalpy charts for moist air are shown in Appendix III. When air at any temperature has a humidity ratio W and the saturated humidity ratio at that temperature is W_s, the *degree of saturation* is defined as

$$\mu = \frac{W}{W_s}$$

Air with a low degree of saturation can be substantially cooled by evaporating water into it. The wet bulb temperature is the temperature that air can be brought down to by adiabatic evaporation from water at the wet bulb temperature. If the water temperature is initially different from the wet bulb temperature, it will soon arrive at the wet bulb temperature if air is continuously blown over it.

In the psychrometric tables in Appendix III, the enthalpy of air is h_a, and the enthalpy of the water vapor at saturation is h_{as}. Thus, for example, at 90°F (305°K) $h_a = 21.625$ Btu/lb (50.30 kJ/kg) dry air is the enthalpy of dry air at 90°F (305°K), and $h_{as} = 34.31$ Btu/lb (79.80 kJ/kg) dry air is the enthalpy of 0.03118 lb (0.01414 kg) of water vapor at saturation. Therefore, the enthalpy of water vapor under these conditions is $34.31/0.03118 = 1100.4$ Btu/lb water vapor.

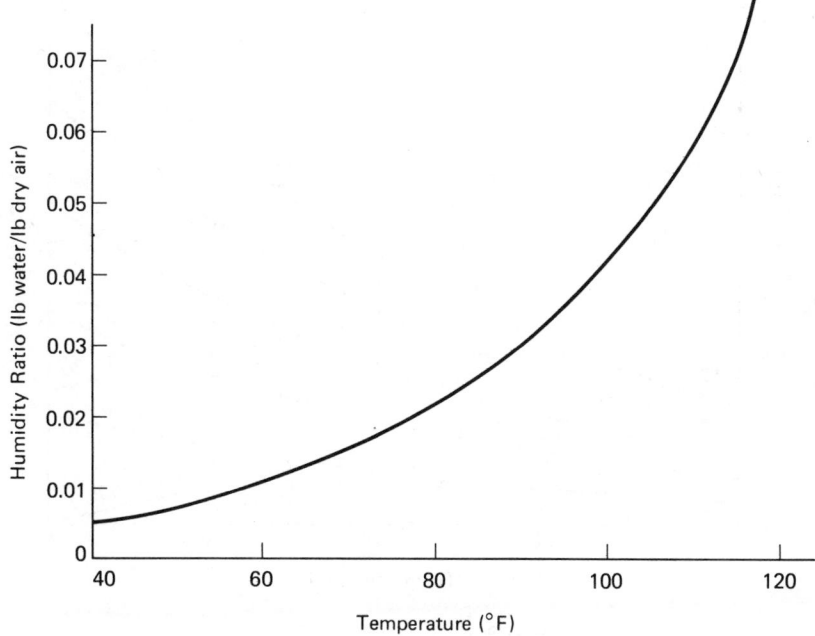

Figure 8.22 Humidity ratio of saturated air.

If dry air at 90°F (305°K) is gradually humidified by evaporation, it cools. The degree of saturation required to cool it to 80°F (300°K) is found from the equation

$$h_{a,90°F} = h_{a,80°F} + \frac{W}{W_s} h_{as,80°F}$$

where W is the humidity ratio required to cool the air to 80°F (300°K), and W_s is the humidity ratio at saturation at 80°F (300°K). Thus,

$$21.625 = 19.221 + \left(\frac{W}{0.02233}\right)(24.47)$$

$$W = 0.00219$$

$$\mu = 0.098$$

Therefore, by humidifying to 9.8% degree of saturation (roughly 10% relative humidity), the air is cooled to 80°F (300°K). The path of continuous humidification is shown on the psychrometric chart of Figure 8.23. A line of constant enthalpy, $h = 21.625$ Btu/lb, intersects the saturation line at $T = 52.5°F$ (285°K). Thus, the dry air at 90°F (305°K) can be cooled by humidification to any temperature above 52.5°F (285°K).

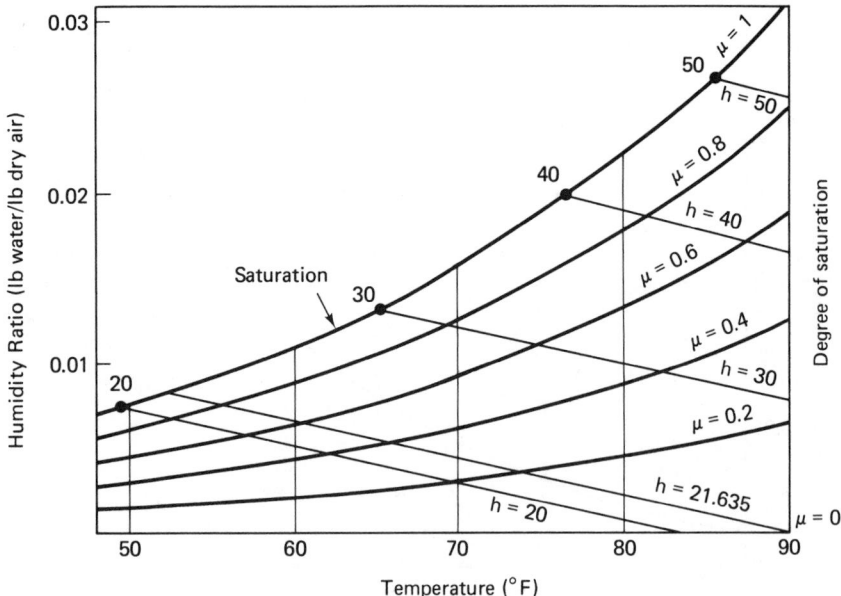

Figure 8.23 Psychrometric chart.

Except for desert areas, dry air is generally not available for humidification cooling. If one started with air at 90°F (305°K) at a degree of saturation of 40%, the air could only be cooled to about 72°F (295°K) by
humidification. Air at 72°F (295°K) and 100% humidity is not very com-
fortable. An air cooling procedure has been developed to dry recirculating
air in a house before evaporative cooling to reduce the temperature and
humidity of the air recirculated to the house. A typical system for doing this
is illustrated in Figure 8.24. The psychrometric chart of the process appears
in Figure 8.25. Ambient air at (5) corresponding to 90°F (305°K) and 50%
degree of saturation is cooled by humidification along a line of constant
enthalpy to (6). A sensible heat exchanger allows this air to absorb heat from
hot dried air at (2) in a counter-flow arrangement. The heated air at (7) is
further heated to point (8) by an external source which could be solar heat.
The hot air at (8) is used to remove moisture from the rotating desiccant
wheel and is exhausted to the atmosphere at (9). In the second loop, return
air from the house at perhaps 75°F (297°K) and 50% degree of saturation is
dried by the desiccant, cooled by sensible heat transfer, and then humidified
to produce cool moist air at about 60°F (289°K) and 80% degree of saturation to be recirculated to the house. The attainable COP and degree of
reliability of this system is not yet known.

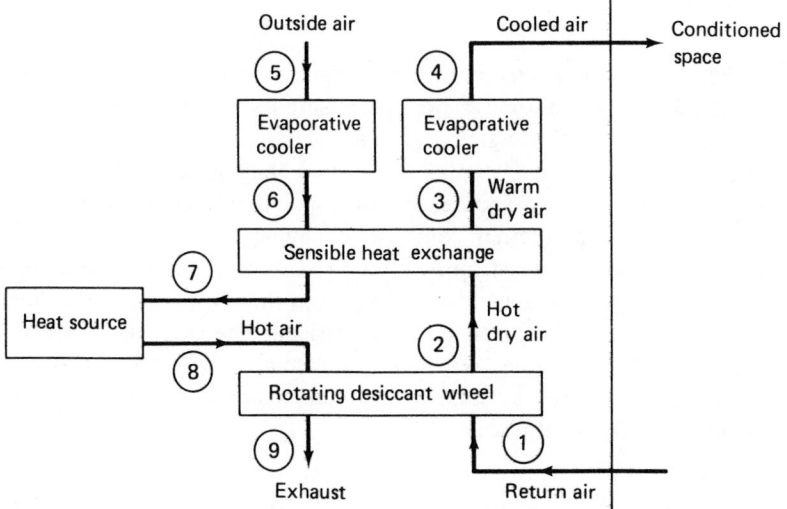

Figure 8.24 Desiccant humidification cooling scheme.

Figure 8.25 Psychrometric plot of desiccant humidification scheme.

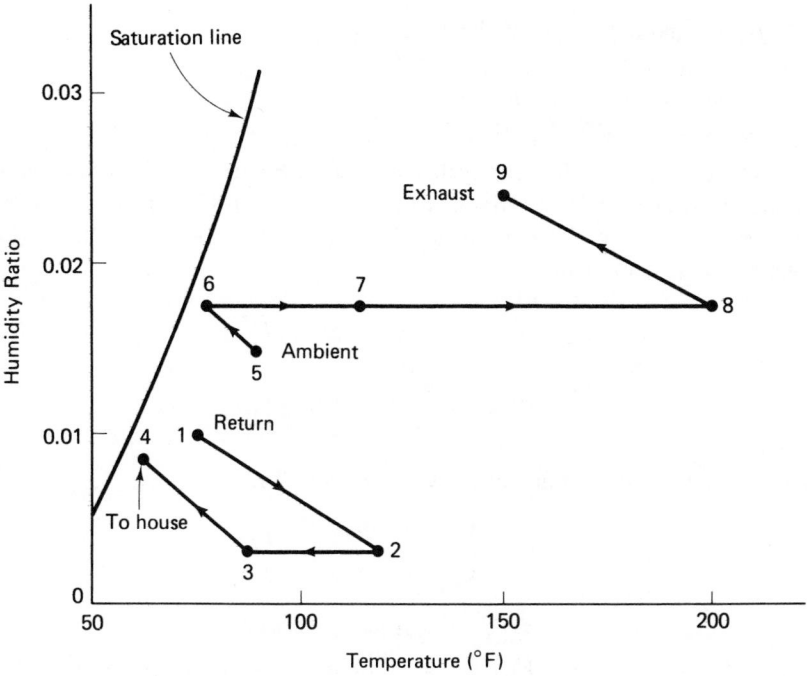

8.9 Advanced Cycles

A number of alternatives to the Rankine cycle exist for conversion of heat into work. The two that appear most promising for use in conjunction with solar energy are the Stirling and Brayton cycles. The Stirling cycle was originally conceived as an "external combustion" engine in which an external combustion chamber furnished heat to an engine that operated analogously, in some ways, to an internal combustion engine. When used with solar energy, either a solar concentrator is focussed directly onto the heat input surface for the engine or solar heat is stored and then conveyed to the engine by a transfer fluid. The Brayton cycle is basically the cycle used in a conventional aircraft jet engine. When used for solar energy applications, air is heated and compressed by solar energy and expanded through the engine to convert to mechanical power.

8.9.1 The Stirling Cycle

To understand the Stirling cycle, it is useful to first review the Carnot cycle. Consider an ideal gas as the working substance. The four steps of the cycle are carried out reversibly and are illustrated in Figure 8.26. These are:

1. Isothermal expansion of the gas at T_2.
2. Adiabatic expansion from T_2 to T_1.
3. Isothermal compression at T_1.
4. Adiabatic compression from T_1 to T_2.

The p-V diagram for this cycle is shown in Figure 8.27. In steps 1 and 3 there is no change in internal energy of the ideal gas working substance because the steps are isothermal, so the heat added is equal to the work done by the system. Thus,

$$Q_1 = W_1 = \int P dV = RT_2 \ln \left(\frac{V_b}{V_a}\right)$$

$$Q_3 = W_3 = \int P dV = RT_1 \ln \left(\frac{V_d}{V_c}\right)$$

Since steps 2 and 4 are adiabatic, $Q_2 = Q_3 = 0$, and

$$W_2 = -\Delta E_2 = -C_v(T_1 - T_2)$$
$$W_4 = -\Delta E_4 = -C_v(T_2 - T_1)$$

The net work done around the cycle is then $W_1 + W_3 + W_2 + W_4 = W_1 + W_3$. It can be shown that $V_a/V_b = V_c/V_d$, and, therefore the net work done

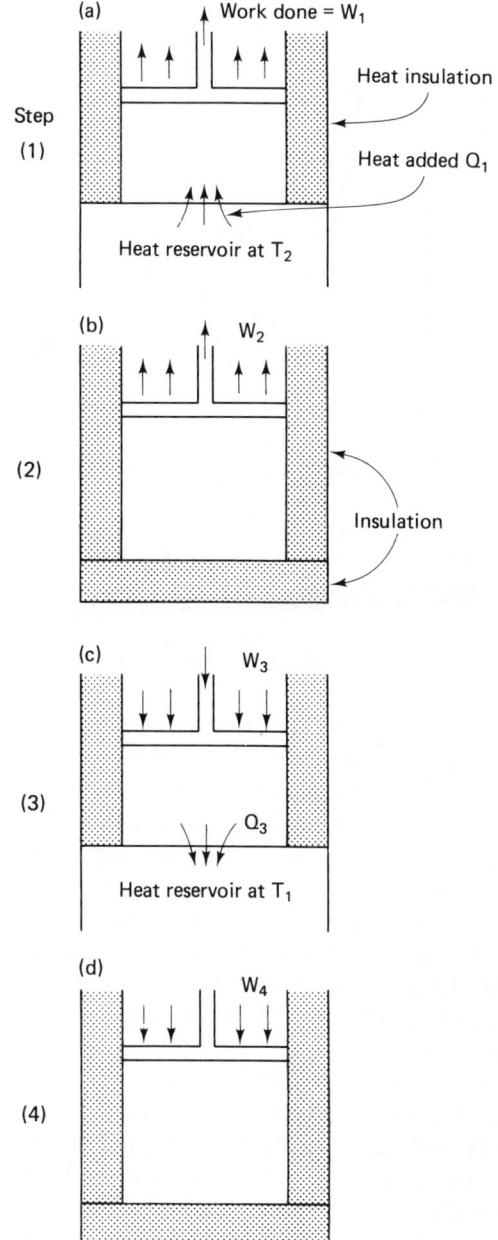

Figure 8.26 Mechanical representation of the Carnot cycle.

285

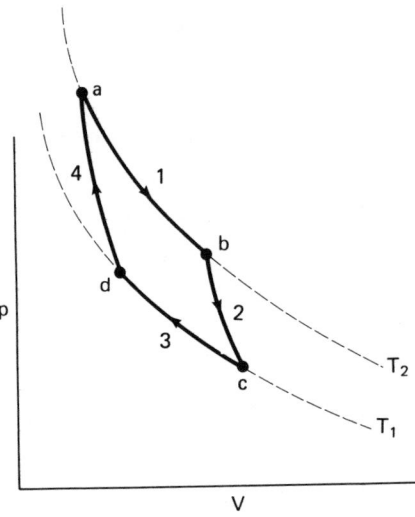

Figure 8.27 The Carnot cycle.

is $R(T_2 - T_1) \ln (V_b/V_a)$. The efficiency of the cycle is the net work divided by the heat input Q_1. Thus,

$$\eta = \frac{R(T_2 - T_1) \ln (V_b/V_a)}{RT_2 \ln (V_b/V_a)} = \frac{T_2 - T_1}{T_2}$$

The problem with the Carnot cycle is that it is difficult, if not impossible, to actually construct and operate. Because of the small area in the p-V plane, the required pressures and swept volumes are quite high.

The Stirling cycle is similar to the Carnot cycle but replaces the two adiabatic steps by constant volume steps, which increases the area in the p-V plane and reduces the required pressure and swept volume. An engine can be built that approximates the theoretical cycle. As in the case of the Carnot cycle, the two constant volume steps cancel out the effect of one another, and the net efficiency is still $(T_2 - T_1)/T_2$. The steps of the Stirling cycle are illustrated in Figure 8.28.[2] Step (1) is the same reversible isothermal expansion as in the Carnot cycle. Step (2) is a reversible constant volume cooling by a series of heat reservoirs from T_2 to T_1. Step (3) is the same isothermal compression as in the Carnot cycle, and step (4) is a constant volume reversible heating from T_1 to T_2. Since steps (2) and (4) are carried out at constant volume there is no work done, and the heat flow is equal to the change in internal energy. For an ideal gas working substance, the specific heat is independent of temperature, and $Q_2 = -Q_4 = C_v(T_1 - T_2)$. These heat flows cancel out, leaving the external heat reservoirs associated with these

[2]G. Walker, *Stirling Cycle Machines*, Clarendon Press, Oxford, 1973.

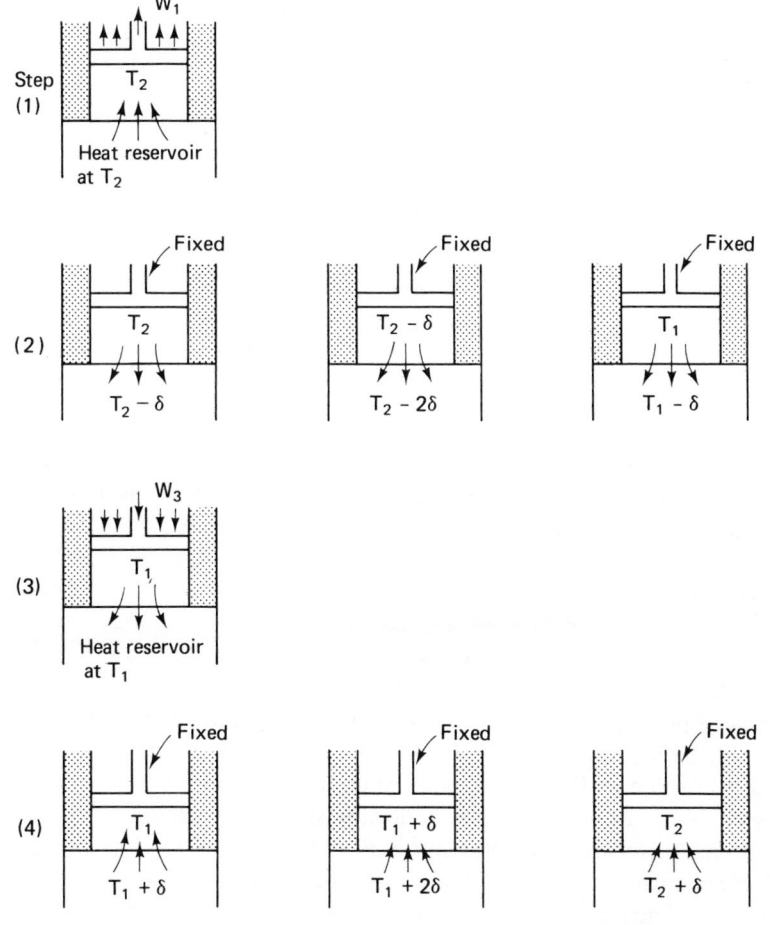

Figure 8.28 Mechanical representation of the ideal Stirling cycle.

steps unchanged after completion of a cycle. The p-V diagram is shown in Figure 8.29. The efficiency is

$$\eta = \frac{RT_2 \ln (V_b/V_a) + RT_1 \ln (V_a/V_b)}{RT_2 \ln (V_b/V_a)} = \frac{T_2 - T_1}{T_2}$$

No exact representation of the Stirling cycle has ever been built; however, several approximations have been constructed and tested.[3] One such

[3]Thermal Power Systems, Advanced Solar Thermal Technology Project, Semi-Annual Progress Report, Jet Propulsion Laboratory, DOE/JPL-1060-7816, November, 1978, Pasadena, CA.

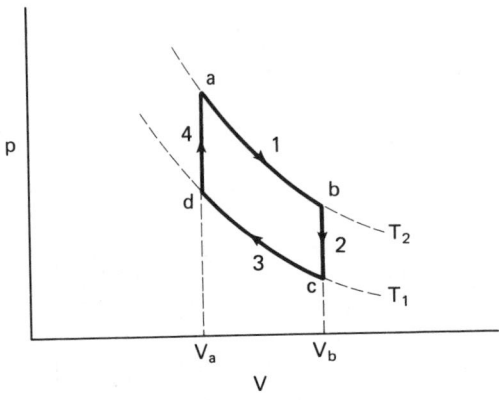

Figure 8.29 p-V diagram for the
Stirling cycle.

engine is illustrated in Figure 8.30. A cylinder is equipped with two pistons. The lower piston is the working piston that delivers work to the user, whereas the upper piston is only used to transfer gas from one region of the cylinder to another. The cylinder is maintained at high temperature near the top, and cooling fluid is circulated near the bottom to remove rejected heat. The transfer piston is used to move gas back and forth between the hot and cold regions when necessary. For the real engine, the steps are not quite as discrete as for the ideal cycle. In the engine shown in Figure 8.30, diagram (a) corresponds to point a in Figure 8.29. The gas is at T_2, and in the following step heat will be supplied at constant temperature, increasing the volume. Diagram (b) in Figure 8.30 corresponds to point b in Figure 8.29. At this point, the gas is fully expanded. In the following step, the working piston reaches bottom and reverses, while the transfer piston moves up, forcing gas from the high temperature region into the low temperature region. The cooling step $b \rightarrow c$ in Figure 8.29 is carried out irreversibly in Figure 8.30 because of the nonhomogeneity in the temperature. Diagram (c) in Figure 8.30 corresponds to point c in Figure 8.29. The final step is heat rejection at T_1 along $c \rightarrow d$. In the actual engine, this is accompanied by transfer of the gas to the hot region so that it is ready for heat gain at point d.

The advantage of the Stirling cycle over the Rankine cycle is that the potential efficiency is somewhat higher for the same maximum temperature. The reason is that all the heat is delivered at the high temperature and rejected at the low temperature, as in the Carnot cycle. The Rankine cycle requires that heat be delivered over a range of temperatures from the condensate temperature up to the vapor temperature. As a result, the average temperature at which heat is delivered is lower in the Rankine cycle. The practical problems of getting good heat transfer between the working gas and

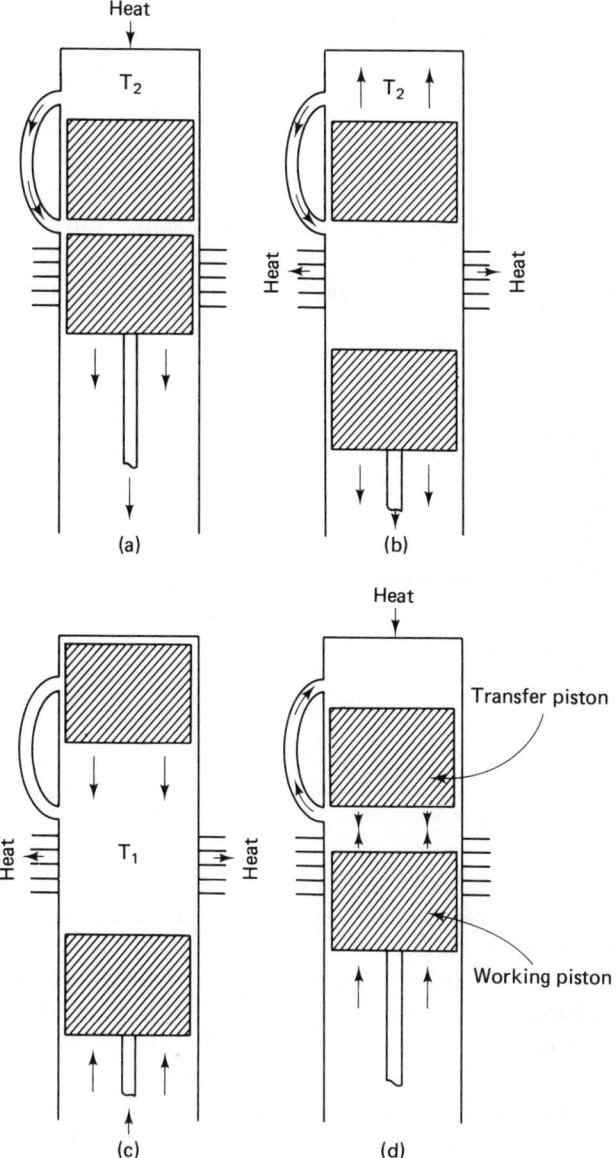

Figure 8.30 A machine to approximate the Stirling cycle.

the cylinder walls may prove to be difficult. It has been proposed[4] that the Stirling cycle be used in conjunction with the parabolic dish solar collector (Sec. 12.3), with the focus of the dish located at the heat input surface of the engine.

8.9.2 The Brayton Cycle

The Brayton cycle is a cycle for utilizing temperatures well in excess of 1000°F (811°K). A p-V diagram for a Brayton cycle is shown in Figure 8.31. A gas is used as a working substance, and the gas is compressed along ab by a compressor mounted on the same shaft as a turbine and generator.

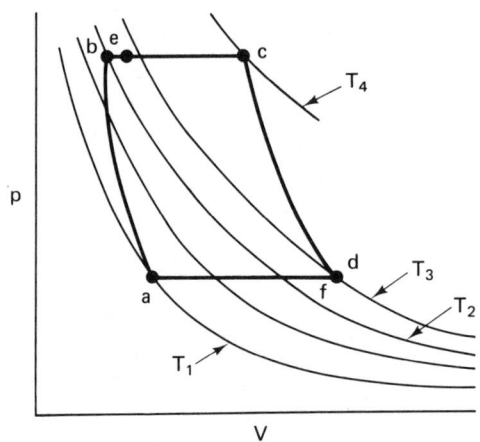

Figure 8.31 p-V diagram for the Brayton cycle.

The compressed gas at b is heated at constant pressure to a high temperature at c. The hot compressed gas is allowed to expand through a turbine along cd, producing enough work to drive the compressor and the generator. The turbine exhaust gas is fed to a heat exchanger for heat rejection prior to recycling back to the compressor. A simple flow diagram for the Brayton cycle is shown in Figure 8.32. In practice, two modifications of the Brayton cycle are made. A heat exchanger (usually called a *regenerator*) is inserted between the turbine exhaust and the heat rejection unit for preheating the compressed gas prior to the main heating step as shown in Figure 8.33. This allows the turbine exhaust, which is hotter than the gas leaving the compressor, to heat the compressed gas from b to e, while the turbine exhaust is cooled from

[4]Focus on Solar Technology: A Review of Advanced Solar Thermal Power Systems, Jet Propulsion Laboratory, DOE/JPL-1060-7815, November, 1978, Pasadena, CA.

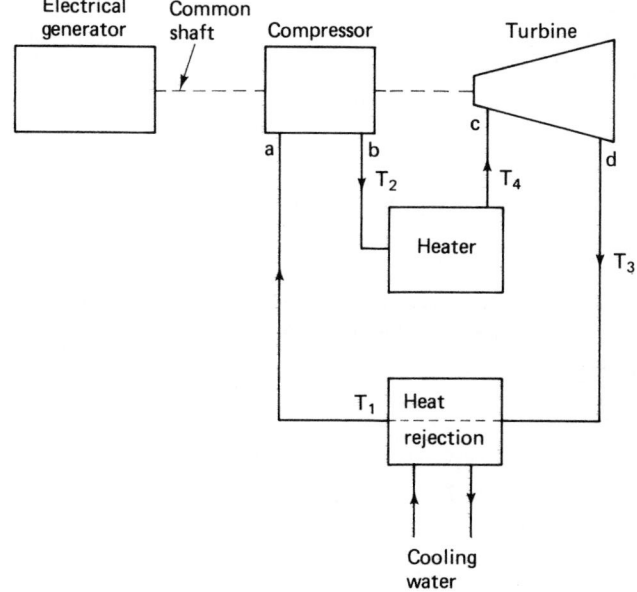

Figure 8.32 A simple Brayton cycle.

Figure 8.33 Modified Brayton cycle.

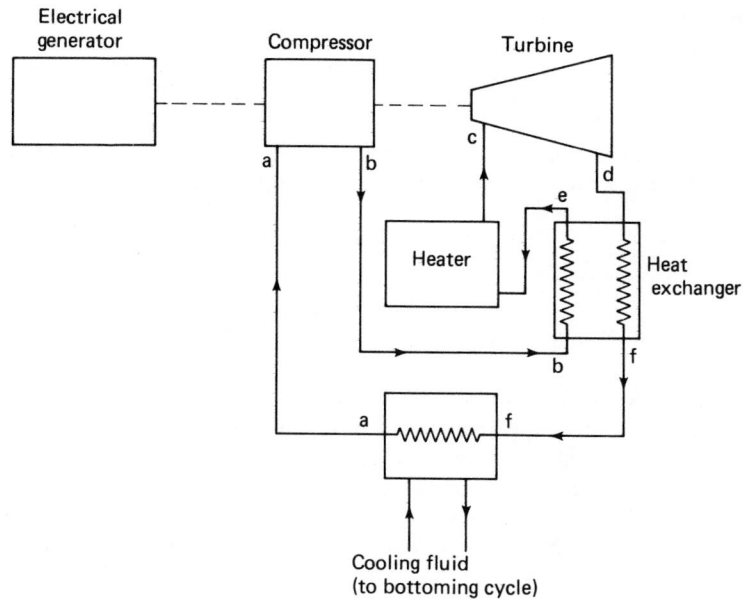

d to f. Since the same amount of work is done but less external heat is required, the cycle efficiency is improved. A second modification is to use the heated cooling fluid from the heat rejection step to supply heat to a bottoming cycle of standard Rankine design.

Worked Examples

1. If a steam Rankine cycle uses saturated steam at 422 psi and has an exhaust temperature of 125°F, what is the theoretical cycle efficiency?

    The inlet steam temperature at 422 psi is 450°F. First calculate the fraction of vapor in the exhaust:

$$S_{V,450} = xS_{V,125} + (1 - x)S_{L,125}$$
$$1.4806 = x(1.9218) + (1 - x)(0.1732)$$
$$x = 0.748$$

The enthalpy of the turbine exhaust is

$$h_{ex} = 0.748h_{V,125} + 0.252h_{L,125}$$
$$h_{ex} = 0.748(1115.4) + 0.252(93.0)$$
$$h_{ex} = 857.8 \text{ Btu/lb}$$

The cycle efficiency is then

$$\eta = \frac{h_{V,450} - h_{ex}}{h_{V,450} - h_{L,125}} = \frac{1205.6 - 857.8}{1205.6 - 93.0} = 0.313$$

2. If irreversibility led to the exhaust containing 80% vapor instead of 74.8% vapor in Example 1, what would the cycle efficiency be reduced to?

$$h_{ex} = 0.8(1115.4) + 0.2(93)$$
$$= 910.9$$
$$\eta = \frac{1205.6 - 910.9}{1205.6 - 93} = 0.265$$

3. An organic Rankine cycle uses R114 as the working fluid. The turbine cycle is arranged as shown in Figure 8.11. The boiler and preheater operate at 222.8 psi to produce saturated vapor, and there is no superheater. Turbine exhaust conditions are maintained by the cooling water, so point d is at 90°F. Points d and e may be treated as having the same temperature. The heat exchanger results in $T_c = T_e + 20°F$. What is the pressure of the turbine exhaust, what are T_b, T_c, and T_f, and what is the theoretical cycle efficiency?

According to tables in Appendix III, saturated liquid at point d at 90°F exists at a pressure of 38.8 psi. To determine point b, we require an isentropic expansion from saturated vapor at point a (220°F, 222.8 psi) to superheated vapor at 222.8 psi where the entropy is $S = 0.1701$ Btu/lb-°F. Following a line of constant entropy $S = 0.17$ down to a pressure of 38.8 psi, T_b is located at 126°F. Since $T_e = T_d$, it follows that $T_c = 110$°F. To calculate T_f, we require that the enthalpy change in going from b to c is equal to the enthalpy change in going from f to e. Therefore,

$$h_f = h_e + h_b - h_c$$
$$= 29.3 + 90.0 - 87.0 = 32.3 \text{ Btu/lb}$$
$$T_f \cong 102°F$$

Theoretical cycle efficiency:

$$\eta = \frac{h_a - h_b}{h_a - h_f} = \frac{100.6 - 90.0}{100.6 - 32.3} = 0.155$$

Thus, in summary, we have:

Point	State	T (°F)	P (psi)	h (Btu/lb)
a	Saturated	220	222.8	100.6
b	Superheated vapor	126	38.8	90.0
c	Superheated vapor	110	38.8	87.0
d	Saturated liquid	90	38.8	29.3
e	Supercooled liquid	90	222.8	29.3
f	Supercooled liquid	102	222.8	32.3

4. Starting with air at 90°F and 30% degree of humidification, what temperature can the air be reduced to by humidification if the maximum degree of humidification is 60%?

An enthalpy balance requires that

$$h_{a,90°F} + 0.3h_{as,90°F} = h_{a,T} + 0.6\,h_{as,T}$$
$$= 21.625 + 0.3 \times 34.31$$
$$= 31.92$$

By trial and error we guess T, and adjust T until the right side of the equation approaches 31.9:

T	$h_{a,T}$	$h_{as,T}$	Right Side
70	16.82	17.27	27.18
75	18.02	20.59	30.37
77	18.50	22.07	31.74
78	18.74	22.84	32.44

Problems

8.1. If a simple steam Rankine cycle operates at 65% of theoretical reversible cycle efficiency, what cycle efficiency can be achieved with saturated steam at 1000 psi and an exhaust at 125°F?

8.2. If the steam in Problem 8.1 is superheated to 800°F, what cycle efficiency should be achieved?

8.3. Suppose a fraction f of the steam flow through the turbine in Problem 8.2 is "bled off" at the point where expansion has taken place to 200 psi. If this steam flow is admixed with condensate return (liquid at 125°F) prior to introduction to the boiler, what cycle efficiency is achievable?

8.4. A Rankine cycle is run with R114 as the turbine fluid at 220°F and with an exhaust temperature of 90°F. What is the efficiency of a reversible cycle?

8.5. Starting with air at 95°F and 0.4 degree of saturation, what is the lowest temperature the air can be cooled to by humidification?

9 | Solar-Assisted Heat Pumps

This chapter concerns heat pump systems and the use of solar energy in conjunction with heat pumps for spacing heating. We show how heat pumps can greatly reduce the operational costs of electric heating. When solar energy is employed in conjunction with a heat pump, the solar system can operate at moderate temperatures, leading to lower costs and higher efficiencies as compared with ordinary solar heating systems. Computer simulation is used to test the performance of various options.

9.1 Conventional Heat Pump Systems

9.1.1 Introduction

When electricity is used to heat a residence or building, one can either use simple resistance heating or a heat pump. Resistance heating tends to be very expensive, and in many areas the cost of electricity per gross heat equivalent is of the order of four times higher than natural gas or fuel oil. However, since the electricity is delivered with nearly 100% efficiency and the fossil fuels are burned with only perhaps 60% efficiency, the price of net energy delivered is typically of the order of slightly less than three times higher with electricity than with gas or oil. The costs are, of course, dependent on region and these figures are only given as an example.

To reduce the operating costs associated with heating with electricity, a heat pump can be employed. A heat pump is simply an air conditioner operated in reverse. Instead of pumping heat from the interior of a house to the exterior as in space cooling, the system is reversed to pump heat from the exterior to the interior. If the coefficient of performance of the heat pump is denoted by COP, the heat delivered to the residence is (1 + COP) times the electrical power input to the heat pump. Most commercial manufacturers report (1 + COP) as the *heating coefficient of performance*. If electricity is to be used for space heating, the choice between resistance heating and a heat pump can be made in terms of the economics of the two systems. A tradeoff study must be made between the extra investment required for a heat pump vs. the electrical power savings each year. If the building is to be air conditioned anyway, the extra investment required to use heat pumps instead of air conditioners is not very great (typically $100/ton).[1] However, use of the heat pumps all winter shortens their life compared to only summer operation for space cooling. In many locations, a fairly convincing argument can be made in favor of heat pumps over resistance heating, especially when conventional air conditioners would have been utilized anyway. In far northern localities where conventional air conditioning is not often employed and prevailing winter temperatures are very low, heat pumps may not be desirable.

9.1.2 Performance and Balance Point

Schematic diagrams of a heat pump in the heating and cooling modes are shown in Figure 9.1. A switchover valve is arranged to reverse flow through the coils so that the indoor coil acts as an evaporator in the cooling mode and as a condenser in the heating mode. In the cooling mode, the heat pump is essentially the same as a conventional air conditioner. However, because the cycle must be used through a wider range of operating conditions than in a simple air conditioner, certain compromises are usually made in refrigerant pressure and design. As a result, heat pumps tend to have lower coefficients of performance than simple air conditioners. It is not clear whether this is the result of current design or fundamental limitations. There does appear to be ample room for improvement of the COP for heat pumps. The performance curves for a typical 10 ton heat pump in the heating mode are shown in Figure 9.2. Smaller units tend to perform less adequately. The heating and cooling performance curves of a commercial 3 ton unit are shown in Figures 9.3 and 9.4. It can be seen that as the ambient temperature deviates further from about 60–70°F (289–294°K), the COP of the unit falls

---

[1]One "ton" of refrigeration is the cooling effect required to produce 1 ton of ice from water at 32°F (273°K) in 24 hr. It is equal to 200 Btu/min = 12,000 Btu/hr = 3.517 kW.

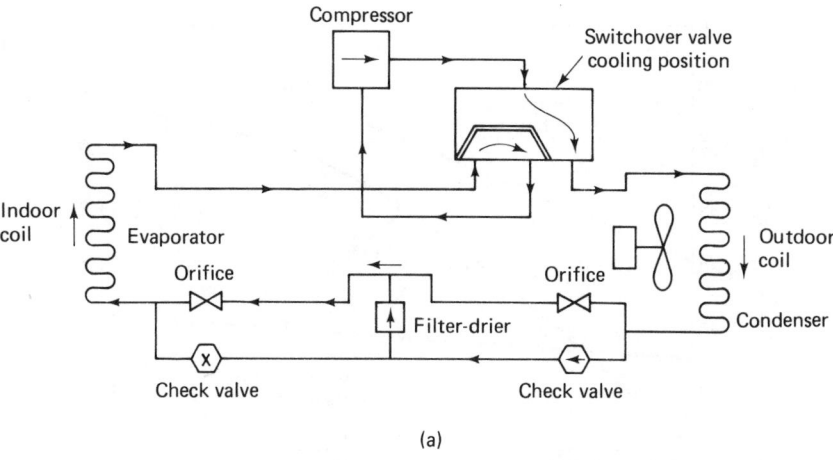

(a)

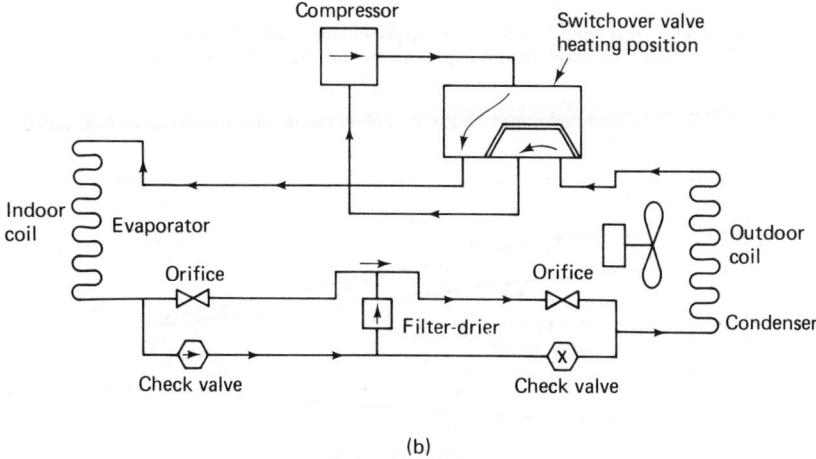

(b)

Figure 9.1 Heating and cooling with a heat pump: (a) cooling with a heat pump; (b) heating with a heat pump.

in both modes. At the same time that the COP falls off, the heat throughput also falls off. In the heating mode, the fall-off in performance with decreasing ambient temperature is quite rapid. As the ambient temperature decreases, the house heating load increases. Thus, a point is reached where the heat pump is barely able to supply the heating load without cycling on and off. This point is called the *balance point*, and when the temperature drops below the balance point, resistance heat must be used to augment the heat pump, which cannot provide the required house load even when it runs continuously. The COP of resistance heat is 0, and the effective COP (heat pumped

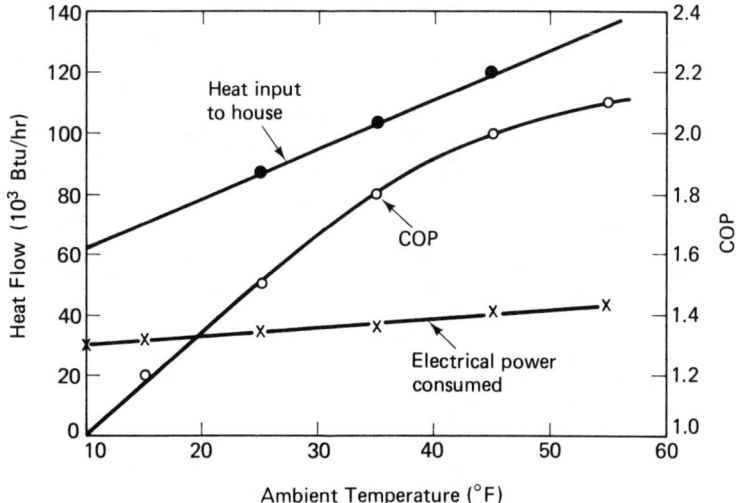

Figure 9.2 Heat input to house, Btu equivalent of electrical power consumed, and COP of a 10 ton air-to-air heat pump as a function of ambient temperature.

Figure 9.3 Heat pumped out of house in cooling mode for 3 ton air-to-air heat pump.

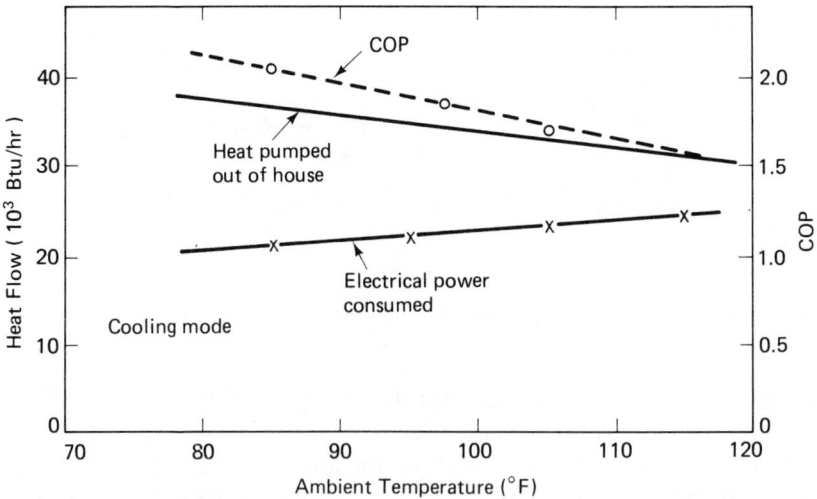

from the ambient plus resistance heat divided by total electrical power for heat pump and resistance heat) falls rapidly when the ambient temperature drops below the balance point. An illustration is provided in Figure 9.5. A 3 ton heat pump is used to heat 1500 ft² (139 m²) of residence with a heat loss coefficient of ∼0.5 Btu/hr-ft²-°F (2.84 W/m²-°K). The heat demanded by the house rises linearly as the ambient temperature drops below 65°F (292°K).

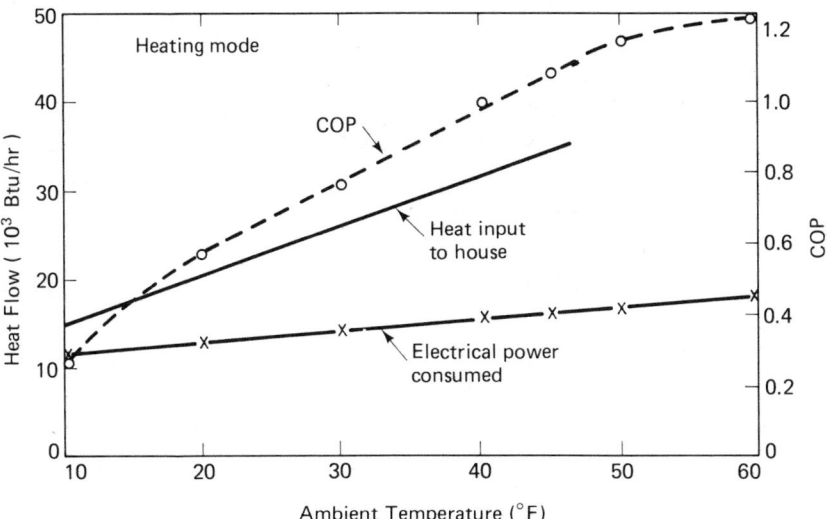

Figure 9.4 Heat input to house, Btu equivalent of electrical power consumed, and COP of a 3 ton air-to-air heat pump as a function of ambient temperature.

Figure 9.5 Effective COP of heat pump when resistance heat (COP = 0) is used to augment shortfall of heat pump below balance point. Heat demand based on (0.5 Btu/hr-ft²-°F) (1500 ft²) (65 − T_{amb}).

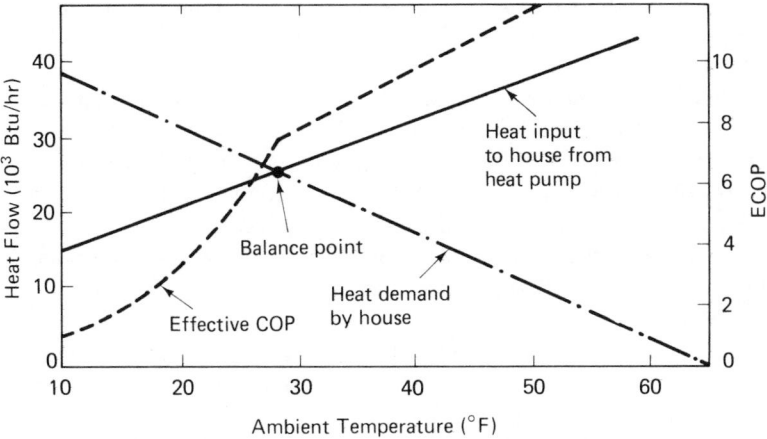

The balance point occurs at 29°F (272°K), and the effective COP drops rapidly at lower temperatures due to greater contribution of resistance heat.

For any heat pump system, we shall define the following quantities:

COP = thermodynamic COP
= (heat pumped from low temperature)/(work done)

HCOP = heating COP

  = (heat delivered)/(work done) = 1 + COP

ECOP = effective COP

  = (heat delivered by heat pump + resistance heat)/(total electricity expended for resistance heat and the heat pump)

If for any period of time E_{hp} energy is used to drive the heat pump and E_R is used for resistance heating, it follows that

$$\text{ECOP} = \frac{(\text{HCOP})(E_{hp}) + E_R}{E_{hp} + E_R} \tag{9.1}$$

9.2 Heat Pumps and Solar Energy

9.2.1 Solar-assisted Heat Pumps

Most of the conventional heat pumps now available are of the air-to-air type, meaning that the outdoor coil is exposed to the air, with a blower circulating ambient air over it, and the inside coil is in an air plenum for heat exchange with a forced-air system. During the late 1970s, a few water-to-air heat pumps were developed in which the outside coil is designed to be immersed in water to exchange heat with circulating water, in a closed loop. This is not really anything new since most commercial and industrial air conditioning units are designed to be water-cooled. Water-to-water heat pumps might be very useful in solar energy applications.

In using heat pumps with solar energy, there are a variety of approaches possible. The major points that must be kept in mind are that solar energy must furnish most of the heat to be pumped, and that the use of a solar-assisted heat pump instead of a simple solar heating system does not materially reduce the amount of heat required to be collected by solar energy. The main function of the heat pump in such applications is to make the solar collected heat readily available to the home heating system, thus allowing the collectors to function at lower temperatures and reducing the costs of collectors and possibly storage. Only the combination of a solar water system with a heat pump will be considered here, although solar air systems can also be coupled to heat pumps.

The heating system described in the computer simulation program in Chapter 7 is typical. Water at a temperature of at least 130°F (328°K) is required to heat the air in the heating plenum, and there is a 30°F (17°K) drop across the heat exchanger. Thus, the real job of the heating system is to heat recirculating water from 100°F (311°K) to 130°F (328°K). For space heating purposes, hot water is useful only when it is above 100°F (311°K).

When using a *solar-assisted* heat pump, the main purpose of the heat pump is to allow the collectors to heat water from, say, 50°F (283°K) to 80°F (300°K), rather than 100°F (311°K) to 130°F (328°K). The heat pump boosts this heat to a usable temperature in the inside coil. During periods of mild weather, storage might conceivably become heated to above 130°F (328°K). Under such conditions it would be wasteful to use the heat pump, so the heat pump can be bypassed. This is sometimes referred to as a *solar augmented* heat pump.

9.2.2 Heat Storage and Heat Pumps

Heat storage can be placed on either side of the heat pump. The optimum choice depends on whether the storage is inside the residence, outside, or buried in the ground. Interior storage may be desirable in areas with high air conditioning loads because the heat pump can be used at night in the cooling mode to produce chilled water in the storage at relatively high COP. This chilled water can be used for space cooling during the day. Storage tanks buried in the ground can be desirable insofar as they pick up heat from the ground, which is substantially warmer than the air in winter. Such storage, if not insulated, may act as an energy collector from the ground even without solar collection. However, when the exterior storage becomes warm, it will lose considerable heat to the ground. Whether the buried storage tank should be insulated or not depends on average water and ground temperatures during the winter. If most of the operating water temperatures are higher than ground temperatures, an insulated tank is desirable. It might be advantageous to use a combination of interior and exterior storage. The purpose of the interior storage is to keep a supply of hot water on hand for heating the house during periods of high demand. The exterior storage is used as a depository for solar heated water and to furnish an inlet temperature to the heat pump as high as possible.

9.2.3 Flow Systems

A simple solar-assisted and augmented water-to-air heat pump system is shown in Figure 9.6. It is not convenient to use this system for hot water, and a separate domestic hot water system, such as is described in Chapter 6, would probably be applied in parallel. This system can be simulated on a computer. The characteristics of the collectors and heat pump would have to be defined. For the house used as an illustration in Chapter 7, the floor area was 2500 ft² (232 m²) and the heat loss coefficient was 0.5 Btu/hr-ft²-°F

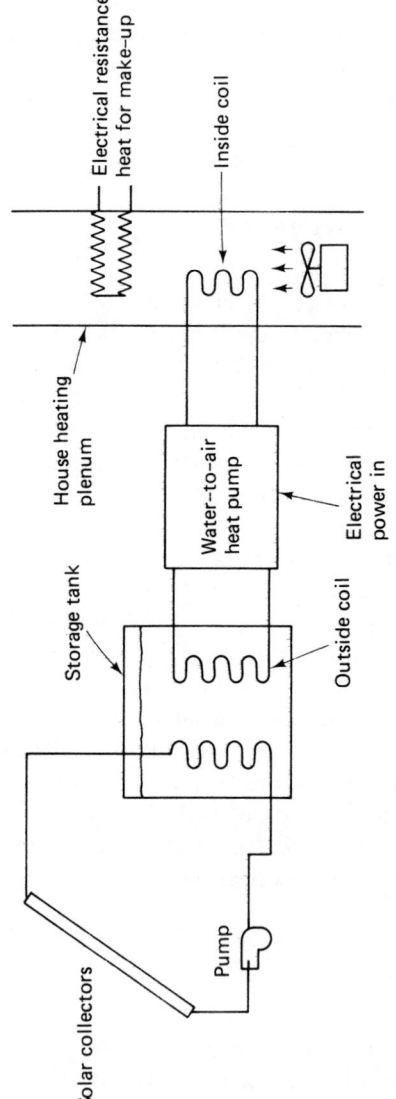

Figure 9.6 Simple heat pump system with exterior storage.

(2.84 W/m²-°K) for each degree below 65°F ambient temperature. The hourly demand for heat is then roughly $1250(65 - T_{amb})$ Btu with T_{amb} in °F. Since solar heated water provides the input to the heat pump, the input to the heat pump is almost always above 40°F (278°K), regardless of the ambient temperature. A design balance point must be selected as the ambient temperature for which the heat pump, with 40°F (278°K) inlet, is just able to provide the heat load to the house. Since there is no problem with icing up, as there would be if heat was being extracted from ambient air, the balance point can be set quite low. For North Central Texas, a 20°F (267°K) balance point is reasonable. This means that the heat pump must draw enough heat from the 40°F (278°K) source to supply the house demand at an ambient temperature of 20°F (267°K). The house demand at $T_{amb} = 20°F$ (267°K) is estimated at 56,250 Btu/hr (16.5 kW). A good 5 ton heat pump should provide this load quite easily. When T_{amb} drops below 20°F (267°K), the excess of house demand above this amount is supplied by electrical resistance heat. The COP of the heat pump will be assumed to be approximated by the function

$$COP = 1.0 + \frac{T_s - 36}{34}$$

where T_s is the storage temperature in °F. The COP reaches 2.2 at $T_s = 76.8°F$ (298.1°K). It will be assumed that when $T_s > 76.8°F$, the COP remains fixed at 2.2. It follows that the maximum HCOP is 3.2. The solar collectors will be assumed to be the same as those used in Chapter 7, i.e., having an efficiency of

$$\eta = 0.7 - \frac{0.9(\Delta T)}{I}$$

with ΔT in °F and I in Btu/hr-ft².

9.2.4 Computer Simulation

The computer simulation of this system may now be carried out. The only differences between this program and that given in Chapter 7 are:

1. The domestic hot water is run on a separate system.
2. No auxiliary make-up heat is required until T_{amb} drops to 20°F (267°K), provided that storage is at least 40°F (278°K).
3. Storage temperatures above 36°F (275°K) will provide an adequate source of heat to drive the heat pump.

At each hour for which there is heat demand by the house, the heat pump draws heat from the storage tank to supply to the house. If storage drops to 36°F (275°K), heating occurs entirely via resistance heating and the heat pump is not used. When the ambient temperature drops below 20°F (267°K), the heat pump can only supply part of the load, and the remainder is made up by electrical resistance heat. These characteristics of the solar-assisted heat pump system can easily be programmed as a modification to the space heating program of Chapter 7. A heat loss term from storage is based on the assumption that the storage tank is buried in the ground and covered with R20 insulation. The ground temperature is taken as tap water temperature. The domestic hot water system is kept separate as in Chapter 6. The solar collector loop functions in the usual way. At each hour where there is a space heat demand, storage supplies heat to the heat pump, provided that storage is above 36°F (275°K); otherwise resistance heat is used. When the ambient temperature drops below 20°F (267°K), resistance heat is used to augment the heat pump, regardless of the storage temperature. Systems with solar augmentation bypass the heat pump when $T_s > 130$°F (328°K).

9.2.5 Hourly Results

When the solar heat pump is run from start-up in January, the results for the first two days are shown in Figure 9.7. It can be seen that the storage tank starts at tap water temperature [assumed to be 55°F (286°K)] and rapidly cools as heat is drawn out by the heat pump on the morning of January 1. As soon as storage drops to below 36°F (6th hour), the heating system goes on to electrical resistance heating. This continues until the solar heat input to the tank raises the storage temperature above 36°F (12th hour), whereupon the heat pump goes back on. Some of the column titles are:

TISP = tank input to heat pump (Btu/hr) for space heating

AUX2 = total electrical energy required for running the heat pump and resistance heat (Btu-equivalents/hr)

FSOL = fraction of space heat load provided from solar energy

SPHT = space heat load (Btu/hr)

COP = thermodynamic coefficient of performance of heat pump = heat pumped/electricity required by pump

ECOP = effective coefficient of performance = heat delivered/electricity expended = 1 + COP

For the first day, 42.4% of the heat required is provided by solar energy, an impressive value considering that storage never became hotter than 80.6°F (300°K). The ECOP for the first day is 688,750/396,790 = 1.74. The second day is similar to the first, with storage providing heat to the heat pump in the early hours until the storage temperature drops to below 36°F (275°K),

Figure 9.7 Hourly printout of results with solar-assisted heat pump.

```
PROGRAM TO EVALUATE PERFORMANCE OF SOLAR ASSISTED HEAT PUMP
```

```
LATITUDE = 33.0
TILT = 33.0
STORAGE = 600.0 GALLONS (UNDERGROUND)
HOUSE AREA = 2500 SQ FT
```

MO	DAY	DECLIN	TTAP
1	1	-22.92	55.0

HOUR	SH	ST	EFF	TA	TS	TSNEW	SHITT	TISP	AUX2	FSOL	SPHT	ECOP
1	0	0	0.0	40	55.0	51.2	0.	19037.	12213.	0.609	31250.	2.56
2	0	0	0.0	39	51.2	47.3	0.	19215.	13285.	0.591	32500.	2.45
3	0	0	0.0	38	47.3	43.4	0.	19263.	14467.	0.571	33750.	2.33
4	0	0	0.0	37	43.4	39.6	0.	19227.	15773.	0.549	35000.	2.22
5	0	0	0.0	36	39.6	35.8	0.	19033.	17217.	0.525	36250.	2.11
6	0	0	0.0	35	35.8	35.8	0.	0.	37500.	0.	37500.	1.00
7	0	0	0.0	35	35.8	35.8	0.	0.	37500.	0.	37500.	1.00
8	0	0	0.0	33	35.8	35.8	0.	0.	40000.	0.	40000.	1.00
9	3	0	0.0	34	35.8	35.8	0.	0.	38750.	0.	38750.	1.00
10	40	73	0.0	38	35.8	35.8	0.	0.	33750.	0.0	33750.	1.00
11	96	163	0.689	42	35.8	42.5	33690.	0.	28750.	0.0	28750.	1.00
12	140	231	0.669	46	42.5	49.2	46306.	12915.	10835.	0.544	23750.	2.19
13	155	256	0.651	48	49.2	56.8	49896.	12356.	8894.	0.581	21250.	2.39
14	176	299	0.635	50	56.8	65.9	56851.	11569.	7181.	0.617	18750.	2.61
15	117	213	0.601	53	65.9	71.6	38452.	9789.	5211.	0.653	15000.	2.88
16	100	217	0.574	52	71.6	76.9	37389.	10918.	5332.	0.672	16250.	3.05
17	63	189	0.531	51	76.9	80.6	30136.	12031.	5469.	0.687	17500.	3.20
18	4	0	0.0	50	80.6	78.0	0.	12891.	5859.	0.687	18750.	3.20
19	4	0	0.0	46	78.0	74.7	0.	16328.	7422.	0.687	23750.	3.20
20	4	0	0.0	44	74.7	71.1	0.	17886.	8364.	0.681	26250.	3.14
21	4	0	0.0	42	71.1	67.2	0.	19270.	9480.	0.670	28750.	3.03
22	4	0	0.0	41	67.2	63.3	0.	19722.	10278.	0.657	30000.	2.92
23	4	0	0.0	40	63.3	59.2	0.	20099.	11151.	0.643	31250.	2.80
24	4	0	0.0	39	59.2	55.2	0.	20390.	12110.	0.627	32500.	2.68

MONTH	DAY	DTOT	DAUX2	DSH	FSOLAR
1	1	688750.0	39679092	688750.0	0.424

MO	DAY	DECLIN	TTAP
1	2	-22.83	55.0

HOUR	SH	ST	EFF	TA	TS	TSNEW	SHITT	TISP	AUX2	FSOL	SPHT	ECOP
1	0	0	0.0	40	55.2	51.3	0.	19059.	12191.	0.610	31250.	2.56
2	0	0	0.0	40	51.3	47.6	0.	18499.	12751.	0.592	31250.	2.45
3	0	0	0.0	39	47.6	43.9	0.	18620.	13880.	0.573	32500.	2.34
4	0	0	0.0	39	43.9	40.3	0.	17936.	14564.	0.552	32500.	2.23
5	0	0	0.0	38	40.3	36.7	0.	17872.	15878.	0.530	33750.	2.13
6	0	0	0.0	37	36.7	33.1	0.	17674.	17326.	0.505	35000.	2.02
7	0	0	0.0	37	33.1	33.1	0.	0.	35000.	0.0	35000.	1.00
8	0	0	0.0	36	33.1	33.1	0.	0.	36250.	0.0	36250.	1.00
9	22	0	0.0	35	33.1	33.1	0.	0.	37500.	0.0	37500.	1.00
10	86	157	0.677	37	33.1	39.5	31789.	0.	35000.	0.0	35000.	1.00
11	141	239	0.653	39	39.5	45.5	46783.	17049.	15451.	0.525	32500.	2.10
12	180	296	0.641	41	45.5	53.6	57021.	16836.	13164.	0.561	30000.	2.28
13	199	328	0.626	43	53.6	62.6	61529.	16571.	10929.	0.603	27500.	2.52
14	193	327	0.609	46	62.6	71.5	59749.	15213.	8537.	0.641	23750.	2.78
15	172	313	0.584	47	71.5	79.5	54868.	15111.	7389.	0.672	22500.	3.04
16	132	286	0.556	48	79.5	86.1	47634.	14609.	6641.	0.687	21250.	3.20
17	73	219	0.498	48	86.1	89.8	32735.	14609.	6641.	0.687	21250.	3.20
18	6	0	0.0	48	89.8	86.8	0.	14609.	6641.	0.687	21250.	3.20
19	0	0	0.0	45	86.8	83.4	0.	17188.	7813.	0.687	25000.	3.20
20	0	0	0.0	43	83.4	79.6	0.	18906.	8594.	0.687	27500.	3.20
21	0	0	0.0	42	79.6	75.6	0.	19766.	8984.	0.687	28750.	3.20
22	0	0	0.0	40	75.6	71.3	0.	21378.	9872.	0.684	30250.	3.17
23	0	0	0.0	38	71.3	66.8	0.	22646.	11104.	0.671	33750.	3.04
24	0	0	0.0	36	66.8	62.0	0.	23774.	12476.	0.656	36250.	2.91

MO	DAY	DECLIN	TTAP
2	24	-9.65	55.0

HOUR	SH	ST	EFF	TA	TS	TSNEW	SHITT	TISP	AUX2	FSOL	SPHT	ECOP
1	0	0	0.0	45	106.9	103.5	0.	17188.	7813.	0.687	25000.	3.20
2	0	0	0.0	43	103.5	99.7	0.	18906.	8594.	0.687	27500.	3.20
3	0	0	0.0	40	99.7	95.4	0.	21484.	9766.	0.687	31250.	3.20
4	0	0	0.0	38	95.4	90.7	0.	23203.	10547.	0.687	33750.	3.20
5	0	0	0.0	35	90.7	85.5	0.	25781.	719.	0.687	37500.	3.20
6	0	0	0.0	34	85.5	80.2	0.	26641.	109.	0.687	38750.	3.20
7	0	0	0.0	33	80.2	74.6	0.	27500.	500.	0.687	40000.	3.20
8	0	0	0.0	32	74.6	69.0	0.	28099.	151.	0.681	41250.	3.14
9	63	92	0.292	32	69.0	55.1	8040.	27364.	886.	0.663	41250.	2.97
10	136	166	0.511	35	65.1	66.0	28883.	24372.	128.	0.650	37500.	2.86
11	194	263	0.559	38	66.0	70.4	44955.	22044.	706.	0.653	33750.	2.88
12	233	313	0.576	43	70.4	77.6	54021.	18373.	9127.	0.666	27500.	3.01
13	256	344	0.572	46	77.6	86.2	58963.	16328.	7422.	0.687	23750.	3.20
14	253	343	0.552	47	86.2	94.5	56747.	15469.	7031.	0.687	22500.	3.20
15	215	298	0.521	50	94.5	101.2	46514.	12891.	5859.	0.687	18750.	3.20
16	180	262	0.483	51	101.2	106.4	37960.	12031.	5469.	0.687	17500.	3.20
17	116	194	0.403	52	106.4	108.9	23506.	11172.	5078.	0.687	16250.	3.20
18	35	0	0.0	52	108.9	106.6	0.	11172.	5078.	0.687	16250.	3.20
19	0	0	0.0	50	106.6	104.1	0.	12891.	5859.	0.687	18750.	3.20
20	0	0	0.0	47	104.1	101.0	0.	15469.	7031.	0.687	22500.	3.20
21	0	0	0.0	44	101.0	97.3	0.	18047.	8203.	0.687	26250.	3.20
22	0	0	0.0	41	97.3	93.2	0.	20625.	9375.	0.687	30000.	3.20
23	0	0	0.0	40	93.2	88.9	0.	21484.	9766.	0.687	31250.	3.20
24	0	0	0.0	38	88.9	84.2	0.	23203.	10547.	0.687	33750.	3.20

Figure 9.8 Hourly printout of results with solar-assisted heat pump on a day when storage is warm.

whereupon the heat pump goes off and resistance heating takes over. As the sun warms storage, the heat pump goes back on. On a bright day with warm storage, the ECOP for the day is greater than 3, as shown for February 24 in Figure 9.8. It can be seen that on this day there is no shortage of available solar energy and storage is at a reasonably high temperature for driving a solar-assisted heat pump. Indeed, one could not do much better than February 24 for a solar-assisted heat pump. Nevertheless, only 68.1% of the space heat is supplied by solar energy due to the electrical requirements of the heat pump. Therefore, the natural size of the collector field will tend to be considerably smaller with a solar-assisted heat pump than with a solar heating system.

9.2.6 Monthly Results

The monthly performance of a solar-assisted heat pump is shown in Figure 9.9. It can be seen that solar energy provides about 68.8% of the space heat demand during the months when there is an excess of solar energy available. This may be regarded as an upper limit comparable to the 100% of a simple solar heating system. If more solar collectors are added, they will only increase the fraction solar during the peak winter months. As a result, the returns from adding more collectors beyond 300 ft² (27.9 m²) are

```
LATITUDE = 33
COLLECTOR AREA = 500 SQ FT
COLLECTOR TILT = 33
STORAGE VOLUME = 1000 GAL
HOUSE AREA = 2500 SQ FT

MONTH    HEAT       SOLAR      FSOLAR    ECOP
         DEMAND     HEAT
         (MILLION   (MILLION
          BTU)       BTU)

  1      19.59      10.98      0.560     2.28
  2      12.80       8.77      0.685     3.17
  3      10.10       6.92      0.685     3.18
  4       1.78       1.22      0.688     3.21
  5       0.66       0.45      0.688     3.20
  6       0.00
  7       0.00
  8       0.00
  9       0.14       0.09      0.687     3.21
 10       1.80       1.24      0.687     3.20
 11       8.03       5.22      0.650     2.85
 12      16.14      10.53      0.653     2.87
```

Figure 9.9 Monthly results for solar-assisted heat pump.

rapidly diminishing, and the natural size of the collectors for a solar-assisted heat pump without augmentation are considerably smaller than for a simple solar heating system. When solar augmentation is used, larger collector fields become more practical. The monthly results for a 500 ft²/1000 gal (46.5 m²/3.78 m³) solar-assisted heat pump with augmentation are as shown in Table 9.1. It can be seen that solar augmentation improves performance during the months of April, May, September, October, and November. On

TABLE 9.1 Monthly performance of solar-assisted and solar-assisted/augmented heat pumps (500 ft², 1000 gal, 45° tilt, 46.5 m², 3.78 m³)

Month	Space Heat Demand (10^6 Btu/mo)	Solar-Assisted Heat Provided (10^6 Btu/mo)	ECOP	Solar-Assisted and Augmented Heat Provided (10^6 Btu/mo)	ECOP
Jan	19.6	11.3	1.75	11.8	1.82
Feb	12.8	8.8	1.96	9.7	2.16
Mar	10.1	6.9	1.99	8.1	2.34
Apr	1.8	1.2	2.37	1.7	3.35
May	0.7	0.5	2.48	0.7	3.47
Jun	0.0	0.0	—	0.0	—
Jul	0.0	0.0	—	0.0	—
Aug	0.0	0.0	—	0.0	—
Sep	0.1	0.1	2.49	0.1	3.62
Oct	1.8	1.2	2.39	1.8	3.59
Nov	8.0	5.3	2.20	6.8	2.82
Dec	16.1	10.7	1.78	11.5	1.91
Year	71.0	46.0		52.2	

a yearly basis, the fraction of required heat supplied by solar energy is raised from 54% to 58% by solar augmentation. A comparison on a monthly basis of the solar augmented and nonaugmented 500 ft²/1000 gal (46.5 m²/3.78 m³) is shown in Figure 9.10. Yearly data for several sizes of installation appear in Table 9.2. It can be seen that solar augmentation greatly improves performance, especially in the larger sizes. A nonaugmented system has an upper limit of 2.2/3.2 = 68.8% for the fraction of space heat that can be provided by solar energy, whereas an augmented system can provide up to 100% of the energy required for space heat. As a result, nonaugmented systems do not perform as well as augmented systems, especially in the larger sizes.

Figure 9.10 Monthly performance of 500 ft²/1000 gal (46.5 m²/3.78 m³) solar-assisted and solar-assisted/augmented heat pumps.

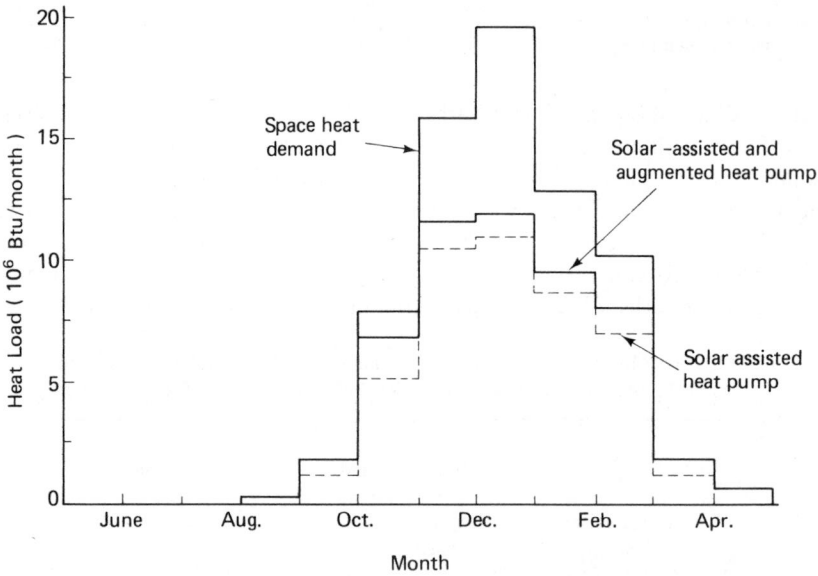

9.3 Comparison of Solar Heating with Solar Heat Pumps

In order to compare with space heating/hot water systems without heat pumps (Chapter 7), a separate domestic hot water system with 100 ft² (9.29 m²) of collector and 120 gal (0.454 m³) of storage will be added to each heat pump design. Such a hot water system can provide 24 × 10⁶ Btu/yr (25.3 × 10⁹ J/yr) out of the 35.7 × 10⁶ Btu/yr (37.7 × 10⁶ J/yr) required. The results for the combined augmented heat pump/hot water systems are shown in

TABLE 9.2   Results of a solar-assisted heat pump to provide space heat to a 2500 ft²
(232 m²) house in Dallas, Texas

Collector Area (ft²)	Storage Volume (gal)	Yearly Space Heat Load (10⁶ Btu)	Solar Heat Provided (10⁶ Btu)	Fraction Solar	Collector Tilt Angle (degrees)	Augmenta-tion Used
200	400	71.0	30.3	0.425	45	No
			31.9	0.448	45	Yes
300	600	71.0	37.2	0.524	33	No
			38.3	0.539	45	No
			41.2	0.580	45	Yes
500	1000	71.0	45.4	0.640	33	No
			46.0	0.648	45	No
			52.2	0.735	45	Yes
750	1500	71.0	48.3	0.680	33	No
			48.5	0.683	45	No
			59.7	0.840	45	Yes

TABLE 9.3   Results for combined solar-assisted and augmented
heat pump/domestic hot water system with 100 ft²
(9.3 m²) of collector and 120 gal (0.454 m³) in the
hot water system

Total Collector Area (ft²)	Total Storage Volume (gal)	Total Heat Required (10⁶ Btu/yr)	Solar Heat Supplied (10⁶ Btu/yr)	Fraction Solar
300	520	107	55.9	0.522
400	720	107	65.2	0.609
600	1120	107	76.2	0.712
850	1620	107	83.7	0.782

Table 9.3. A comparison of these results with those of Chapter 7 (including heat loss from storage) is shown in Figure 9.11. The heat pump provides more solar heat than the heating/hot water system.

The economics of a solar-assisted heat pump should be discussed in terms of a reference base of a house heated by electricity with a heat pump and no solar assistance. A calculation has been performed for such a system, using an air-to-air heat pump with an assumed COP given by

$$COP = 1 + \frac{T_{amb} - 42}{34} \tag{9.3}$$

where T_{amb} is the ambient temperature in °F. This function goes to 0.706 at $T = 32$°F. This is the assumed balance point. For ambient temperatures

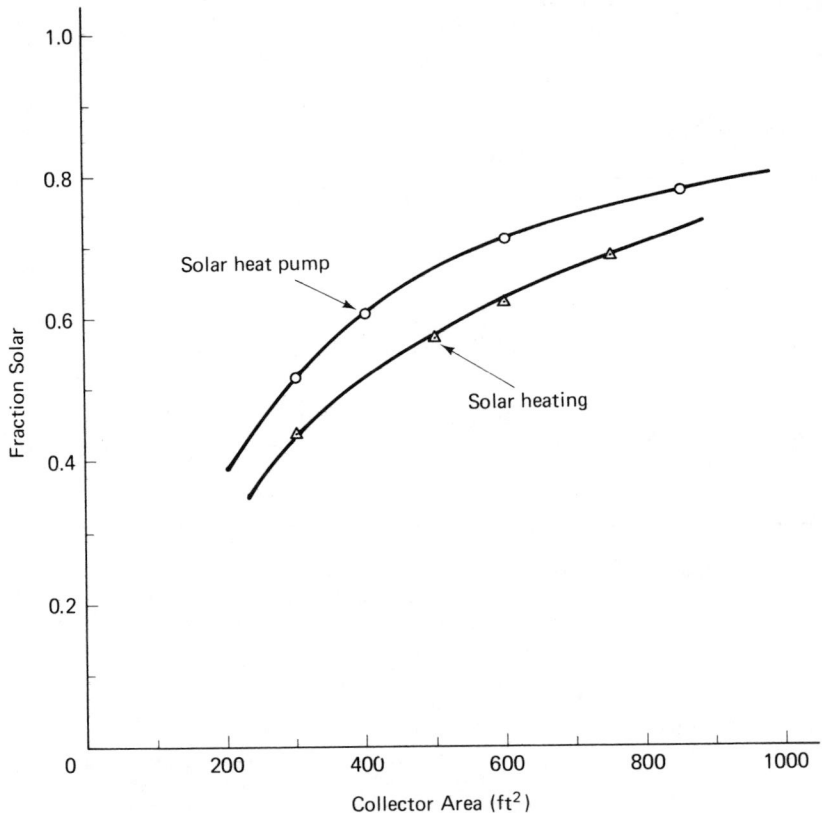

Figure 9.11 Comparison of yearly performance of solar-assisted/augmented heat pump with solar heating system.

lower than 32°F (273°K), the assumed COP is

$$\text{COP} = 1 + \frac{T_{\text{amb}} - 37}{17} \tag{9.4}$$

This goes to 0 at $T_{\text{amb}} = 20$°F. The heating COPs are, of course, 1 plus the COPs. The assumed balance point is 32°F (273°K), and increasing amounts of resistance heat must be used as T_{amb} is reduced below 32°F (273°K). The results for such a system are shown in Table 9.4. The effective ECOP is 1.90 for the year, and the heat pump reduced electrical requirements from 71.0 × 10^6/3413 = Btu/yr = 20,800 kWh per yr with resistance heat, to 10,920 kWh per yr with the heat pump. This system may now be used as a reference system for comparison with solar-assisted/augmented heat pumps as in Table 9.5. Compared to the saving of nearly 10,000 kWh per yr in going from

TABLE 9.4 Space heating results from operating a heat pump with ambient air as a heat source

Month	Space Heat Demand (10^6 Btu/yr)	Heat from Ambient (10^6 Btu/yr)	Electrical Energy Used (kWh)	Average ECOP
Jan	19.6	8.4	3280	1.75
Feb	12.8	6.3	1910	1.96
Mar	10.1	5.0	1490	1.99
Apr	1.8	1.0	220	2.37
May	0.7	0.4	80	2.48
Jun	0.0	0.0	0	—
Jul	0.0	0.0	0	—
Aug	0.0	0.0	0	—
Sept	0.1	0.06	20	2.49
Oct	1.8	1.1	220	2.39
Nov	8.0	4.4	1070	2.20
Dec	16.1	7.0	2660	1.78
Year	71.0	33.7	10,920	1.90

TABLE 9.5 Electrical energy requirements of solar-assisted/ augmented heat pump compared with electrical energy requirements of a nonsolar heat pump

Collector Area (ft^2)	Storage Volume (gal)	Electrical Energy Required (solar assist) (kWh/yr)	Electrical Energy Required (no solar assist) (kWh/yr)	Electrical Energy Saved by Solar (kWh/yr)
200	400	11,460	10,920	−540
300	600	8730	10,920	2190
500	1000	5510	10,920	5410
750	1500	3310	10,920	7610

resistance heat to a heat pump, use of solar assist provides less savings. Thus, a solar-assisted heat pump may look good when compared with resistance heat, but it does not compare so well with a nonsolar heat pump.

The economics of the heat pump systems will be estimated on the basis that costs are the same as for the solar heating systems in Chapter 7, except that collectors for heat pump systems will be valued at $7 per ft^2 ($75 per m^2) instead of $10 per ft^2 ($108 per m^2). With this assumption, the costs are listed in Table 9.6. There is no net present value to any of the installations, and the pay-back periods are greater than 15 years. By comparison, the installation of an air-to-air heat pump without solar assistance and augmenta-

TABLE 9.6 Economics of solar-assisted/augmented heat pumps compared with a reference system of a nonsolar heat pump. Collectors are valued at $7/ft² ($75/m²) and scenario III in Chapter 4 is used. Electricity is valued at $0.04/kWh in the first year

Collector Area (ft²)	Storage Volume (gal)	kWh/yr Saved by Solar	Estimated Cost of Solar System ($)	NPV $ (15 yr life)	PBP (yr)
200	400	−540	3900	—	—
300	600	2190	4825	−3623	>15
500	1000	5410	6581	−3696	>15
750	1500	7610	8686	−4217	>15

tion saves 20,800 − 10,920 = 9880 kWh per year at a cost of perhaps $2000. This leads to a net present value of $10,490 and a pay-back period of less than 5 years. It may be concluded that in a mild climate like Dallas a heat pump is a good investment, but solar assistance is not yet worthwhile. If the cost of the heat pump is taken as the excess above an ordinary heater/air conditioner, the pay-back period for a nonsolar heat pump is shortened to 1 year. On the other hand, in a very cold climate, solar assistance might be necessary to maintain a moderate temperature at the evaporator of the heat pump to improve the COP.

9.4 Passive Cooling of the Heat Pump Condenser

In climates where the summer nights are cool, water from the storage tank can be circulated to the solar collectors at night to cool off the storage system. The storage tanks then act as a reservoir of cold water for cooling the condenser of the heat pump in the air conditioning mode. This keeps the COP of the air conditioner high at very little additional operating expense. If the nights were reliably cool enough, this might even obviate the need for a cooling tower.

Worked Examples

1. A heat pump is operating in the heating mode with a COP = 1.3 at an ambient temperature below the balance point. If resistance heat must be used for 30% of the load, what is the effective COP, and what fraction of the load must be supplied by electricity?

Let the load be 100 units. Then 30 units are supplied by resistance heat; 70 units are supplied by the heat pump.

Since the COP $= 1.3$, the electricity required for the heat pump is

$$\frac{1}{2.3} \times 70 = 30.4 \text{ units}$$

The total electricity required is 60.4% of the load. The effective COP is

$$\text{ECOP} = \frac{\text{heat delivered}}{\text{electricity required}} = \frac{100}{60.4} = 1.66$$

2. Flat plate solar collectors are available with an efficiency curve

$$\eta = 0.7 - \frac{0.8(T - T_{amb})}{I}$$

where the temperatures are in °F, and I is the intensity on the collectors in Btu/hr-ft². The average value of I during winter operation is 230, and the average value of $T_{amb} = 40°F$. If the collectors are used in a solar space heating system, the average value of T is 140°F, whereas when used in a solar-assisted heat pump system, the average value of T is 90°F. What is the difference in efficiency of the two systems? What fraction of the collector area required for solar space heating is required for a solar-assisted heat pump with the same solar heat gain?

Using the solar heating system, the average collector efficiency is

$$\eta = 0.7 - \frac{0.8(140 - 40)}{230} = 0.352$$

Using the solar-assisted heat pump,

$$\eta = 0.7 - \frac{0.8(90 - 40)}{230} = 0.526$$

The solar heat pump would deliver the same amount of solar gain with 0.352/ 0.526 $\times$ 100 = 66.9% of the collector area required for the solar heating system.

Problems

9.1. A heat pump has a thermodynamic COP given by the function COP $= 1 + (T_{amb} - 40)/20$ for temperatures above 20°F.

(a) What is the heating COP at $T_{amb} = 35°F$?
(b) If the heat pump operates at $T_{amb} = 25°F$ and supplies two-thirds of the house demand, and the other one-third of the house demand is supplied by resistance heat, what is the effective COP?

9.2. A house of 2000 ft^2 has a heat loss coefficient of 0.4 Btu/hr-°F per ft^2 of floor area. If the electrical power consumed by a heat pump is 4.5 kW independent of ambient temperature, and the COP is $1 + (T_{amb} - 40)/20$, at what value of T_{amb} is the balance point reached? What is the effective COP at 10°F below the balance point?

9.3. A house in Texas requires 10,000 kWh for cooling (with a compression chiller) in the summer and 18,000 kWh for resistance heating in the winter. If a new installation is being installed, what are the economics of using a heat pump with an effective COP of 1.8 for the winter instead of resistance heat, if the summer cooling COP of the heat pump is the same as COP of the compression chiller? The heat pump costs an extra $1600, and electricity is worth $0.04 per kWh in the first year.

9.4. A house is located in Chicago, Illinois, where the monthly average temperatures for October through May are 55°F, 40°F, 28°F, 25°F, 32°F, 36°F, 49°F, and 58°F, respectively. If the effective COP of a heat pump is given by the function

$$\text{ECOP} = 3.5 \qquad\qquad T_{amb} > 70°F$$

$$\text{ECOP} = 2 + \frac{T_{amb} - 40}{20}, \qquad 70°F \geq T_{amb} \geq 20°F$$

$$\text{ECOP} = 1 \qquad\qquad T_{amb} < 20°F$$

what is the average effective COP for the entire heating season? If solar assistance can maintain storage (on average) 30°F warmer than ambient, what fraction of the total electricity expended in the nonsolar heat pump system can be saved by use of solar assistance? Compare this with Dallas, Texas, where the average temperatures for the same time period are 63°F, 54°F, 47°F, 45°F, 49°F, 55°F, 62°F, and > 65°F. Is solar assistance to heat pumps a better investment in Chicago or Dallas?

9.5. A house in Chicago requires 30,000 kWh of resistance heating for an average winter season. Using the average effective COP for the winter in Chicago with and without solar assistance from Problem 9.4, estimate the electricity saved by using a solar-assisted heat pump. If capital is borrowed at 6% after taxes for 15 years, what maximum cost must the solar-assisted heat pump have to produce a positive net present value? What cost will lead to a pay-back period of 8 years?

10

Intermediate Temperature Collectors

This chapter concerns the theory and design of solar collectors for producing heat in the range 200–330°F for application to areas such as solar air conditioning systems and industrial process heat. In general, the heat losses associated with flat plate collectors must be reduced to operate effectively at these elevated temperatures. This can be achieved by evacuating the collectors or by using concentrators. Concentrators can be run in three modes: fixed, periodic adjustment, or continuous tracking. To operate in fixed or periodic adjustment modes, it is desirable to maintain as large an acceptance aperture for any concentration. The theory of the relation between aperture and concentration is explored in some detail. The "ideal" concentrator is one that provides the maximum aperture for any concentration. The CPC collector is an example of such an ideal concentration scheme, which is discussed in great detail. Other collection schemes that may be more practical are also discussed. The parabolic trough is covered in some detail.

10.1 Introduction

Intermediate temperature collectors are usually solar collectors designed primarily to provide hot fluid to drive solar heating/cooling/hot water systems. In order to power solar cooling systems, operating temperatures in the range 200–330°F (367–439°K) are desirable during the summer, depending on the type of cooling system used. One of the best solar cooling approaches involves the use of a two-stage absorption chiller, which requires at least

300°F (422°K) input temperature. During the winter it is usually not necessary for an intermediate temperature collector to produce temperatures
higher than 120–180°F (322–355°K) for space heating, depending on the
application.

10.2 Definition of Concentration

Flat plate collectors can only function as intermediate temperature collectors
if the space between the absorber plate and the cover glazing is evacuated to
reduce heat loss. An evacuated flat plate collector has several disadvantages,
and most intermediate temperature collectors are designed to concentrate
the incoming sunlight onto a small receiver to reduce the surface area which
loses heat to the surrounding air. The degree of concentration is a critical
factor in solar collector design and has been defined in a variety of confusing
ways in the field. The *theoretical concentration* C_t of a lens or reflector can
be defined as the ratio of the area of the entrance aperture to the area of an
exit aperture containing the refracted or reflected rays. Although this definition is conceptually simple and appealing, it is not convenient to use if the
receiver is not planar. In general, we define the *effective concentration* C_e
of a solar collector as the ratio of the entrance aperture area to the area of
the receiver that is not well insulated. This definition can be ambiguous and
difficult to interpret in some cases. Nevertheless, it does emphasize the most
important aspect of concentrating collectors, which is the area on the receiver
that is exposed for heat loss. The effective and theoretical concentrations
of solar collectors are based on the assumption of 100% transmission by
lenses, 100% reflection by reflectors, and 100% absorption by receiving
surfaces. Since the object of a concentrating collector is to increase the intensity on a receiving surface, imperfect optical characteristics reduce the final
intensity proportional to the optical efficiency. Therefore, a third definition
of concentration is the *useful concentration* C_u, which is the product of C_e
times the overall optical efficiency η_{opt}. The overall optical efficiency is the
product of transmissivities, reflectivities, and absorptivities for light interact-
ing with all surfaces between the entrance aperture and receiver.

   To illustrate these definitions, consider the following examples. The
flat plate collector illustrated in Figure 10.1 has $C_t = C_e = 1$ because the
bottom of the collector plate is heavily insulated. If the underside of the collector plate was not insulated, C_e would be approximately 0.5 because both
the top and bottom surfaces of the receiver plate would lose heat, although
C_t would remain equal to unity. The optical efficiency of the flat plate collector is probably about 0.7, and thus $C_u \cong 0.7$ for a typical flat plate col-

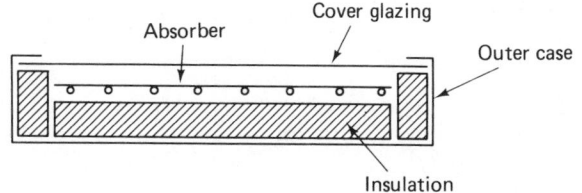

Figure 10.1 Cross section of flat plate collector.

Figure 10.2 Front reflecting parabolic troughs.

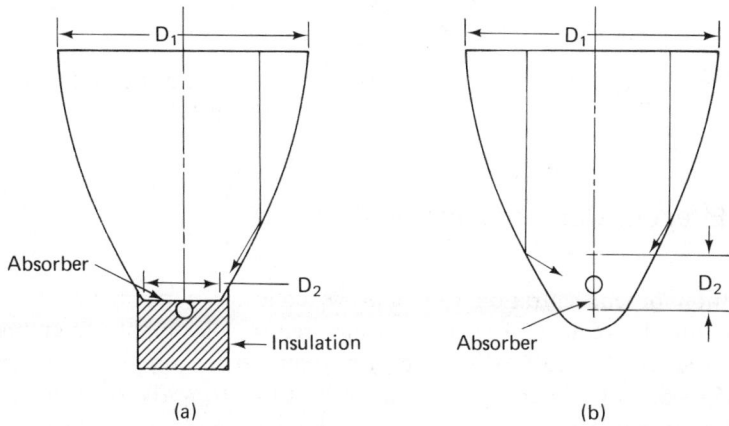

lector. For the front reflecting parabolic trough shown in Figure 10.2(a), the width of the entrance slot is D_1 and the width of the exit slot is D_2. Therefore, $C_t = C_e = D_1/D_2$. The C_u is equal to D_1/D_2 times the reflectivity of the reflecting surface. The alternative receiver design shown in Figure 10.2(b) has both sides of the receiver exposed to heat loss and, therefore, $C_e = D_1/(2D_2)$. However, if the backs of the reflectors are insulated and there is a glazing over the front aperture, the rate of heat loss can be reduced. Nevertheless, the design shown in Figure 10.2(b) is inherently less satisfactory than that in Figure 10.2(a) because the light is concentrated onto a surface with twice the heat loss area. Finally, consider the cross section of a parabolic trough illustrated in Figure 10.3. The receiver is covered with heavy insulation on the top and is exposed to the reflector underneath. The theoretical concentration is $(W - D_2)/D_1$. However, if $\theta = 120°$, the curved receiver surface has an area of $\pi D_1/3$ per unit length, and therefore

$$C_e = \frac{(W - D_2)}{(\pi D_1/3)} \tag{10.1}$$

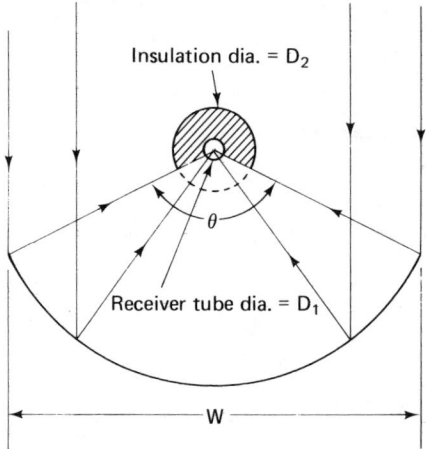

Insulation dia. = D_2

θ

Receiver tube dia. = D_1

W

Figure 10.3 Back reflecting parabolic trough.

10.3 Effect of Concentration on Efficiency

Evacuation or concentration can improve collector performance at high temperatures by reducing heat losses compared to solar input per unit area of the receiver. The rate of solar energy input to the receiver is $I_0 C_e \eta_{opt}$, where I_0 is the solar intensity. The rate of heat loss from the receiver can be written approximately $B(T_{abs} - T_{amb})$ over limited ranges of T_{abs}, where T_{abs} is the absorber temperature, T_{amb} is the ambient temperature, and B is a linearized approximate heat transfer coefficient per unit area of the receiver. As T_{abs} increases, B also increases due to the fourth power dependence of radiative losses on temperature. The net rate of energy gain by fluid in the receiver is the difference between solar input and heat loss, and the net gain divided by $C_e I_0$ is the efficiency:

$$\eta = \frac{I_0 C_e \eta_{opt} - B(T_{abs} - T_{amb})}{I_0 C_e} \qquad (10.2)$$

$$\eta = \eta_{opt} - \frac{B(T_{abs} - T_{amb})}{C_e I_0} \qquad (10.3)$$

This has the same general form as the expression for efficiency of flat plate collectors. The concentration acts as a multiplier for the solar intensity. However, the transmissions and/or reflections required to produce C can reduce η_{opt} and in no case can η exceed η_{opt}.

10.4 Relation between Concentration and Acceptance Aperture

Rabl[1] has written an important paper on concentrating collectors. He shows that there is an inverse relationship between the concentration and the acceptance aperture of a collector system. The higher the concentration, the smaller the range of acceptance angles by a collector. The acceptance aperture of a collector is an important property because it determines the type and accuracy of tracking that is required and because it determines the ability of the collector to receive circumsolar or diffuse light. Therefore, to achieve elevated temperatures, the choice of an approach for concentration can have significant impact on tracking requirements. Rabl[2] was concerned with the limitations on acceptance angle produced by concentration by various techniques, whether some forms of concentrators are by nature more ideal than others, and what levels of concentration can be achieved with fixed non-tracking collectors.

There are three levels of operation of solar collectors:

1. Continuously tracking
2. Periodic (seasonal) adjustment of tilt
3. Fixed position

It is clear that collectors with fixed position are by nature cheapest and easiest to operate. However, it is difficult to achieve the required acceptance aperture for reasonable collection during the various seasons unless the concentration is low. Collectors with high concentration require continuous tracking. There is, however, a middle ground where the concentration is high enough to achieve moderately elevated temperatures (200–300°F = 367–422°K) and yet have an aperture large enough to require only periodic seasonal adjustment of the tilt. In designing such collectors, it is desirable to obtain as high an acceptance angle as is possible for any concentration, so as to reduce the number of such seasonal adjustments to a minimum. On the other hand, the economics of such a collector may not be good, depending on the specifics of the design. To achieve the maximum concentration for fixed aperture, the amount of reflector surface required may become excessive. Therefore, collectors with maximum aperture may not be very desirable in a practical sense.

[1] A. Rabl, "Comparison of Solar Concentrators," *Solar Energy*, **18**, 93, 1976.
[2] Ibid.

10.4.1 Maximum Concentration for Fixed Aperture

We now proceed to give several basic theorems following the derivations given by Rabl. The first theorem is:

I. The maximum possible concentration achievable with a collector that only accepts all incident light rays within the half-angle θ is $1/\sin^2 \theta$ for a three-dimensional collector and $1/\sin \theta$ for a two-dimensional collector.

To prove this theorem, Rabl set up a thought-experiment as shown in Figure 10.4. An isotropically radiating sphere of radius r is surrounded by

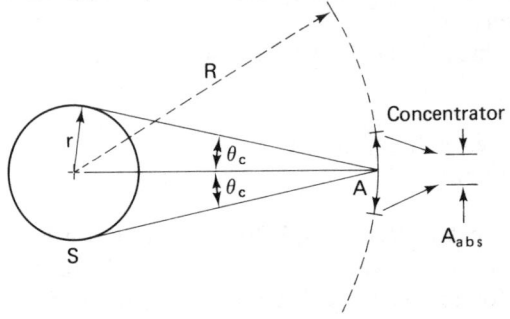

Figure 10.4 Radiation transfer from source S through aperture A of concentrator to absorber A_{abs}. By permission, *Solar Energy*, (A. Rabl, "Comparison of Solar Concentrators," *Solar Energy*, **18**, 93, 1976).

a sphere of radius R which has an aperture of area A that admits light to a concentrator behind A. The concentrator focuses the light entering through A onto a smaller area A_{abs}, corresponding to a theoretical concentration of $C_t = A/A_{abs}$. It is assumed that the concentrator focuses all light entering A within the half-angle θ of the normal to A. If A/R^2 is small, the light from the source S incident on A is uniformly distributed over all angles between 0 and θ, and there is no light from S outside of θ. Extraneous radiation is eliminated by assuming black walls at absolute zero temperature. The procedure is now to calculate the rates of emission of radiant energy by the source and the absorber, and then to note that the radiant transfer between the source and absorber must be equal if they are both at the same temperature. The source emits the radiant power

$$Q_S = 4\pi r^2 \sigma T_S^4 \tag{10.4}$$

where σ is the Stefan-Boltzmann constant, of which a fraction

$$F_{S \to A} = \frac{A}{4\pi R^2} \tag{10.5}$$

hits the aperture. With the assumed perfect concentrator optics, no radiation is lost between the aperture and the absorber. Thus, the power radiated from the source to the absorber is

$$Q_{S \to \text{abs}} = Q_S F_{S \to A} = A \frac{r^2}{R^2} \sigma T_S^4 \tag{10.6}$$

The absorber radiates an amount

$$Q_{\text{abs}} = A_{\text{abs}} \sigma T_{\text{abs}}^4 \tag{10.7}$$

and the fraction of this radiation, $E_{\text{abs} \to S}$, which reaches the source cannot exceed unity. Hence, the radiative power transfer from the absorber to the source is

$$Q_{\text{abs} \to S} = E_{\text{abs} \to S} A_{\text{abs}} \sigma T_{\text{abs}}^4 \tag{10.8}$$

with

$$E_{\text{abs} \to S} \leqq 1 \tag{10.9}$$

If the source and absorber are at the same temperature, the second law of thermodynamics requires that there cannot be any net heat transfer between these bodies. Thus, if we set $T_{\text{abs}} = T_S$, it follows that

$$Q_{S \to \text{abs}} = Q_{\text{abs} \to S} \tag{10.10}$$

Therefore,

$$A \frac{r^2}{R^2} = E_{\text{abs} \to S} A_{\text{abs}} \tag{10.11}$$

from which we deduce that the theoretical concentration satisfies

$$C_t = \frac{A}{A_{\text{abs}}} = \frac{R^2}{r^2} E_{\text{abs} \to S} = \frac{E_{\text{abs} \to S}}{\sin^2 \theta} \tag{10.12}$$

Since the maximum possible value of $E_{\text{abs} \to S}$ is unity, the concentration must satisfy

$$C_t \leqq \frac{1}{\sin^2 \theta} \tag{10.13}$$

Collectors for which the equal sign holds are called *ideal* by Rabl. Rabl indicated that even though this proof rests upon a specific geometry, it is nevertheless completely general because, if a concentrator with greater concentration could be found from a different geometry, this concentrator could be used in the geometry of Figure 10.4 to produce a violation of the second law of thermodynamics.

For a two-dimensional collector with cylindrical symmetry, Figure 10.4 may still be used, except that S and A are sections of cylinders instead of spheres. The calculation then proceeds analogously, except that r^2/R^2 is replaced by r/R, and the final result is

$$C_t \leq \frac{1}{\sin \theta} \tag{10.14}$$

10.5 Maximum Achievable Temperature

Rabl also elucidated the question of how high a temperature can be achieved at the absorber of a concentrator which concentrates light emitted by a black body source at T_S. It is clear from first principles that $T_{abs} \leq T_S$, for if T_{abs} were greater than T_S, this would imply spontaneous heat flow from S to a hotter body, which is a violation of the second law. A more detailed analysis is made by considering only direct rays from the source, assuming that the temperature of the rest of the universe, collectively called the *ambient*, is $T_{amb} = 0$. For a real collector, the power absorbed by the absorber is

$$Q_{S \to abs} = \eta_{opt} A \frac{r^2}{R^2} \sigma T_S^4$$

$$= \eta_{opt} A \sin^2 \theta T_S^4 \sigma \tag{10.15}$$

where η_{opt} is the optical efficiency (product of transmissivities, reflectivities, and absorptivities). If the source is the sun, the appropriate half-angle is $\theta \approx \frac{1}{4}°$. The radiative power loss from the absorber is

$$Q_{abs \to amb} = \epsilon_{abs} A_{abs} \sigma T_{abs}^4 \tag{10.16}$$

where ϵ_{abs} is the effective emissivity of the absorber over the spectral wavelength region characteristic of a body emitting at T_{abs}. If the efficiency of the collector for capturing incident radiation as useful energy in a transfer fluid is called η, the heat balance on the absorber is

$$Q_{S \to abs} = Q_{abs \to amb} + \eta Q_{S \to abs} \tag{10.17}$$

assuming there are no conductive or convective heat losses from the absorber. Thus,

$$(1 - \eta)\eta_{opt} \sigma A \sin^2 \theta T_S^4 = \epsilon_{abs} A_{abs} \sigma T_{abs}^4 \tag{10.18}$$

This may be written

$$T_{abs} = T_S\left[(1 - \eta)\frac{\eta_{opt}}{\epsilon_{abs}}C_t \sin^2\theta\right]^{1/4} \qquad (10.19)$$

Following Rabl, if we denote $C_{ideal} = 1/\sin^2\theta$, it follows that

$$T_{abs} = T_S\left[(1 - \eta)\frac{\eta_{opt}}{\epsilon_{abs}}\frac{C_t}{C_{ideal}}\right]^{1/4} \qquad (10.20)$$

The maximum possible value of C_t is C_{ideal}. The term η_{opt} always includes an absorptivity of the absorber. As T_{abs} approaches T_S, this absorptivity manifestly approaches ϵ_{abs}, and therefore $\eta_{opt}/\epsilon_{abs}$ cannot exceed unity as T_{abs} approaches T_S. Thus, the maximum value of T_{abs} is T_S. But this can only occur if the optics of the concentrator are perfect and no useful heat is removed from the absorber. If the collector operates at, say $\eta = 0.5$, this implies that the maximum possible absorber temperature is $T_{abs} = T_s(0.5)^{1/4} = 0.841T_S$. A reasonable practical value of η_{opt} is 0.7, and reasonable collector optics might yield $C_t/C_{ideal} \approx 0.7$. Thus, a reasonable upper limit to T_{abs} (neglecting conductive/convective losses) is

$$T_{abs} = T_S \times (0.5 \times 0.7 \times 0.7)^{1/4} \cong 0.707T_S \qquad (10.21)$$

Considering the sun as a black body at about 6000°K, the maximum T_{abs} is about 4200°K. The assumption that $T_{amb} \approx 0$ is a good one if the absorber is this hot. A concentration of about $0.7/\sin^2(\frac{1}{4}°) \cong 31,500$ is required to achieve this temperature. This applies to a three-dimensional concentrator with no convective/conductive losses. Equation (10.19) is general for a spherical source, regardless of whether the concentrator is two or three dimensional. When a two-dimensional concentrator is used, $C_{ideal} = 1/\sin\theta$, and therefore Equation (10.20) is replaced by

$$T_{abs} = T_S\left[(1 - \eta)\frac{\eta_{opt}}{\epsilon_{abs}}\frac{C_t}{C_{ideal}}\sin\theta\right]^{1/4} \qquad (10.22)$$

For a perfect two-dimensional concentrator with no useful heat removal, the maximum temperature achievable is, therefore,

$$T_{abs} = T_S (\sin\theta)^{1/4} \qquad (10.23)$$

which for the sun reduces to

$$T_{abs} = 6000\left(\frac{1}{213}\right)^{1/4} = 1570°K \qquad (10.24)$$

For a real collector, assuming $(1 - \eta) = 0.5$, $\eta_{opt}/\epsilon_{abs} = 0.7$, and $C_t/C_{ideal} = 0.7$, we find a maximum possible temperature of about 1100°K, corresponding to a concentration of $0.7 \times 213 = 149$, neglecting convective/conductive losses.

10.6 Tracking and Nontracking Modes

It was mentioned earlier in this section that concentrating collectors can be operated in continuously tracking, periodic seasonal adjustment, or fixed orientation modes. For continuously tracking collectors, intermediate temperatures ($200–300°F = 367–422°K$) can most easily be achieved with two-dimensional concentrators. For periodic adjustment or fixed orientation, the concentrators manifestly must be two dimensional, with the long axis in the east-west direction. Therefore, the remainder of this chapter concerns only two-dimensional concentrators. For continuously tracking systems, the acceptance aperture need only be made large enough to include the solar disc, plus errors in mirror or lens construction and errors in collector steerage orientation. This easily allows concentration ratios in the range 5–15 even with relatively crude construction and tracking systems, and intermediate temperatures are readily achievable. The exact degree to which C_t approaches C_{ideal} is not of great importance, and any number of approaches for concentration are possible.

10.6.1 Collectors with Periodic Adjustment or Fixed Orientation

For concentrating systems with periodic adjustment or fixed orientation, there is a great advantage in having as large an acceptance aperture as possible for fixed concentration, since this affects the number of adjustments required in the periodic case and the yearly energy collected in the fixed case. The fixed orientation case is very appealing because it eliminates the moving parts and controls of a tracking system, thus reducing investment and operating costs and greatly simplifying the collector so as to reduce maintenance requirements. The flat plate collector is usually mounted in a fixed orientation mode. The difficulty is that as one employs more concentration than a flat plate collector, the acceptance angle of the collector goes down, and therefore the yearly excursions of the sun's elevation place stringent limits on how much concentration can be used. At solar noon, the range of solar elevations during the year is $\pm 23.45°$. Thus, it is clear that a collector with fixed orientation must have an acceptance aperture half-angle of at least $23.45°$. On the equinoxes, the solar elevation remains constant all day at an angle $(\pi/2 - L)$ above the southern horizon, where $L = $ latitude. Thus, the natural orientation of a fixed orientation collector is at this angle. As the year proceeds, the solar elevation moves further from $(\pi/2 - L)$ as one approaches the solstices. At solar noon on the winter or summer solstices, the solar elevation is $23.45°$ above or below $(\pi/2 - L)$. Thus, a collector with fixed orientation must have an acceptance half-angle of at least $23.45°$ to receive light at solar noon on all days of the year. On the solstices, the sun moves higher (lower) as time moves away from noon in the summer (winter). At about $3\frac{1}{2}$ hr from noon, the excursion in elevation from $(\pi/2 - L \pm 23.45)°$ is about $12°$. Therefore, to guarantee 7 hr of collection on all days of the year, an acceptance half-angle of about $35.5°$ is required. To guarantee 6 hr of collection on all days of the year, an acceptance half-angle of about $30°$ is required. The maximum concentrations possible for fixed collectors with these angular apertures are $C_t = 1.7$ for 7 hr of collection, and $C_t = 2$ for 6 hr of collection. If a collector does not approach the ideal concentration in such a design, the actual concentration will not be much greater than unity and the advantage over a flat plate collector will be minimal. Rabl has emphasized the CPC concept (see Sec. 10.9) for this application to fixed orientation concentrating collectors, since full CPC concentrators have the ideal efficiency. However, such concentrators require large amounts of reflector surface and utilize reflections at glancing incidence, leading to high cost and poor performance. The CPC concentrators may be truncated, but this reduces the concentration at fixed aperture, and it is difficult, if not impossible, to obtain a significant concentration for a real fixed orientation collector. Some manufacturers have begun production of fixed truncated CPC collectors with a concentration of about 1.8. This concentration will improve performance slightly at elevated temperatures but will be accomplished by a loss in optical efficiency compared to flat plate collectors. It seems doubtful whether they are viable.

For concentrators with periodic seasonal tilt adjustment, a number of options are available for collector design. Rabl, in comparing such collectors, has again emphasized the CPC concept because its large aperture for fixed concentration tends to require fewer periodic adjustments than other collector designs. Actually, this is an illusion; the CPC collector offers no great advantage over other collector designs. Rabl has placed primary emphasis on the concentration achievable for any fixed acceptance aperture. However, when the ratio of reflector areas is compared for CPC concentrators and other concentrators, the advantages of the CPC collector diminish rapidly. Since the CPC concept involves surfaces at grazing angles to the incident solar rays, reflectivities are not very high. In short, it is not very practical to select a collector on the basis of abstract mathematical formulations. Bannerot[3] has compared the performance and reflector surface require-

[3]R. B. Bannerot, "The Radiative Characteristics of Non-tracking Moderately Concentrating Solar Energy Collectors", *Proc. ERDA on Concentrating Collectors*, Atlanta, GA, September, 1977.

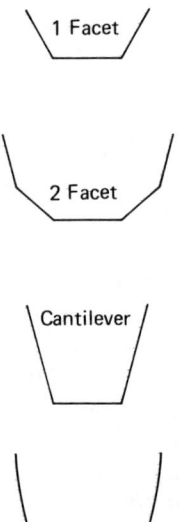

Figure 10.5 Concentrator shapes used in comparison. Adapted from "The Radiative Characteristics of Nontracking Moderately Concentrating Collectors" by R. B. Bannerot in "Solar Concentrating Collectors," Georgia Institute of Technology, Atlanta, GA, Sept. 26, 1977.

ments of several simple collector geometries (including the CPC) which might be used in periodically adjusted concentrating collectors. The concentrators he considered are illustrated in Figure 10.5. Bannerot showed that other geometries, perhaps with lower fabrication cost and higher reflectivities, can approach the truncated CPC collector in performance (defined by ratio of reflector area to aperture area required to achieve some concentration ratio at fixed aperture). His results for collectors with 9° half-angle of acceptance are shown in Figure 10.6. For concentration ratios under about 1.8, a simple one-facet design is almost as good in performance as the CPC and is surely cheaper. For concentration ratios between 1.8 and 2.5, other designs can almost approach the truncated CPC in performance. In more recent work, Bannerot[4] has shown that when the single reflection criterion is relaxed, the performance of simple faceted collectors closely rivals or exceeds the truncated CPC collector. The number of periodic adjustments required for various degrees of concentration are discussed in Sec. 10.9.6.

10.7 Evacuated Flat Plate Collectors

An evacuated flat plate collector acts like any ordinary flat plate collector except that the heat loss coefficient is reduced. A "soft" vacuum of about 0.001 atm will essentially eliminate convective heat transfer from the col-

---

[4]R. B. Bannerot, University of Houston, private communication, March 1980.

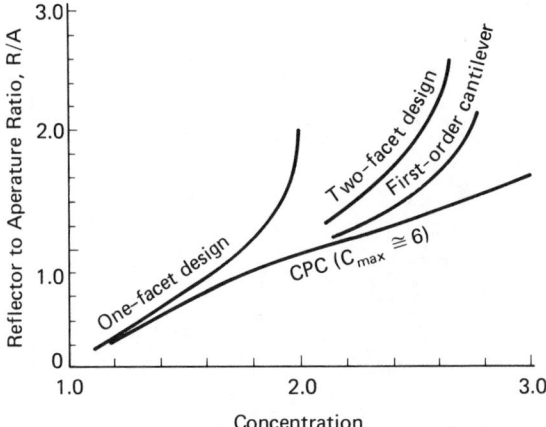

Figure 10.6 Performance comparison of various concentrator shapes for a 9° acceptance half-angle. Adapted from "The Radiative Characteristics of Nontracking Moderately Concentrating Collectors" by R. B. Bannerot in "Solar Concentrating Collectors," Georgia Institute of Technology, Atlanta, GA, Sept. 26, 1977.

lector plate to the case enclosing the plate, whereas a "hard" vacuum ($< 10^{-6}$ atm, or < 0.1 P) is required to eliminate conduction as well. It is probably only feasible to consider a soft vacuum for flat plate collectors. By eliminating convection, such collectors improve the performance of a flat plate collector at elevated temperatures. It is estimated that perhaps 50% of the heat loss from a conventional flat plate collector is due to convection when the absorber plate is about 100°F (55°K) higher than ambient (see Chapter 4). An actual evacuated collector has been marketed by Solar Systems, Inc., of Tyler, Texas. Their collector is illustrated in Figure 10.7. It has a single cover glazing supported by numerous pegs (the total force on the 32 ft² (2.97 m²) cover is about 4500 lb (641 kN). The manufacturer reports an efficiency curve of

$$\eta = 0.71 - 0.43 \frac{\Delta T}{I} \qquad (10.25)$$

with I in Btu/hr-ft² and ΔT in °F. The heat loss coefficient of 0.43 is considerably better than the heat loss coefficient expected for the best single-glazed collectors (~ 0.8 Btu/hr-ft²-°F). On the other hand, a conventional collector with no vacuum but with two glazings and a selective surface should be capable of achieving

$$\eta = 0.75 - 0.6 \frac{\Delta T}{I} \qquad (10.26)$$

Under typical operating conditions, $\Delta T/I$ lies in the range 0.2 to 0.5. In this

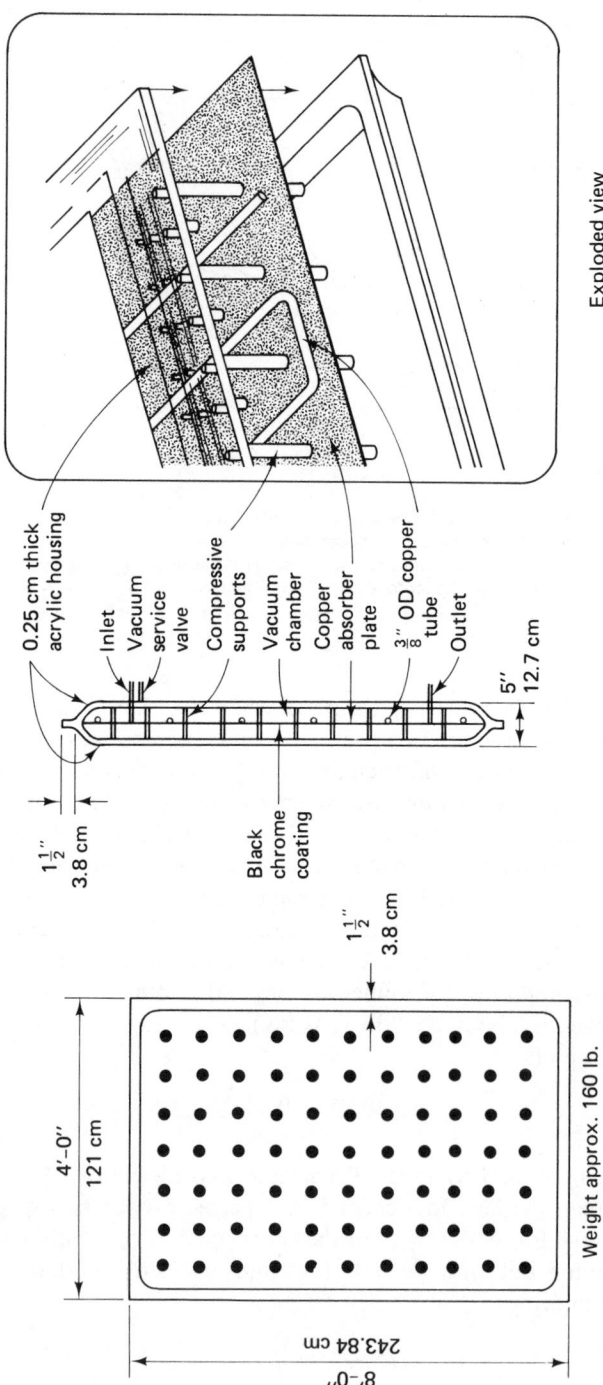

Exploded view

0.25 cm thick acrylic housing

Inlet
Vacuum service valve
Compressive supports
Vacuum chamber
Copper absorber plate
$\frac{3}{8}$" OD copper tube
Outlet

5"
12.7 cm

$1\frac{1}{2}$"
3.8 cm

Black chrome coating

$1\frac{1}{2}$"
3.8 cm

4'-0"
121 cm

243.84 cm

8'-0"

Weight approx. 160 lb.

Figure 10.7 Evacuated flat plate collector. Courtesy of Solar Systems, Inc., Tyler, TX.

Figure 10.7   (Continued)

range, the evacuated collector ranges in efficiency from 0.62 to 0.50, whereas
the conventional collector ranges from 0.63 to 0.45. The performance is
quite close under these conditions, and the extra cost and maintenance
required for evacuated flat plate collectors does not appear to be justified.
On the other hand, for higher temperature applications where $(T_{abs} - T_{amb}) >$
150°F (83°K), the operating range of $\Delta T/I$ could be 0.5 to 1.0, where the
evacuated collector has great advantages in performance.

10.8 Evacuated Tubular Collectors

Several of the large manufacturers of fluorescent lights have utilized their
highly developed technologies for producing evacuated glass tubes for the
development of tubular solar collectors. The tubes are arranged in parallel
rows to form an array that has the outward appearance of a flat plate collector module. Each tube consists of an outer glass tube of perhaps 10 cm
diameter and an inner glass tube of about 8 cm diameter. The inner tube is
coated with a selective absorbing surface, and a hard vacuum is maintained

in the annular space between the tubes. The inner tube is lined on the inside with a metal tube to conduct heat to water tubes in the interior. A diagram of a tubular collector is shown in Figure 10.8 and a photo of an assembled collector in Figure 10.9. A 90° reflector is provided with each tube. These

Figure 10.8 Diagram of evacuated tubular collector. Courtesy of Owens-Illinois, Inc., Toledo, Ohio.

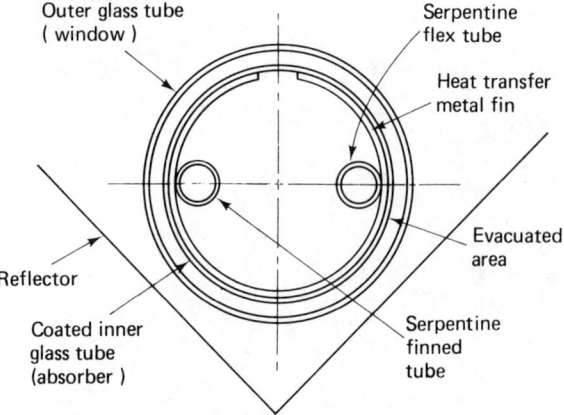

Figure 10.9 Photograph of evacuated tubular collector. Courtesy of Owens-Illinois, Inc., Toledo, Ohio.

reflectors increase the solar input to the tubes and reduce the cost of the array. Figure 10.10 illustrates the concentrating effect of a 90° reflector. The performance curve of a typical tubular collector is shown in Figure 10.11; the performance is very good. If the initial cost and maintenance requirements of these collectors can be reduced, they may become important for intermediate temperature applications.

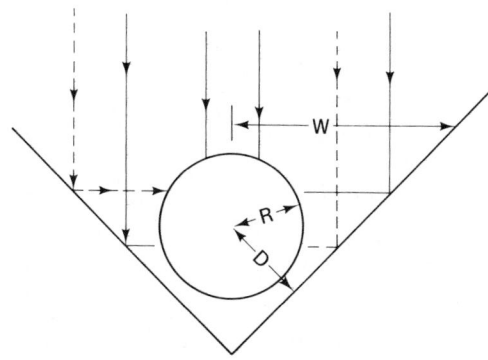

Figure 10.10 Concentration produced by right angle reflector.

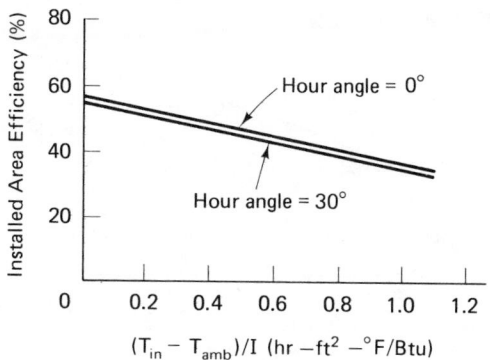

Figure 10.11 Performance curve of evacuated tubular collector. Courtesy of Owens-Illinois, Inc., Toledo, Ohio.

10.9 The CPC Collector

10.9.1 Introduction

The purpose of a concentrating collector is to concentrate the solar irradiation falling on a large area onto a smaller area so that the hot receiver will have less surface area to produce heat losses, thus allowing operation at a higher temperature with reasonable efficiencies. Usually, it is not necessary that the concentrating system produce a perfect image of the sun. Suppose one has a device with an entrance slot of width D_1 facing the sun. Parallel rays from the sun impinge on this slot at an angle θ to the normal to the slot. It is desired to concentrate these rays onto the smaller surface of width D_2, a distance L behind D_1. It is of no consequence what angles the rays make with the lower slot as long as they pass through the lower slot (see Figure 10.12). It is evident that some sort of parabolic configuration is required.

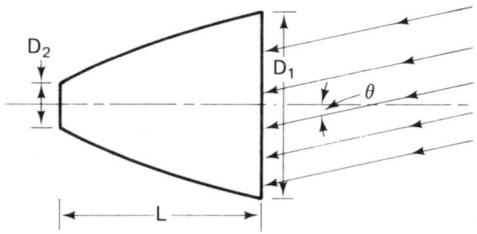

Figure 10.12   Rays impinging on a collector at angle θ.

Some of the focusing properties of parabolas include:

1. Light parallel to the axis of a parabola becomes focused at a point (called the *focus*), regardless of which point along the parabola the reflection occurs, as shown in Figure 10.13.

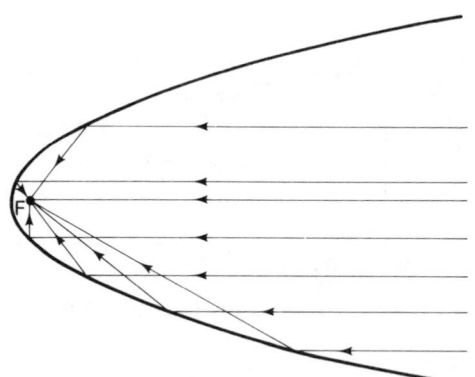

Figure 10.13   Focusing property of a parabola for coaxial rays.

2. Light rays making an angle θ with the parabola axis do not focus at a single point (see Figure 10.14).
3. Light rays at glancing angles, as shown in Figure 10.15, do not focus at a point but do pass through the line segment FB drawn through the focus F after one, two, or more reflections. The case of two reflections is illustrated as the line labelled "2 Ref" in Figure 10.15.

Winston[5] has proposed a simple but ingenious procedure for maximizing the light collection by a system with apertures D_1 and D_2 and length L.

[5]R. Winston, "Solar Concentrators of a Novel Design," *Solar Energy*, **16**, 89, 1974.

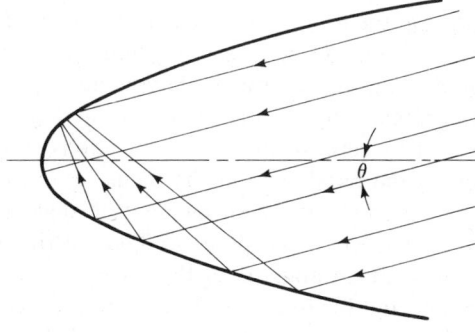

Figure 10.14 Focusing of a parabola for nonaxial rays.

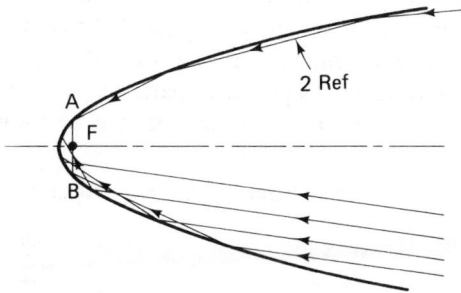

Figure 10.15 Single and double reflections of slightly off axis rays.

The basis of Winston's approach is to make the curved surfaces of Figure 10.12 from parabolas that are tilted relative to the main collector axis, rather than using a parabola with its axis along the collector axis. The tilted parabolic surfaces produce glancing collisions of light rays with the surfaces, resulting in complete collection of reflected rays through the second slot under some circumstances. A tilted parabolic surface is shown in Figure 10.16. The collector axis is line FC, and the parabola axis is line FD. The angle between these axes is θ. The line AC is half the entrance slot width, and BF is the full exit slot width. Light rays that enter the collector parallel

Figure 10.16 Relation between collector axis and parabola axis in CPC collector.

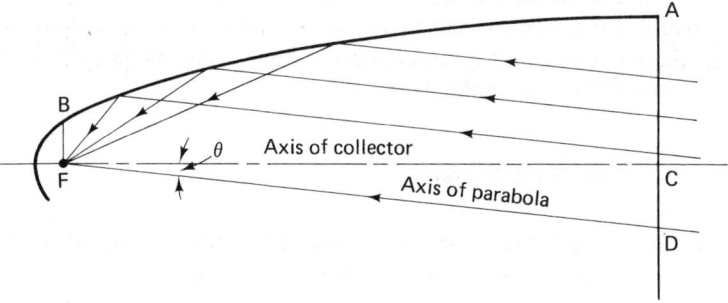

to the parabola axis pass through the focus F. These rays enter at an angle θ to the main collector axis. For light rays that impinge on AC at angles less than θ to the collector axis, there is no single point of focus. However, such rays are glancing rays as described in Figure 10.15. Therefore, these rays are bent by zero, one or more reflections so that they pass through line BF between the parabola and the focus. This condition is illustrated in Figure 10.17 for a light ray JHE incident at angle ϕ, which is smaller than θ and which intersects BF at point E. Thus, if the parabola is tilted at angle θ so that light at an angle θ to the collector axis is parallel to the axis of the parabola and focuses at F, then light rays incident at an angle ϕ (less than θ) will focus at point E on BF. Hence, all light rays impinging at angles less than θ will pass through the second slot, and the collector is a total collector for rays within the aperture angle θ. To complete the design, a second parabola is placed symmetrically on the other side of the collector axis as shown in Figure 10.18. The focus of parabola #1 is point A, and the focus of parabola #2 is at C. The collector axis passes midway between the two foci.

Figure 10.17   Ray tracings in a CPC collector.

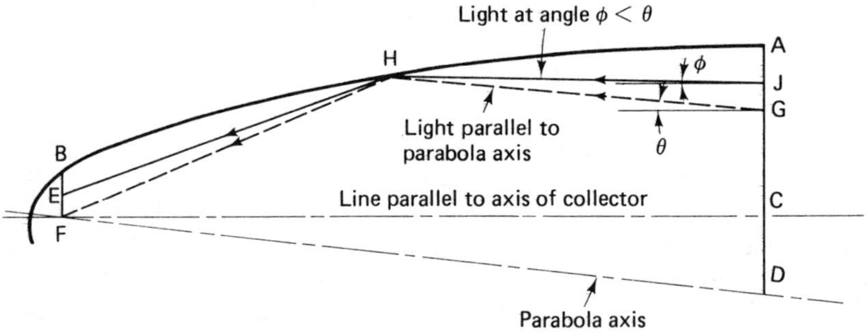

The parabolas must be terminated at the points where they become parallel to the collector axis, since further continuation would actually decrease the entrance slot width. In actual practice, it is almost surely desirable to terminate the parabolas even sooner because there is very little gain in effective aperture for the outer parts of the parabolas. Under any circumstances, there is no advantage in continuing the parabolas past the points of intersection with the parabolic axes.

10.9.2 Coordinate Systems

We now proceed to define the required parabolas in detail. In Figure 10.19, four coordinate systems are defined. The parabola $CBAE$ defines the entrance slot half-width CD, the exit slot full-width AP, the parabolic

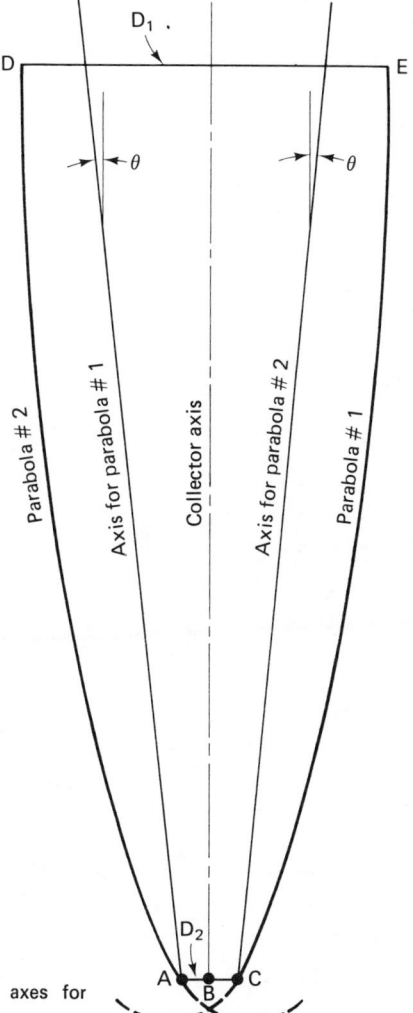

Figure 10.18 Parabola axes for the CPC collector.

axis EP, the parabola focus P, and the collector axis OD. The (x, y) coordinates are natural to the collector axis, the (x_1, y_1) coordinates are natural to the parabolic axis, and the (x_2, y_2) coordinates are simply the (x_1, y_1) coordinates translated down from the focus to the bottom of the parabola. In addition, a fourth coordinate system (x_3, y_3) is defined with origin at P but axes parallel to (x, y). It is desired to express the equation of the parabola in the coordinate system (x, y) natural to the collector axis. The parabola equation is most readily expressed in terms of (x_2, y_2) coordinates as

$$y_2 = bx_2^2 \tag{10.27}$$

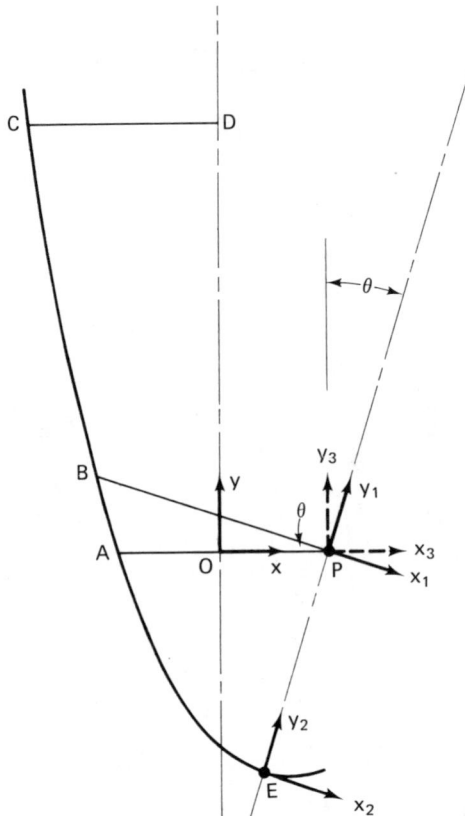

Figure 10.19 Coordinate systems.

where the constant b determines the width PB as $1/(2b)$ and the focus line PE as $1/(4b)$. The transformation from (x_1, y_1) to (x_2, y_2) is simply a linear translation, and is given by

$$x_2 = x_1$$
$$y_2 = y_1 + \frac{1}{4b} \tag{10.28}$$

The linear transformation from (x, y) to (x_3, y_3) is

$$x_3 = x - \frac{D_2}{2} \tag{10.29}$$
$$y_3 = y$$

The transformation from (x_3, y_3) to (x_1, y_1) is a rotation of axes, and is

$$x_1 = x_3 \cos \theta - y_3 \sin \theta$$
$$y_1 = x_3 \sin \theta + y_3 \cos \theta \tag{10.30}$$

Therefore, the transformation from (x, y) to (x_1, y_1) can be taken by steps: $(x, y) \rightarrow (x_3, y_3) \rightarrow (x_1, y_1)$, and is

$$x_1 = \left(\frac{x - D_2}{2}\right) \cos \theta - y \sin \theta$$

$$y_1 = \left(\frac{x - D_2}{2}\right) \sin \theta + y \cos \theta$$

(10.31)

Finally, by taking the steps: $(x, y) \rightarrow (x_1, y_1) \rightarrow (x_2, y_2)$ we obtain

$$x_2 = \left(\frac{x - D_2}{2}\right) \cos \theta - y \sin \theta$$

$$y_2 = \left(\frac{x - D_2}{2}\right) \sin \theta + y \cos \theta + \frac{1}{4b}$$

(10.32)

The parabola is thus given by

$$\left(\frac{x - D_2}{2}\right) \sin \theta + y \cos \theta + \frac{1}{4b} = b\left[\left(\frac{x - D_2}{2}\right) \cos \theta - y \sin \theta\right]^2 \quad (10.33)$$

In particular, point C, which has coordinates $(x, y) = (-D_1/2, L)$, lies on the parabola, and use of these values leads to the quadratic equation in b:

$$(\bar{D} \cos \theta + L \sin \theta)^2 b^2 + (\bar{D} \sin \theta - L \cos \theta)b - \tfrac{1}{4} = 0 \quad (10.34)$$

where $\bar{D} = (D_1 + D_2)/2$. Thus, for any arbitrarily chosen values of D_1, D_2, L and θ, the above equation defines b (and therefore the parabola) required for complete collection. However, for some choices of the parameters a real solution for b may not exist.

A more valuable approach is to select arbitrary values of D_2 and θ dictated by design requirements. Then L is treated as a variable and is adjusted to maximize D_1, thus finding the parabola with maximum entrance width aperture for fixed exit aperture and fixed angular acceptance aperture. Consider Figure 10.20 in which BK is the collector axis, FG is the parabola axis, F is the focus of parabola JA, AK is half of the entrance slot width, and JF is the full exit slot width.

10.9.3 Theorems in Geometry

To proceed further, we use two theorems provable by geometry. The first theorem is that *if one draws the line FA in Figures 10.20 or 10.21 at an angle 2θ to the parabola axis (or θ to the collector axis), the slope of the parabola at point A will be parallel to the collector axis.* The importance of this theorem is that it provides the construction that gives the maximum D_1

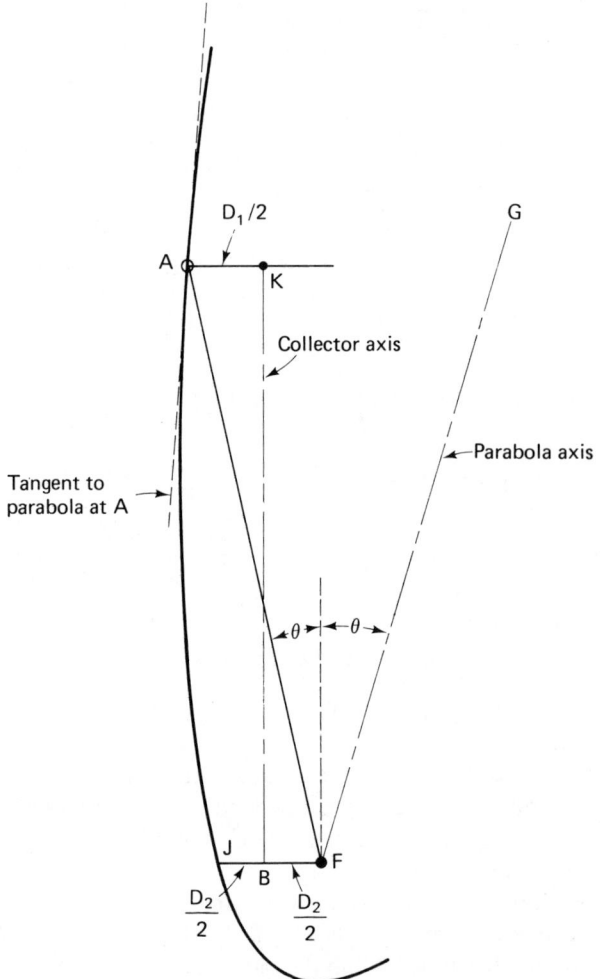

Figure 10.20 Geometrical construction.

for fixed D_2 and θ, since the slope at point A is parallel to the collector axis. To prove this theorem, we use Figure 10.21. A coordinate system is chosen with x horizontal and y vertical. The equation of the parabola is

$$y = bx^2 \tag{10.35}$$

Line AF is written as

$$y = \tan\left(\frac{\pi}{2} + 2\theta\right)x + \frac{1}{4b} \tag{10.36}$$

since point F is $(0, 1/4b)$, and the slope of line AF is $\tan\left(\frac{\pi}{2} + 2\theta\right)$. Point A

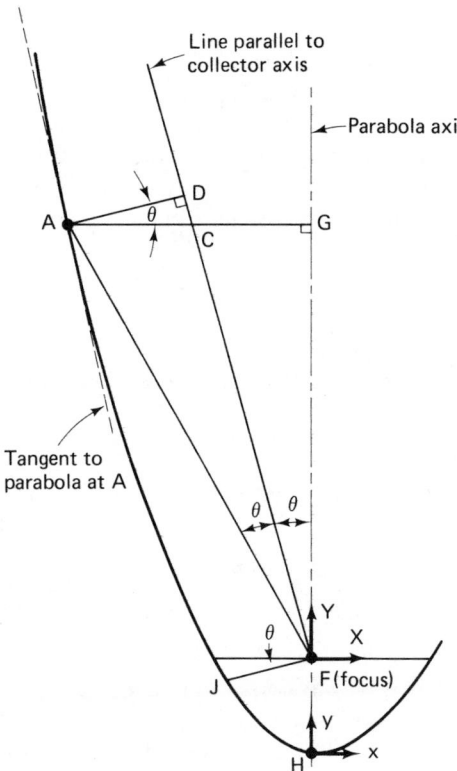

Figure 10.21 Geometrical construction.

is determined as the intersection of line AF with the parabola:

$$bx_A^2 = \tan\left(\frac{\pi}{2} + 2\theta\right)x_A + \frac{1}{4b} \qquad (10.37)$$

The slope at A is

$$\left(\frac{dy}{dx}\right)_A = 2bx_A \qquad (10.38)$$

The slope of line CF is

$$t = \tan\left(\frac{\pi}{2} + \theta\right) = -\cot\theta \qquad (10.39)$$

Solving Eq. (10.37) for x_a, we find that

$$x_A^2 - \frac{T}{b}x_A - \frac{1}{4b^2} = 0 \qquad (10.40)$$

where $T = \tan\left(\frac{\pi}{2} + 2\theta\right) = -\cot 2\theta$, and therefore $\qquad (10.41)$

$$2bx_A = T - \sqrt{1 + T^2}$$
$$= -\cot 2\theta - \csc 2\theta = -\cot \theta \qquad (10.42)$$

Therefore $2bx_A = t$, and the slope of the parabola at A is proven to be parallel to the collector axis.

The second theorem required is the statement that *for the maximum entrance aperture*,

$$D_2 = D_1 \sin \theta \qquad (10.43)$$

Winston proved this with a phase space approach, but we shall be content with a simple geometrical proof. Consider Figure 10.21. We require to prove that

$$JF = 2(AD) \sin \theta \qquad (10.44)$$

The capitalized (X, Y) coordinate system is used. The equations of various lines are then

$$
\begin{aligned}
JF: & \quad Y = (\tan \theta)X \\
FA: & \quad Y = -(\cot 2\theta)X \\
FD: & \quad Y = -(\cot \theta)X \\
AD: & \quad Y = (\tan \theta)X + \beta
\end{aligned}
\qquad (10.45)
$$

where β is the value of Y at the point where AD (extended) intersects the parabola axis. The equation of the parabola is

$$Y = bX^2 - \frac{1}{4b} \qquad (10.46)$$

Point J is found by finding the intersection of JF with the parabola. The result is

$$
\begin{aligned}
X_J &= \frac{1}{2b}\left(\frac{\sin \theta - 1}{\cos \theta}\right) \\
Y_J &= \frac{1}{2b}\frac{\sin \theta(\sin \theta - 1)}{\cos^2 \theta}
\end{aligned}
\qquad (10.47)
$$

The distance JF is then readily calculated to be

$$JF = \sqrt{X_J^2 + Y_J^2} = \frac{1}{2b}\left(\frac{1}{1 + \sin \theta}\right) \qquad (10.48)$$

Point A is found as the intersection of AF with the parabola. The result is

$$
\begin{aligned}
X_A &= -\frac{1}{2b}\cot \theta \\
Y_A &= \frac{1}{4b}(\cot^2\theta - 1)
\end{aligned}
\qquad (10.49)
$$

Since point A lies on line AD, it can be used to evaluate β from Equation 10.45. The result is

$$\beta = \frac{1}{4b} \csc^2 \theta \tag{10.50}$$

Point D is the intersection of AD with FD. Thus,

$$X_D = -\frac{1}{4b} \cot \theta$$

$$Y_D = \frac{1}{4b} \cot^2 \theta \tag{10.51}$$

The length of AD is

$$AD = \sqrt{(X_D - X_A)^2 + (Y_D^2 - Y_A^2)}$$

$$AD = \frac{1}{4b} \csc \theta \tag{10.52}$$

Therefore

$$\frac{AD}{JF} = \frac{1 + \sin \theta}{2 \sin \theta} = \frac{1}{2}\left(1 + \frac{1}{\sin \theta}\right)$$

$$\frac{D_1 + D_2}{2D_2} = \frac{1}{2}\left(1 + \frac{1}{\sin \theta}\right)$$

$$\frac{D_1}{D_2} = \frac{1}{\sin \theta} \tag{10.53}$$

This proves the required theorem.

10.9.4 Design Equations

From simple geometry (Figure 10.20), it follows that

$$\tan \theta = \frac{D_1 + D_2}{2L} \tag{10.54}$$

Combining this with Eq. 10.53 we obtain

$$D_1 = D_2 \csc \theta \tag{10.55}$$

$$L = \frac{D_2(1 + \csc \theta)}{2 \tan \theta} \tag{10.56}$$

These equations indicate the maximum entrance aperture D_1 and the length L for any arbitrarily chosen combination of D_2 and θ.

10.9.5 Truncated CPC Collectors

Although it may be desirable from a theoretical point of view to maximize D_1, it may not be an optimum design in a practical sense. Near the entrance aperture, the parabolas are nearly parallel to the collector axis, and therefore the sheet metal sides of the parabolic surfaces must be extended a considerable length to increase the aperture a small amount. It may, therefore, be economical to truncate the parabolas before the points where they become parallel to the collector axis. For a small loss in aperture, a considerable saving in sheet metal cost might be effected. For full nontruncated parabolas, a plot of L/D vs. D_1/D_2 is the straight line in Figure 10.22 labelled

Figure 10.22 Ratio of L/D_1 vs. D_1/D_2 at fixed θ for various degrees of truncation.

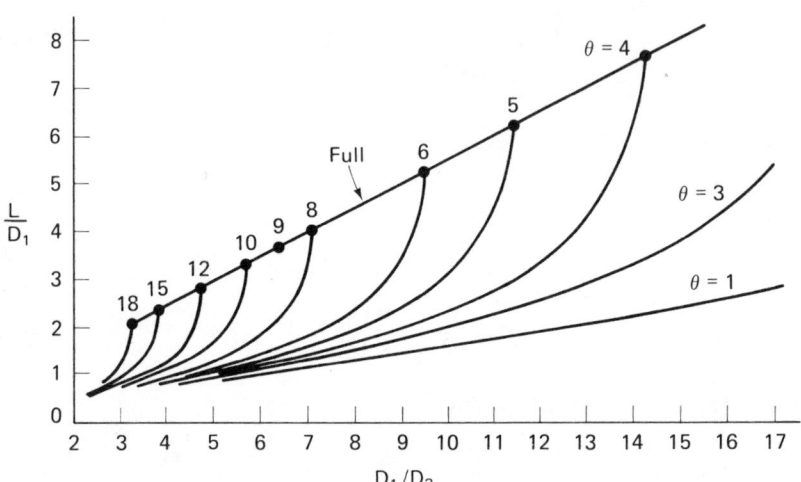

"Full." The numbers along this line refer to the half acceptance angle θ in degrees. By truncating the parabolas at various lengths, the curves shown for constant θ are obtained. The horizontal scale in this figure is the concentration ratio D_1/D_2, and the vertical scale is the ratio of collector length to entrance aperture width. It can be seen that the curves are nearly vertical where they intersect the "Full" line. Therefore, as one gradually truncates the parabolas from full, there is a rapid decrease in L, accompanied by only a small loss in concentration. For example, consider a full parabola with entrance half-aperture of 4°. The design point is $L/D_1 = 7.7$ and $D_1/D_2 = 14.4$. If the length of the parabola is truncated to half the full length, keeping D_1 fixed, the value of D_1/D_2 is reduced to 12.7. Thus, by using half the sheet metal the collector suffers a minor loss in concentration from 14.4 to 12.7.

The actual shapes of truncated CPC collectors are shown in Figure 10.23 for a fixed concentration of 3 and variable half-angle of acceptance.

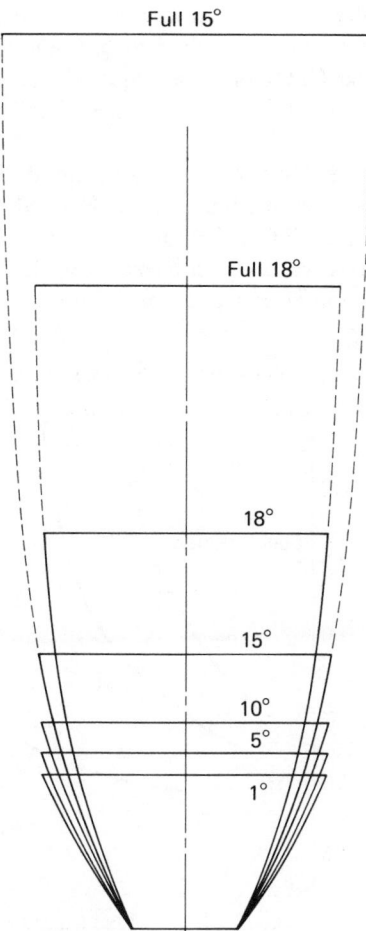

Figure 10.23 Shapes of CPC collectors for various apertures at a concentration of 3.

In designing truncated CPC collectors, as one approaches the full parabola, large amounts of additional sheet metal are required to produce small gains in concentration. In Figure 10.23, full and truncated parabolas for $\theta = 18°$ and $\theta = 15°$ are plotted. It can be seen that, in these cases, the loss in concentration produced by severe truncation is minimal. In any specific application, the best truncated CPC collector should be chosen to meet the requirements of the job. *The full CPC collector will never be an economic design.* In Table 10.1, parameters are presented for a reasonable selection of truncated CPC collectors, which have been chosen to reduce sheet metal requirements with minimum loss of concentration. The collectors are numbered 1 through 10, and have concentrations ranging from 3 to 12. The corresponding parameters for the third nontruncated collector has $L/D_2 = 8.25$ as compared to 19.2 for the full parabola. The concentration of the truncated collector is 5 as compared to 5.76 for the full parabola. In this same table, some figures are given as to the number of adjustments of the orientation of the CPC collector required per year to guarantee 6, 7, or 8 hr of complete collection per day. These numbers result from a calculation to be explained next.

Another way of visualizing the design considerations for a CPC collector is shown in Figure 10.24. Here, the length per unit entrance aperture of truncated CPC collectors vs. θ, the half acceptance angle, is given for various fixed values of concentration. It shows the loss in aperture produced by truncation at constant concentration.

Figure 10.24 L/D_1 vs. θ for various concentration ratios.

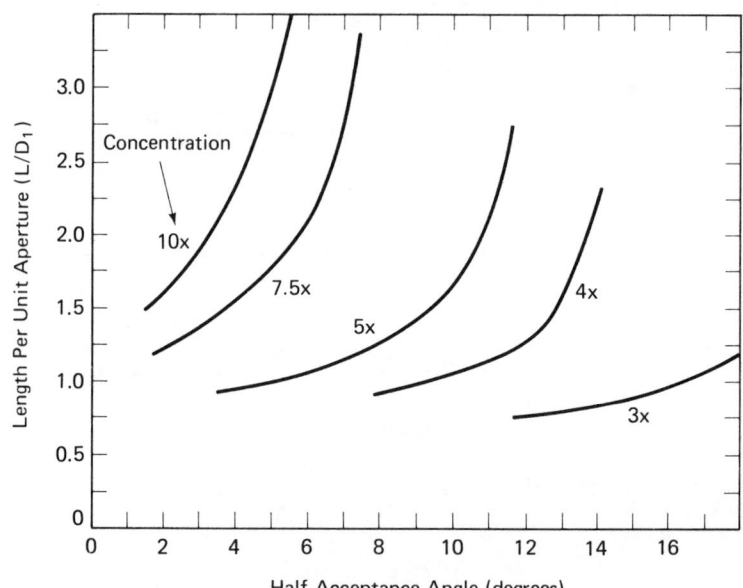

10.9.6 Periodic Adjustments

The CPC collectors must be reoriented periodically during the year to keep the sun within their ranges of acceptance. Collectors with larger acceptance angles naturally need fewer adjustments during the year. We shall consider CPC collectors in an east-west orientation and rotate the axis of the collectors periodically so that the angle α that the axis makes with a horizontal plane has various fixed values for various periods during the year. A very simple description of the apparent movement of the sun about the

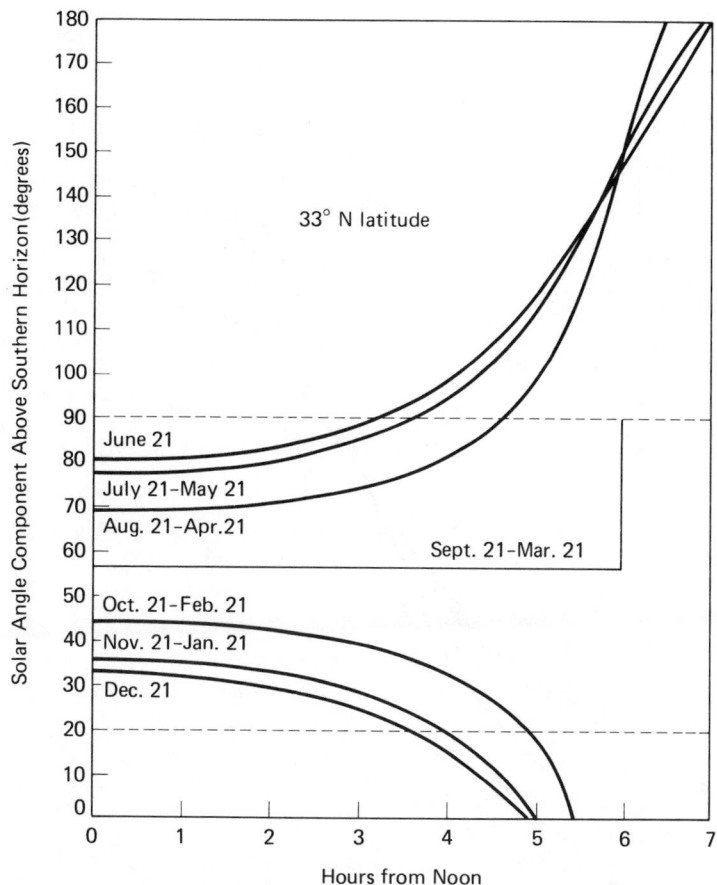

Figure 10.25 Solar angle χ above southern horizon on various days as a function of hour.

earth for a latitude of 33°N is shown in Figure 10.25, which is a modified form of Figure 2.28. The angle χ is the angle between a horizontal plane and a plane passing through an east-west line in the horizontal plane that also passes through the sun. Clearly, we want to periodically adjust the orientation angle α so that it is within θ of χ for best collection.

In this work, the days of the year are numbered from December 21 as day 1, so that the day of the solar year is 10 greater than the actual calendar day of year. From Figure 10.25 the solar elevation angle χ can be determined for various numbers of hours away from noon for various days of the year. These results are plotted in Figure 10.26. The curves are symmetrical about day 182.6. The day of the year with the greatest variation in

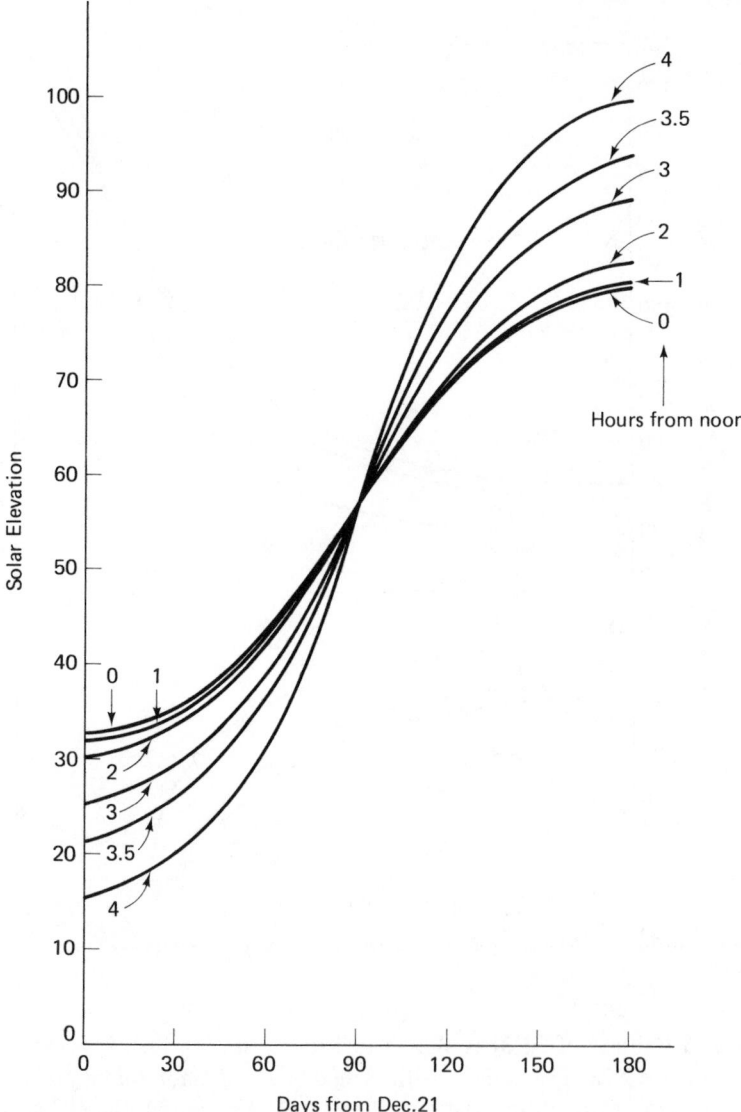

Figure 10.26 Variation of solar elevation χ with hours from noon at various days of the year.

χ during the middle of the day is June 21. December 21 also has a large variation. On March 21 and September 21, the solar elevation χ is constant all day. Since the change in χ over a 3 hr period measured from noon on June 21 is 9°, it is clear that any CPC with θ less than 4.5° cannot collect more than 6 hr of sunlight on that day with a fixed orientation. Similarly, a

CPC with θ less than $10°$ cannot collect more than 8 hr of sunlight on June 21 unless it is made to track periodically during the day.

To determine how many adjustments are necessary, the plots shown in Figures 10.27, 10.28, and 10.29 have been prepared. These indicate the range of variation of χ over 6, 7, and 8 hr periods centered on noon. The optimum setting of the orientation tilt angle of the CPC for daily tracking is the midpoint of the range of χ on any day. For periodic tracking, we begin by

Figure 10.27 Optimum setting of CPC orientation and range of solar elevation for 6 hr minimum collection per day.

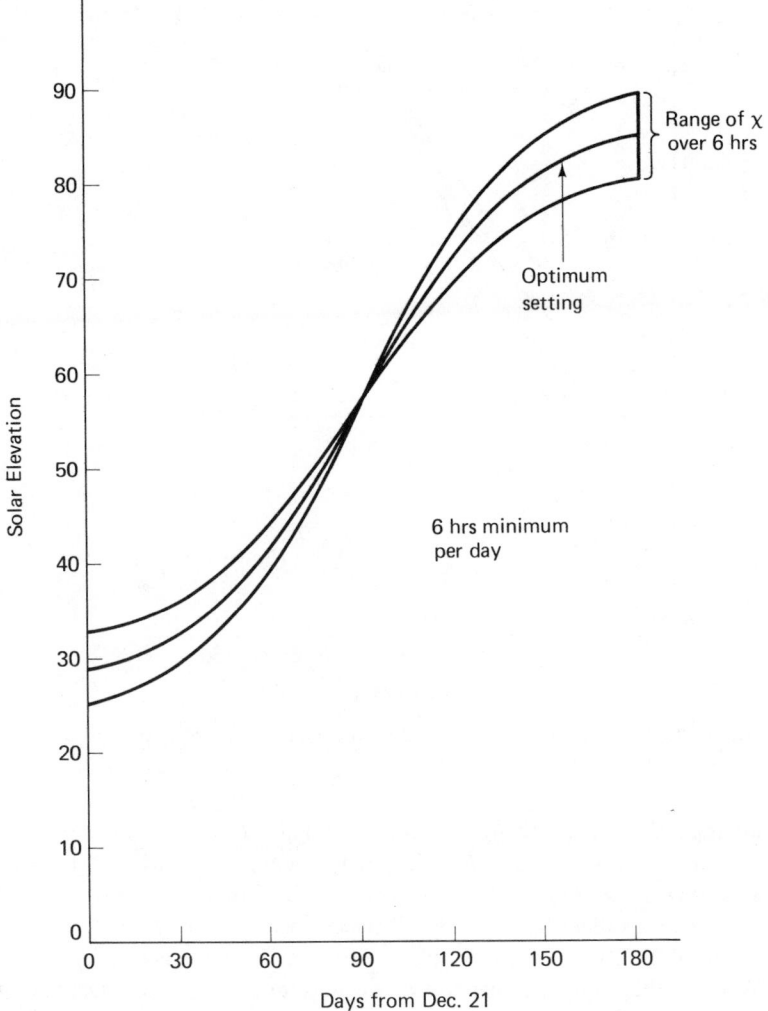

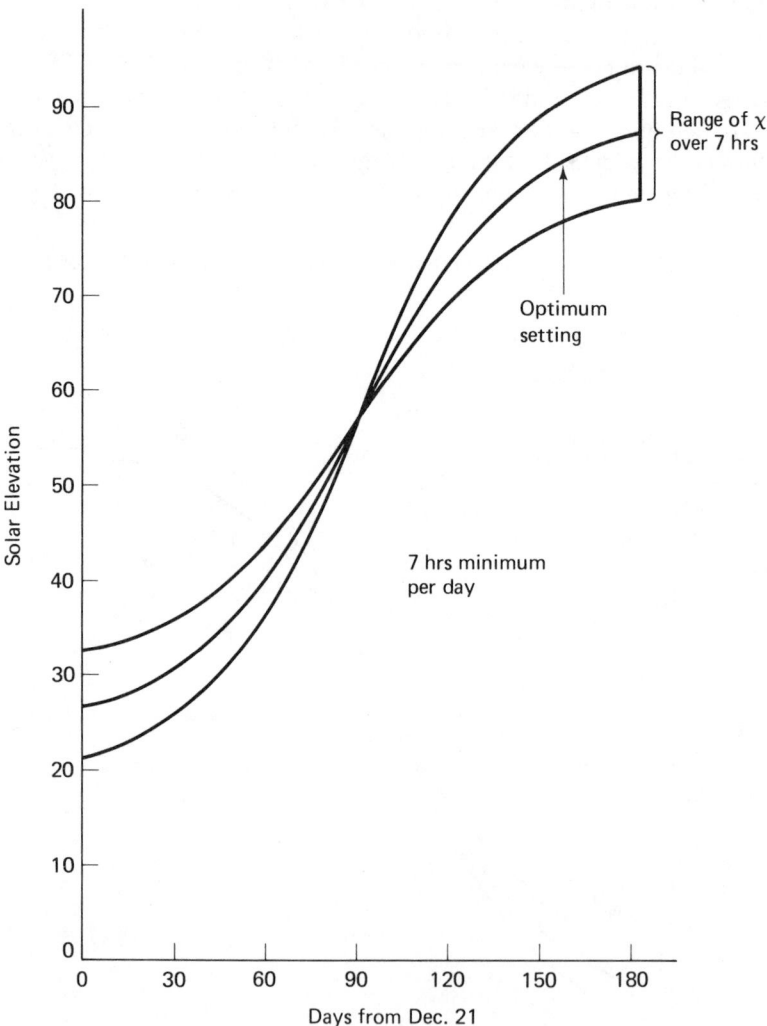

Figure 10.28 Optimum setting of CPC orientation and range of solar elevation for 7 hr minimum collection per day.

choosing a midwinter setting such that $\alpha \cong \chi_m + \theta$, where χ_m is the minimum value of χ on December 21 (day = 1). We have written $\cong$ because in practice we allow about $0.5°-1°$ of variance in this relation and make α up to a degree smaller than $\chi + \theta$. With this value of α, we determine from Figures 10.27, 10.28, or 10.29, as the case may be, what is the maximum day of the year this angle will just collect all the sunlight over the time span under consideration. At that day, the angle α must be readjusted so that it is again

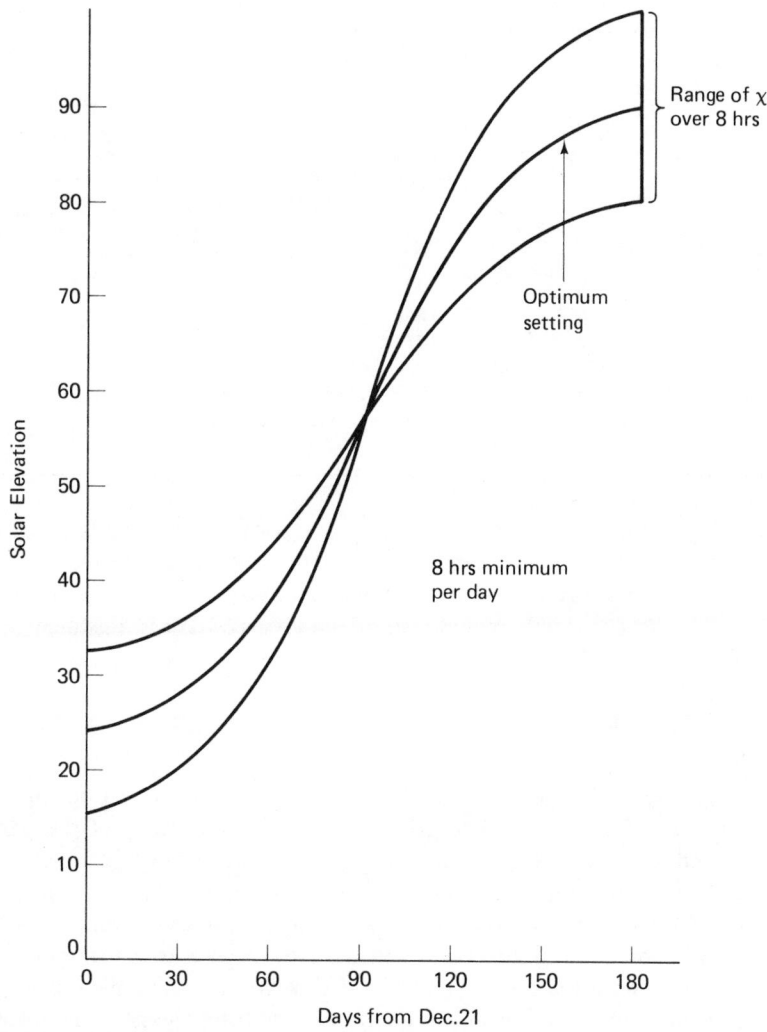

Figure 10.29 Optimum setting of CPC orientation and range of solar elevation for 8 hr minimum collection per day.

slightly greater than $\chi + \theta$ on that day. This process is repeated until the entire span to June 21 is covered. Symmetry is used to extend the results to the second half of the year. The results are shown in Table 10.2 and summarized in Table 10.1. In Table 10.1, it is stated that, for concentrations of 3, 4, 5, 6, 8, and 10, one requires 2, 6, 8, 10, 14, and 22 resettings of α during the year for a minimum of 6 hr of collection every day. For longer periods of collection, the number of adjustments increases sharply. The

TABLE 10.1   A selection of truncated CPC collectors

CPC I.D. No.	Conc.	Half-angle	L/D_1	L/D_2	Conc. Full	L/D_2 Full	Number Adj. per Year for		
							6 hr	7 hr	8 hr
							Collection		
1	3	18	1.20	3.60	3.24	6.52	2	4	4
2	4	12	1.25	5.00	4.81	13.7	6	8	14
3	5	10	1.65	8.25	5.76	19.2	8		
4	6	8	1.80	10.8	7.19	29.1	10	18	
5	8	6	2.35	18.8	9.57	50.3	14		
6	8	3	1.50	12.0	19.1	192.			
7	10	5	3.00	30.0	11.5	71.3	22		
8	10	3	2.00	20.0	19.1	192.			
9	12	4	3.30	39.6	14.3	110.			
10	12	1	1.90	22.8	57.3	1670.			

actual days (measured from December 21 as day 1) and the angles are shown in Table 10.2. For a concentration of 3, 2–4 adjustments are required per year. For a concentration of 10, at least 22 adjustments are required, and higher summer efficiencies would result from virtually daily adjustment.

10.9.7 Summary

The CPC collector offers the highest possible light collection for fixed apertures D_1, D_2, and length L, assuming perfect reflectivity of the parabolic reflectors. In actual practice, the angles of incidence of incoming rays to the walls of the reflectors tend to be at glancing angles, which produces lower reflectivities than in collectors with more normal incidence between solar rays and reflector surfaces. Furthermore, the fact that maximum collection is produced for fixed L may be a deceptive feature. While it is important to collect light over as large an entrance aperture θ as possible for fixed D_1 and D_2, there is no particular reason to minimize L except insofar as it reduces the amount of reflector surface area required. However, the very concept of the CPC collector has the major disadvantage that the reflecting surfaces tend to be along the collector axis, resulting in large amounts of surface being required. This disadvantage is especially bad in the full CPC collector but is also a problem in the truncated CPC collector. The main possibility of utility for a CPC collector is to obtain a modest concentration (in the range 2–5) without requiring continuous tracking. It is not clear whether such a system is desirable compared to a simple back-reflecting parabolic trough.

TABLE 10.2 Days of the solar year for readjustment of CPC collectors for various minimum durations of collection per day (see Table 10.1 for characteristics of various CPC I.D. numbers)

CPC ID No. = 1		CPC ID No. = 2		CPC ID No. = 4		CPC ID No. = 5	
Day	Angle	Day	Angle	Day	Angle	Day	Angle
6 hr minimum		6 hr minimum		6 hr minimum		6 hr minimum	
90	73	66	53	48	42	36	42
272	41	100	72	74	54	70	50
		142	85	96	67	89	62
7 hr minimum		222	72	120	77	106	70
		264	53	147	85	123	76
57	51	299	36	217	77	138	81
102	78			244	67	162	85
262	51	7 hr minimum		268	54	202	81
308	32			290	42	226	76
		57	46	316	33	241	70
8 hr minimum		90	68			258	62
		104	74	7 hr minimum		275	50
68	53	138	87			294	42
105	80	226	74	30	34	328	31
259	53	260	68	51	41		
297	30	274	46	71	52	CPC ID No. = 7	
		308	32	93	68		
CPC ID No. = 3				116	75	6 hr minimum	
		8 hr minimum		130	80		
6 hr minimum				144	83	25	33
		45	36	158	86	40	37
57	46	69	48	172	87	54	42
87	64	93	69	192	86	66	47
117	77	117	79	202	83	79	55
150	85	135	85	220	80	94	64
214	77	156	89	234	75	109	70
247	64	208	85	248	68	121	75
277	46	229	79	271	52	132	78
307	34	247	69	293	41	153	83
		271	48	313	34	165	84
		295	36	334	29	199	83
		319	28			211	78
						232	75
						243	70
						255	64
						270	55
						285	47
						298	42
						310	37
						324	33
						339	30

10.10 Fresnel Lens Collectors

10.10.1 Introduction

According to the laws of optics a standard circular lens will focus coaxial parallel light rays to a point, and a cylindrical lens will focus coaxial parallel rays to a line (see Figure 10.30). There are two problems associated with the use of such lenses. One is that due to the necessary curvature the lenses must be quite thick, which makes them heavy and expensive. The other is that spherical aberration begins to limit their performance as the curvature becomes extreme.

Figure 10.30 Spherical and cylindrical lenses.

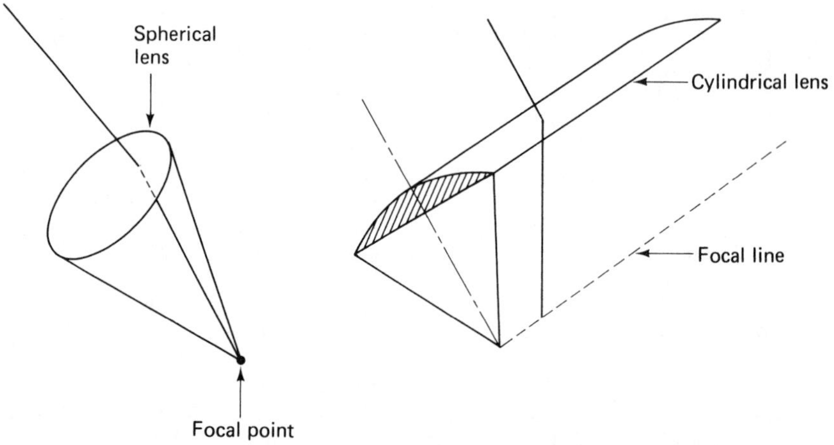

A *Fresnel lens* is a device for producing the same kind of focusing action, but with much less material and at a fraction of the cost of a standard lens. The Fresnel lens consists of a series of segments called *facets*, which are tilted to refract the incident light to a focal cone. It is not necessary for the facets to have curved surfaces if the width of the facets is smaller than the receiver. The comparison of a Fresnel lens with a standard lens is shown in Figure 10.31. Such Fresnel lenses are typically made of plastic and are relatively inexpensive. Each facet is tilted at an angle appropriate to focus coaxial light onto a receiver. The transition region at the junction of each pair of facets does not focus parallel light and is thus a net loss. Therefore, only a fraction of the area of a Fresnel lens is useful for concentration (see Figure 10.32).

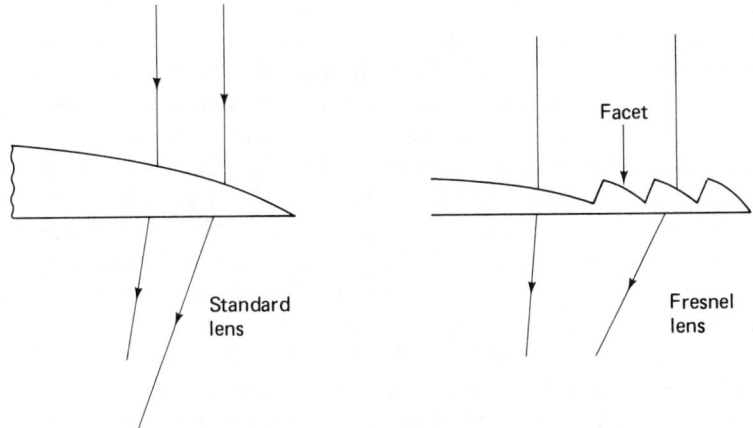

Figure 10.31 Standard and Fresnel lenses.

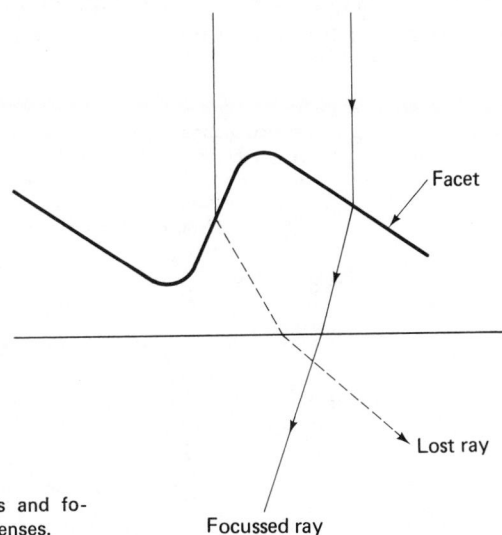

Figure 10.32 Lost rays and focused rays with Fresnel lenses.

10.10.2 Design Consideration: East–West Orientation

Nelson, Evans, and Bansal[6] have studied the performance of linear
Fresnel lens concentrators. A comparison of the transmission was made
when the facets are placed on the front and back surfaces of the lens. It was

[6]D. T. Nelson, D. L. Evans, and R. K. Bansal, "Linear Fresnel Lens Concentrators," *Solar Energy*, **17**, 285, 1975.

found that losses at the edges of the facets are greater when the facets are on the front surface. Since it is desirable to place the facets on the back surface to minimize the adherence of dirt, it appears on both grounds that the facets should be placed on the back surface. Nelson, Evans, and Bansal examined the performance of a linear Fresnel lens with main axis in the east-west direction, with seasonal adjustment of tilt but no tracking. They found, as might be expected, that as the time progresses away from solar noon, the position of the refracted image from a facet moves both along the lens axis and also transverse to it. The radiation entering each side of the Fresnel lens moves in opposite directions, so there is a smearing of the image at off-noon hours. This effect occurs even on the equinoxes when the solar elevation is constant all day (Figure 10.33). On days other than the equinoxes, this effect is magnified and compounded by the variation in solar elevation as time varies from solar noon. Because of this effect, the performance of a seasonally adjusted east-west aligned linear Fresnel lens will suffer as time progresses away from solar noon, and the best that one can hope for is a limited number of hours of collection centered on solar noon. The wider the absorber, the more collection time will result, at the cost of reduced concentration.

Figure 10.33   Linear Fresnel lens concentrator, indicating the paths taken by solar rays at noon and noon + 2 hr on an equinox day.[6] By permission of Pergamon Press Ltd.

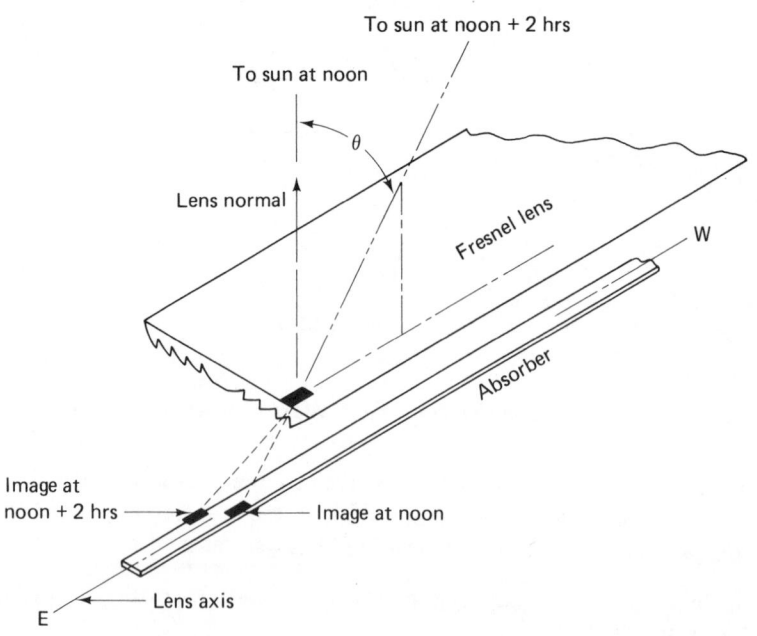

The design of the facets and the placement of the absorber can be achieved in two ways. Each of the facets can be designed to project refracted light onto the center of the absorber at solar noon, or the facets can be arranged to project refracted light onto the outer edge of the absorber at solar noon. With the latter design, the image tends to move across the absorber as time progresses away from solar noon on the equinox days. However, the design in which the images are centered on the absorber at solar noon is better at the solstices because, when the compounding effect of variation of solar altitude with hour is taken into account, centering the image at noon keeps the image on the absorber longer. In general, as the absorber is made wider it should be moved closer to the lens to obtain optimum performance.

It was found that at noon the amount of power received by the absorber is about 85% of the solar power incident at the lens. However, after about 1 hr from noon the spreading of the image rapidly reduces the collection efficiency. At 2 hr from noon only about 50% of the incident solar intensity reaches the absorber. And for an entire day at the equinoxes, as the theoretical concentration ratio is reduced from about 12 to about 6, the fraction of daily solar intensity received at the absorber varies from 0.48 to 0.59.

The number of seasonal adjustments required was investigated by studying the effect of a tilt angle between the collector axis and the solar elevation at noon. It was found that when the tilt ranges from $-1.6°$ to $+1.6°$ the daily collection efficiency stays within a 10% variation. The rate of change of solar elevation is greatest at the equinoxes, when it varies by about 0.4° per day. Thus, by readjusting the concentrator alignment every 8 days near the equinoxes, variations in efficiency will be kept within 10%. Near the solstices, readjustments can be made less frequently.

An estimate of the effects of tracking were made. The effect of tracking is to make the performance each day approximate the performance of a non-tracking system on the equinoxes when the solar elevation is constant. On a yearly basis, for a particular design, seasonal adjustment produces a collection efficiency of 50%, whereas tracking results in 56% efficiency. The performance falls off fast enough at off-noon hours so as to make tracking of little help. In other words, the spreading of the image at off-noon hours is the major problem, not the change in solar elevation. A typical curve of power incident on the absorber is shown in Figure 10.34.

10.10.3 Performance in the North–South Orientation

Fresnel lens collectors can also be mounted with the main axis in the north-south direction. In this configuration, the north end must be elevated so as to tilt the collector. The collector must track the diurnal path of the

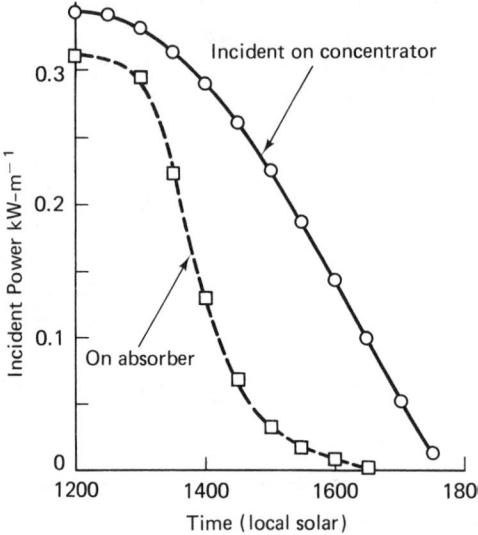

Figure 10.34 Power (in kW) delivered to the absorber per meter of length vs. time of day for a linear Fresnel lens concentrator.[6] The solid curve represents the power incident upon the concentrator. Data are for March 21. By permission of Pergamon Press Ltd.

sun during the day. If the tilt is fixed, it is one-axis tracking. The tilt may also be seasonally adjusted or tracked to make the collector two-axis tracking. If the tilt is fixed, the optimum angle for year-round performance is the latitude angle. At this tilt angle, the solar rays make an angle of 23.45° with a normal to the lens at the solstices even at solar noon. According to Nelson, Evans, and Bansal, a substantial amount of beam spread occurs at this angle, and one might expect a 25% reduction in collection efficiency at noon with such a system, compared to the efficiency of a seasonally adjusted or two-axis tracking system. The advantage of a two-axis tracking system over a seasonally adjusted system would only be felt at several hours from noon near the solstices, when substantial variations occur in solar elevation. This only results in minor improvements in performance. A two-axis tracking north-south system should have an optical efficiency of about 80% all day for every day of the year. The yearly optical efficiencies of other systems are shown in Table 10.3, on the basis of rough estimates. The improvement of

TABLE 10.3   Relative yearly optical efficiencies of
             Fresnel lens systems

Lens Configuration	Relative Efficiency
Two-axis tracking (N/S)	1.0
Seasonally adjusted (N/S)	0.95
Fixed tilt (N/S) one-axis tracking	0.70
One-axis tracking (E/W)	0.60
Seasonally adjusted (E/W)	0.50

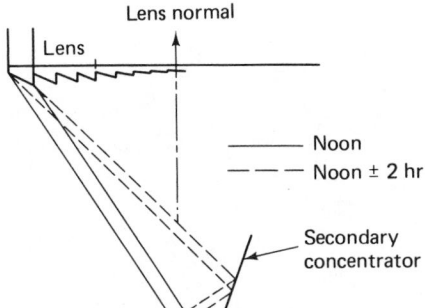

Lens normal

Lens

———— Noon

— — — Noon ± 2 hr

Secondary
concentrator

Figure 10.35 Illustration of secondary concentrator used in conjunction with Fresnel lens.[6] By permission of Pergamon Press Ltd.

north-south one-axis tracking over east-west one-axis tracking is much less than for parabolic troughs because of the poor off-axis performance of Fresnel lens collectors. Fresnel lens collectors with secondary concentration, as shown in Figure 10.35, may offer some substantial advantages.

10.11 Parabolic Troughs

10.11.1 Efficiency vs. Concentration

The parabolic trough can be used for both intermediate and high temperature applications. A parabolic trough is illustrated in Figure 10.36. A crucial aspect of any concentrating collector is the design concentration ratio. In particular, one should know what concentration ratios are required to produce reasonable collection efficiencies at various operating temperatures. Consider a receiver consisting of a metal tube of diameter D_R and a reflector of width W. The ratio of the area of the entrance aperture to the area of receiver per unit length is the effective concentration

$$C_e = \frac{W}{\pi D_R} \qquad (10.57)$$

If the receiver is simply a flat plate of width W_R and is heavily insulated on top, πD_R should be replaced by W_R. If the solar intensity falling on the entrance aperture of the trough is N, the rate of heat gain by the receiver is

$$H_g = \eta_{\text{opt}} N W \qquad (10.58)$$

per linear ft of collector, where η_{opt} is the optical efficiency. The rate of heat loss per linear foot of collector for a cylindrical receiver is

$$H_L = U(T_c - T_a)(\pi D_R) = \frac{U(T_c - T_a)W}{C_e} \qquad (10.59)$$

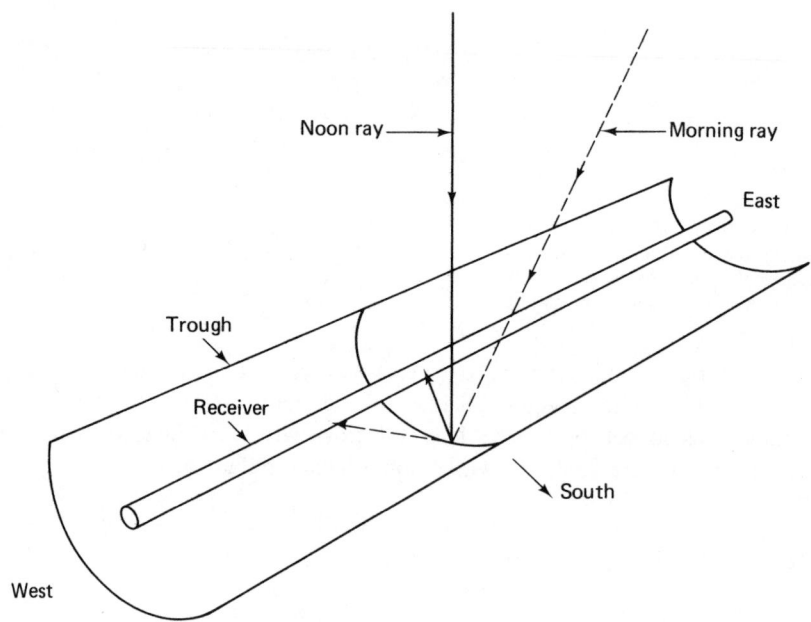

Figure 10.36 Parabolic trough with main axis in EW direction.

where U is a linearized overall heat loss coefficient, and $T_c - T_a$ is the difference between the receiver and ambient temperatures. The net rate of heat gain by the fluid in the receiver, per linear foot, is

$$H_f = H_g - H_L = \eta_{opt} NW - \frac{U(T_c - T_a)W}{C_e} \qquad (10.60)$$

At solar noon in bright sun we may assume $N \cong 280$ Btu/hr-ft^2, and a reasonable value for η_{opt} is 0.65. The linearized heat loss coefficient U will increase as T_c increases, due to the fourth-power radiative power law. For intermediate temperatures, a reasonable value is 1.0 Btu/hr-ft^2-°F (5.68 W/m^2-°K). The efficiency of the collector is

$$\eta = \frac{H_f}{NW} = \eta_{opt} - \frac{U(T_c - T_a)}{C_e N} \qquad (10.61)$$

and, with the approximate values chosen,

$$\eta \cong 0.65 - \frac{(T_c - T_a)}{280 C_e} \qquad (10.62)$$

To achieve 50% collection efficiency under bright sunny conditions, it is required that

$$C_e = \frac{(T_c - T_a)}{(.15)(280)} = \frac{T_c - T_a}{42} \tag{10.63}$$

It can be seen that at $T_c - T_a \approx 100°F$ (55°K), the required value of $C_e \approx$ 2.5. At $T_c - T_a \approx 200°F$ (111°K), the required $C_e \approx 5$. As $T_c - T_a$ increases, U also begins to increase, and the required concentration goes up faster than this linear prediction. Furthermore, these estimates are for bright sunlight near solar noon when $N \approx 280$ Btu/hr-ft² (883.4 W/m²-°K). Under other less ideal conditions, higher concentrations are required. To operate in the range of $\sim 150°F$ (83°K) above ambient under a moderate range of conditions, a concentration ratio of at least 5 is recommended.

10.11.2 The East–West Orientation

Parabolic troughs can be mounted in several different ways. The most common method is to mount the collectors horizontally with the long axis in the east-west direction as shown in Figure 10.36. Except on March and September 21, the altitude of the sun above the southern horizon varies during the day. Therefore, the troughs must be continuously rotated during the day to stay in focus. At solar noon, the rays impinge on the collector in a plane perpendicular to the main axis. At any other time, the rays are focused onto the receiver at points "downstream" of where they are reflected from the reflector. Therefore, there are end losses at any time other than solar noon. The distance downstream can be evaluated by resolving a vector representing the direction of a solar ray into components parallel and perpendicular to the main axis. The component parallel to the main axis can be treated as a simple reflection along the axis, whereas the component perpendicular to the main axis can be treated as if it were the solar ray at solar noon, producing the focusing action. At any arbitrary time, the effective solar intensity falling on the aperture of the trough is the product of the direct normal intensity and the cosine of the angle θ between the rays and a line perpendicular to the plane of the entrance aperture as shown in Figure 10.37. The end losses can be reduced by making the linear arrays long compared to the focal distance. However, the cosine factor remains a source of loss at off-noon hours. The solar energy available to an east-west oriented parabolic trough on the 21st day of December, March, June, and September are illustrated in Figures 10.38 to 10.40, assuming clear weather in North Central Texas. The curves of direct normal solar intensity have been multiplied by the cosine of θ in Figure 10.35. It can be seen that the period of major collectable solar energy is within about a 3 hr period on either side

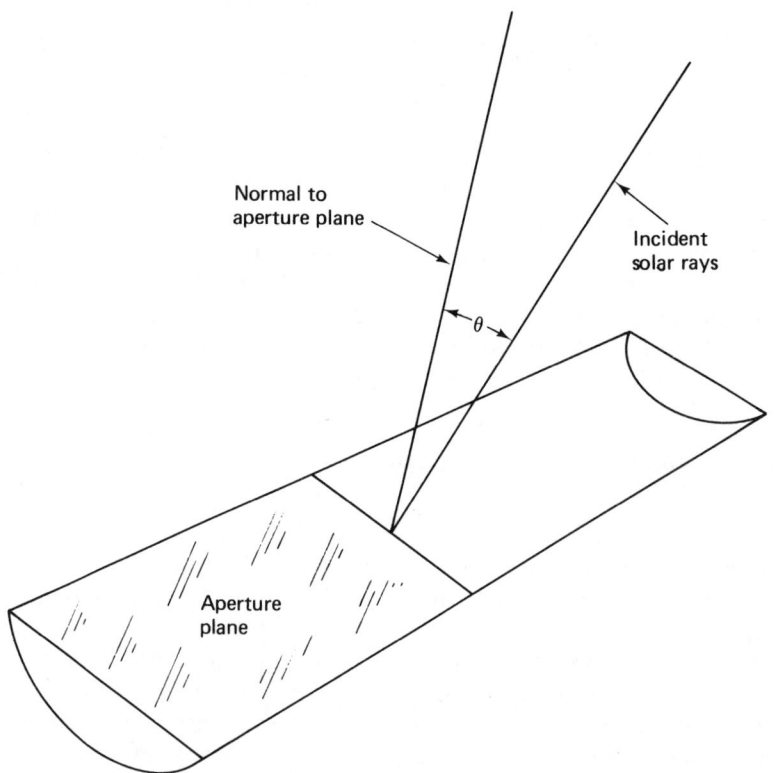

Figure 10.37 Angle between normal to trough and incident solar rays.

of solar noon. East-west parabolic troughs can be made with fixed receivers and reflectors that rotate about the focal line. This has the advantage of fixed rigid plumbing, but the tracking system tends to be more complicated and the required aperture of the receiver is larger. Another approach is to build the receiver/reflector units as rigid structures that rotate to track the sun. The tracking is then simpler, but flexible couplings are required for the fluid connections to the receiver. Flexible couplings can be a potential source of problems. In Figure 10.41 the two methods of tracking are illustrated. The east-west alignment of parabolic troughs has the advantages that the collectors can be mounted horizontally on the ground with low-cost supports and that alignment and tracking are relatively simple. The main disadvantage is the cosine loss at off-noon hours. One other consideration for east-west aligned troughs is that for moderate operating temperatures one can operate a trough at moderate concentration (~ 5) and use a fixed nontracking trough. The tilt of the trough about the main axis must be adjusted periodically during the year at, roughly, monthly intervals. In this mode of operation the

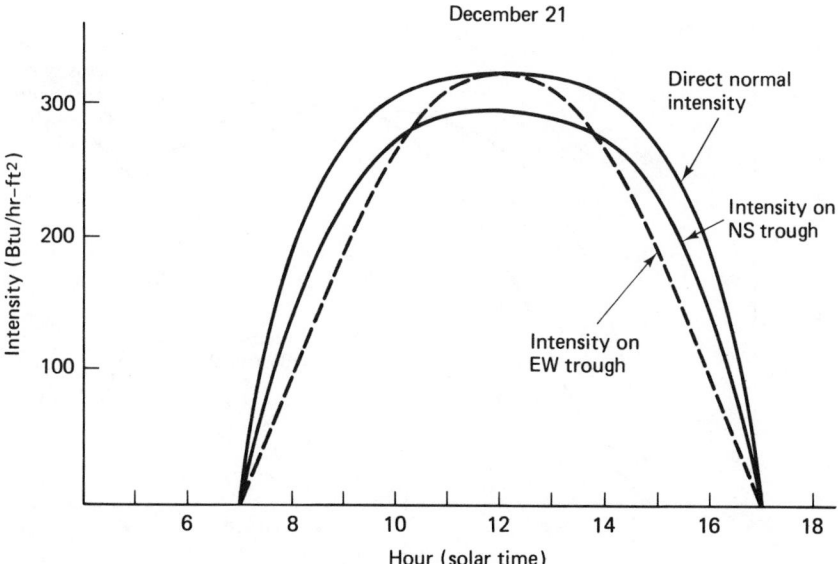

Figure 10.38 Estimated intensities on parabolic troughs on December 21 at 33°N latitude.

Figure 10.39 Estimated intensities on parabolic troughs on June 21 at 33°N latitude.

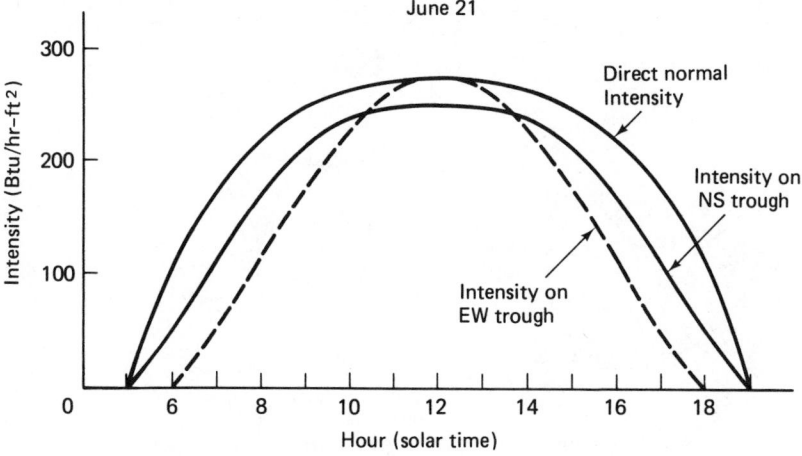

use of a parabolic trough is simplest and cheapest. The range of solar elevations during the day is shown in Figures 2.28 and 10.26, 10.27, 10.28, and 10.29. On March and September 21, the trough is in focus all day long with a single fixed tilt. The days of greatest variation of solar elevation during the day are June 21 and December 21, and the hourly variation of solar angles

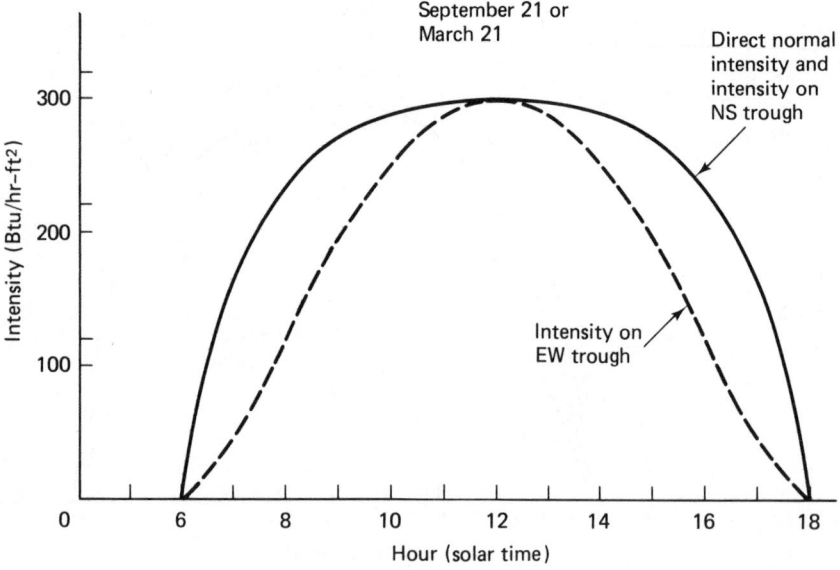

Figure 10.40 Estimated intensities on parabolic troughs on the equinoxes at 33°N latitude.

Figure 10.41 Methods of tracking parabolic troughs.

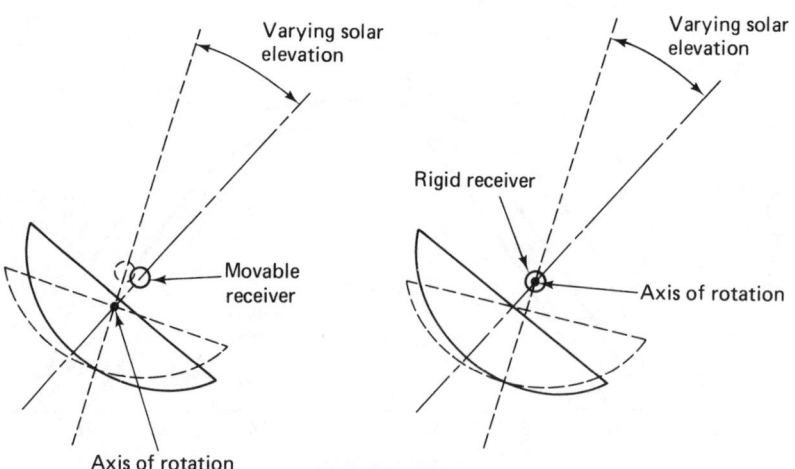

during a day does not change much for about 6 weeks on either side of these dates. On December 21, the variation in solar elevation is over a 4° range for a 4 hr period centered on solar noon and over an 8° range for a 6 hr period centered on noon. On June 21, the variation is 4° over 4 hr and 9° over 6 hr. At any time of the year, it is desirable to adjust the tilt so that focus is perfect at about 1 to 1½ hr from solar noon so that the variation in solar elevation from solar noon to 2 or 3 hr from solar noon keeps the reflected rays on the receiver. During the periods when the solar elevation changes most rapidly from day to day (spring and fall) the tilt of the fixed troughs must be adjusted more often. This approach is mainly feasible for concentration ratios C_e in the range 3–5 on flat roofs that are easily accessible for adjustment.

10.11.3 The North–South Orientation

Parabolic troughs can also be mounted with the long axis in the north-south direction. When this orientation is used, tracking is accomplished by rotation of the trough about the long axis through 180° in a day as shown in Figure 10.42. The tilt angle is a very important parameter for north-south

Figure 10.42 Parabolic trough with long axis in the NS direction.

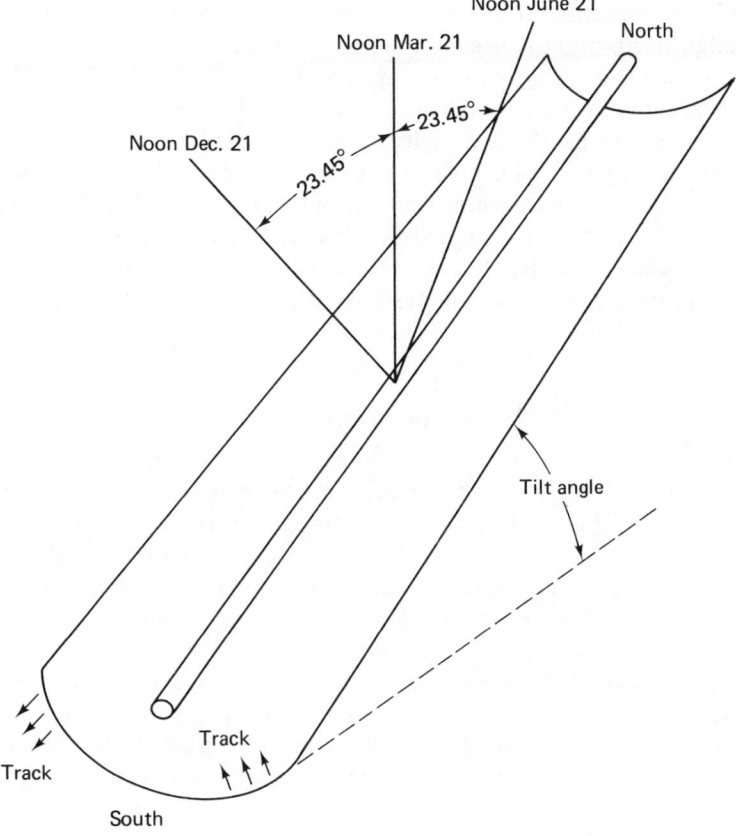

parabolic troughs. A very reasonable compromise for year-round operation is to tilt the trough at the latitude angle. With this tilt, the solar rays at noon approach the trough 23.45° too high in midsummer, 23.45° too low in midwinter, and are perpendicular in spring and fall. As a result, the solar intensity available to the north-south trough at solar noon on December 21 and June 21 is cos (23.45°) times the solar noon availabilities to an east-west trough. On the other hand, the large cosine factor at off-noon hours is not present in north-south troughs as it is with east-west troughs. For a load that peaks in summer (or winter) the tilt angle of the troughs should be less than (or greater than) the latitude angle. The ideal trough would have adjustable tilt angle. If the tilt angle could be adjusted about 12 times a year, it would closely approximate the performance of a two-axis tracked trough. The relative performance of east-west and north-south troughs for 4 days in the year are illustrated in Figures 10.38, 10.39, and 10.40. In general, the area under the north-south curves is higher than under the east-west curves, especially near the equinoxes. However, because the north-south arrays are usually tilted, the substructure supports and piping runs are more expensive. In some cases, the additional cost of the north-south mountings outweighs the gain in collection of energy, and the east-west orientation is preferred. Although north-south collectors mounted horizontally are equal in cost to east-west collectors, they perform very poorly in the winter and are not very useful except in cases where the load peaks strongly in the summer.

Tracking of parabolic troughs can be achieved by a simple mechanism illustrated in Figure 10.43. A shadow band, consisting of a circular band of sheet metal, is placed above two photoconductive cells. When the trough is perpendicular to the sun, the shadow of the band is exactly between the two cells, and when out of balance one of the cells is shaded. This unbalances an electrical bridge circuit which drives a servo motor to seek a reorientation to bring the cells into balance.

10.12 Segmented Mirror Reflectors

An alternative to the parabolic trough is a segmented mirror system, which replaces the large curved surface of the trough with a series of narrow segments that are tilted to approximate the surface angles of a trough. A typical design is illustrated schematically in Figure 10.44. The principal hope of this design is that the sum total of costs of narrow mirror segments can be fabricated for less than a large curved trough. The mirror segments can be made flat so that the concentration is simply the number of slats (less one for the receiver shadow). Of course, if the slats are tilted at an angle to the sun's rays, there is a cosine loss as illustrated in Figure 10.44. The receiver for a segmented mirror reflector system with flat slats can be similiar to an

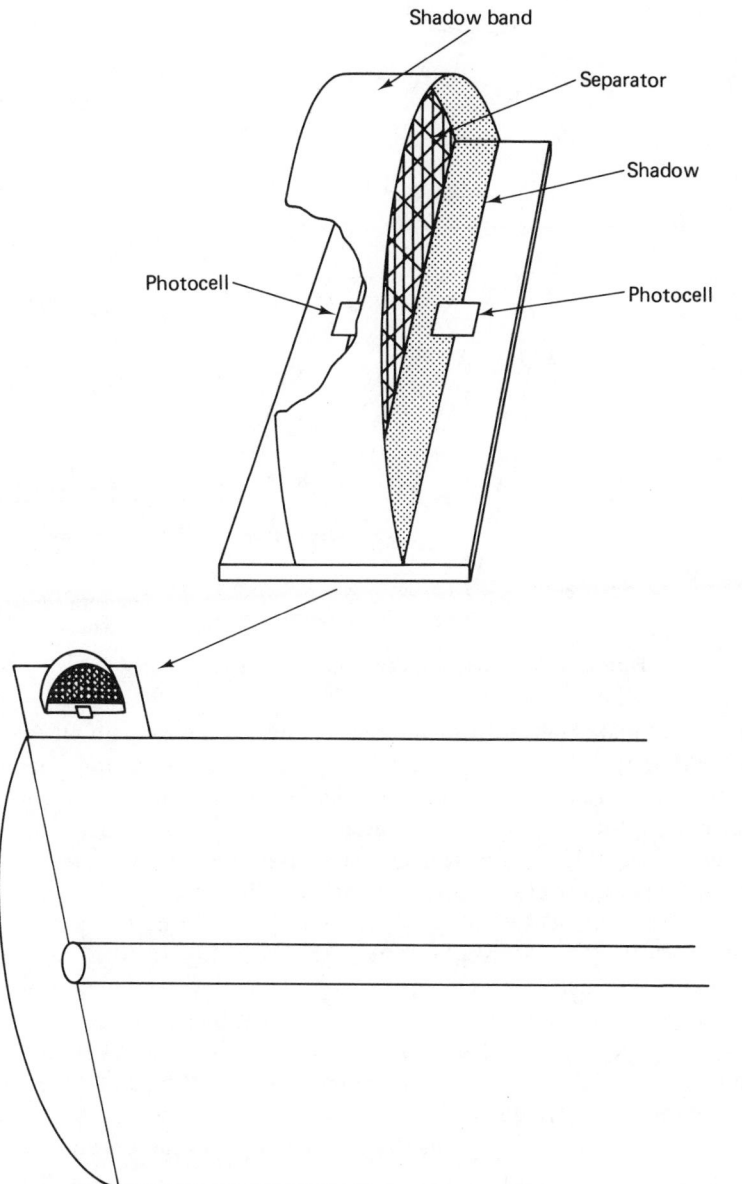

Figure 10.43 Shadow band for tracking of parabolic trough.

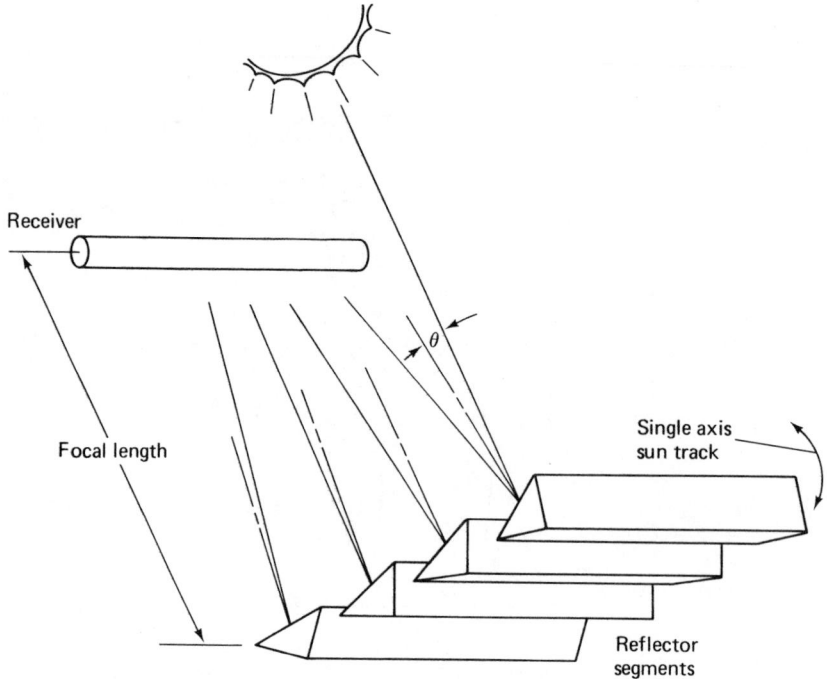

Figure 10.44 Schematic design of segmented mirror collector.

ordinary flat plate collector which faces downward. By using about 6–8 slats, it is possible to achieve the elevated temperatures required for solar space cooling. For additional concentration, the individual slats can be curved as circular cylindrical surfaces to concentrate the reflected light onto the receivers. Use of 10 mirror segments and a secondary concentration of 3 should produce an overall theoretical concentration of about 30.

Losses in segmented mirror collectors are caused by a variety of contributing effects. The receiver shadow is shown in Figure 10.45, where it is clear that some reflected rays from slat 10 are blocked by slat 9, thus blocking some rays from reaching the receiver. When a slat blocks incident rays from an adjacent slat, that is called *shading*, and when a slat blocks reflected rays from an adjacent slat, that is known as *screening*. Shading and screening are illustrated in Figure 10.46. End effect losses occur at hours other than solar noon (see Figure 10.47). Losses also occur due to cosine factors and imperfect reflectivities.

The advantages of the segmented mirror reflector are that the receiver is fixed and the mirror slats are small and presumably relatively inexpensive. On the other hand, the mechanism and the cost of labor for its assembly can be expensive; it is not clear whether this approach is viable.

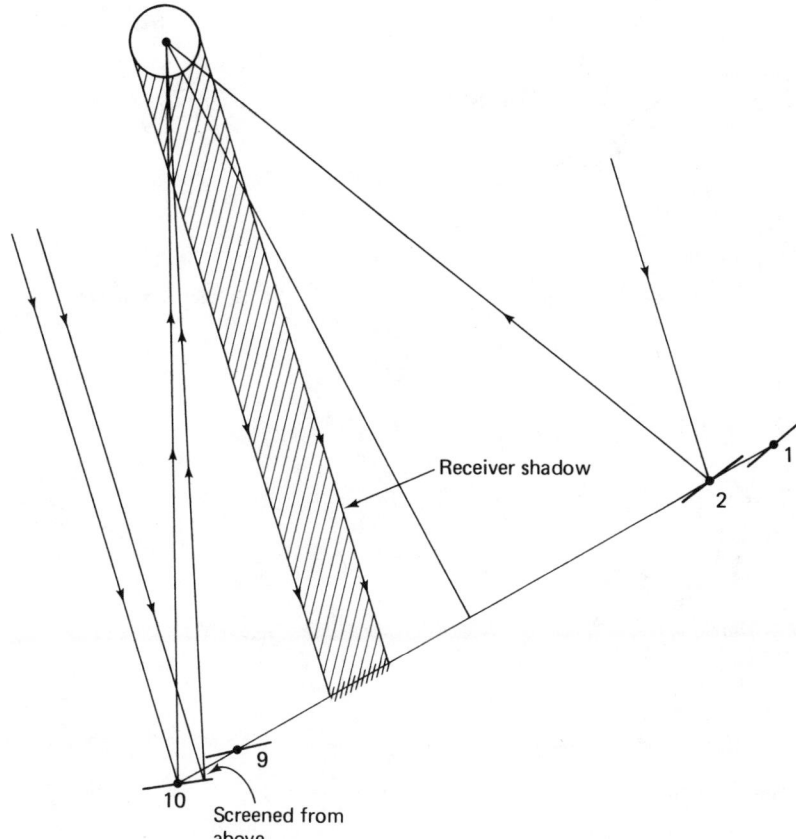

Figure 10.45   Solar rays and receiver shadow on April 20 at noon at 33°N latitude.

10.13 Summary

To produce intermediate temperatures (200–330°F = 367–439°K) some alternative method must be used to reduce the heat losses that would be disastrous with ordinary flat plate collectors. Two methods for doing this are (1) evacuation of the receiver and (2) concentration of the irradiation onto a smaller receiver. Evacuation has been used in a flat plate collector and in tubular collectors. Such collectors can achieve intermediate temperatures with reasonable efficiency. However, the initial costs tend to be high and maintenance costs are, at best, uncertain.

Concentration can be achieved with lenses or reflectors. Reflector systems appear to have advantages over refractors. Reflector concentrators can be operated in fixed, periodically adjusted, or continuous tracking modes.

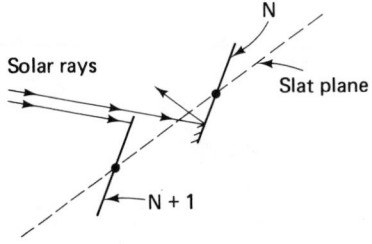

(a) Shading from below

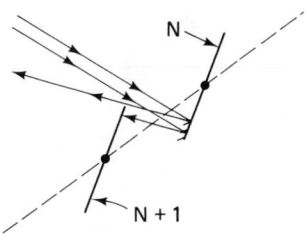

(b) Screening from below

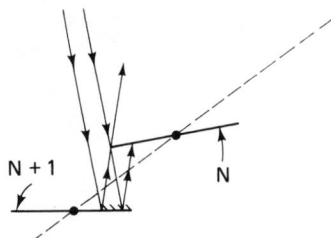

(c) Screening from above

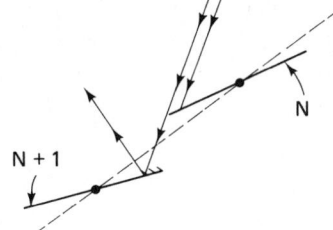

(d) Shading from above

Figure 10.46 Illustration on shading and screening in segmented mirror collectors.

Figure 10.47 Illustration of end effect loss.

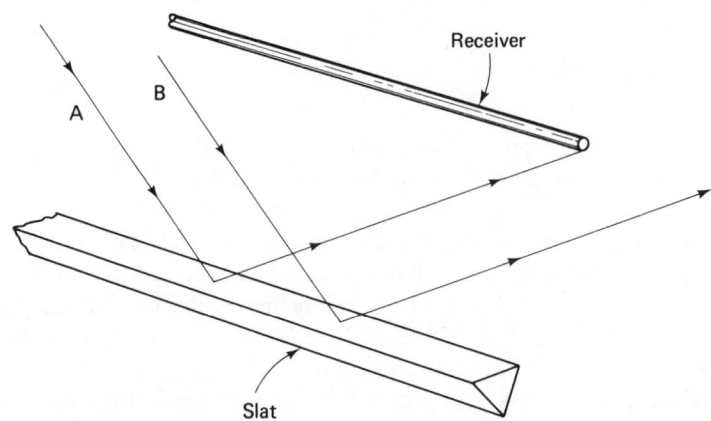

When operating in a fixed or periodically adjusted mode, the relationship between concentration and acceptance aperture becomes important. Because of the change in the elevation of the sun during the day and the seasonal changes in this daily pattern, it is important to achieve as high an acceptance angle as possible for any concentration. The maximum possible aperture for any concentration can be derived by thermodynamic analysis. A physical collector that achieves this ideal relationship is the CPC collector. Unfortunately, the CPC collector has certain physical disadvantages that may negate its theoretical superiority. The maximum possible concentration that can be achieved with a fixed collector that can collect for 6 hr per day on all days of the year is about 2. When periodic adjustment is used, concentrations up to about 5 are feasible. These concentrators should be adequate for intermediate temperatures.

Several concentrating systems which employ continuous tracking can be used. Although these systems have good performance characteristics, they tend to be expensive, have moving parts, and may require considerable maintenance. They have the advantage of not requiring hand readjustment at periodic intervals.

Worked Examples

1. A large dish collector has a concentration of 1000, $\eta_{opt} = 0.7$, and $\epsilon_{abs} = 0.3$. What absorber temperature can be achieved at 50% collection efficiency if only radiative losses from the absorber are included?

$$T_{abs} = T_s \left[(1 - 0.5)\frac{0.7}{0.3} \times 1000 \times \sin^2 (0.25°) \right]^{1/4}$$

$$T_{abs} = 2316°K$$

2. A concentrator has a half angle of acceptance of 8°. Use Figure 10.27 to determine how many readjustments are required per year for 6 hr of daily collection.

On December 21, set the collector tilt at 33°. This will suffice until the variation of χ during the day exceeds 41°, which occurs around February 8. The collector tilt is reset to 49° on that date. This will suffice until the variation of χ during the day exceeds 57°, which occurs around March 20. The collector tilt is reset to 65°. This is continued, leading to the following result:

Setting	Date	Collector Tilt Angle
1	12–21	33°
2	2–8	49°
3	3–20	65°
4	4–17	81°
5	8–7	65°
6	9–22	49°
7	10–21	33°

Problems

10.1. (a) A parabolic trough has an aperture of 24 in. (0.61 m) and a receiver, suspended above the trough, consisting of a simple metal tube of 2 in. (5.1 cm) diameter. If the product of reflector reflectivity and receiver absorptivity is 0.7, what is the effective concentration and the useful concentration? (b) If the upper part of the receiver tube is covered with 2 in. (5.4 cm) thick insulation around an angle of 240°, how are these values modified?

10.2. A parabolic trough as described in Problem 10.1(a) is operating with inlet water at 160°F and outlet water at 200°F when the direct solar intensity on the trough aperture is 250 Btu/hr-ft² (788 W/m²) and the ambient air is at 60°F (289°K). If the heat loss coefficient from the receiver is 0.3 Btu/hr-°F per linear foot (0.52 W/°K-m), estimate the collection efficiency.

10.3. If an ideal dish concentrator has $\eta_{opt}/\epsilon_{abs} = 0.7$ and an acceptance aperture of 1°, what is the maximum absorber temperature achievable at 50% collection efficiency if conductive/convective heat losses are neglected?

10.4. If the collector described in Problem 10.3 is operated at 1500°K, what is the maximum efficiency that can be achieved?

10.5. An ideal two-dimensional concentrator has an acceptance aperture of 2°. What absorber temperature can be achieved at 50% collection efficiency if $\eta_{opt}/\epsilon_{abs} = 0.7$ and conductive/convective heat losses are neglected?

10.6. Calculate the solar elevation χ at 3 hr from noon on December 21 and June 21 at your latitude, and from that determine the minimum aperture that a fixed collector must have to guarantee 6 hr of collection on all days of the year. What is the concentration of an ideal concentrator with this aperture?

10.7. From an inspection of Figure 10.9, what is the effective concentration as a function of the ratio D/R of a collector consisting of a tube in front of a 90° reflector?

10.8. Calculate the hourly variation of solar power input to the aperture plane of a parabolic trough on December 21 at your latitude for the following cases:
(a) North-south tilted at the latitude.
(b) North-south horizontal.
(c) East-west.
Use the following data on hourly direct normal intensities:

Hours from Noon	(Btu/hr-ft²)	(W/m²)
0	300	946
1	290	914
2	270	851
3	210	662
4	120	378
5	0	0

10.9. If a parabolic trough is operated under conditions such that the efficiency is given by

$$\eta = 0.6 - \frac{30}{N_a}$$

where N_a is the direct intensity on the aperture plane of the trough in Btu/hr-ft² (use Problem 10.8), estimate the hourly and daily heat collection for the three orientations of Problem 10.8.

10.10. A parabolic trough is mounted in the east-west orientation. If the maximum distance from the reflector to the receiver is 3 ft (0.914 m), what is the width of the line image due to the finite size of the sun (0.5°) at solar noon and 3 hr from noon on December 21, March 21, and June 21 at your latitude?

10.11. If the tracking errors for a parabolic trough are about ±0.5°, how wide must the receiver be on an east-west trough to guarantee complete collection at 3 hr from solar noon on March 21, taking into account both tracking errors and the finite solar disc? The maximum distance from trough to receiver is 3 ft (0.914 m).

10.12. If the trough of Problem 10.11 extends over a 180° sector, what limitations do the finite size of the sun and the tracking errors place on the achievable concentration if complete collection is to be achieved at 3 hr from solar noon on March 21?

11

Space
Heating/Cooling
Systems

This chapter deals with space cooling by means of solar energy. Cooling loads are estimated by analogy with space heating. The major illustration is in terms of a residential system with a high-quality flat plate collector driving a single-stage absorption chiller. Emphasis is placed on alternate methods for using back-up energy. It is concluded that electricity from a central power plant driving a compression chiller is more efficient than burning natural gas or oil to drive an absorption chiller. The solar space cooling system needs further cost reduction to become competitive.

11.1 Estimation of Cooling Loads

In Sec. 7.1, we discussed the variation of heating and cooling loads with season, day, and hour. The heating and cooling loads are correlated with degree-days, and on a seasonal or monthly basis there is good proportionality between them. The proportionality constant can be estimated from the construction of the exterior building surfaces, and there is good correlation between such calculations and observed loads. However, on a daily basis the correlation between the load and degree-days is not so good. At any hour, the correlation between heating load and $65 - T_{amb}$, where T_{amb} is in °F, is made obscure by thermal inertia, variations in wind and solar conditions, and other factors. The analysis of a residence in Dallas, Texas, in Sec. 7.1 (as evidenced in Figures 7.1 through 7.4) leads to the conclusion (for that case) that the rate at which heat had to be supplied in winter, or removed in summer, is as tabulated in Table 11.1. It can be seen that the summer load

TABLE 11.1   Comparison of winter and summer loads for a residence in Dallas, Texas

	Heat Supplied to Residence in Winter (Btu/ft^2)	Heat Removed from Residence in Summer (Btu/ft^2)
Daily Basis (per degree-day)	14	16
Hourly Basis (per degree difference from 65°F)	0.6	0.7

per degree-day or per degree-hour is higher than the winter load. This condition is presumably due to direct solar gain and the need for dehumidification in the summer. The monthly, daily, and hourly loads can be estimated by using the winter heating values and then adding about 15% for dehumidification. The validity of this procedure depends on prevailing humidities, and the figure given applies only to North Central Texas. Some handbooks recommend adding about 30% for dehumidification and direct solar gain.

11.2 Options for Solar Space Cooling

11.2.1 Absorption Chillers

To cool an interior space using solar energy, there are two approaches that appear worth considering.[1] One is to produce solar heated water to drive an absorption chiller. With this approach, one can either heat water to ~300°F (422°K) to drive a two-stage absorption chiller or heat water to ~200°F (367°K) to drive a one-stage absorption chiller. The two-stage chiller has the advantage of a higher COP but requires a more sophisticated solar collector and storage system to achieve higher temperatures. At present, two-stage chillers are commercially available only in very large sizes. Single-stage chillers are available in a wide range of sizes. However, there is at present only a single one-stage chiller available for the residential market, and it is quite expensive.

11.2.2 Rankine Cycle/Compression Refrigeration Cycle

A second approach is to use a small Rankine cycle turbine with an organic working fluid to drive a conventional compression-type chiller. The problem with this approach is that small turbines are not commercially

---

[1] The method of humidification of desiccant dried air described in Chapter 8 has not been sufficiently developed yet to consider as a third approach here.

available at present. Another problem is that small turbines tend to be irreversible and lead to poor cycle efficiencies. The product of turbine cycle efficiency and compression refrigeration COP is an overall efficiency that may be compared with the COP of an absorption refrigeration system. With a 200°F (367°K) hot temperature and 90°F (305°K) cooling water, a one-stage absorption chiller should be able to operate at a COP of about 0.6. Under the same conditions, an organic turbine can probably operate at about 9% cycle efficiency. If the COP of the compression chiller is approximately 4, this leads to an overall efficiency of about 0.36, which is lower than for the absorption chiller. There is hope that cycle efficiencies of small turbines can be improved. The theoretical limit of a turbine operating between 200°F (367°K) and 90°F (305°K) is about 12.7%.

11.2.3 Methods for Use of Back-up Energy

A major advantage of the compression-type chiller is that when the solar system is depleted and conventional make-up energy is used, the combination of electricity from a central power plant driving a chiller with high COP is a very desirable one. The alternative with an absorption chiller is to burn gas or oil to supply heat to the absorption chiller, which is less desirable because it uses scarce fuels and tends to be more expensive. A substantial fraction of the electricity from a central power plant can be derived from coal or nuclear energy rather than oil or gas. Consider the diagrams shown in Figure 11.1. These diagrams illustrate the energy flows from primary fuel to the final removal of one unit of heat from a building, using a compression chiller, a one-stage absorption chiller, and a two-stage abscrption chiller. It can be seen that the use of absorption chillers involves greater use of conventional fuel to remove one unit of heat from a building and that locally burned fuel must be scarce oil or gas. Furthermore, considerably more heat must be dissipated in the cooling tower. In terms of cost, if electricity is valued at $11 per 10^6 Btu ($11.60 per 10^9 J) and gas at $3.70 per 10^6 Btu ($3.90 per 10^9 J) for delivered heat, the cost of pumping 10^6 Btu out of a residence is calculated to be about $3.50 using a compression chiller, $6.00 using a single-stage absorption chiller, and $3.70 using a two-stage absorption chiller. Particularly note that when a single-stage absorption chiller is used the cost is nearly double that of an electrically driven chiller per Btu pumped out of the house. Therefore, if a house is solar energized to provide 50% of the required cooling via a single-stage absorption chiller, and the other 50% is made up by burning gas or oil to drive the absorption chiller, the yearly operating cost will be nearly the same as using no solar installation and simply us ng a compression chiller with electricity from a central power plant. Yet, at present, the only commercially available way to use solar

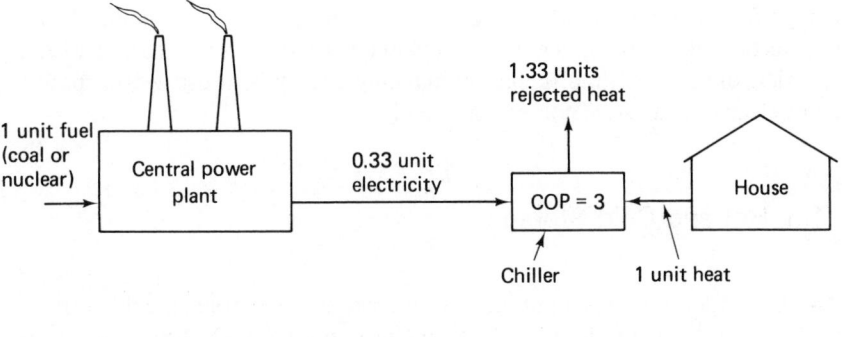

(a)

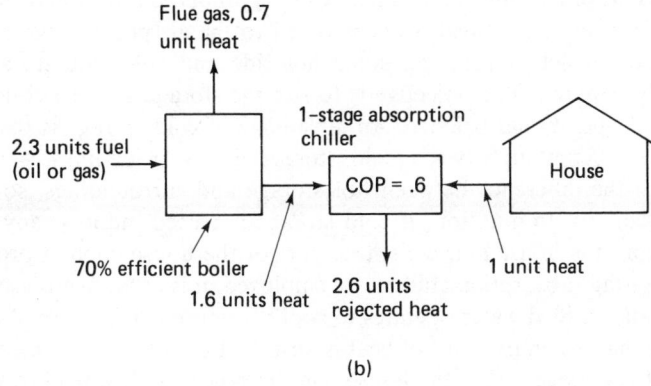

(b)

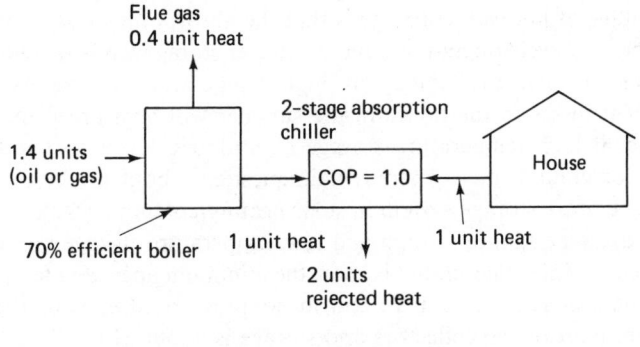

(c)

Figure 11.1 Comparison of energy flows in three approaches for supplying make-up cooling: (a) compression chiller; (b) one-stage absorption chiller; (c) two-stage absorption chiller.

energy for cooling is via absorption chillers. Thus, it may be concluded that two-stage chillers are much more advantageous than single-stage chillers, and that the best method to supply make-up energy is to use central power plant electricity to drive compression chillers.

11.3 Hot and Cold Storage

Another important consideration is whether to use hot storage, cold storage, or a combination of the two. In using hot storage, solar heated water is stored in tanks from which it is drawn as required by the absorption chiller. In using cold storage, solar-generated hot water is fed into the absorption chiller as fast as practicable, and chilled water is produced and stored. As the house requires cooling, chilled water is piped to the air plenum, where the air is cooled. In actual practice, some hot side and cold side storage is undoubtedly required. The objective is to size the storage and the chiller to achieve good operational balance. An advantage of cold storage is that the temperature differential between cold storage and surroundings is much smaller than the difference between hot storage and surroundings, so heat losses are reduced. In addition, if cold storage is placed indoors, any heat leaks are from the house and constitute part of the house cooling process. When single-stage absorption chillers are employed, it is desirable to use cold storage because chilled water operates to cool the house with a COP of unity in the sense that, when one unit of heat is absorbed by the chilled water, one unit of heat is removed from the house. On the other hand, when hot water storage is employed, for every unit of heat removed from the hot water to feed the chiller with COP ≈ 0.6, only 0.6 unit of heat is removed from the house. Therefore, the hot water storage requires considerably more capacity. The advantage of hot water storage is that the absorption chiller works best when it does not cycle on and off, and hot water storage tends to smooth out fluctuations in solar availability. If high temperature storage is located indoors, heat losses to the house in the summer will be a problem. On the other hand, if high temperature storage is outdoors, losses to the ambient will be a problem in winter. There remain questions about the best place for a high temperature storage system in solar heating/cooling systems.

The dynamic ranges of high and low temperature storage are not very great. When a single-stage chiller is used, the minimum operating temperature for the generator is about 180°F, and in nonpressurized systems the upper temperature limit of the collectors and storage is about 210°F (372°K). The solar energy system may therefore be viewed as heating water from about 180°F (355°K) to about 210°F (372°K). Once storage drops to below 180°F

(355°K) it is nearly useless for chilling. The chilled water works best for space conditioning when it is cooled to about 40°F (278°K). By the time it climbs to about 57°F (287°K) it has lost most of its potency for cooling. Thus, the chiller may be viewed as a device for cooling 57°F (287°K) water to 40°F (278°K). When the chilled water exceeds 57°F (287°K), make-up energy will have to be used for space cooling. Therefore, the storage requirements for space cooling are considerably higher than for space heating, where the dynamic range of temperatures is much greater.

11.4 Flow Diagrams

A space heating/cooling/hot water system using a compression chiller for back-up is illustrated in Figure 11.2. In the cooling mode, solar heated water is used to drive the single-stage absorption chiller whenever T_1 is high enough. In order to reduce cycling of the absorption chiller, a reasonable control strategy would be to require T_1 to rise to 190°F (361°K) before turning on the chiller if the chiller is off, and to keep the chiller running until T_1 drops below 180°F (355°K). The chiller functions whenever the solar heat input is large enough to drive it. Cooling of the air plenum is obtained by circulating chilled water from the cold storage tank through the solar cooling coil. As long as cold storage is below some preset value (perhaps $\sim 57°F$), cooling is achieved by this procedure without use of back-up. When cold storage exceeds this temperature limit, cooling is achieved with a compression chiller, using Freon, that operates from electricity as a back-up power source. Hot water is produced by circulating transfer fluid with pump $P1$ through a water preheat tank. During the heating season, pump $P1$ is actuated whenever $T_1 > T_2 + 10°F$. In the cooling season, to avoid depleting storage, $P1$ should only be actuated when the absorption chiller is not running.

Other variations of this design can be utilized. In particular, storage bypass and recirculation modes can be added, as in the case of space heating.

It is also possible to design a system which uses only a single absorption chiller for operation from solar energy and from back-up energy. Such a design is shown in Figure 11.3. With this design, in the cooling mode the absorption chiller goes on whenever $T_1 > 57°F$. Thus, if $T_c > 57°F$ and $T_1 < 180°F$, the back-up auxiliary heater is turned on to boost the inlet temperature to the generator of the absorption chiller. This system has the advantage of using only a single chiller. However, the use of back-up fuel is very inefficient because the product of the back-up heater efficiency and the COP of the chiller is of the order of 0.3 to 0.4, which means that it takes about three units of fossil energy to remove one unit of heat from the house.

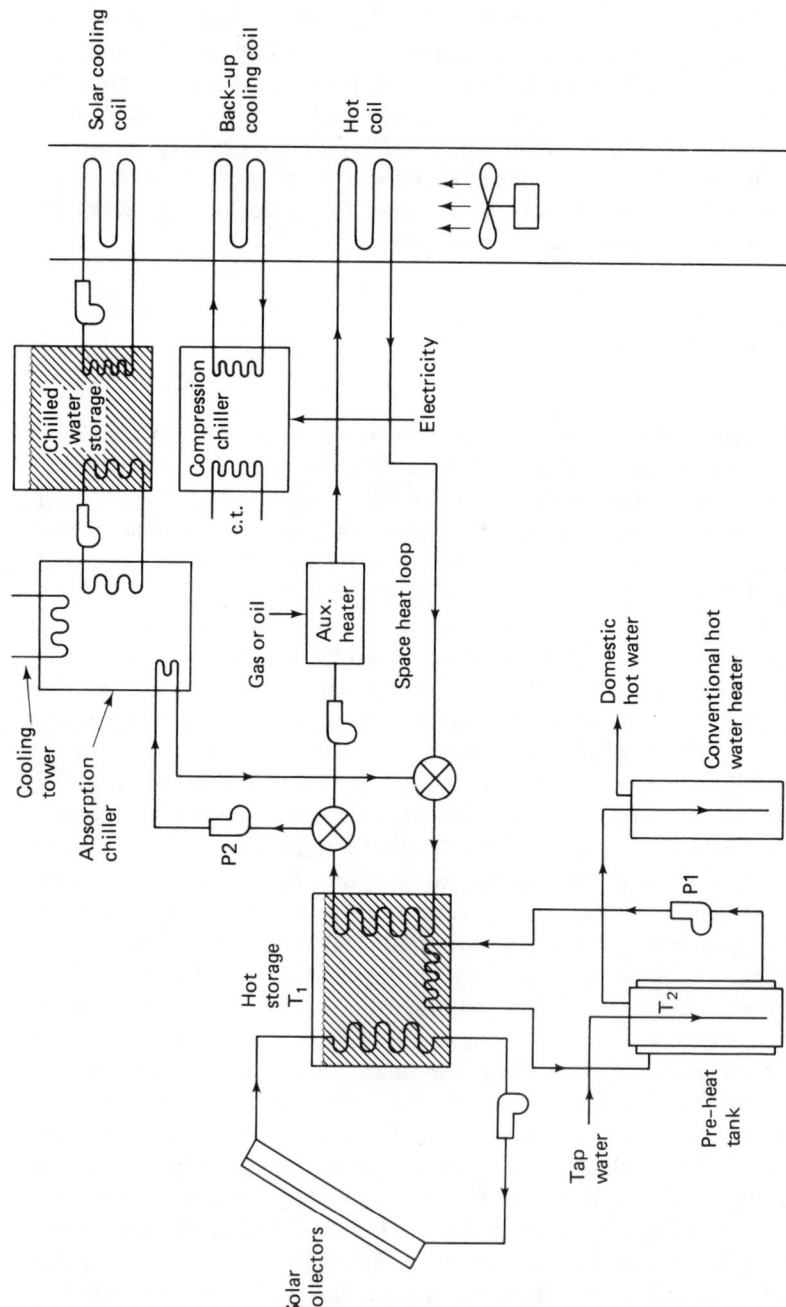

Figure 11.2 Schematic diagram for space heating/cooling/hot water system with compression chiller back-up.

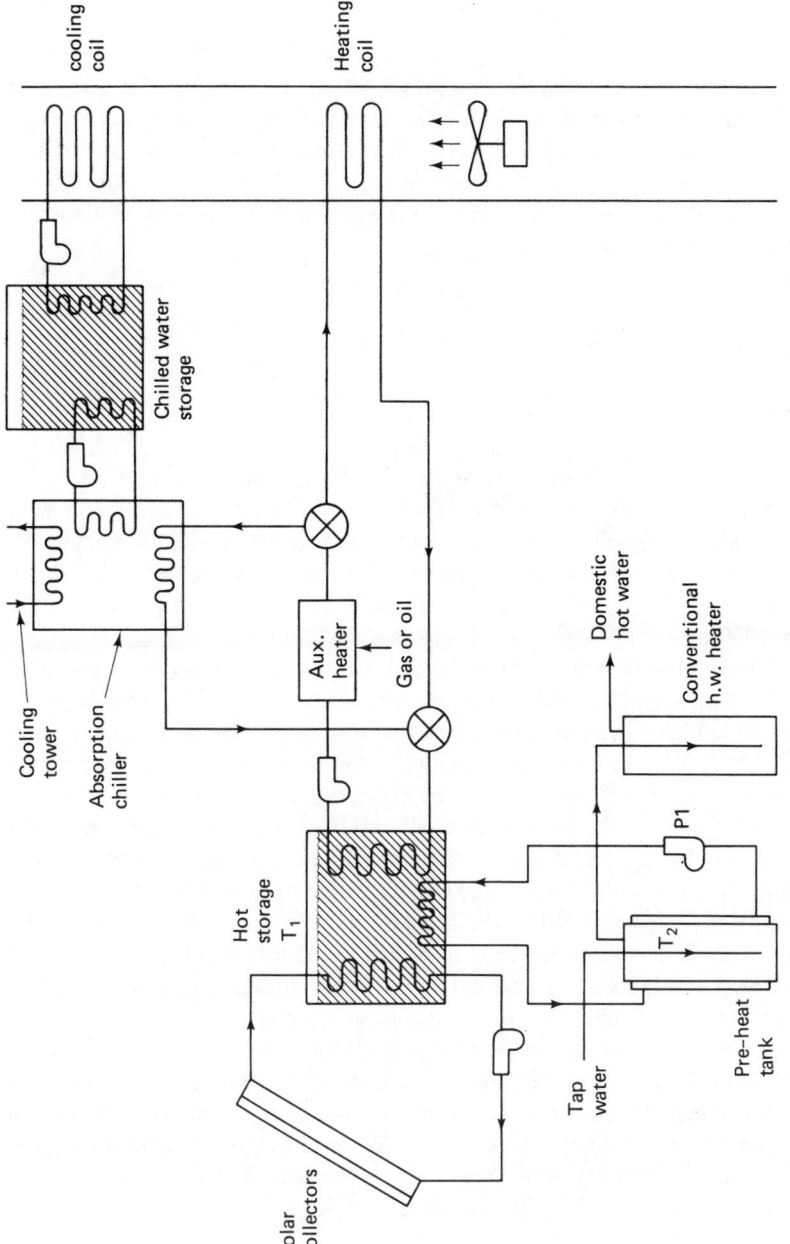

Figure 11.3 Schematic diagram for space heating/cooling/hot water system with fossil fuel burner driving an absorption chiller for back-up.

379

11.5 Computer Simulation

The space heating/cooling system shown in Figure 11.2 can be simulated with a computer program similar to that used in Chapter 7 for space heating. During the winter heating season (roughly from October 20 to April 15 in Dallas, Texas), essentially the same program described in Figure 7.16 is used. The only change is that a more efficient flat plate collector is assumed, with an efficiency

$$\eta = 0.7 - \frac{0.6(\Delta T)}{I} \tag{11.1}$$

instead of the function

$$\eta = 0.7 - \frac{0.9(\Delta T)}{I} \tag{11.2}$$

used in Chapter 7. (Note: I is in Btu/ft^2-$°$F and ΔT in $°$F.) Performance would be considerably better if parabolic troughs were used to drive a two-stage absorption chiller, which was not practical for residential systems as of the mid-1970's. However, for large-scale commercial systems (hundreds of tons of cooling) such an approach might be close to competitive by the early 1980s.

During the months of May through October, the program is modified to include an absorption chiller system. The space cooling demand is assumed to be 20% higher, at any hour when the ambient temperature is x degrees above 65$°$F, than the space heat demand would be when the ambient temperature was x degrees below 65$°$F. The 20% figure is a guess as to the effects of extra solar gain in the summer and dehumidification. The hourly heat load is called *SPHT* in the program and is negative during the air conditioning season. The solar collectors charge the hot water storage tank. The following assumptions are made. When the hot water storage tank reaches 190$°$F, the absorption chiller goes on. The chiller stays on as long as the temperature of hot storage remains at or above 180$°$F. When storage drops below 180$°$F, the chiller goes off and stays off until storage once again reaches 190$°$F. The maximum temperature permitted in the collector/storage loop is assumed to be 230$°$F (383$°$K). The operation of the absorption chiller is simulated as follows. When the chiller is operating continuously (it has run during the previous hour), the COP is taken as 0.6. When the chiller was off during the previous hour, the COP is taken as 0.4 due to start-up inefficiency. The continuous operation of the chiller removes

$$Q_c = 36,000 + (TS - 195)800 \text{ Btu/hr} \tag{11.3}$$

from cold storage during any hour where TS is the hot storage temperature in $°$F at the beginning of the hour. During a start-up hour, this is multiplied by 0.6. The heat removed from the hot storage tank during any hour is Q_c/COP and is denoted as *TIAC* (tank input to air conditioner) in the computer program.

The cooling of the building can be achieved by two mechanisms. When chilled water is available in cold storage at temperatures below 57°F, chilled water is circulated through a coil in the forced air plenum to remove heat from the house. During any hour when the temperature of the cold tank begins the hour at TC $\leq$ 57°F, the net heat input to the cold tank (CTI) is the sum of the heat gain from the air plenum (− SPHT) and the heat removed by the absorption air conditioner $-Q_c$. When TC exceeds 57°F, no chilled water is circulated from storage to the air plenum, and a conventional electrically driven compressor-type air conditioner with COP = 2.8 cools the air in the house, removing SPHT Btu from the house for the hour. In this case, the net heat input to the cold tank (CTI) is simply $-Q_c$.

11.6 Design Results

The performance of the solar air conditioning system is illustrated in Figure 11.4 for a system with 1000 ft² (28.3 m²) of collector area, 1500 gal (42.5 m³) of hot storage, 1000 gal (28.3 m³) of cold storage, and a collector tilt angle of 23°. The vertical columns are HOUR, SH = Btu/hr-ft² solar intensity on a horizontal surface, ST = intensity on tilted collectors, EFF = collector efficiency, TA = ambient temperature in °F, TS = hot storage temperature (°F) at beginning of hour, TSNEW = hot storage temperature at end of hour, TC = cold storage temperature at beginning of hour, TCNEW = cold storage temperature at end of hour, SHITT = solar heat input to hot storage tank (Btu), TIAC = hot water tank input to absorption chiller (Btu), AUXEL = electrical energy required to drive the make-up air conditioner (units are Btu; divide by 3413 to convert to kWh), CTI = cold tank input (Btu) = net heat flow to cold storage, SPHT = heat demand by house (Btu); negative for space cooling load. These results are plotted in Figure 11.5.

An examination of Figures 11.4 and 11.5 shows that June 1 begins with hot storage slightly cooler than the 180°F required to operate the absorption chiller. Cold storage is below 57°F, so the small cooling load in the early morning is supplied from cold storage. However, by the end of the second hour on June 1, TC rises above 57°F. At this point, the electrically driven compression chiller takes over the space cooling requirements. The system remains in this mode for several hours. There is a dip in TS in the 7th and 8th hours due to hot water usage. During the 9th hour, there is a net solar gain which increases TS. As TS continues to increase due to solar gain, it finally passes 190°F at the end of the 11th hour. Thus, the absorption chiller

Figure 11.4 Performance of a solar system for space cooling and hot water on two selected days.

6 1

HOUR	SH	ST	EFF	TA	TS	TSNEW	TC	TCNEW	SHITT	TIAC	AUXEL	CTI	SPHT
1	0	0	0.0	73	178.1	178.1	54.8	56.3	0.	0.	0.	12000.	-12000.
2	0	0	0.0	72	178.1	178.1	56.3	57.5	0.	0.	0.	10500.	-10500.
3	0	0	0.0	72	178.1	178.1	57.5	57.5	0.	0.	3750.		-10500.
4	0	0	0.0	71	178.1	178.1	57.5	57.5	0.	0.	3214.	0.	-9000.
5	0	0	0.0	70	178.1	178.1	57.5	57.5	0.	0.	2679.	0.	-7500.
6	0	0	0.0	69	178.1	178.1	57.5	57.5	0.	0.	2143.	0.	-6000.
7	0	0	0.0	69	178.1	177.4	57.5	57.5	0.	0.	2143.	0.	-6000.
8	70	58	0.0	71	177.4	176.6	57.5	57.5	0.	0.	3214.	0.	-9000.
9	141	129	0.198	75	176.6	178.3	57.5	57.5	25547.	0.	5357.	0.	-15000.
10	204	196	0.363	78	178.3	183.7	57.5	57.5	71161.	0.	6964.	0.	-19500.
11	241	237	0.410	81	183.7	191.2	57.5	57.5	97355.	0.	8571.	0.	-24000.
12	274	273	0.434	84	191.2	196.3	55.2	55.2	118321.	49388.	10179.	-19755.	-28500.
13	263	262	0.412	84	196.3	199.7	55.2	54.1	107882.	61781.	0.	-8569.	-28500.
14	240	236	0.379	85	199.7	201.2	54.1	53.0	89503.	66242.	0.	-9745.	-30000.
15	223	214	0.350	87	201.2	201.4	53.0	52.0	75081.	68257.	0.	-7954.	-33000.
16	230	210	0.344	87	201.4	201.3	52.0	51.0	72522.	68511.	0.	-8107.	-33000.
17	184	152	0.222	88	201.3	197.8	51.0	50.2	33769.	68464.	0.	-6579.	-34500.
18	120	72	0.0	87	197.8	192.0	50.2	49.6	0.	63796.	0.	-5277.	-33000.
19	49	0	0.0	86	192.0	187.2	49.6	49.3	0.	56011.	0.	-2106.	-31500.
20	0	0	0.0	83	187.2	182.8	49.3	49.0	0.	49536.	0.	-2721.	-27000.
21	0	0	0.0	78	182.8	178.9	49.0	48.2	0.	43754.	0.	-6753.	-19500.
22	0	0	0.0	76	178.9	178.6	48.2	50.2	0.	0.	0.	16500.	-16500.
23	0	0	0.0	74	178.6	178.2	50.2	51.8	0.	0.	0.	13500.	-13500.
24	0	0	0.0	72	178.2	177.9	51.8	53.1	0.	0.	0.	10500.	-10500.

Figure 11.4 (Continued)

		6	2										
HOUR	SH	ST	EFF	TA	TS	TSNEW	TC	TCNEW	SHITT	TIAC	AUXEL	CTI	SPHT
1	0	0	0.0	72	177.9	177.9	53.1	54.3	0.	0.	0.	10500.	-10500.
2	0	0	0.0	70	177.9	177.9	54.3	55.2	0.	0.	0.	7500.	-7500.
3	0	0	0.0	71	177.9	177.9	55.2	56.3	0.	0.	0.	9000.	-9000.
4	0	0	0.0	70	177.9	177.9	56.3	57.2	0.	0.	0.	7500.	-7500.
5	0	0	0.0	68	177.9	177.9	57.2	57.2	0.	0.	1607.	0.	-4500.
6	0	0	0.0	68	177.9	177.9	57.2	57.2	0.	0.	1607.	0.	-4500.
7	0	0	0.0	66	177.2	177.2	57.2	57.2	0.	0.	536.	0.	-1500.
8	69	57	0.0	69	177.2	176.6	57.2	57.2	0.	0.	2143.	0.	-6000.
9	144	132	0.198	73	176.6	178.4	57.2	57.2	26100.	0.	4286.	0.	-12000.
10	208	200	0.362	76	178.4	183.9	57.2	57.2	72365.	0.	5893.	0.	-16500.
11	265	261	0.426	78	183.9	192.5	57.2	57.2	111108.	0.	6964.	0.	-19500.
12	297	295	0.443	81	192.5	198.6	57.2	54.8	130873.	50958.	8571.	-20383.	-24000.
13	304	302	0.438	82	198.6	203.7	54.8	53.2	132474.	64752.	0.	-13351.	-25000.
14	311	306	0.433	83	203.7	208.3	53.2	51.2	132526.	71578.	0.	-15947.	-27000.
15	293	281	0.407	85	208.3	210.9	51.2	49.2	114503.	77678.	0.	-16607.	-30000.
16	253	231	0.346	86	210.9	210.5	49.2	47.2	80121.	81195.	0.	-17217.	-31500.
17	197	162	0.210	85	210.5	206.1	47.2	45.1	34089.	80654.	0.	-16852.	-30000.
18	129	78	0.0	85	206.1	199.5	45.1	43.3	0.	74813.	0.	-14888.	-30000.
19	53	0	0.0	84	199.5	193.8	43.3	42.0	0.	65948.	0.	-11069.	-28500.

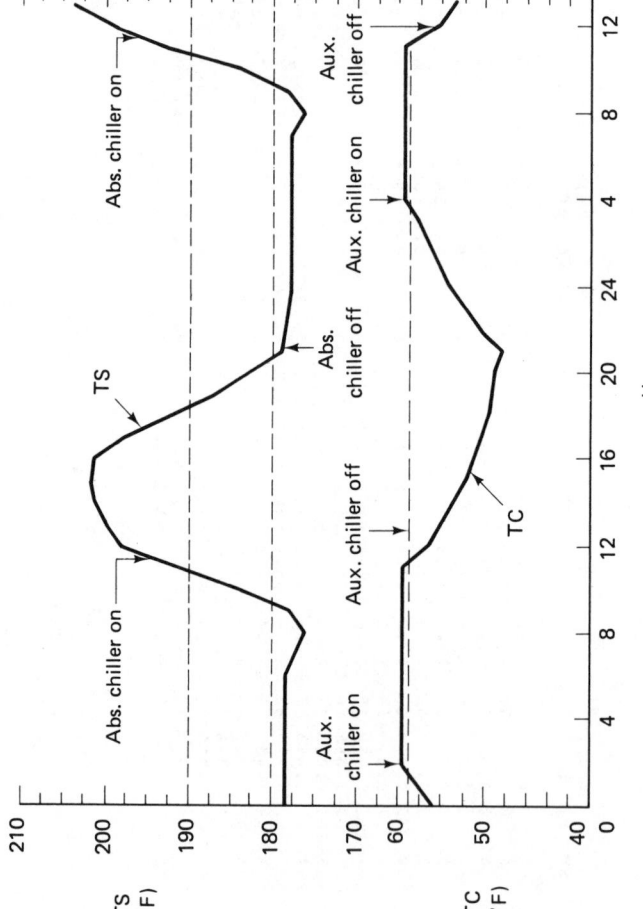

Figure 11.5 Temperatures of hot and cold storage for the days shown in Figure 11.4.

goes on during the 12th hour, operating under start-up conditions. During the 12th hour, operation of the absorption chiller reduces TC to below 57°F. Therefore, the compression chiller goes off after this hour, and cooling is achieved by circulation of chilled water to the air plenum. As the day progresses, TS rises and passes through a maximum. When TS finally drops to below 180°F during the 21st hour, the absorption chiller goes off at the end of that hour. However, since TC remains below 57°F, space cooling continues from circulation of stored chilled water to the air plenum. During the 4th hour of the next day (June 2), TC finally rises above 57°F, and therefore the compression chiller takes over the cooling load during the 5th hour. As the sun supplies energy to storage, TS rises and exceeds 190°F during the 11th hour of June 2. The absorption chiller therefore goes back on during the 12th hour. One hour of operation of the absorption chiller brings TC below 57°F, and at the end of the 12th hour the compression chiller goes off and space cooling is achieved by circulation of chilled water to the air plenum. It was not shown in these figures, but during each day the domestic hot water demand was met 100% by withdrawing heat from hot storage.

On a monthly basis, the performance of this system, 1000 ft² (92.9 m³) at 23°, 1500 gal (42.5 m³) hot storage, 1000 gal (28.3 m³) cold storage, is illustrated in Table 11.2. It can be seen that this system displaces 86.1×10^6 Btu (90.8×10^9 J) of thermal load and 4353 kWh of electricity during a year.

TABLE 11.2 Monthly performance of a solar heating/cooling/hot water system with 1000 ft² (92.9 m²) of collectors at 23° tilt, 1500 gal (5.67 m³) of hot storage, and 1000 gal (3.78 m³) of cold storage for a 2500 ft² (232 m²) house in Texas

Month	Space Heat			Domestic Hot Water			Space Cooling		
	Req'd (10^6 Btu)	Solar	Fraction Solar	Req'd (10^6 Btu)	Solar	Fraction Solar	Req'd (kWh)	Solar	Fraction Solar
1	19.6	11.0	0.561	3.4	2.9	0.868			
2	12.8	11.1	0.653	3.1	3.0	0.964			
3	10.1	9.4	0.932	3.2	3.2	0.991			
4	1.8	1.8	1.000	2.9	2.9	1.000			
5				3.0	3.0	1.000	1145	754	0.659
6				2.6	2.6	1.000	1751	1077	0.615
7				2.7	2.7	1.000	2347	1134	0.483
8				2.7	2.7	1.000	2328	1207	0.518
9				2.6	2.6	1.000	1609	976	0.607
10				3.0	3.0	1.000	924	603	0.653
11	8.0	6.3	0.787	3.1	2.9	0.961			
12	16.1	11.8	0.733	3.3	3.0	0.915			
Year	68.4	51.4	0.751	35.7	34.7	0.972	10104	5751	0.569

TABLE 11.3 Performance of solar heating/cooling/hot water systems on an annual basis in Texas

Collector Area (ft²)	Tilt Angle (°)	Hot Storage (gal)	Cold Storage (gal)	Space Heat			Domestic Hot Water			Space Cooling		
				Req'd (10^6 Btu/yr)	Solar (10^6 Btu/yr)	Fraction Solar	Req'd (10^6 Btu/yr)	Solar (10^6 Btu/yr)	Fraction Solar	Req'd (kWh/yr)	Solar (kWh/yr)	Fraction Solar
600	23	900	600	68.4	35.3	0.52	35.7	33.7	0.94	10,100	3250	0.32
800	23	1200	800	68.4	44.5	0.65	35.7	34.3	0.96	10,100	4540	0.45
1000	23	1500	1000	68.4	51.4	0.75	35.7	34.7	0.97	10,100	5750	0.57
	33	1500	1000	68.4	55.0	0.80	35.7	34.9	0.98	10,100	5350	0.53
1200	23	1800	1200	68.4	56.2	0.82	35.7	35.0	0.98	10,100	6850	0.68

Most of the electricity displaced is during peak hours of the peak season. Valuing the heat at $3.67 per 10^6 Btu ($3.48 per 10^9 J) delivered, and the electricity at $0.04 per kWh, the annual saving at these energy prices is $316 in heat and $174 in electricity.

A series of such runs were made for various design parameters, and the results are reported on an annual basis in Table 11.3. Note that varying the collector tilt angle from 33° (the latitude) to 23° reduces the heating performance but improves the cooling performance. The loss of 3.8×10^6 Btu (4.01×10^9 J) valued at $14 nearly balances the gain of 400 kWh valued at $16. Thus, on a yearly basis, tilt angles from about 20–35° are equally desirable. Variation of the volume of cold storage had only a small effect on system performance.

It will be assumed that the cost of this system may be approximated by:

Collectors	$12 per ft² ($129 per m²)
Hot storage	$(500 + 0.5v)$ (v in gal)
Cold storage	$(400 + 0.4v)$ (v in gal)
Absorption chiller	$3000
Other fixed costs	$3000

Then the value of fuel saved and the cost of the installation may be calculated and is shown in Table 11.4. Using scenario III of Chapter 5, the net present values and pay-back periods can be calculated; the results are given in Table 11.5. The economics are not very encouraging. The tax credit

TABLE 11.4 Estimated cost and value of energy for solar heating/cooling/hot water systems in Texas

System	Collector Area (ft²)	Tilt Angle (°)	Hot Storage (gal)	Cold Storage (gal)	Value of Heat Displaced	Value of Electricity Displaced	Cost of System ($)
1	600	23	900	600	$253	$130	$14,800
2	800	23	1200	800	$289	$182	$17,400
3	1000	23	1500	1000	$316	$230	$20,100
4	1200	23	1800	1200	$335	$274	$22,700

TABLE 11.5 Economics of residential solar space heating/space cooling/ domestic hot water systems in Texas

System	Collector Area (ft²)	Cost ($)	Value of 1st yr Energy Savings ($)	NPV (15 yr)	PBP
1	600	14,800	383	−11,813	>15 yr
2	800	17,400	471	−13,150	>15 yr
3	1000	20,100	546	−15,123	>15 yr
4	1200	22,700	609	−17,594	>15 yr

of \$2,200 allowed by the Energy Act of 1978 would improve the economics of these residential systems somewhat. However, the economics are considerably better for commercial installations. For the systems described in Tables 11.4 and 11.5, a commercial investor can deduct all payments of principal (as depreciation) and interest. If the company is in a 50% tax bracket, that implies that the net costs are roughly 50% of those shown in Tables 11.4 and 11.5.

Worked Examples

1. A house has a summer cooling load which may be expressed as the requirement that 70×10^6 Btu must be removed from the house per summer season. An electrically driven conventional compressor-type chiller can be employed with a COP = 2.5. What is the annual cost of electricity (at \$0.04 per kWh) for air conditioning?

$$\text{Cost} = \frac{70 \times 10^6 \text{ Btu}}{3413 \text{ Btu/kWh}} \times \frac{1}{2.5} \times \$0.04 = \$328$$

2. A solar energy air conditioning system is used on the house in Example 1, which utilizes an absorption chiller with an average COP of 0.5. If the solar energy system displaces 60% of the conventional air conditioning load, and auxiliary make-up energy is supplied in the form of electrically heated water to the absorption chiller, how much electricity is saved per summer season with the solar system?

Assume a COP of 0.5 for the absorption chiller.

$$\text{Heat required for absorption chiller} = \frac{70 \times 10^6}{0.5} = 140 \times 10^6 \text{ Btu}$$

$$\text{Make-up energy required} = 0.4 \times 140 \times 10^6 = 56 \times 10^6 \text{ Btu} = 16,400 \text{ kWh}$$

The electrical requirements in the absence of the solar energy system are

$$\frac{70 \times 10^6}{3413} \times \frac{1}{2.5} = 8,200 \text{ kWh}$$

Thus, the solar energy system actually requires more electricity than the conventional system. Because of the low COP of the absorption chiller (0.5) compared to the high COP of the conventional chiller (2.5), each unit of electricity used as make-up in the solar system removes one-fifth as much heat as each unit of electricity used in the compression chiller.

3. If make-up energy is supplied to the system described, in Example 2, in the form

of a separate compression chiller running directly from electricity, how much electricity is saved from use of the solar energy system?

$$\text{Make-up energy} = 0.4 \times 8200 \text{ kWh} = 3280 \text{ kWh}$$

$$\text{Electricity saved} = 0.6 \times 8200 \text{ kWh} = 4920 \text{ kWh}$$

4. Using the results of Example 3 and scenario III of Table 5.2 and a 15 year loan at an after-tax interest rate of 7%, what is the maximum cost that additions to a solar heating/hot water system (extra collectors, absorption chiller, cold storage, labor, etc.) for air conditioning can have to produce a pay-back period of 6 years? Is this realistic?

The value of the first year's saving in electricity is

$$4920 \times 0.04 = \$196.80$$

The annual payments are \$107.88 per \$1,000 borrowed. Let the cost be C. A tax credit of

$$\$600 + 0.2(C - 2000)$$

reduces the amount borrowed to

$$0.8C - \$200$$

For a 6 year pay-back,

$$\$196.80 R_6 = (0.8C - \$200)0.10788 \times T_{15}$$

$$C = \left(\frac{\$196.80 \times \$11.05}{0.10788 \times \$10.38} + 200\right) 1.25 = \$2677$$

As of 1979, it is doubtful whether such an addition for solar air conditioning could be made for less than \$6,000–\$8,000. Extra collectors, storage, an absorption chiller, a cooling tower, pumps, controls, and labor would far exceed \$2677.

Problems

11.1. In the February/March 1976 issue of *Solar Engineering*, it is reported that the City of Dallas spent \$70,000 to solar energize a fire station with 1200 sq ft of collector and a 2000 gal storage tank. Heating, hot water, and air conditioning (through a 3 ton Arkla absorption chiller) were provided. The City of Dallas figured a 21 year pay-back period based on 10% compounded annual escalation of electricity costs. Comment on this analysis. Use the following data:

Month	Average Daily Insolation on a Surface Tilted at the Latitude (Btu/ft^2)	Average Daily Collection Efficiency
January	1300	0.48
February	1650	0.47
March	1680	0.45
April	1650	0.44
May	1750	0.43
June	1800	0.43
July	1800	0.43
August	1900	0.43
September	1750	0.44
October	1580	0.45
November	1400	0.47
December	1220	0.48

Hot water requirement = 200 gal/day

Electricity savings in first year = \$12 per 10^6 Btu displaced

11.2. A house requires 1000 ft^3 of gas for heating on a winter day with 20 heating degree-days. During the summer, an absorption air conditioner is used on a day with 30 cooling degree-days. How much gas is required to fire the air conditioner?

11.3. A gas-fired absorption chiller has a COP of 0.5, and a conventional electrically driven compression chiller has a COP of 2.0. If approximately 3000 Btu/hr = 1 kW and electricity is about three times as expensive as gas, which type of chiller is more economical to run?

11.4. If a two-stage absorption chiller were available with a COP = 1.0, how would this change the comparison in Problem 11.3?

12

Concentrating Collectors for High Temperatures

In this chapter we describe concentrating methods for achieving high temperatures ($> 422°$K). One method is simply an extension of an approach used for intermediate temperatures (367–$439°$K), namely, linear arrays. This class of collectors includes parabolic troughs and segmented mirror arrays. The segmented mirrors can utilize movable mirrors and a fixed receiver, or fixed mirrors and a movable receiver. The hourly performance of a linear array on bright clear days at various seasons is discussed.

Greater concentration (and therefore higher temperatures) are achievable with "dish"-type concentrators. However, these require two-dimensional tracking, which can be expensive. Dish concentrators can use a movable paraboloid or a fixed spherical dish and a movable receiver.

Very high concentrations can be achieved with central receiver systems. These systems use large fields of mirrors on pedestals which move so as to concentrate the reflected light onto a fixed receiver atop a tower above the field.

12.1 Introduction

Concentrators for producing high temperatures [greater than about $300°$F ($422°$K)] may be classified according to basic principles of geometry. Linear arrays are concentrators with one long axis, such as a parabolic trough, that are basically designed to track along one axis. However, when mounted in a north-south orientation, adjustment of the tilt corresponds to tracking along a second axis. A second approach is use of bowl geometry, which is best typified by a parabolic dish or paraboloid of revolution or by a spherical dish. This collector is capable of higher concentration than the linear arrays but requires more complex tracking and mounting. The approaches involving linear arrays or dish-type geometries are usually referred to as *distributed fields* because the collectors are made in the form of modules that are repeated many times in a field. In both the linear array and dish-type geometries, Fresnel lenses may be used in place of reflectors. Although the development of Fresnel lens collectors has lagged behind reflective collector development, there is a considerable potential in both approaches. One major detriment of distributed fields is that the entire field is full of plumbing lines, which are undesirable because of the possibility of pressure drops, leaks, cost, service and maintenance, and large surface areas for heat loss. The use of distributed fields without field plumbing has been proposed by using engines directly coupled to the receivers of dish collectors.[1] The problems of field plumbing can also be avoided by a third general approach: the central receiver concept. In this approach, there is one central place where fluid is exposed to concentrated sunlight from many fixed tracking reflectors (heliostats); thus, the problem of having a field full of plumbing is avoided, allowing much higher concentrations and temperatures. The central receiver concept has been primarily considered with the receiver on a tower surrounded by reflectors on pedestals that rotate so as to focus light onto the receiver. It has also been considered in the form of a large fixed reflective bowl of several hundred feet in diameter, with a movable receiver that is moved to track the image of the sun.

Although there is little doubt that, technically, the various collector concepts can be made to work, the estimates of cost based on large-scale production are conjectural. There appear to be many hidden costs involved in most designs. Some are expensive to ship to the installation site. Others can only be shipped in disassembled form and are very expensive to install in the field. Ground preparation and mountings and pedestals are often very expensive. Maintenance costs can be high—especially for cleaning mirrors. It remains to be seen whether concentrating collectors for high temperatures will become economically competitive sources of solar energy.

12.2 Linear Arrays

12.2.1 Parabolic Troughs

The parabolic trough is the prime example of a linear array. It was discussed previously in Sec. 10.11. When used for higher temperatures, this type of trough must operate in a continuous tracking mode. The Honeywell

---

[1]A Preliminary Assessment of Small Steam Rankine and Brayton Point Focusing Solar Modules, Jet Propulsion Lab, DOE/JPL-1060-16, March 1, 1979; Techno-Economic Projections for Advanced Small Solar Thermal Electric Power Plants to Years 1990–2000, Jet Propulsion Lab, DOE/JPL 1060-4, November 15, 1978.

Corporation began developing a parabolic trough collector around 1973 with support from the National Science Foundation. Support was later transferred to ERDA and then to the Department of Energy. Tests were carried out at Desert Sunshine Tests, Inc., located in Arizona. A small unit was built and tested at temperatures as high as 572°F (573°K). In more recent work (1977–78), Honeywell redesigned for a half-trough, consisting of half of a parabolic trough as shown in Figure 12.1. Honeywell claims that the primary advantage of the half-trough is its almost flat profile, which allows it to be stowed (when not in collection mode) horizontally facing downward to reduce wind forces and the adhesion of dew and dust. The receiver is a steel tube with a selective coating mounted behind etched low-iron glass and covered around 3/4 of the periphery by calcium silicate insulation. The receiver is fixed relative to the concentrator and rotates with it. The optical

Figure 12.1 The half parabolic trough. The assembly consisting of half-trough and receiver can rotate about the pivot point as a rigid unit. In the "stow" position, the half-trough points down and the receiver is at ground level.

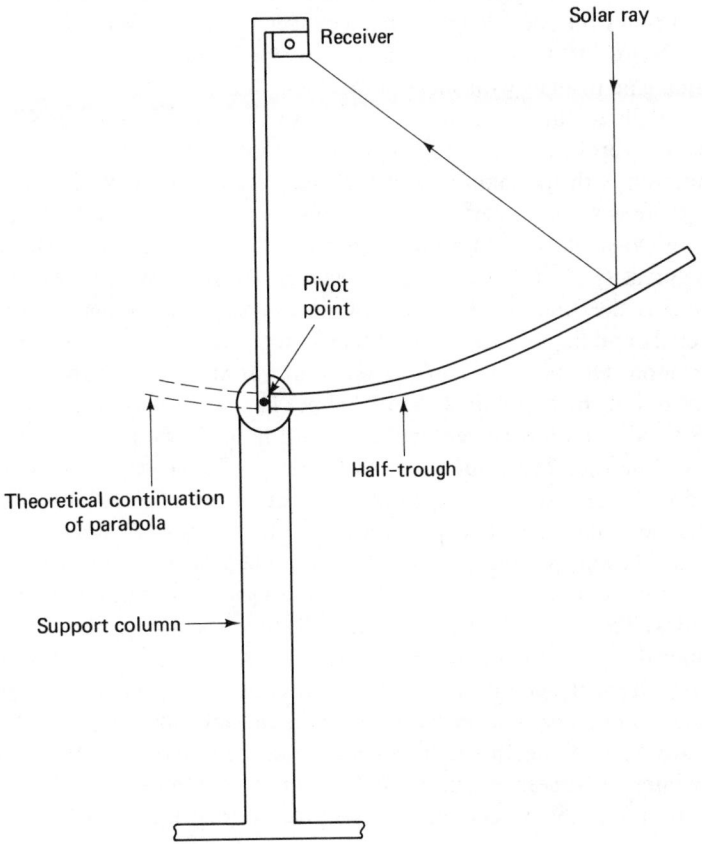

efficiency is predicted to be about 70% with clean mirrors. At 400°F (478°K) heat losses reduce the efficiency to about 50%. At 600°F (589°K), the efficiency is expected to be in the 30–40% range.

At the Conference on Solar Concentrating Collectors held at the Georgia Institute of Technology in September, 1977, several other parabolic collectors were described. The Jacobs-Del Company has selected a relatively small aperture of 2 ft (0.6 m) in order to minimize installation labor. Of course, this increases the amount of field plumbing as compared with the use of troughs having a larger aperture. Jacobs-Del also used a fixed receiver and rotated the concentrator about the receiver axis with a sleeve and worm drive. Acurex, Hexcel, and Solar Kinetics are among the manufacturers that also market parabolic troughs for operation in the range 300–500°F (422–533°K). Many thousands of square feet of these collectors have been installed from 1978–1980 in various solar applications. Troughs are now available in widths of 7 ft from Solar Kinetics of Dallas, Texas.

To reduce end effects at off-noon hours, it is usually necessary to string a number of modules in a line so that the length of the array is at least several times the focal length. This arrangement makes it difficult to mount the collector string in any other way except to lay it horizontally on the ground. The troughs with shortest focal length are most amenable to mounting at a tilt to the horizontal, but even so the substructure costs are quite high. The problem with horizontal mounting is that it is best suited to orientation with the main axis in the east-west direction. With this orientation, there is a loss in performance at off-noon hours due to the angle θ that the sun's rays make with a line perpendicular to the aperture. The effective solar intensity at any hour is the direct normal intensity N times $\cos \theta$. The angle θ is the angle between the line to the sun and an east-west line. On March 21 and September 21, the angle θ increases by 15° for each hour from solar noon. On other days of the year, the variation of θ with hour is nonuniform, but the approximation of 15° per hr is typically good in estimating $\cos \theta$ to within a few percent within several hours of noon. This approximation will be used here. Summer and winter performance are fairly well balanced with the east-west orientation. Any other orientation of the main axis is also possible. The extreme opposite is the north-south orientation. With this orientation, performance is strongly peaked in the summer because the array is not far from perpendicular to the sun in midsummer at moderate latitudes. Therefore, the collectors are effective for a much larger part of a summer day than for the east-west orientation. In the winter, however, the low rays from the sun in the southern sky come in at a glancing angle, even at solar noon, resulting in very poor winter performance. At orientations part way between north-south and east-west, the difference between summer and winter performance is part way between the extremes.

It is desirable to estimate the relative performance of a long string of parabolic troughs on a clear day as a function of season. This will be illustrated for a site at 40°N latitude, corresponding to the middle of the United States. The effective solar intensity at any hour is as follows:

$$N_{eff} = N \cos \theta \tag{12.1}$$

The efficiency of a solar collector will generally be of the form

$$\eta = A - \frac{B \, \Delta T}{C N_{eff}} \tag{12.2}$$

where A is the optical efficiency, B is a heat loss coefficient for the receiver, C is the concentration, and ΔT is the temperature difference between the receiver and the ambient. Based on the limited data that are available on parabolic trough performance, reasonable figures are

$$A = 0.68$$

$$\frac{B}{C} = 0.09 \text{ Btu/hr-°F-ft}^2 \text{ (0.51 W/°K-m}^2\text{) of aperture}^2$$

For purposes of illustration, ΔT will be assigned as 500°F (533°K). Thus, we will use

$$\eta = 0.68 - \frac{45}{N \cos \theta}$$

with N in Btu/hr-ft². We may now estimate the energy collected at any hour by multiplying η by N_{eff} at that hour. The values of N are estimated for clear days from evidence given in Chapter 3, and θ is calculated based on the assumption that θ increases by 15° for each hour from noon. The heat gain per unit area of aperture is denoted as Q. The appropriate calculations are shown in detail for 40° latitude in Tables 12.1, 12.2, and 12.3, and the results are plotted in Figure 12.2. The east-west orientation collects roughly equal amounts of heat on clear days for all days of the year, whereas the north-south orientation at 40° latitude collects very strongly in the summer and very weakly in the winter.

To evaluate the best collector orientation one should take into account:

1. The user requirements as a function of season.
2. Solar availability as a function of season.

The user requirements will tend to dictate how the collectors should be oriented. This must be coupled to information on solar availability. If

²Recent developments by Solar Kinetics Inc. of Dallas, TX indicate that values $A = 0.7$ and $B/C = 0.06$ Btu/hr-ft²-°F are routinely achieved with commercial troughs.

TABLE 12.1 Performance of E/W and N/S oriented horizontal parabolic troughs on a clear day (Dec. 21) at 40°N lat. (Units of N, N_{eff}, and Q are Btu/hr-ft².)

Hour	N	East-West				North-South			
		$cos\,\theta$	N_{eff}	η	Q	$cos\,\theta$	N_{eff}	η	Q
7	0								
8	148	0.500	74	0.072	5.3	0.446	66	0.000	0.0
9	244	0.707	173	0.420	72.7	0.446	109	0.266	28.9
10	277	0.866	240	0.493	118.3	0.446	124	0.316	39.0
11	290	0.966	280	0.519	145.3	0.446	129	0.332	42.9
12	295	1.000	295	0.527	155.5	0.446	132	0.338	44.5
13	290	0.966	280	0.519	145.3	0.446	129	0.332	42.9
14	277	0.866	240	0.493	118.3	0.446	124	0.316	39.0
15	244	0.707	173	0.420	72.7	0.446	109	0.266	28.9
16	148	0.500	74	0.072	5.3	0.446	66	0.000	0.0
17	0								
Total					838.7				266.1

TABLE 12.2 Performance of E/W and N/S oriented horizontal parabolic troughs on a clear day (Mar. 21 or Sep. 21) at 40°N lat. (Units of N, N_{eff}, and Q are Btu/hr-ft².)

Hour	N	East-West				North-South			
		$cos\,\theta$	N_{eff}	η	Q	$cos\,\theta$	N_{eff}	η	Q
6	0								
7	133	0.259	34	0.000	0.0	0.766	102	0.239	24.4
8	225	0.500	123	0.314	38.6	0.766	172	0.418	72.0
9	252	0.707	178	0.427	76.0	0.766	193	0.447	86.2
10	276	0.866	239	0.492	117.5	0.766	211	0.467	98.5
11	281	0.966	271	0.514	139.3	0.766	215	0.471	101.2
12	284	1.000	284	0.522	148.1	0.766	218	0.474	103.2
13	281	0.966	271	0.514	139.3	0.766	215	0.471	101.2
14	276	0.866	239	0.492	117.5	0.766	211	0.467	98.5
15	252	0.707	178	0.427	76.0	0.766	193	0.447	86.2
16	225	0.500	123	0.314	38.6	0.766	172	0.418	72.0
17	133	0.259	34	0.000	0.0	0.766	102	0.239	24.4
18	0					0.766	0		
Total					890.9				867.7

the average levels of direct normal intensity are only slightly lower in winter than in summer as in Central Texas, then applications to a summer (or winter) peaking load should be oriented north-south (or east-west). On the other hand, in localities where winter levels of direct normal intensity are very small, it might be desirable to emphasize summer collection with a north-south orientation. Except in the case of a strong summer peaking load,

TABLE 12.3 Performance of E/W and N/S oriented horizontal parabolic troughs on a clear day (June 21) at 40°N lat. (Units of N, N_{eff}, and Q are Btu/hr-ft^2.)

		East-West				North-South			
Hour	N	$cos\,\theta$	N_{eff}	η	Q	$cos\,\theta$	N_{eff}	η	Q
5	0								
6	136	0.000				0.959	130	0.334	43.4
7	204	0.259	53	0.000	0.0	0.959	196	0.450	88.3
8	232	0.500	117	0.295	34.6	0.959	223	0.478	106.6
9	256	0.707	181	0.431	78.1	0.959	246	0.497	122.3
10	262	0.866	227	0.482	109.4	0.959	251	0.501	125.7
11	265	0.966	256	0.504	129.1	0.959	254	0.503	127.7
12	266	1.000	266	0.511	135.9	0.959	255	0.504	128.4
13	265	0.966	256	0.504	129.1	0.959	254	0.503	127.7
14	262	0.866	227	0.482	109.4	0.959	251	0.501	125.7
15	256	0.707	181	0.431	78.1	0.959	246	0.497	122.3
16	232	0.500	117	0.295	34.6	0.959	223	0.478	106.6
17	204	0.259	53	0.000	0.0	0.959	196	0.450	88.3
18	136	0.000				0.959	130	0.334	43.4
19	0								
Total					838.3				1356.4

Figure 12.2 Expected heat collection of a horizontal string of parabolic troughs per day per ft^2 of aperture on clear days.

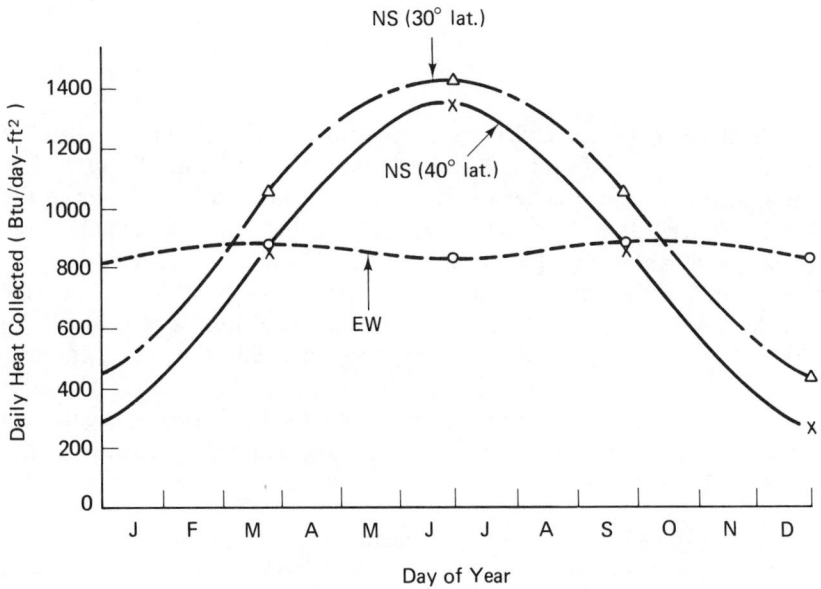

the east-west orientation is generally preferred. At lower latitudes (30°N) the north-south orientation gains in effectiveness compared to the east-west orientation, especially in winter. This is illustrated in Table 12.4, which may be compared with Table 12.1. The variation with day of the year for 30°N latitude is compared with the variation for 40°N latitude in Figure 12.2. It can be seen that, for horizontal arrays, the north-south orientation gains relative to the east-west orientation as one approaches the equator.

TABLE 12.4   Performance of a N/S trough on a clear day (Dec. 21) at 30°N lat. (Units of N, N_{eff}, and Q are Btu/hr-ft².)

Hour	N	$\cos \theta$	N_{eff}	η	Q
7	0	0.552	0		
8	148	0.552	82	0.131	10.8
9	244	0.552	135	0.347	46.8
10	277	0.552	153	0.386	59.0
11	290	0.552	160	0.399	63.8
12	295	0.552	163	0.404	65.8
13	290	0.552	160	0.399	63.8
14	277	0.552	153	0.386	59.0
15	244	0.552	135	0.347	46.8
16	148	0.552	82	0.131	10.8
17	0	0.552	0		
Total					426.6

12.2.2 Segmented Mirrors

Segmented mirror collectors were mentioned briefly in Sec. 10.12. A segmented mirror collector is being manufactured by Suntec Systems, Inc., and named the "Slats" collector. It has the general structure shown in Fig. 10.42, with ten 1 ft wide reflectors, each of which produces a secondary concentration of about 4: 1 by means of a curved surface. The focal length is about 3 m, and this implies that a row length should be at least 25 m long. French, Mooney, McDowell, and Uselton[3] reported results of field tests on such a collector. The reported efficiencies were quite low, the optical efficiency being 0.594, and the heat loss about 430 Btu/hr per linear ft (362 W/m) of receiver at operating temperatures near 450°F (505°K). This particular collector was one of the first manufactured, and had some defective mirror segments. The receiver design had many defects and has since been discarded.

[3]R. L. French, L. G. Mooney, J. H. McDowell and R. B. Uselton, Concentrator Collector Testing at the Solar Engineering Test Module, *Proc. ERDA on Concentrating Solar Collectors*, Georgia Inst. of Technology, September, 1977.

It is not clear how well such collectors can perform at high temperatures. However, it seems probable that the segmented mirror collector should, in general, perform about as well as a parabolic trough collector. The advantages of the segmented mirror system are that:

1. The receiver is fixed and can be plumbed with rigid connections.
2. The aperture can be much wider than for a parabolic trough with low wind resistance, thus requiring less plumbing in the field.
3. The mirror segments are light and individually replaceable.

The major question is whether the installed cost of a segmented mirror collector array can be lowered to reasonable levels.

A novel approach to segmented mirror collectors was developed by Russell et al.[4] In this approach, the mirror segments are held fixed, and the receiver is moved to track the sun. This is just the opposite of the segmented mirror approach described earlier. The advantage of the fixed mirror concentrator is that the mirrors and supports are fixed and can be manufactured as a rigid unit, whereas the lighter receiver is the only moving part. It has been proposed that these reflectors could be molded in the ground, with the mirrors mounted directly on forms made of cement or asphalt.[5]

The principle on which the fixed mirror collector works is an ingenious optical property of tilted flat mirrors that is illustrated in Figure 12.3. If a set of mirror segments is arranged around part of a circle so that the tilt of each mirror is one-quarter of the angle θ shown in Figure 12.4, then coparallel rays of light will be reflected by the mirrors to an element of the same circle. Therefore, the receiver needs to be tracked around part of the same circle on which the mirror segments are deployed. In actual construction, the receiver swivels about the center of the reference circle as shown in Figure 12.5.

To show the focusing property of this system, an analysis will be given corresponding to solar noon. The appropriate analysis for off-noon hours merely involves resolution of the incident solar ray vector into components along the main collector axis and perpendicular to the main axis. The perpendicular component obeys the laws described, and the parallel component merely produces a reflected ray displaced "downstream" along the collector.

Consider the diagram of Figure 12.6. Solar rays along the direction $\hat{s}$ impinge on two mirror facets at points P_1 and P_2. The normals to the mirror facets are along $\hat{n}_1$ and $\hat{n}_2$, and the reflected rays are along $\hat{r}_1$ and $\hat{r}_2$ and cross at P on the circle. It is desired to evaluate θ, the angle locating P. Unit vectors

[4]J. L. Russell, E. P. DePlomb, and R. K. Bansal, "Principles of the Fixed Mirror Concentrator," *General Atomic Rept. GA-A12902*, May 31, 1974.

[5]J. L. Russell, "Central Station Solar Power," *Power Engineering*, November, 1974.

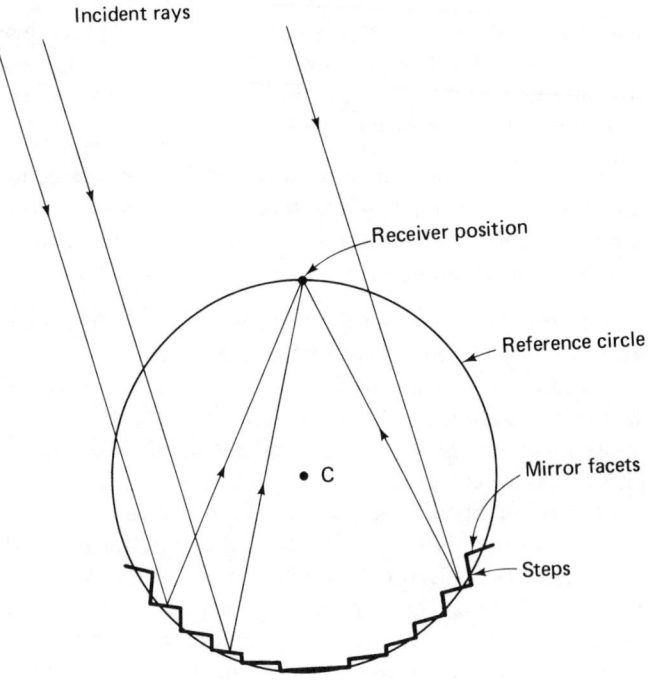

Figure 12.3 Focusing property of fixed mirror concentrator.

Figure 12.4 Illustration of the mirror facet located at angle θ from the center line, tilted at $\theta/4$ to horizontal.

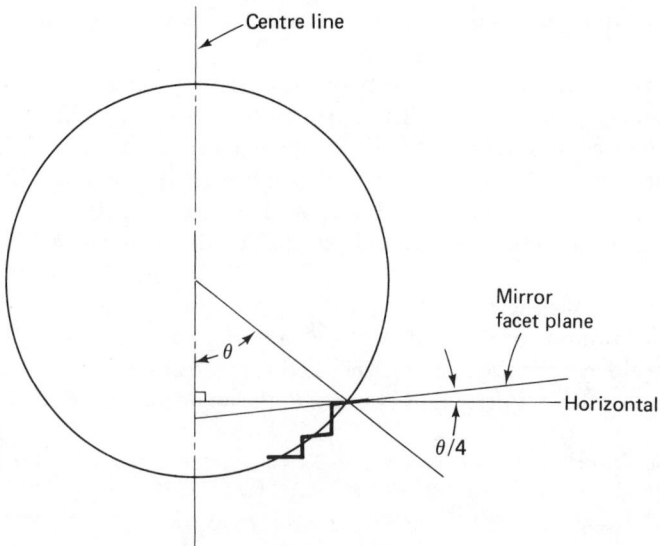

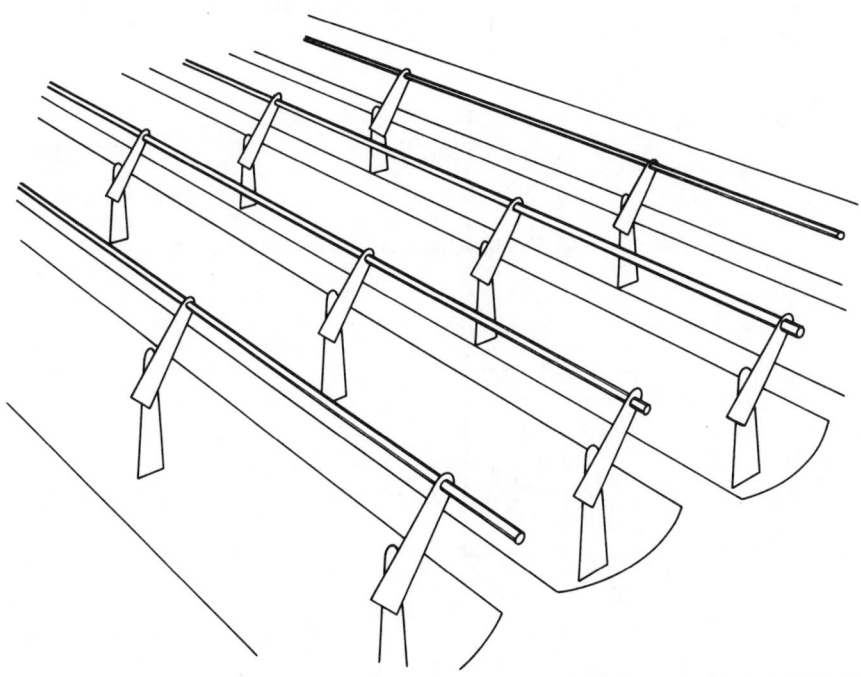

Figure 12.5 Swivel motion of the receiver at various solar elevation angles.

Figure 12.6 Reflection of rays at two points around the design circle.

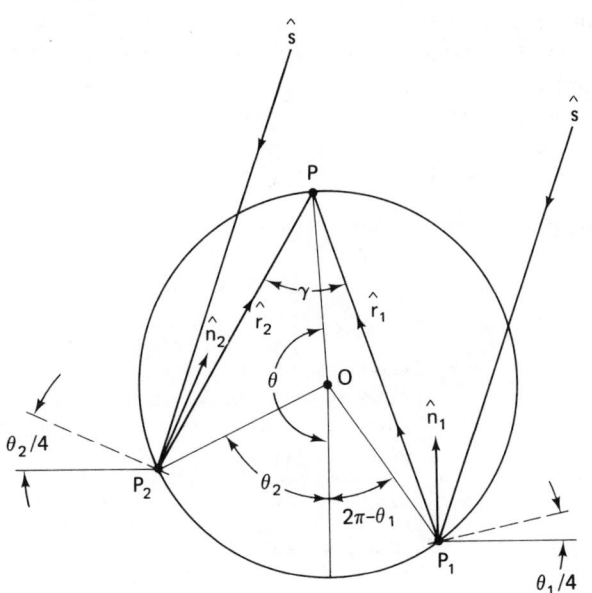

401

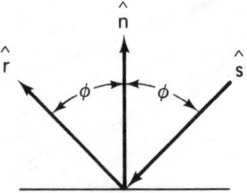

Figure 12.7 Unit vectors for incident solar ray, reflected ray, and normal to surface.

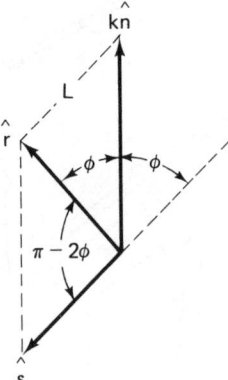

Figure 12.8 Vector diagram for constructing $\hat{r}$ from $\hat{s}$ and a vector of length k along $\hat{n}$.

are defined in Figure 12.7 along the incident solar ray $\hat{s}$, the reflected ray $\hat{r}$, and the normal to the mirror facet $\hat{n}$. The angle of incidence equals the angle of reflectance (equals ϕ). A vector diagram is then constructed as shown in Figure 12.8, in which unit vectors $\hat{r}$ and $\hat{s}$ are used directly, but unit vector $\hat{n}$ is multiplied by a constant k to make it of length k. The constant k is chosen to complete a parallelogram so that line L is parallel to $\hat{s}$. It therefore follows that

$$\hat{r} = \hat{s} + k\hat{n} \tag{12.3}$$

To find k, the law of cosines is applied, and

$$k^2 = 1^2 + 1^2 - 2 \cos(\pi - 2\phi)$$
$$k^2 = 2 + 2 \cos(2\phi) = 4 \cos^2 \phi$$
$$k = 2 \cos \phi \tag{12.4}$$

Thus,

$$\hat{r} = \hat{s} + 2 \cos \phi \hat{n} \tag{12.5}$$

and

$$\cos \phi = -\hat{s} \cdot \hat{n} \tag{12.6}$$

This gives the reflected ray $\hat{r}$ as a function of the solar ray $\hat{s}$ and the tilt of the mirror facet as expressed by $\cos \phi$ and $\hat{n}$. The mirror facet is located from the center of the circle by the angle θ. The vector $\hat{n}$ can be written

$$\hat{n} = - \sin \left(\frac{\theta}{4}\right)\hat{x} + \cos \left(\frac{\theta}{4}\right)\hat{y} \tag{12.7}$$

The angle ϕ depends on the angle θ as well as the angle β that the sun makes with a vertical line as shown in Figure 12.9. Thus,

$$\phi = \beta + \frac{\theta}{4} \tag{12.8}$$

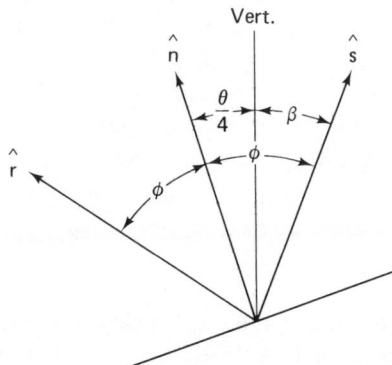

Figure 12.9 Evaluation of ϕ in terms of θ.

Let us now find the angle between the two vectors $\hat{r}_1$ and $\hat{r}_2$

$$\cos \gamma = \hat{r}_1 \cdot \hat{r}_2$$
$$= \hat{s} \cdot \hat{s} + 4 \cos \phi_1 \cos \phi_2 \hat{n}_1 \cdot \hat{n}_2 - 2 \cos^2 \phi_1 - 2 \cos^2 \phi_2$$
$$\hat{n}_1 \cdot \hat{n}_2 = \frac{\sin \theta_1}{4}\frac{\sin \theta_2}{4} + \frac{\cos \theta_1}{4}\frac{\cos \theta_2}{4}$$
$$= \cos \left(\frac{\theta_1 - \theta_2}{4}\right) = \cos (\phi_1 - \phi_2)$$
$$= \cos \phi_1 \cos \phi_2 + \sin \phi_1 \sin \phi_2$$
$$\cos \gamma = 1 + (1 + \cos 2\phi_1)(1 + \cos 2\phi_1) + \sin 2\phi_1 \sin 2\phi_2$$
$$\quad -(1 + \cos 2\phi_1) - (1 + \cos 2\phi_1)$$
$$\cos \gamma = \cos 2\phi_1 \cos 2\phi_2 + \sin 2\phi_1 \sin 2\phi_2$$
$$\cos \gamma = \cos (2\phi_1 - 2\phi_2) = \cos \left(\frac{\theta_1 - \theta_2}{2}\right) \tag{12.9}$$

Thus, the angle between the reflected rays $\hat{r}_1$ and $\hat{r}_2$ is $\theta_2 + (2\pi - \theta_1)$, and the angle P_2PP_1 is half of the angle P_2OP_1 in Figure 12.6. Since the angle between the reflected rays is constant, regardless of the sun angle β, it can be shown by geometry that point P must lie on the same circle as P_1 and P_2. To understand this, consider the construction shown in Figure 12.10. Points P_1, P_2 are on a circle, and radii r are drawn from the center to each point.

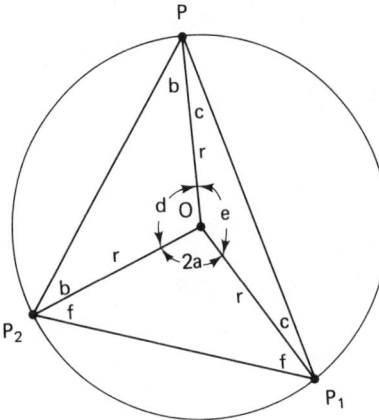

Figure 12.10 Geometrical construction.

It is desired to show that if point P is also on the circle, then the sum of angles $b + c$ equals a. Each of the triangles is isosceles. Therefore,

$$f + f + 2a = 180°$$
$$2b + 2c + 2f = 180°$$

Hence,

$$b + c = a$$

Since it was proved that this relation holds for the reflected rays, it follows that the intersection of the reflected rays lies on the circle.

12.3 "Dish"-type Concentrators

A major problem with linear arrays is that it is difficult to achieve the very high concentration that is required for high temperatures. When linear arrays are mounted in the east-west orientation, the performance falls off rapidly at more than about 2 hr from noon. When mounted in the north-south orientation, the performance is quite good when tilted at about the latitude angle, but the costs for supporting structures may be excessive. When mounted horizontally in the north-south orientation, the summer performance is excellent, but winter performance at moderate latitudes is very poor. Dish-type collectors have the advantage of higher concentration and much greater utilization of the solar intensity at off-noon hours. There are two major difficulties when dish collectors are used. One is that two-dimensional tracking is required, and this is inherently more complicated, expensive, and unwieldy than the one-dimensional tracking used for linear arrays. Another problem with dish collectors is that when used as a distributed field of modular dishes, the heat losses in the field can be great at high operating temperatures. To avoid heat transfer losses, a Stirling or Brayton power conversion cycle can mounted on each dish.[6] The receiver would actually be the hot end of the Stirling engine cylinder. The great advantage is that there is no extended network of plumbing for hot fluids. However, integration of the electrical outputs of the engines could be a problem. The alternative is to use a single very large dish in a fixed configuration with a tracking receiver. It is not clear whether such approaches have performance/cost characteristics that are viable.

An intriguing approach has been proposed by J. Reichert,[7] of Texas Tech University, in which a large dish of about 200 ft diameter is built into the ground to resemble a football stadium. Unlike the smaller dishes, which are paraboloids of revolution, the large dish has been proposed as a spherical mirror. The paraboloid has the optical property that it will focus parallel light to a single focal point. However, the paraboloid must be rotated to make it coaxial with the solar rays. This is clearly impossible for large dishes. The dish proposed by Reichert utilizes a fixed spherical lens and a movable receiver, and has been called the *FMDF* (fixed mirror/diffuse focus) *collector*. It is similar to the SRTA collector[8] concept (stationary reflector/tracking absorber).

In principal, the optics of a paraboloid are far superior to the optics of a spherical lens. However, the paraboloid must be continuously tracked to stay coaxial with the solar rays. A fixed paraboloid is not very useful as a solar concentrator. Even though a spherical mirror has aberration, it does possess one major virtue which is that (in the words of Reichert) "... a sphere, viewed from any angle still looks like a sphere." Thus, as the angle of the sun's rays change, the position of the focal region will change, but

[6]A Preliminary Assessment of Small Steam Rankine and Brayton Point Focusing Solar Modules, Jet Propulsion Laboratory, DOE/JPL-1060-16, March 1, 1979; Techno-Economic Projections for Advanced Small Solar Thermal Electric Power Plants to Years 1990–2000, Jet Propulsion Laboratory, DOE-JPL 1060-4, November 15, 1978.

[7]J. D. Reichert, "The Crosbyton Solar Power Project," *Proc. ERDA on Concentrating Solar Collectors*, Georgia Inst. of Technology, September, 1977.

[8]J. F. Kreider and F. Kreith, *Solar Heating and Cooling*, McGraw-Hill, New York, 1975, pp. 87–97.

the basic optical properties of the fixed spherical mirror will remain un-changed. Furthermore, a certain amount of spreading of the solar image on the receiver may not be a disadvantage if adequate fluxes can be achieved over a suitable area. Indeed, if a paraboloid is used it might be difficult to build a receiver that can absorb the concentrated solar intensity without major problems in obtaining materials suitable for construction of the receiver.

The focal properties of a spherical reflector are illustrated in Figure 12.11. For rays not very far from being perpendicular to the reflector, the reflected rays approach a good focus at a point half way to the center of the circle. As the rays are displaced from perpendicularity to the spherical reflec-tor, the aberration increases, with the rays passing through a cone between the nominal focus and the reflector as shown in Figure 12.12. For rays that are at near grazing angles to the spherical reflector, multiple reflections will occur before the reflected rays reach the base of the focal cone. In actual

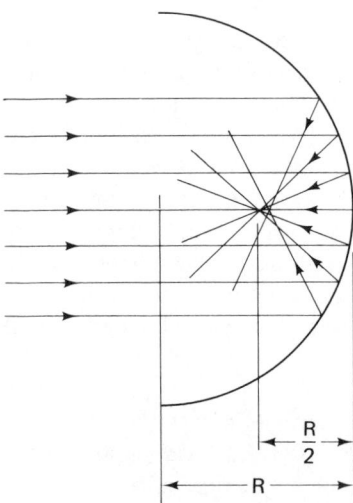

Figure 12.11 Focusing property of a spherical lens.

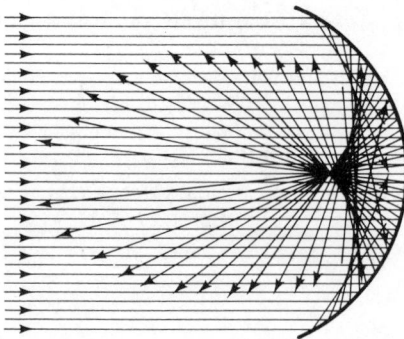

Figure 12.12 Cone of focus of a spherical lens of large aperture. Adapted from F. W. Sears, *Optics*, Fig. 5.4, Addison-Wesley Publishing Co., 1949, by permission of the publisher.

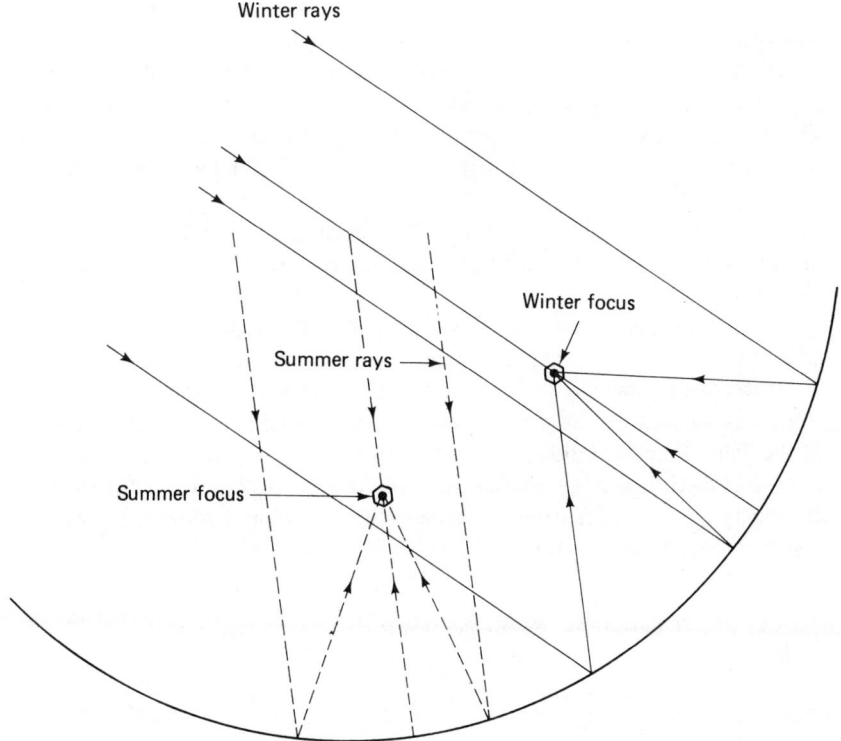

Figure 12.13 Placement of tip of collection cone in summer and winter. The collection cone must also move out of the plane of the paper at off-noon hours.

operation, the cone must be moved so that it is always coaxial with the incident solar rays. The placement of the cone during summer and winter is illustrated in Figure 12.13. The tilt of the reflector will determine the relative performance during summer and winter. Reichert has selected a tilt of 15° for use in Texas, where the electrical load peaks in late summer. A winter peaking utility could require much higher tilt angles.

12.4 Central Receiver Systems

12.4.1 General Description

One of the major problems associated with large fields of distributed collectors is the extensive network of piping for conveyance of transfer fluid from the collector field to the heat conversion system. When linear arrays are used, there are vast lengths of connecting pipes as well as receiver tubes, all of which are expensive, sources of heat loss, and potential sources of leakage or accident. Most transfer fluids at high temperatures (pressurized water, oil, liquid metal, molten salts) are quite dangerous, and this network exposed to lightning, weather, vandals, and accidents poses a great risk. The cost of purchasing and installing receivers and field piping represents one of the major costs in a distributed field of collectors. One possibility for easing this problem is to use a field of modular dish-type collectors with small Stirling cycle engines attached to each dish. For large-scale power projects another way to avoid the field piping problem is to use a central receiver system.

A central receiver system is a distributed set of fixed mirrors (with adjustable orientation) that can concentrate their reflected rays onto a receiver located atop a nearby tower as shown in Figures 12.14 and 12.15. A power station is located on the ground, either directly under the tower or adjacent to the field. Transfer fluid is pumped up to the receiver, where it is heated and then returned down to the power conversion system for conversion to electricity. A line focus central receiver has also been proposed, representing a sort of super linear array as shown in Figure 12.16.

Figure 12.14   Perspective sketch of central receiver system.

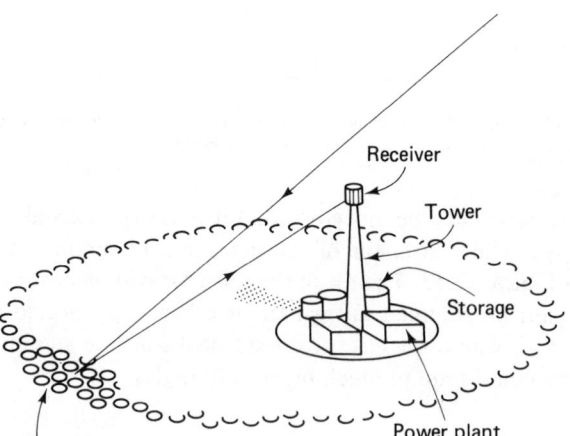

Each of the fixed mirrors is called a *heliostat*. The heliostats must be capable of rotation about two perpendicular axes. The initial central receiver systems are small pilot plants rated from 0.4 to 10 MW electric output. (A modern conventional central power plant is usually rated at 500 to 2000 MW.) The Department of Energy is considering a number of designs of

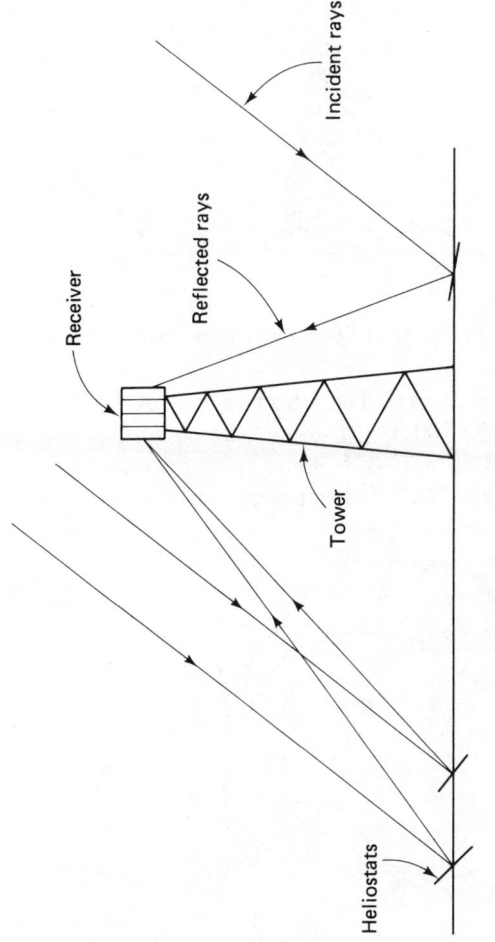

Figure 12.15 Schematic drawing of a central receiver system.

409

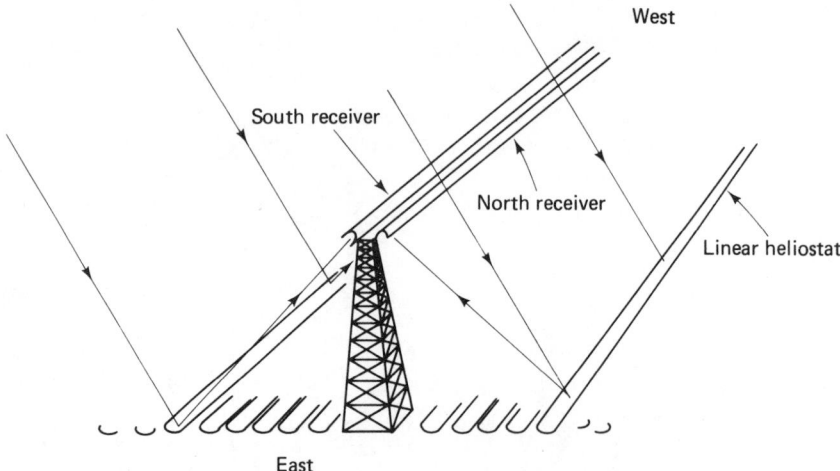

Figure 12.16   Concept for linear central receiver.

heliostats for these plants.[9] The reflective area of a heliostat is projected to be in the range 400–800 ft² (37–74 m²). Some designs use metallized plastic reflectors encased in a plastic dome for protection from the weather as illustrated in Figure 12.17. Other designs use silvered glass mirrors without

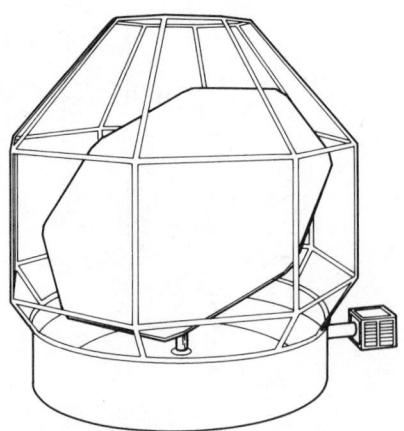

Figure 12.17 Heliostat in covered enclosure. Adapted from "Solar Central Receiver Systems," Sandia Laboratories, Livermore, CA, April 1978.

[9]"Solar Central Receiver Systems," U.S. Dept. of Energy, San Francisco Operations Office, April, 1978; "Highlights Report, Solar Thermal Conversion Program, Central Power Projects," *Sandia Rept. SAND 77-8011*, March, 1977; "Department of Energy Large Solar Central Power Systems Semiannual Review," *Sandia Rept. SAND 78-8511*, November, 1978; "Focus on Solar Technology: A Review of Advanced Solar Thermal Power Systems," *Jet Propulsion Laboratory Rept. DOE/JPL-1060-7815*, Document 5102-96, November, 1978.

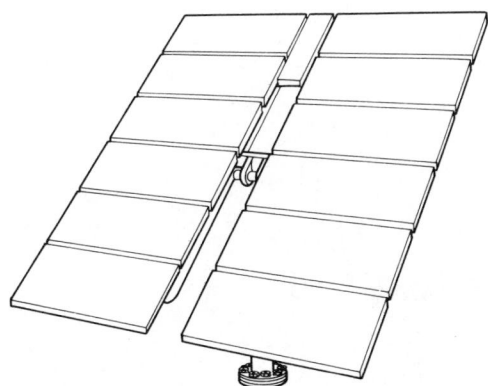

Figure 12.18 Exposed heliostat. Adapted from "Solar Central Reciever Systems," Sandia Laboratories, Livermore, CA, April 1978.

covers as shown in Figure 12.18. Steerage of the heliostats can be accomplished by computer control of servo motors according to stored geometrical algorithms (open loop). The accuracy of the procedure can be checked periodically by checking the position of the reflected image on the tower. Closed loop steerage could be used by placing a sight tube in front of the heliostat aimed at the tower, with a quadrant detector. However, this approach still requires open loop steerage for course seeking and has generally been discarded.

The field layout requires specification of:

1. Tower position and height
2. Field shape
3. Density of fill of heliostats

In general, the higher the tower, the more horizontal the mirrors can be, and thus cosine losses are reduced. However, the problem of spreading of the reflected beam becomes worse and costs rise sharply with increased tower height. A tower height of about 330 ft (101 m) has been chosen for a 10 mW pilot plant, and 200 ft (61 m) for a 5 mW pilot plant by the Department of Energy. Tower heights of over 1000 ft are indicated for plants in the 200+ mW range. For locations in the 30°–40°N latitude range, the prevailing angles of solar incidence lie mainly in the southern sky. Therefore, the tower is usually placed near the south end of the field and the receiver points downward and to the north as shown in Figure 12.19. This allows the mirrors to remain, on average, more perpendicular to the sun than if the tower were centrally located. The "fill factor," or fraction of the field covered by mirror surface, is an important variable that should be optimized. If the heliostats are placed too close to one another, a considerable amount of shading and screening will occur, especially at several hours from noon. If the heliostats are the major cost of the entire system, fill factors of 40–50% may be optimum, since additional heliostats would produce diminishing returns.

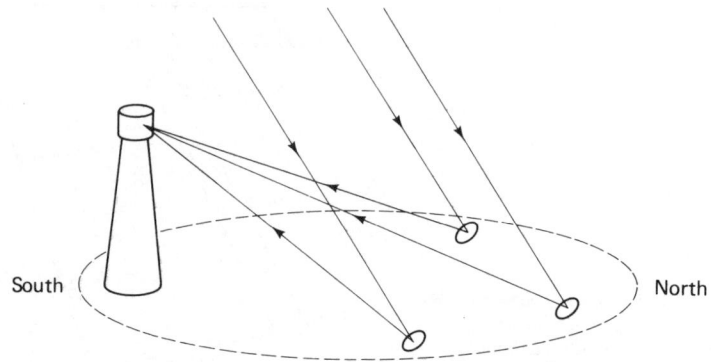

Figure 12.19   Placement of tower at south end of field.

The design of the receiver is a critical issue that is still undergoing reevaluation. Receivers are called *external* if there is an external network of tubes containing the heat transfer fluid and *internal* if a cavity is used to trap radiation onto an internal network of tubing for heat transfer. The problem is to maintain adequate flow and heat transfer surface to eliminate nonuniformities and local overheating. Typical external and internal receivers are shown in Figure 12.20.

Figure 12.20   Typical external and internal receivers. Adapted from "Solar Central Receiver Systems," Sandia Laboratories, Livermore, CA, April 1978.

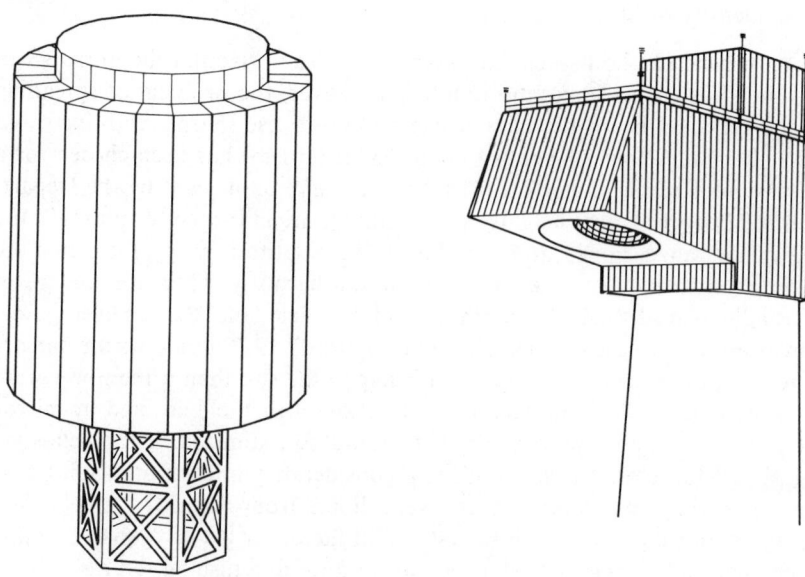

12.4.2 Power Cycles

Several power cycles for use with central receivers have been considered by the Department of Energy. These include:

1. Liquid sodium transfer fluid heated to $\sim 1150°F$ (894°K) in the receiver, used to produce $\sim 1000°F$ (811°K) superheated steam at ~ 2000 psi (13,790 kP) by heat exchange. Conventional Rankine cycle runs on steam.
2. Hot air Brayton cycle. Air is introduced into the receiver at $\sim 1150°F$ (894°K), where it is heated to $\sim 1700°K$ (1200°K). The general layout is similar to that shown in Figure 8.32.
3. The heat transfer fluid is water which is boiled at 1450 psi (9998 kP) and superheated to $\sim 950°F$ (783°K) in the receiver. Steam can either be used directly or can be condensed to give up its heat to a storage system (molten salt or oil/rock) at a lower temperature. Steam produced from storage will be degraded to about 700°F (644°K).

The ultimate purpose of the solar central receiver system is to heat the working fluid in the turbine as it goes through its cycle of temperatures. Typically, this fluid enters the turbine at $\sim 1000°F$ (811°K), and heat is rejected at $\sim 100°F$ (311°K). The central receiver system is an ideal system for high temperature heating from, perhaps, 600° to $> 1000°F$ (589° to 811°K). However, a less sophisticated collector system could be used for the preheating of condensate return. It is instructive to consider the heat fluxes required at the various temperatures encountered. For example, in system #1 described above, a preliminary layout[10] is shown in Figure 12.21. The sodium cycles between 600°F (589°K) and 1150°F (894°K), whereas the water in the turbine cycle varies between $\sim 100°F$ (311°K) and 1000°F (811°K). The heat required to heat 1 lb (0.454 kg) of water from liquid at 100°F (311°K) to superheated steam at 1000°F (811°K) and 2000 psi (13,790 kP) is broken down as follows:

(a) preheat water from 100°F to 635.8°F: 603 Btu (636 kJ);
(b) boil water at 635.8°F: 463 Btu (488 kJ);
(c) superheat steam to 1000°F: 329 Btu (347 kJ).

Thus, only ~ 800 Btu (844 kJ) of the ~ 1400 (1477 kJ) Btu total needs to be supplied by the central receiver system, and ~ 600 (633 kJ) Btu could be supplied by less sophisticated systems. It might be useful to consider a

[10]"Liquid Metal Cooled Solar Central Receiver Feasibility Study and Heliostat Field Analysis," University of Houston, McDonnell-Douglas, Atomics International, and Rocketdyne, Highlights Report—Central Power Projects, *Sandia Rept. SAND 77-8011*, March, 1977.

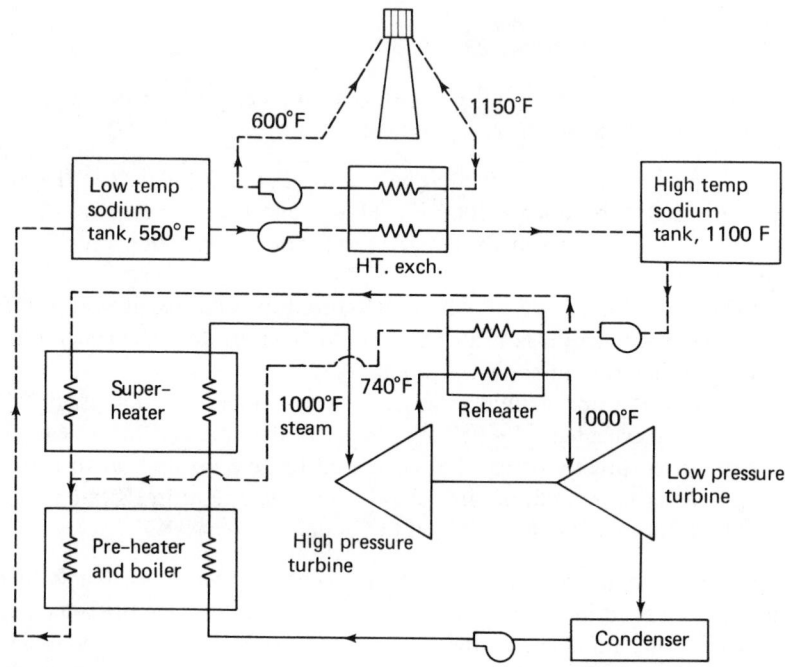

Figure 12.21 Diagram of a central receiver power plant using liquid sodium as the transfer fluid. For simplicity, tap offs from the turbines for boiler feedwater preheating are not shown. Dashed lines are sodium; solid lines are water.

Figure 12.22 Possible arrangement for splitting the collector field into high, medium, and low temperature central receiver

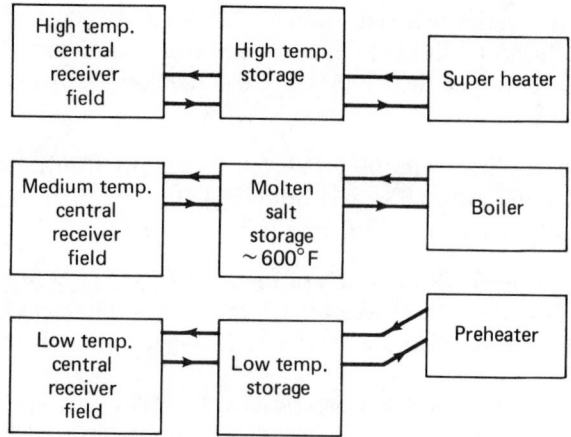

crude central receiver system for preheating from 100°F (311°K) to ~600°F (589°K) adjacent to a more sophisticated central receiver system applying heat at ~600°F (589°K) and for final heating from 600°F (589°K) to 1000°F (811°K). The system diagram might look like Figure 12.22. One difficulty with all these systems is the inadequate state of the art of energy storage. In system #1 described above, energy storage in tanks of liquid sodium is prohibitively expensive. Sodium has a low specific heat, and the volumes required are enormous. Not only is the cost high, but such storage is very dangerous. System #3 suffers from the fact that energy is degraded down from ~950°F (783°K) to ~730°F (661°K) when it is put into storage. System #2 will require development of reversible storage systems to operate in the 1500°F–2000°F (1089°–1366°K) range.

Cost estimates for central receiver systems are still very preliminary. Until the energy storage problem is dealt with, such systems must be regarded as augmenting conventional energy sources, rather than replacing them.

Worked Examples

1. A parabolic trough is operated at 45°N latitude on December 21 with the long axis in the east-west direction. What are the appropriate cosine factors between the sun's rays and the aperture plane at 0, 1, 2, 3, 4 and 4.29 hr from solar noon? Compare with the approximation of 15° per hr from solar noon.

   From the results of Problem 2.3, using

$$\cos \theta_h = \cos L \cos d \cos h + \sin L \sin d$$

$$\cos \phi = \frac{\sin d - \sin L \cos \theta_h}{\cos L \sin \theta_h}$$

$$\tan \chi = \frac{-\cos \theta_h}{\cos \phi_h}$$

The values of χ at each hour are found. The angle θ_t is the angle that the solar rays make with the aperture plane, which is tilted up toward the south at $90° - \chi$. Thus,

$$\cos \theta_t = \cos (L + \chi - 90°) \cos d \cos h + \sin (L + \chi - 90°) \sin d$$

For $L = 45$, $d = -23.45°$, we find that

$$\cos \theta_t = \cos (\chi - 45°) \cos (-23.45°) \cos h + \sin (\chi - 45°) \sin (-23.45°)$$

The results are as follows:

(Δh = hours from solar noon)	χ	$\cos \theta_t$	$\cos [15° (\Delta h)]$
0	21.55°	1.000	1.000
1	20.81°	0.971	0.966
2	18.39°	0.889	0.866
3	13.47°	0.761	0.707
4	4.05°	0.607	0.500
4.29	0.00°	0.562	0.433

It can be seen that the approximation that $\cos \theta_t \cong \cos [15°(\Delta h)]$ works fairly well within 2 hr of noon but underestimates $\cos \theta_t$ at times further displaced from noon.

2. If the array length is 100 ft and the focal distance is 5 ft, what is the fractional loss due to end effects for the system in Example 1 above?

The end effect loss fraction is $5 \tan \theta_t / 100 = 0.05 \tan \theta_t$. Thus, we obtain the following data:

Hours from Solar Noon	Fractional Loss Due to End Effects
0	0.000
1	0.012
2	0.026
3	0.043
4	0.065
4.29	0.074

3. A parabolic trough operates in Fort Worth, Texas, on a clear day in late December. Calculate the fraction of daily direct normal intensity that can be collected based on the following data:

Collector Efficiency (Fraction of Intensity on Aperture Plane That Is Collected in Fluid)	Intensity on Aperture Plane (W/m^2)
0.31	200
0.44	300
0.52	400
0.56	500
0.59	600
0.61	700
0.62	800
0.625	900
0.63	1000

Hours from Solar Noon	Direct Nornal Intensity (W/m^2)	Cosine Factor
0.5	951	0.992
1.5	934	0.936
2.5	847	0.830
3.5	612	0.686
4.5	134	0.531

Solution: The daily direct normal intensity can be approximated by the sum of the values at the center points of 1.0 hr intervals measured from solar noon. Thus, the daily direct normal intensity is $2(951 + 934 + 847 + 612 + 134)$ $= 6.96$ kWh/m². The collected energy is obtained for each interval by multiplying the direct normal intensity by the cosine factor to obtain the direct intensity on the aperture plane. This is used to evaluate the collector efficiency. The results are as follows:

Hour Interval Centered on __ hr from Solar Noon	Direct Intensity on Aperture Plane (W/m^2)	Efficiency	Collected Energy (kWh/m^2)
0.5	943	0.63	0.594
1.5	874	0.62	0.542
2.5	703	0.61	0.429
3.5	420	0.52	0.218
4.5	71	0.00	0.000

The daily collected energy is then $2(0.594 + 0.542 + 0.429 + 0.218) =$ 3.57 kWh/m². The daily collection efficiency based on daily direct normal intensity is then

$$\eta = \frac{3.57}{6.96} = 0.51$$

4. If the daily collection efficiency based on direct normal intensity ranges from 0.51 on December 21 to 0.44 on March 21 to 0.42 on June 21, estimate the yearly collected energy of an east-west oriented parabolic trough in Fort Worth, Texas, based on the solar availability given in Table 3.4.

From Table 3.4 and the estimated collection efficiencies, we obtain:

Month	Direct Normal Intensity (Btu/ft^2-day)	Days in Month	Average Collection Efficiency	Monthly Heat (Btu/ft^2)
January	1090	31	0.49	16,560
February	1260	28	0.47	16,580
March	1405	31	0.44	19,160
April	1375	30	0.43	17,740
May	1520	31	0.42	19,790
June	1850	30	0.42	23,310
July	1940	31	0.42	25,260
August	1840	31	0.43	24,530
September	1590	30	0.44	20,990
October	1520	31	0.47	22,150
November	1135	30	0.49	16,680
December	1130	31	0.51	17,870
Year				240,600

5. If a solar process heat system is installed in Fort Worth, using parabolic troughs and no storage, what is the maximum price of collectors per ft^2 of aperture to obtain a 6 year pay-back period? Assume depreciation over 20 years and 1% annual maintenance charges. The company is in the 40% income tax bracket. Use the following information:

Fuel displaced = natural gas (present cost = $3.25 per 10^6 Btu delivered).
Scenario III in Table 5.2 for future prices.
Loan at 12% amortized level payments over 5 years.

Solution: Value of fuel saved in the first year per ft^2 of aperture equals

$$240,600 \text{ Btu} \times \$3.25 \times 10^{-6} \text{ per Btu} = \$0.782$$

Let the initial cost per ft^2 of aperture be C. There is a 20% investment tax credit, so the net cost to a factory may be taken as $0.8C$.

The after-tax interest rate is equivalent to $0.6 \times 12\% = 7.2\%$. From Table 5.2, yearly payments on a 5 year loan at 7.2% are approximately $238.80 per $1000 borrowed. The value of T_5 from Table 5.2 is 4.33. In year-zero dollars, total payments are, therefore, $4.33 \times \$238.80 = \$1034$ per $1000 borrowed, or $1.034C$.

Maintenance costs are assumed to be initially 1% per year and increasing yearly with inflation. Thus, total maintenance costs over 6 years are 6% of the cost, or $0.06C$ in year-zero dollars.

Depreciation credit per year is $0.40 \times \frac{1}{20} \times 0.8C$. In year-zero dollars, after 6 years, the depreciation credit is

$$0.40 \times \tfrac{1}{20} \times 0.8C \times T_6 = 0.081C$$

For a 6 year pay-back, we require that

$$0.782R_6 = 1.034C + 0.06C - 0.081C$$
$$C = \frac{0.782 \times 11.05}{1.013} = \$8.53$$

Thus, a 6 year pay-back is achievable for a system costing \$8.53 per ft^2 of aperture. In 1979, costs for a solar industrial process heat system, including design, construction, and testing, were around \$30 per ft^2. Therefore, solar industrial process heat was still a factor of approximately 3.5 too expensive to compete with natural gas at \$3.25 per 10^6 Btu delivered.

Problems

12.1. A parabolic trough is operated at 40°N latitude in the east-west orientation. What are the appropriate cosine factors between the sun's rays and the normal to the aperture plane at 0, 1, 2, 3, 4, ... hr from solar noon on December 21, March 21, and June 21?

12.2. What are the average cosine factors for a fixed mirror concentrator of the type illustrated in Figures 12.3 through 12.5 for the conditions mentioned in Problem 12.1? Discuss the relative aperture available in the two systems.

13

Storage Techniques

The simplest method for storing heat collected by solar collectors is to allow the temperature of a reservoir to rise as sensible heat is injected. In many applications, it is desirable to supply heat and remove heat from storage nearly isothermally, which can be done with a two-tank system with flow regulation to achieve a constant temperature rise across the collector. Unfortunately, this system doubles the tankage required. Stratified thermocline storage, in which hot fluid is stored above cold storage in the same tank may be a good compromise. Phase change storage has the potential to store a considerable amount of heat isothermally, but certain difficulties are encountered. Pumped hydro storage is an ideal storage system for large-scale solar power generation in localities where the terrain permits this approach. The ultimate storage procedure will involve storage in a chemical reaction, most probably dissociation of water. The hydrogen so produced could be stored without loss for an indefinite period.

13.1 Sensible Heat

The simplest method for storing energy is to store sensible heat by raising the temperature of a heat reservoir. As heat is removed, the temperature of the reservoir drops. The main problem with sensible heat storage is that very large masses of storage material are required to hold useful heat. The amount of heat that can be stored is proportional to the range over which the storage temperature can be varied. In almost any conceivable solar application, wide excursions in storage temperature are very undesirable. For a space heating system with a nominal desired working temperature of 130°F (328°K), the only way to store heat when storage is already at 130°F (328°K) is to allow storage to rise in temperature; then the fluid in the collectors also rises in temperature and, as a result, the efficiency decreases. Yet, in many applications, it is the total amount of heat above some minimum temperature that is of importance, and elevated temperatures do not supply more effective heating. Conversely, when storage drops below 130°F (328°K) and solar heat becomes available, it must bring storage back up to 130°F (328°K) before the need for auxiliary make-up energy is removed.

A single storage tank for sensible heat storage "floats" in temperature, depending on the system conditions. For large-scale commercial installations, it may be desirable to use a two-tank system in which each tank is of volume V, and the total volume of water is V. The remaining volume V is filled with air. One tank is designated the cold tank at T_C, and the other is the hot tank at T_H. The solar collectors are operated so as to heat water from T_C to T_H, and the user system removes heat from hot storage at T_H and returns water to cold storage at, roughly, T_C. This system avoids the excursions in storage temperature characteristic of one-tank systems, but it requires sophisticated flow control systems and double the tankage volume.

The amount of heat that can be stored by any medium is proportional to its specific heat. On a unit weight basis, water is one of the best storage media with a specific heat of 1 Btu/lb-°F (4.19 kJ/kg-°K). Heat transfer oils have specific heats of about 0.5, rocks have about 0.2, inorganic salts have approximately 0.2, and metals have 0.1 to 0.2. However, when the density is taken into account, the

$$\text{volume heat capacity} = (\text{specific heat})(\text{density})$$

ranges from 62.5 for water, to about 63 for iron, to about 20 for oil, to about 35 for rock, all in Btu/ft³-°F.

13.2 Stratified Thermocline Sensible Heat Storage

13.2.1 Introduction

One method for making heat more available within the context of sensible heat storage is to use stratification. Since hot liquid has a lighter density than cool liquid, one can arrange a vertical cylindrical storage tank with the liquid from the collector output introduced into the top of the tank and liquid from the bottom of the tank connected to the collector inlet. If the tank is designed to minimize mixing, substantial temperature gradients can be maintained across the tank. As a result, heat remains available at high temperatures because only part of storage is heated. By withdrawing liquid from the top of the tank for user applications, hotter fluid is available than if storage were mixed. If stratification worked perfectly, there would be a narrow transition zone, between the hot and cold zones, that would move down and up in the storage tank as storage was charged or discharged.

It is not clear what degree of stratification can be achieved in practice. The effectiveness of stratification that can be achieved depends on whether the system is static or dynamic. In a static system one may consider a storage tank that is carefully prepared to have a sharp transition zone between the hot and cold regions as shown in Figure 13.1. Due to conduction through the walls and the liquid, the transition zone gradually spreads as shown in Figure 13.2. By neglecting the effect of the metal walls and neglecting eddy currents, the spreading of the transition zone can be estimated in terms of a simple model of a solid that starts out at an initial time with a sharp dividing line between hot and cold zones. For such a model, the temperature profile as a function of time can be calculated.

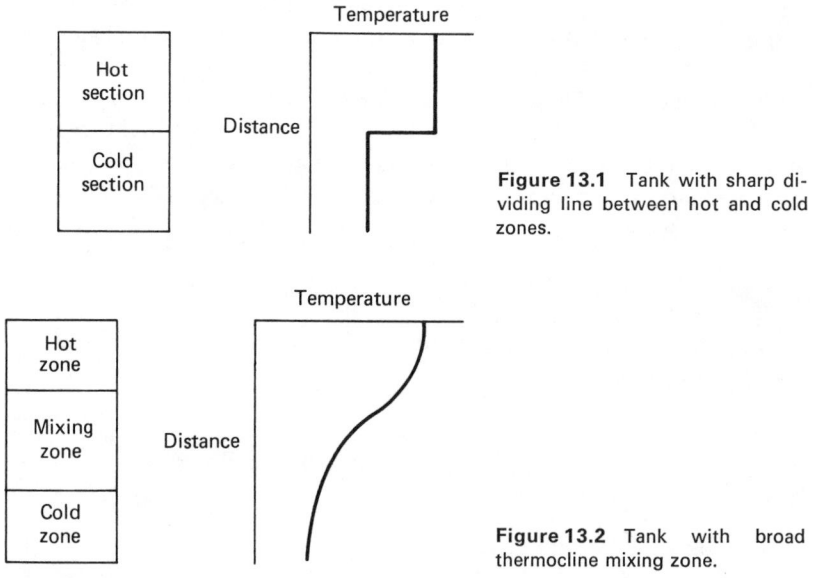

Figure 13.1 Tank with sharp dividing line between hot and cold zones.

Figure 13.2 Tank with broad thermocline mixing zone.

13.2.2 One-dimensional Model

The heat transfer in a static stratified thermocline storage tank takes place through the thermocline from hot water to cold water and through the walls. The heat transfer through the walls can lead to heat flow from the hot water to the cold water as well as from the water to the surroundings. An idealized calculation can be carried out in which heat flow through the walls is neglected and only conduction through the thermocline is included. This should be a rough approximation for well-insulated tanks of large diameter. Consider a vertical column with a temperature distribution that is initially T_2 in the upper half and T_1 in the lower half. At any time subsequent to $t = 0$, the temperature profile is $T(x, t)$, where x varies from 0 to $+L$. The equation of heat flow is[1]

$$\frac{\partial T}{\partial t} = \alpha^2 \frac{\partial^2 T}{\partial x^2} \tag{13.1}$$

where

$$\alpha^2 = \frac{\kappa}{c\rho} \tag{13.2}$$

and κ is the thermal conductivity, c is the specific heat, and ρ is the density. As $t \rightarrow \infty$, $T(x, t)$ must approach $(T_1 + T_2)/2 = T_{av}$. If we write

$$T(x, t) = T_{av} + U(x, t) \tag{13.3}$$

then $U(x, t)$ is a function that varies from $-\delta$ for $x < L/2$ to $+\delta$ for $x > L/2$ at $t = 0$, where $\delta = (T_2 - T_1)/2$. Thus, we must solve

$$\frac{\partial U}{\partial t} = \alpha^2 \frac{\partial^2 U}{\partial x^2} \tag{13.4}$$

where $U(x, 0) = -\delta$ for $x < L/2$ and $U(x, 0) = +\delta$ for $x > L/2$. Equation (13.4) can be solved by assuming a solution in the form

$$U(x, t) = F(x)G(t)$$

This leads to two total differential equations:

$$\frac{dG}{dt} = \alpha^2 kG \tag{13.5}$$

$$\frac{d^2 F}{dx^2} = kF \tag{13.6}$$

where k is a constant of separation. The solution of Eq. (13.5) is

$$G(t) = e^{k\alpha^2 t} \tag{13.7}$$

[1]S. Whitaker, *Elementary Heat Transfer Analysis*, Pergamon, Oxford, 1976.

and since $G(t)$ cannot be allowed to increase indefinitely, k must be negative. If we write $p = -k$, then

$$G(t) = e^{-p\alpha^2 t} \tag{13.8}$$

The solution of Eq. (13.6) must be found subject to the boundary condition that at $t = 0$, $F(x) = -\delta$ for $x < L/2$ and $F(x) = \delta$ for $x > L/2$. The solution to Eq. (13.6) is sinusoidal if $k < 0$ and, in order to fit the initial step function, a Fourier series must be constructed from solutions. The step function at $t = 0$ can be fitted[2] by the series

$$F(x) = \begin{cases} -\delta & (x < L/2) \\ +\delta & (x > L/2) \end{cases}$$

$$= \frac{-4\delta}{\pi}\left(\cos\frac{\pi x}{L} - \frac{1}{3}\cos\frac{3\pi x}{L} + \frac{1}{5}\cos\frac{5\pi x}{L} - \ldots\right) \tag{13.9}$$

in which the value of p associated with each term in the series is $(n\pi/L)^2$, where $n = 1, 3, 5, \ldots$ for successive terms. When the appropriate time dependence is attached to each term, the result is

$$U(x, t) = \frac{4\delta}{\pi}\left[\cos\left(\frac{\pi x}{L}\right)e^{-\alpha^2\pi^2 t/L^2} - \frac{1}{3}\cos\left(\frac{3\pi x}{L}\right)e^{-9\alpha^2\pi^2 t/L^2} + \ldots\right] \tag{13.10}$$

Thus,

$$T(x, t) = \left(\frac{T_1 + T_2}{2}\right)$$

$$+ \frac{4}{\pi}\frac{(T_2 - T_1)}{2}\sum_{n=1,3,5\ldots}(-1)^{[(n-1)/2]}\frac{1}{n}\cos\left(\frac{n\pi x}{L}\right)e^{-n^2\pi^2\alpha^2 t/L^2} \tag{13.11}$$

When $t \to 0$, all the exponentials approach 1, and the term under the summation sign goes to the step function $\pi/4$ for $x < L/2$ and $-\pi/4$ for $x > L/2$. Thus, $T(x, 0)$ goes to T_1 for $x < L/2$ and T_2 for $x > L/2$ at $t = 0$. At large t, the exponentials with large n will be small, and the term with $n = 1$ will dominate the sum. Thus, for large t only the term $n = 1$ need be retained:

$$\lim_{t \to \infty} T(x, t) = \frac{(T_1 + T_2)}{2} + \frac{4}{\pi}\frac{(T_2 - T_1)}{2}\cos\left(\frac{\pi x}{L}\right)e^{-\pi^2\alpha^2 t/L^2} \tag{13.12}$$

When water is the conducting medium, the constants are $\kappa = 0.000101$ Btu/ft-sec-°F, $\rho = 62.4$ lb/ft³, $c = 1$ Btu/lb-°F, and $\alpha^2 = 0.00583$ ft²/hr $(1.5 \times 10^{-7}$ m²/sec). A sample calculation was performed for a cylindrical tank of 20 ft (6.1 m) height. The tank is assumed to have perfect stratification

[2]H. W. Reddick and F. H. Miller, *Advanced Mathematics for Engineers*, Wiley, New York, 2nd ed., 1947, p. 196.

at time $t = 0$, with temperature T_2 above and T_1 below. The profiles of temperature vs. position in the tank for subsequent times are shown in Figure 13.3. It can be seen that the rate of expansion of the thermocline region is very slow. Thus, in the absence of mixing effects, heat conduction across a thermocline in water is quite slow.

Figure 13.3 Spreading of a thermocline in water, treating the liquid as if it were a solid and neglecting wall effects.

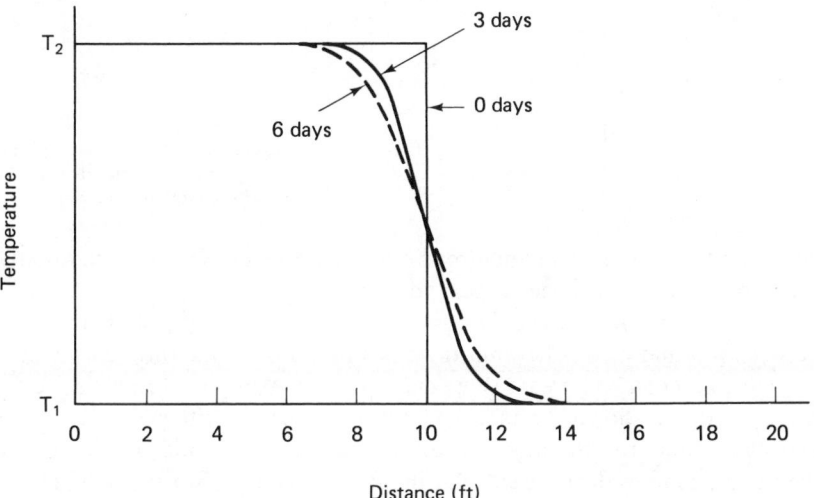

More sophisticated transient heat conduction calculations can be carried out. The transient thermal behavior of a cylinder with initial stratification and an insulating barrier on the outside wall has been extensively studied.[3] It was found that when convective mixing effects are neglected, heat loss through the insulation tends to dominate over heat transfer across the thermocline. Based on only conductive heat transfer, the thermoclines are quite stable.

13.2.3 Problems with Thermoclines

The effect of conduction through the walls will be to greatly speed up the mixing process. If the range of temperature transition in the metal walls increases more rapidly than the range in the liquid, convection currents will be set up when cold liquid contacts a warm wall and rises or when warm

[3]M. A. Abdoly and D. Rapp, unpublished work, University of Texas at Dallas, P.O. Box 688, Richardson, TX, 75080, May, 1979.

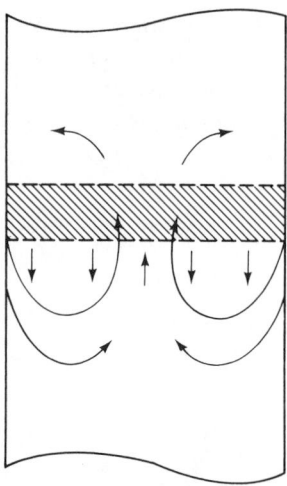

Figure 13.4 Eddy currents set up during charge cycle due to hot liquid contacting cold walls.

water contacts a cold wall and drops (see Figure 13.4). No good quantitative measurements of this effect appear to exist.

In the dynamic case there is initially a sharp line of demarcation between hot and cold zones, but there is a liquid flow rate through the tank. If mixing and conduction are slow, one may hope that the thermocline region will ride up and down the tank as fluid is pumped in either direction. One may then allow the thermocline to rise when user demand dominates and the thermocline to drop when solar input dominates. Unfortunately, mixing can become serious in dynamic situations because hot liquid is continually contacting cold walls when the thermocline drops, and vice versa.

One approach that can be used to reduce the mixing losses in stratified storage is to pack the tank with objects to restrict free convection. This approach has been used for hot oil storage in which the tank is packed with rocks. The thermoclines spread at a moderate rate as they move up and down the tank. This constant tendency toward degradation of the thermocline is a major problem for stratified storage. The rock-packed tank would not be a very good idea if water is the fluid because the volumetric heat capacity of water is almost twice that of rock, and therefore twice the volume of rock is required to store the same amount of heat as a given volume of water. Oil, on the other hand, has a low volumetric heat capacity, like rock, and therefore can be utilized with rock-packed tanks. There is some evidence, however, that exposure of oils at high temperature to the large surface area of the rocks can lead to rapid degradation of the oil.

The primary difficulty with stratified thermocline storage is that the mixing problem goes on continuously at rates which may cause the thermocline to spread considerably in a day or two. The picture of a narrow thermocline moving up and down the tank may be totally unrealistic. After an initially narrow thermocline is established, a few movements up and down the tank could conceivably cause the thermocline to spread over most of the tank. Thus, if one were storing, say, 140°F (333°K) water on top of 100°F (311°K) water, after a few passes up and down the tank, a large volume of stored water might be in the 110°F (317°K) to 130°F (328°K) range. This would have to be passed through the solar heaters to bring it back up to 140°F (333°K) so as to reestablish a sharp thermocline. Thermocline management could become a nuisance. When this is coupled to the fact that not all the volume in the thermocline tank is useful due to the thickness of the thermocline and the need for flow diffusers at the ends, stratified storage may not yet be a practical reality. Some data obtained by Rocketdyne, Inc., on discharging heat from a rock-packed thermocline storage system have been reported recently.[4] The tank was about 40 ft (122.2 m) high and was initially charged with oil at ~580°F (578°K). Cooler oil at ~420°F (489°K) was pumped in through the bottom of the tank at a rate of about 10 linear feet (3.05 m) per hour. The temperature profiles down the tank were measured approximately every half hour during the 4 hr discharge time. These profiles are shown in Figure 13.5. Initially, the thermocline is about 2 ft (0.61 m) thick. The thermocline gradually spreads as discharge proceeds, reaching about 8 ft (2.44 m) after 2 hr, and about 15 ft (4.57 m) after 4 hr. Only about 3 hr of discharge at temperatures within 10°F of 580°F (578°K) can be

Figure 13.5 Measured temperature distributions at 0.5 hr intervals in a 40 ft (12.2 m) high rock-packed cylindrical tank when 420°F (489°K) oil is passed up from the bottom through the tank, initially at 580°F (578°K).

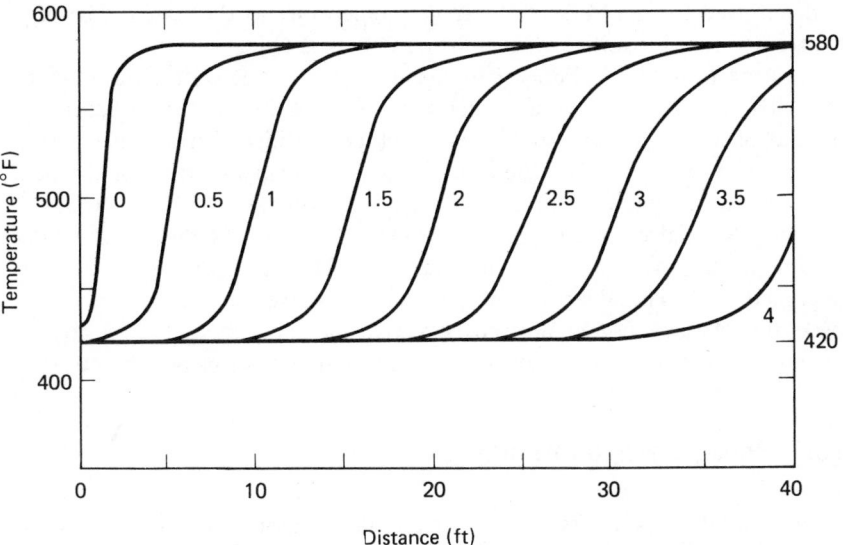

[4]Highlights Report, Solar Thermal Conversion Program, Central Power Projects, Washington DC, December, 1976, *Sandia Report SAND 77*–8011, March, 1977.

achieved, with the last hour producing gradually degraded heat at temperatures ranging from 570°F (572°K) down to 470°F (517°K). If the purpose of the solar energy collector system is to heat oil from 420°F (489°K) to 580°F (578°K), then the storage system allows roughly 80% recovery of the deposited heat, with the remaining 20% degraded. If it is assumed that 20% of the heat from the solar collectors is degraded during the process when storage is charged, then heat recovery from the storage unit at design temperature is about 65%. The remaining heat is not lost, but must be upgraded by solar heating.

13.2.4 Two-tank Systems

In some solar energy systems, the control system varies the flow rates to maintain a constant temperature differential between the low and high sides of storage. This approach is used primarily for large commercial applications. In such cases, there is a question of whether to use two isothermal tanks or a single thermocline tank. When two isothermal tanks are used, two tanks, each of volume V, are required to hold a volume V of liquid. When storage is "full of heat," the high temperature tank is full of liquid, and the low temperature tank is empty. When storage has no available heat, all the liquid is in the low temperature tank. All stages of partial distribution of liquid between the tanks are, of course, permitted. The advantage of the two-tank system is that there is no degradation of heat. Liquid is added to and removed from a tank of the same temperature as the liquid. The only problem is that the initial expense is high because twice the tankage volume is required to hold a volume of liquid. The great objective for thermocline stratified storage is to contain a volume V of liquid in a storage volume V instead of $2V$. It remains to be seen whether stratified thermocline storage can be made to work. For the example illustrated in Figure 13.5, the heat recovery at design temperature is only approximately 65%. A two-tank system with total volume $65/50 = 1.3$ times as large as the thermocline tank volume would store as much heat and would not require upgrading of degraded heat. The absence of thermocline management operations might make the two-tank system more cost effective due to its inherent lower operating costs. Further study of thermoclines in dynamic systems is required.

13.3 Phase Change Storage

When a solid melts, it absorbs a relatively large amount of heat at constant temperature. The reverse process, crystallization, liberates heat at constant temperature. Phase change storage therefore offers the possibility of storing moderate amounts of heat over a very small temperature range. Furthermore, the relatively high heats of melting of some crystals allows a relatively large amount of heat to be stored in a small volume when compared to sensible heat storage. Phase change storage consists of a container which has an internal network of heat exchange tubing, which is filled with a material that melts at an appropriate temperature. A heat transfer fluid (usually oil or water) is circulated through the tubing to exchange heat with the phase change material. When storage is "full of heat," the material is completely molten and further addition of heat will cause a temperature rise of the molten material. When storage is "empty," it is completely crystallized.

For any particular application, one should select a substance that has a melting point in the desired range of temperature and which has as high a heat of fusion and as low a cost as possible. In addition to pure substances, eutectic mixtures of substances are good candidate materials. When a mixture of substances is heated, it is usually found that the mixture begins to melt at one temperature, but that as heat is added the temperature rises as more solid melts, until all the solid is melted. Thus, the mixture melts over a range of temperatures. The melting range depends on the mixture of components that is used. In some cases, a eutectic mixture of a particular percentage of each component can be found which has the property that it melts at a single temperature, like a pure substance. Such eutectic mixtures are very important because they allow blending of substances to achieve a desired melting point.

The heats of fusion of inorganic salts tend to be quite high, but the melting points also tend to be very high. Organic substances tend to have lower melting points and lower heats of fusion. The desirable melting points for various applications are roughly as follows:

1. Cold storage for space cooling $40°–45°F$ ($278°–280°K$)
2. Warm storage for forced air heating systems $85°–95°F$ ($303°–308°K$)
3. Hot storage for water space heating systems:
   (a) Residential $110°–130°F$ ($317°–328°K$)
   (b) Commercial $130°–150°F$ ($328°–339°K$)
4. Single-stage absorption chiller systems $190°–220°F$ ($361°–378°K$)
5. Two-stage absorption chiller systems $300°–350°F$ ($422°–450°K$)
6. Process steam or electrical power systems $> 350°F$ ($> 450°K$)

Historically, much of the original effort in phase change materials has been centered on applications for space heating. More recently, considerable emphasis has been placed on high temperature applications. The advantage of phase change storage is afforded by the following example.

Each square foot of solar collector may be expected to collect about 900 Btu (950 kJ) on a clear day. If water storage is used, then for each gallon of storage the temperature rise will be 900/(8.3) = 108°F (60°K). To keep the temperature rise to within, say 40°F (22°K), one should have 2.7 gal (0.01 m³) of storage per ft² (0.093 m²) of collector. In principle, fusion materials can be found with heats of fusion of the order of 50 Btu/lb (116 kJ/kg). Assuming a density of twice that of water, such a substance could store 5 × 2 × 8.3 = 830 Btu per gal (0.232 kJ/cm³) with no temperature rise. Thus, 1 gal of fusion material could store about as much heat isothermally as 2.5 gal (0.00095 m³) of water with a 40°F (278°K) temperature rise. Whether such fusion storage can be made practical is uncertain at this time.

There are several major problems associated with phase change storage. These include:

1. Materials are expensive.
2. Materials are corrosive.
3. Supercooling of liquid.
4. Heat exchangers required in storage.
5. Materials must be kept anhydrous.

Many of the phase change materials are both expensive and corrosive. These are problems that can only be considered on a case-by-case basis. A general problem that occurs in many phase change systems is supercooling. When the molten material is cooled to the melting point, the material will begin crystallizing if enough time is allowed. However, in practice, many materials tend to supercool below the melting point as liquids in a nonequilibrium state. This reduces their effectiveness as phase change materials. Supercooling can be reduced by adding seed crystals to initiate crystallization, but this is difficult to do in practice. Certain additives such as silicates are said to thicken the melt and improve crystallization when cooled, but the effectiveness of this approach is doubtful. In cases where the substance is a salt-hydrate, the solution can separate into phases of varying water content with a range of melting points. Another problem is that large heat exchange surfaces must be provided inside the storage tank for phase change storage, both in the loop to the solar collectors and in the loop to the user application. As heat is supplied to a crystallized phase change tank, crystals melt in localized regions around these heat exchange surfaces. Unless the heat exchange surface is large, melting of the bulk of the crystals away from the heat exchangers can be rather slow. In the reverse process, when cooling the molten salt, crystals will form around the heat exchange surfaces and tend to prevent the rest of the melt from being cooled, except by slow conduction processes. Many of the phase change materials are inorganic salts which are quite hygroscopic. They must be kept anhydrous because water will make them more corrosive and will broaden the melting range.

It is clear that phase change storage has the potential to store a much larger amount of heat per unit volume that sensible heat storage and to supply that heat at nearly constant temperature. On the other hand, there are severe problems associated with phase change storage, and these must be overcome if it is to become a commercial reality.

13.4 Pumped Hydro Storage

Pumped hydro storage is an important possibility for solar electrical power systems located in terrain suitable for storing large amounts of water at elevations differing by at least 100 ft (30.5 m).

In the electrical generation system, when solar-generated power is available from the collector field, water is pumped from the lower pool to the upper pool. Electricity is then generated by releasing water from the upper pool through a penstock to drive a hydroelectric generator. The design requires that (1) pools be of sufficient size so that during nights and periods of inclement weather (when solar-powered pumping cannot be done) the discharged water does not seriously overload the lower pool or drain the upper pool, and, conversely, that (2) during extended periods of good weather the lower pool is not drained and the upper pool is not overfilled. An additional consideration is that rain tends to raise the levels of the pools, and evaporation and drainage tends to empty the pools.

In the usual applications of pumped water storage, the purpose is not to make use of solar energy but to use off-peak load energy to pump water to a higher elevation, and then at times of peak load requirements to discharge the water through the system to generate electricity. In a conventional pumped storage system, nights or weekends are the typical pumping times, and midday is the typical discharge and generation time. In contrast, in the solar energy system pumping would be performed during daylight periods of sufficient solar intensity to drive the pump against the necessary head to the upper pool.

To calculate the water flow rate through the hydroelectric generator necessary for any arbitrary electrical output, the required turbine horsepower is determined first. Assuming a generator efficiency η_g of 0.97, the required horsepower per megawatt of electrical output is

$$\frac{1 \text{ MW}}{0.97} \frac{1341 \text{ hp}}{\text{MW}} = 1382 \cong 1400 \text{ hp}$$

The rate of water flow Q to produce an amount of power depends on the turbine efficiency η_t and the head of water, in feet, between the lower and upper pools H, and is given by the equation

$$Q = \frac{\text{hp} \times 550 \text{ ft-lb/sec}}{\eta_t \rho H} \text{ ft}^3/\text{sec}$$

where ρ is the water density (62.4 lb/ft³). Assuming a turbine efficiency of 0.9, the required flow rate to produce 1400 hp is

$$Q = \frac{1400 \times 550}{0.9 \times 62.4 \times H} = \frac{13,704}{H} \text{ ft}^3/\text{sec}$$

The number of days of storage provided by a given pool is the useful volume of the pool divided by the total flow per day. The useful pool volume is estimated as the volume above the midheight of the dam retaining the pool. The effective level of a pool is estimated as three-quarters of the dam height, and the head available between pools is taken as the difference in levels as estimated by this method. These assumptions are illustrated in Figure 13.6.

Figure 13.6 Useful volumes and difference in head of pools. By permission, American Technological University, Killeen, TX.

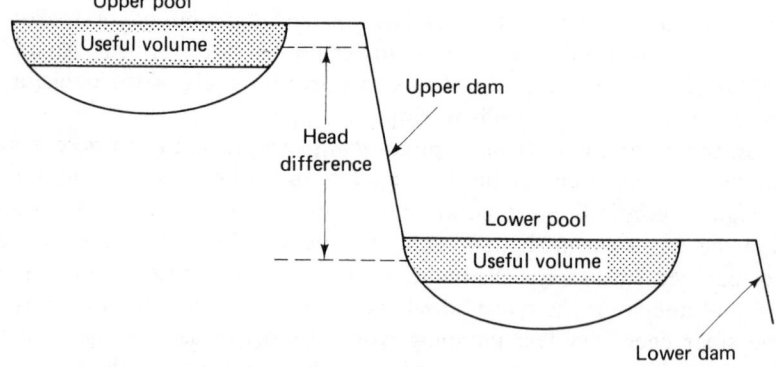

For a volume V (ft³) and a head H between upper and lower pools, the storage capacity C of a pool in MW-days is

$$C = \frac{V}{Q \times 86,400 \text{ sec/day}}$$

$$= \frac{V}{13,704/H \times 86,400 \text{ sec/day}} \cong 8.45 \times 10^{-10} \, HV$$

The volume V required for a given number of MW-days storage is

$$V = \frac{C}{8.45 \times 10^{-10} H}$$

For example, for a head H of 100 ft, the volume required for one MW-day is

$$V = \frac{1}{8.45 \times 10^{-10} \times 100} = 11.83 \times 10^6 \text{ ft}^3 \ (0.334 \times 10^6 \text{ m}^3)$$

or 11.83 million cubic feet of water.

13.5 Chemical Storage

One of the major impediments to large-scale use of solar energy as a prime (rather than augmentory) fuel is the lack of adequate technology for energy storage. It has been conjectured that in order for solar energy to become a prime energy source, large areas will have to be set aside for solar energy collection, and that the collected energy must be stored in the form of a chemical reaction, most probably

$$2H_2O \longrightarrow 2H_2 + O_2$$

The hydrogen could be stored at high pressure and transported via pipeline. It could become a universal fuel that is burned in residences, industry, and transportation.

The question arises as to the best approach for utilizing solar energy for dissociation of water. One method is to produce electricity, and dissociate water by electrolysis. The electricity can be generated by solid-state direct conversion devices or by solar thermal conversion. In 1978, solid-state devices were generally less than 12% efficient and were about 30 times more expensive than conventional sources of electricity. Solar thermal conversion to electricity is capable of higher efficiency, but costs appear to be high. The technology for electrolysis of water is fairly well established. Efficiencies of 70–80% can be realized for conversion of water to hydrogen by electricity. The overall efficiency for conversion of solar energy into dissociated water is the product of efficiencies of solar collection/conversion to electricity/electrolysis. Using solar cells, this efficiency is about $0.12 \times 0.70 = 0.084$. Using solar thermal conversion at 70% collection efficiency, it could be about $0.7 \times 0.3 \times 0.7 = 0.15$. Bockris[5] has pointed out that the electrolysis of water requires both electricity and heat to proceed. At 298°K, 83.7% of the total energy required for dissociation is needed in the form of electricity, and 13% is needed as heat. A minimum voltage of 1.23 V is required for dissociation, but this will cause the cell to cool as the reaction absorbs heat. At 1.47 V, the electrolysis will be isothermal, the additional

---

[5]J. O. M. Bockris, "On Methods for the Large-Scale Production of Hydrogen from Water," *Hydrogen Energy*, ed. T. N. Veziroglu, Plenum, New York, 1974.

0.24 V providing the required heat. In principle, it would be desirable to supply 1.23 V as electricity and 0.24 V as heat. However, the *rate* of electrolysis would be very slow due to overpotentials. The overpotential problem can be greatly reduced by carrying out the electrolysis at high temperature ($\sim 1000°F = 811°K$). Furthermore, at high temperatures the electrical requirements decrease and the heat requirements increase. For example, at 1000°F (811°K), only about 1.04 V is required in the form of electricity, whereas the equivalent of about 0.46 V is required as heat. It should be feasible to run such a cell on about 1.1 V of electricity with 0.4 V of heat supplied. If a supply of heat were available at 1000°F (811°K), the electrolysis would then be represented (per unit of energy expended) as

$$2H_2O + \begin{cases} 0.27 \text{ unit heat} \\ 0.73 \text{ unit electricity} \end{cases} \longrightarrow 2H_2 + O_2$$

The overall process could be written as

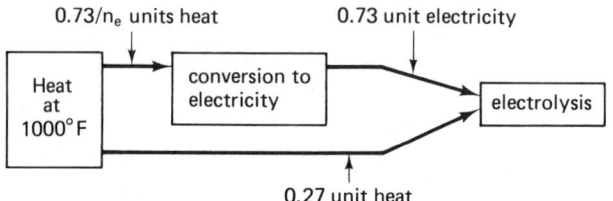

where η_e is the efficiency of conversion of heat at 1000°F (811°K) to electricity. If, for example, $\eta_e \cong 0.365$, it takes 2.27 units of heat to supply a unit of energy to the electrolysis reaction by this means. If all the energy fed to the electrolysis reaction were in the form of electricity (as in electrolysis at room temperature) the heat required would be $1/\eta_e = 2.74$ units of heat per unit of electricity supplied to the cell. Thus, the use of high temperature electrolysis at $\sim 1000°F$ leads to a thermal efficiency of $1/2.27 = 0.441$, as opposed to 0.365 for electrolysis at room temperature.

   Once the hydrogen is formed, it can be stored, transported by pipeline, and burned or converted to electricity at local sites. Conversion to electricity might take place in fuel cells, which could run at greater than 50% conversion efficiency.

   Another approach for conversion of water to hydrogen is by thermochemical splitting, which has the advantage, in principle, that the energy input to splitting can be utilized with a higher fraction in the form of heat than in electrolysis. Direct thermal splitting via the reaction $2H_2O \longrightarrow 2H_2 + O_2$ requires very high temperatures and low pressures to achieve adequate conversion. At 1 atm pressure, the fractional conversion to $H_2 + O_2$ is only about 5.8% at 2000°K.[6] The use of a closed cycle scheme for thermochemical hydrogen production offers the advantage, in principle, that most of the energy of reaction can be supplied in the form of heat at much lower temperatures than those required for direct thermal splitting. A thermochemical cycle is a series of reactions for which the overall net effect is $2H_2O \longrightarrow 2H_2 + O_2$, but reaction intermediates are involved which are regenerated around the cycle. For example, one of the many cycles that has been studied is

$$H_2O + Cl_2 \longrightarrow 2HCl + \tfrac{1}{2}O_2 \qquad (973°K)$$

$$2VCl_2 + 2HCl \longrightarrow 2VCl_3 + H_2 \qquad (298°K)$$

$$4VCl_3 \longrightarrow 2VCl_4 + 2VCl_2 \qquad (973°K)$$

$$2VCl_4 \longrightarrow 2VCl_3 + Cl_2 \qquad (298°K)$$

Heat may be absorbed or rejected in each step, and work may also be required. It is presumed that the heat can be stored and reused, and that the net work around the cycle is zero. The only net heat input is allowed to be q_2 at T_2, and the only net heat output is q_1 at T_1 as shown in Figure 13.7.

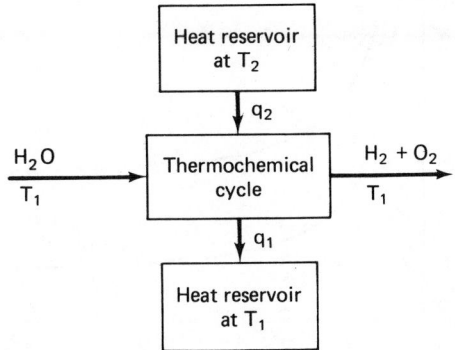

Figure 13.7 Thermochemical cycle.

The H_2O enters the cycle at T_1 and pressure p_1, and the H_2 and O_2 leave at T_1 and p_1. The energy required to dissociate H_2O is $E_{diss} = 3794$ Btu/lb (8840 kJ/kg). The question now arises as to what is the ideal minimum requirement of q_2 to supply 3794 Btu/lb (8840 kJ/kg) to the reaction, leaving $q_2 - 3794 = q_1$ rejected at T_1? This problem has been studied by several authors,[7] and it can be shown that, for an ideal reversible system, the ratio

[6]R. E. Chao and K. E. Cox, "An Analysis of Hydrogen Production Via Closed-Cycle Schemes," *Hydrogen Energy*, ed. T. N. Veziroglu, Plenum, New York, 1974.

[7]J. E. Funk and R. M. Reinstrom, *I.&E.C. Proc. Des. Develop.*, 5, 336, 1966; J. B. Pangborn and J. C. Sharer, "Analysis of Thermochemical Water Splitting Cycles," *Hydrogen Energy*, ed. T. N. Veziroglu, Plenum, New York, 1974.

q_2/E_{diss} is given by

$$\frac{q_2}{E_{\text{diss}}} = \frac{q_2}{\Delta G}\frac{T_2 - T_1}{T_2}$$

where ΔG is the change in free energy associated with the dissociation reaction. Since $\Delta G = 3150$ Btu/lb (7340 kJ/kg), it follows that the heat efficiency η is

$$\eta = \frac{q_2}{E_{\text{diss}}} = 1.20\frac{T_2 - T_1}{T_2}$$

where the temperatures are absolute. This equation only holds for an ideal reversible system. For a fixed value $T_1 = 298°$K, the results are shown in Figure 13.8. It can be seen that as T_2 increases, $\eta = q_2/E_{\text{diss}}$ increases. At around 2600°F (1700°K), q_1 becomes negative, meaning that heat is absorbed at both the heat reservoirs, leading to values of η in excess of unity. Since the heat efficiency of electrolysis is almost certain to be below 50%, the

Figure 13.8   Heat flows and heat efficiency of ideal water splitting cycle.

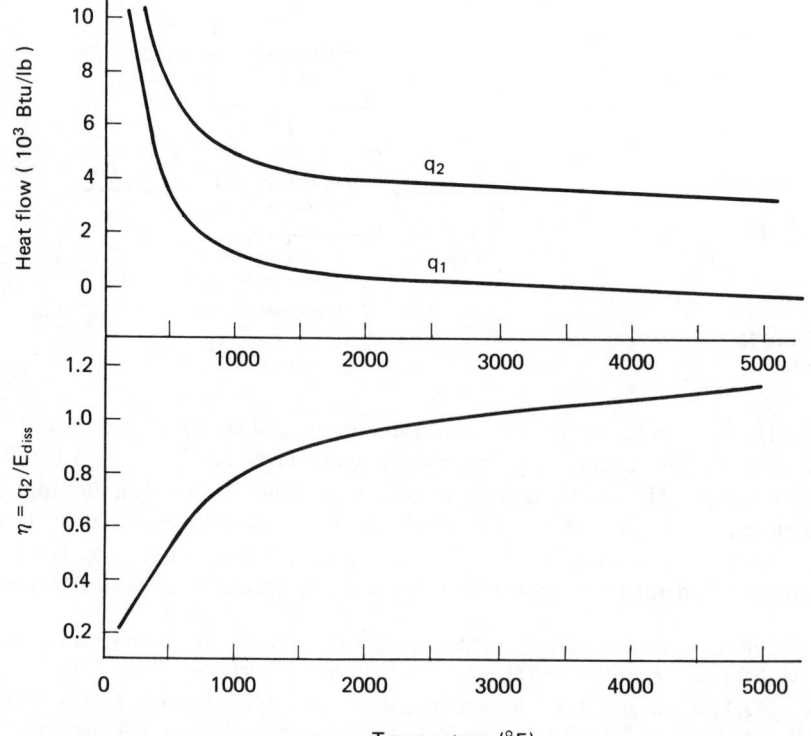

appeal of the thermochemical cycle is obvious. No real cycle can approach these ideal limits. When real cycles are analyzed[8] it is found that work is required in some steps, heat is released in some steps, and work is required to separate the products. Assuming reasonable fractions of heat recovery, and conversion efficiencies for heat into work, the achievable thermal efficiencies for various cycles appear to be, at most, in the range achievable with electrolysis.

However the hydrogen is made[9], there is little doubt that chemical storage via dissociation of water is the key that might unlock solar energy to be an ultimate major prime source of energy for our future.

Worked Examples

1. If a square foot of solar collector can collect about 1000 Btu on a clear day, what volume of water storage is required to keep the temperature rise at 30°F?

$$1000 = V\rho c_p(\Delta T)$$

$$V = \frac{1000}{62.4 \times 1 \times 30} = 0.534 \text{ ft}^3 = 4 \text{ gal}$$

2. A perfectly insulated stratified thermocline tank of 20 ft height is filled with water at 50°F in the lower half and water at 150°F in the upper half. After 10,000 hr, what are the temperatures 5 ft above and below the center of the tank?

Use Eq. (13.12). The appropriate constants lead to:

$$T(x, t) = 100 + \frac{200}{\pi}e^{-1.438} \cos\left(\frac{\pi x}{20}\right)$$

$$T(x, t) = 100 + 15.11 \cos\left(\frac{\pi x}{20}\right)$$

$$T(5, 10{,}000) = 100 - 10.68 = 89.32°F$$

$$T(15, 10{,}000) = 100 + 10.68 = 110.68°F$$

It can be seen that heat flow through the thermocline by conduction is very slow, and that convection currents and losses through the walls will determine the actual rate of dissipation of the thermocline.

[8] J. B. Pangborn and J. C. Sharer; also J. E. Funk, W. L. Conger, and R. H. Carty, *Hydrogen Energy*, ed. T. N. Veziroglu, Plenum, New York, 1974.

[9] J. H. Kelley and E. A. Laumann, "Hydrogen Tomorrow" JPL Rept. 5040-1, Jet Propulsion Laboratory, Pasadena, CA, December, 1975.

3. A solar electric power plant is designed to produce 50 MW of electricity continuously. If pumped hydro storage is used, what storage volume is required in the upper pool to provide for a 10 day outage of solar availability? There is a 400 ft head difference between upper and lower pools.

According to Sec. 13.4, the volume in ft^3 required to supply C MW-days of electricity with a head difference H is

$$V = \frac{C}{8.45 \times 10^{-10} H}$$

To provide 50 MW for 10 days requires 500 MW-days of storage. Thus,

$$V = \frac{50}{8.45 \times 10^{-10} \times 400} = 1.48 \times 10^8 \text{ ft}^3$$

Problems

13.1. A 9 in. inside diameter plastic tube about 6 ft long is mounted vertically and is initially filled halfway with cold water at 65°F. Hot water at 114°F is carefully poured on top of the cold water to form a thermocline. The actual data[10] of temperature profiles down the tube (at a radius of about 3 in. from the center) are shown in Figure 13.9 for various time periods subsequent to initial charging. The tube was wrapped with crude insulation, so heat flow took place both from the tube to the room (~ 70°F) and from hot water to cold water across the thermocline. The following analysis is desired:

(a) If heat flow from the upper part of the tank is mostly to the surrounding air, show that if one balances the temperature drop of a slab of water of height h against a heat loss from the circumferential area of the slab, the differential equation for the temperature of the water is

$$\frac{dT}{dt} = -\frac{UA}{MC}(T - T_a)$$

where U is the heat loss coefficient from water to the air, A is the area $\pi D h$, M is the mass of water $(\pi D^2/4)h \cdot \rho$, and C is the specific heat of water.

(b) Show that in integrated form this leads to

$$\frac{T(t) - T_a}{T(0) - T_a} = e^{-0.769(U/D)t}$$

with t in hours, U in Btu/hr-ft^2, and D in inches.

[10]M. A. Abdoly and D. Rapp, unpublished work, University of Texas at Dallas, 1979.

(c) Assume $T_a = 70°$F and plot $\ln\left[\dfrac{T(t) - 70}{T(0) - 70}\right]$ vs t, and show that U is equal to 0.239.

(d) Plot the expansion of the thermocline vs. time. Define the width of the thermocline to be the distance between points 80% of the temperatures of the tank ends.

Figure 13.9

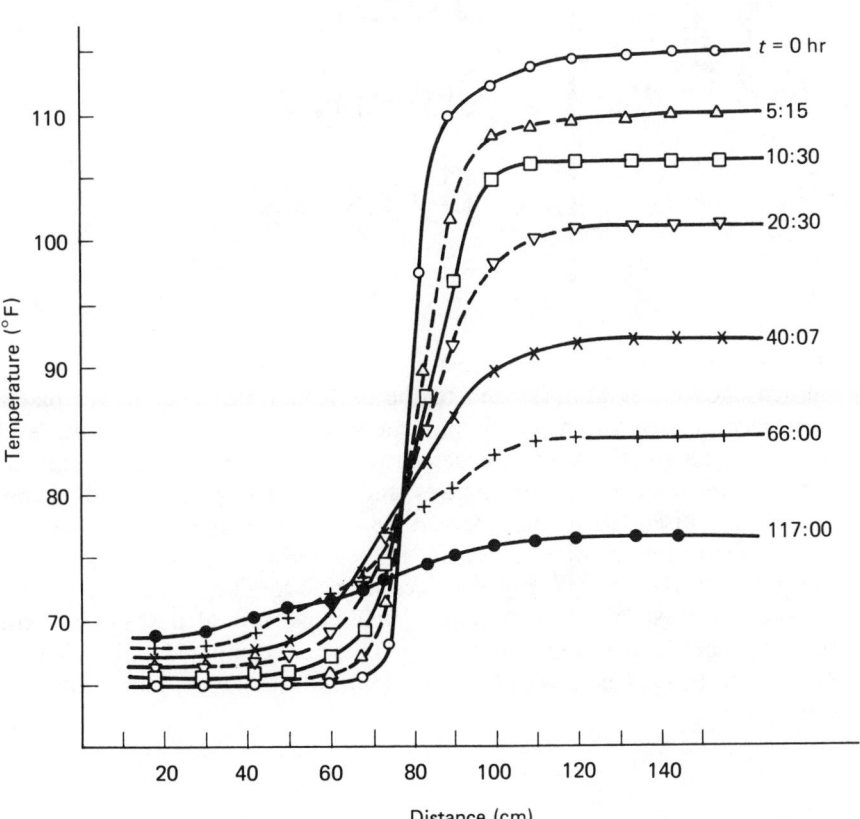

14 | Solar Total Energy Systems

In most central electrical power plants, about two-thirds of the raw energy burned ends up as heat rejected to the environment. Local power plants situated near users can make use of some of this rejected heat for low-level heat applications. Such an arrangement is called a *total energy system* or *cogeneration* system. The advantages and disadvantages of total energy systems are discussed in this chapter. Solar total energy systems have an appeal in some cases where there are no economies of scale for expansion of the solar collector field beyond a certain point. When solar total energy systems are used for space cooling of buildings, several possibilities exist for the procedure for producing cooling. These are compared and evaluated. Other aspects, such as choice of working fluid, collector type, and orientation, are discussed.

14.1 Total Energy Systems

During the past 50 years, there has been a growing use of electricity for energy. The use of energy in all forms has increased at an annual rate of about 3%, but the use of electrical energy has increased at about double that rate. As a result, in 1978 a point was reached where about 25% of the raw energy used in the United States was for production of electricity. According to most projections, this figure will increase to around 40–50% by the year 2000. In general, when heat is converted to electricity in central power plants, only about one-third of the heat is converted to electricity that is available to consumers; the remainder is either rejected to cooling towers or ponds or rivers or dissipated in transmission lines.

It has occurred to many engineers that if local electrical power plants could be build adjacent to areas of demand for low-level heat, the heat rejected from such "total energy systems" could be distributed to such users. This approach represents a large potential saving of energy because roughly two-thirds of the energy released in the burning of fuel appears in the power station exhaust. A total energy system is illustrated schematically in Figure 14.1. Such an approach is not entirely feasible for modern American society.

Figure 14.1 Conventional vs. total energy system: (a) conventional utility system; (b) total energy system.

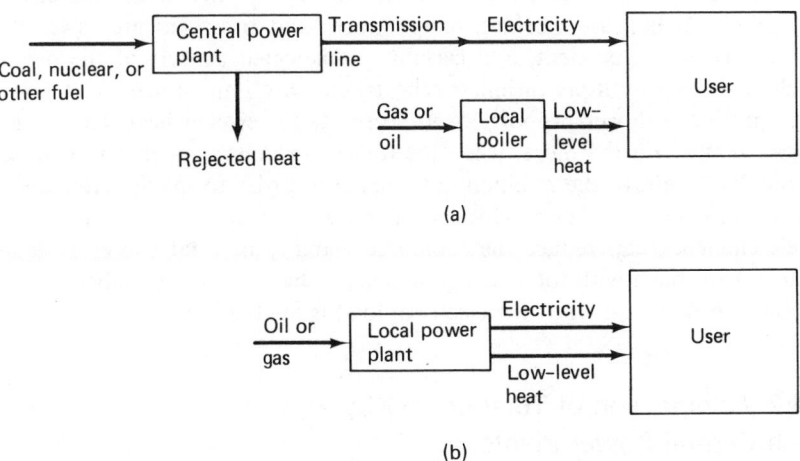

House heating systems rarely are fed by hot water, and transmission of hot water from the power plant to residences can be expensive and undependable. Furthermore, there are tremendous economies of scale involved in the use of large power plants that would be lost if many small local power plants were used. When coal and nuclear fuels are used, small power plants become rather impractical. Management and operational costs are also very high in small power plants. As a result of all these factors, such total energy systems are quite rare. It appeared at one time that high-efficiency fuel cells available in small sizes might make total energy systems more feasible, but the shortage of natural gas has made this approach less desirable. If synthetic gas can be manufactured from coal, this may still be an appealing approach. The ideal potential users for total energy systems are industrial, commercial, or residential complexes that have the following characteristics:

1. A central energy facility for energy distribution to outlying buildings or areas.

2. A reasonable balance between electrical and thermal loads over most of the year.
3. Sufficient size so that certain fixed costs do not become oppressive compared to size-dependent costs.

One of the problems in the use of total energy systems for residential applications, such as large complexes of apartments, dormitories, or barracks, is that the thermal loads are usually difficult to balance on a yearly basis. In cold climates, the thermal loads are small from late spring to early fall. In hot climates, there are peaks in thermal load in summer and winter that are not necessarily balanced. In all cases of residential application, there are periods (spring and fall) where there is no user for rejected low-level heat.

Total energy systems usually have lower cycle efficiencies than central power plants because working temperatures and pressures are lower, the turbine is not as efficient, and certain sophisticated aspects of the overall cycle involving multiple turbines, reheats, etc., are impractical. As a result, they produce a lower proportion of electricity to rejected heat than central power plants. Furthermore, when the turbine exhaust temperature is raised about 100°F above the minimum attainable in order to make better use of the exhaust heat, the cycle efficiency is reduced even further. The lowered cycle efficiency can reduce the economic viability of total energy systems. Another problem with total energy systems is that control and efficiency are difficult to maintain when the electrical load is fluctuating.

14.2 Comparison of Total Energy Sytems with Central Power Plants

The economics of operation of total energy systems vs. central power plants may be illustrated by an oversimplified example. Suppose electricity is worth 3 units of money per Btu equivalent and heat is worth 1 unit of money per Btu equivalent. If a central power plant can convert one-third of each unit of fuel to electricity, the value of electricity produced is 1 unit of money per unit of fuel expended. If a total energy system has a cycle efficiency of 20%, and because of load fluctuations only 70% of that is usable, the value of electricity produced is $3 \times 0.2 \times 0.7 = 0.42$ units of money per unit of fuel expended. The value of the rejected heat is estimated on the basis that if account is taken of thermal load variations during the year, perhaps 50% of the rejected thermal energy is usable. Thus, the value of heat produced is $1 \times 0.8 \times 0.5 = 0.40$ unit of money per unit of fuel expended. The value of electricity plus heat produced by the total energy system is less than the value of electricity produced by the central power plant. Admittedly, this is an oversimplified example, but it does show that the "free" rejected heat from the total energy system does not necessarily represent a saving over central power plants. Since the central power plants can (at least partially) burn coal or nuclear fuel and total energy systems require scarce oil or gas, this is another disadvantage of the use of total energy systems. If total energy systems can be made efficient enough, and if the electrical and thermal loads can be well matched to the system, it is possible that they could be competitive. However, the cost of physical plant per unit of energy tends to be much higher with total energy systems, and maintenance and operations are probably higher. A complete back-up system is probably needed to deal with periods when the system is down.

14.3 Solar Total Energy Systems

14.3.1 Introduction

It was shown in Sec. 14.2 that when conventional energy is used, a total energy system will not compete favorably with a central power plant unless its efficiency is high and the thermal and electrical loads are well matched. The advantages of a local total energy system, which appear to be great at first glance, tend to fade under closer examination.

When solar energy is used as the main energy source, it may be more advantageous to use a total energy system approach. In some solar energy systems there is a natural size, such as in central receiver systems. In distributed fields of solar collectors such as parabolic troughs, there are no large economies of scale involved in the use of very large fields. Indeed, the cost of piping runs and pressure drops in such piping would tend to favor moderate rather than large collector fields. For a moderate-sized collector field (say 2 to 50 acres of ground coverage) it would appear to make sense to consider placing the field adjacent to a central energy facility which distributes utilities to a building complex, in a total energy system mode. One thing is clear. If such a system were to be used to produce electricity as a primary goal, it would be foolhardy not to use the rejected heat from the power plant for low-level heating purposes. On the other hand, if the cycle efficiency is too low, it might be advisable to forego the desire to produce electricity and only produce low-level heat from solar energy. In other words, it is not clear whether one should attempt to produce electricity from solar energy via a heat engine, but if that were to be done, it should be part of a total energy system to make use of rejected heat. The arguments for and against total energy systems are rather different when solar energy is involved because a high-efficiency solar powered central power plant is usually not a viable option for comparison. This leads to the conclusion that if electricity is to be produced from solar energy, it should probably be part of a total energy system.

14.3.2 Solar Total Energy vs. Solar Heating/Cooling

The question of whether to use a solar total energy system or a solar heating/cooling/hot water system in any application is an important one. Solar heating/cooling/hot water systems have been described in Chapter 11. A typical solar total energy system is shown schematically in Figure 14.2. A solar collector field supplies fluid at as high a temperature as possible to high temperature storage. This fluid is circulated to a heat exchanger to generate steam (or organic vapor) to drive the turbine. The turbine is mechanically linked to an electrical generator, and the exhaust from the turbine is cooled with water circulated from low temperature storage. The exhaust temperature is maintained between $140°F$ and $300°F$, depending on the low-level heat application. The condensate is recirculated to the steam generator. Heat stored in the low temperature storage can be used for space heating, space cooling, domestic hot water production, or other low-level heat applications.

In a solar heating/cooling system, a lower temperature solar collector field is used, and the output from the field is routed directly to low temperature storage. Electricity is purchased from a central power plant. A solar total energy system requires a considerable extra investment for high temperature solar collectors, high temperature storage, and the Rankine cycle power train (steam generator, turbine, generator, condenser, pump). Whether the electricity produced justifies this investment or not must be examined in each case. Present indications are that the extra costs associated with the Rankine cycle are not justified unless fairly high cycle efficiencies can be produced ($\sim 25\%$). First generation designs with high temperatures near $550°F$ have cycle efficiencies in the $10-15\%$ range. If the working temperatures can be increased, the prospects for solar total energy systems will improve.

14.3.3 Operation of the Vapor Generator

The steam (or vapor for organic cycles) generator can be subdivided into preheater, boiler, and superheater sections as shown in Figure 14.3. The hot fluid from storage at T_1 is cooled to T_4 at the exit, while the turbine fluid flows counterflow and is heated from liquid at T_5 to vapor at T_8. Since the turbine fluid is heated by the storage fluid, at no point in the heat exchange can the temperature of the turbine fluid exceed the temperature of the

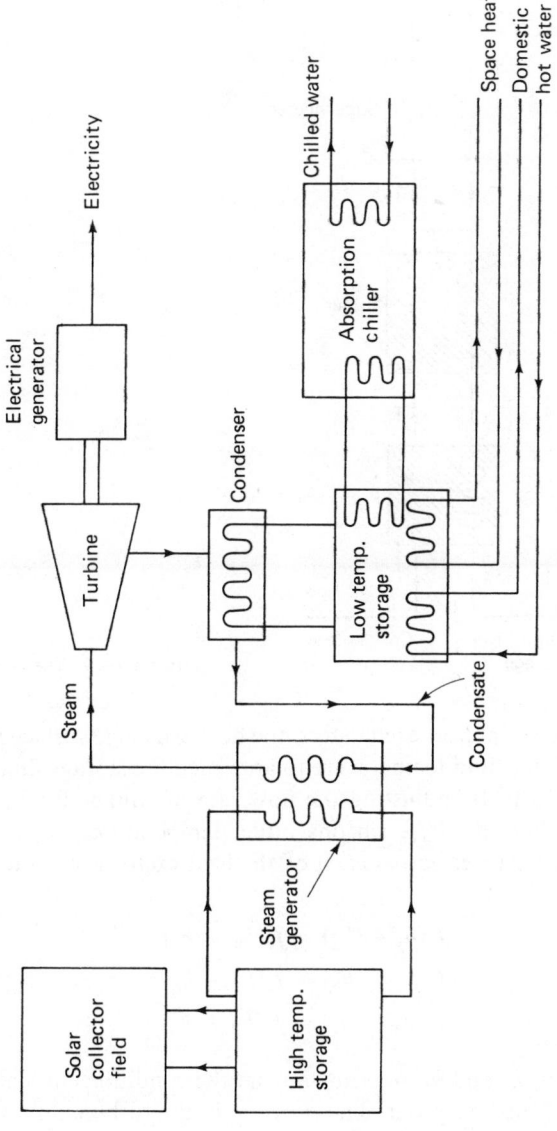

Figure 14.2 Typical solar total energy system.

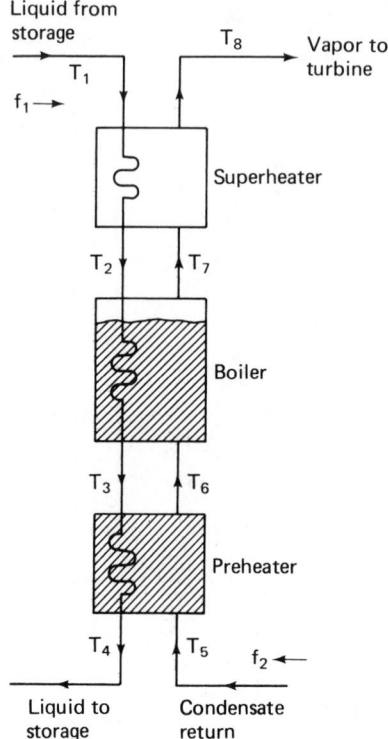

Figure 14.3 The vapor generator.

storage fluid. In actual heat exchangers, the heat exchange surface is designed to allow the turbine fluid temperature to approach the storage fluid to within about 10°F to 20°F. By adjusting the flow rate of storage fluid f_1 for fixed flow rate of turbine fluid f_2, various outlet temperatures (T_4 and T_8) will result. The heat balances across each of the heat exchangers in terms of the enthalpies are:

$$f_1(h_1 - h_2) = f_2(h_8 - h_7) \qquad (14.1)$$

$$f_1(h_2 - h_3) = f_2(h_7 - h_6) \qquad (14.2)$$

$$f_1(h_3 - h_4) = f_2(h_6 - h_5) \qquad (14.3)$$

It is assumed that h_1 and h_5 are known from the conditions in high temperature storage and the condenser. The pressure in the turbine fluid loop determines the boiling point, and therefore h_6 and h_7 correspond to saturated liquid and saturated vapor, respectively, at this pressure. The ratio f_1/f_2 can be treated as a single variable, and thus there are nine variables, of which four are immediately known, and there are three equations. The ratio f_1/f_2 can be treated as an independent parameter which takes on various values, and thus, of the nine variables, five are known, and there are three equations.

There is one more condition required to solve the problem. This condition is the amount of heat exchanged across the exchangers. Fluid being cooled in the primary loop when flowing countercurrent to the liquid being heated in the secondary loop can never be cooled down to the temperature of the fluid in the secondary loop at any place in the heat exchangers. The larger the heat exchange surface, the closer the secondary loop temperatures can approach the primary loop temperatures. In actual work, one usually designs for about a 20°F approach as a minimum difference. This temperature difference is sometimes referred to as a "pinch" temperature. Consider the extreme case where $f_1/f_2 \gg 1$. Then T_4 will be only slightly less than T_1, and T_8 will approach T_1 to within the "pinch" difference. Further increase in f_1/f_2 will not increase T_8 much but will only make T_4 approach T_1 more closely. Consider the diagram in Figure 14.4. Because $f_1/f_2 \gg 1$, the "pinch"

Figure 14.4 Pattern of temperatures at high values of f_1/f_2.

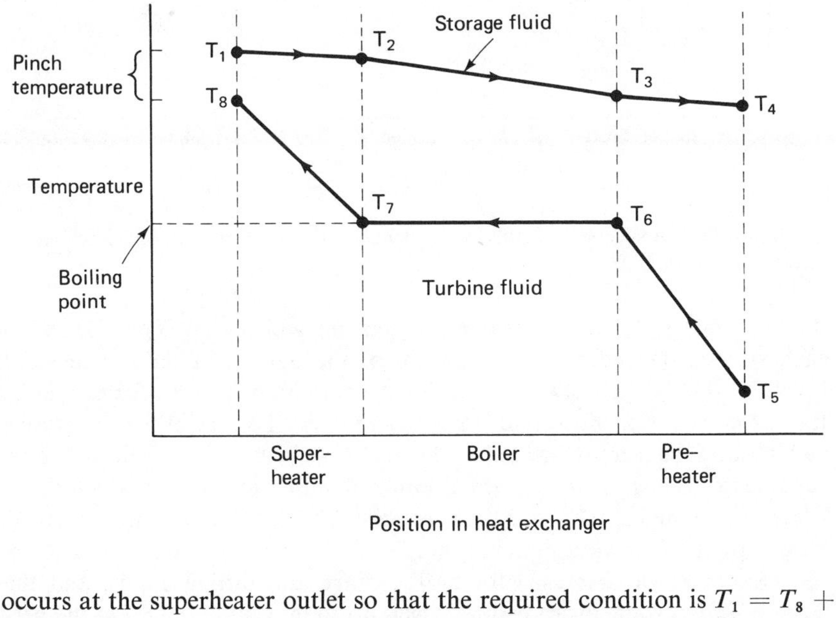

occurs at the superheater outlet so that the required condition is $T_1 = T_8 + 20°F$. As the ratio f_1/f_2 is reduced, the temperature pattern of the turbine fluid remains unchanged, while the pattern for the storage fluid begins to drop toward that of the turbine fluid as shown in Figure 14.5 as lines A and B. Eventually, when f_1/f_2 is reduced to a critical value, T_3 will be reduced to $T_6 + 20°F$ as shown by line B in Figure 14.5. Further reduction of f_1/f_2 will cause the temperature T_8 to drop as shown as curve C in Figure 14.5. When f_1/f_2 is reduced to a second critical value, T_8 is just equal to T_7 and no superheat occurs. Further reduction in f_1/f_2 will result in T_6 dropping below the boiling point so that some of the preheat stage will occur in the boiler. Thus,

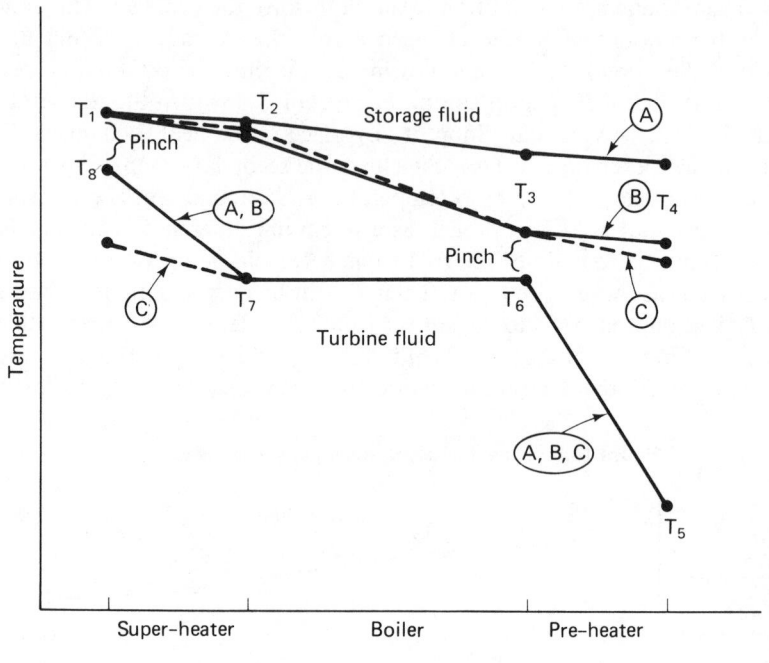

Figure 14.5 Temperature pattern as a function of f_1/f_2. Curves A, B, C represent decreasing f_1/f_2 ratio.

the minimum f_1/f_2 to be considered is that for which $T_3 - T_6 = 20°F$. For each system, depending on the nature of the fluids and the temperatures involved, the location of the "pinch" is the condition required for evaluating the unknowns. For most fluids (especially water) the heat of vaporization is so high that the heat exchanged in the boiler is much greater than that in the superheater. Thus, there is a broad range of values of f_1/f_2 for which superheated steam at $T_8 = T_1 - 20°F$ is produced. This discussion may be made clearer by the following specific example.

Suppose water is used for both storage and turbine fluids, and that $T_1 = 500°F$ at some overpressure > 680 psi to prevent boiling. Let the pressure in the turbine loop be 247.1 psi, corresponding to $T_6 = T_7 = 400°F$, and let $T_5 = 140°F$. Then, according to steam table data,

$$h_1 \cong 487.7 \text{ Btu/lb}, \qquad h_7 = 1202.0$$
$$h_6 = 375.1, \qquad h_5 = 108.0$$

The first critical ratio f_1/f_2 is found by requiring that $T_8 = T_1 - 20°F$ *and* $T_3 = T_6 + 20°F$. It is found that this occurs at $f_1/f_2 = 9.67$. For $f_1/f_2 > 9.67$, $T_8 = T_1 - 20°F$, and $T_3 > T_6 + 20°F$. The second critical ratio is found by requiring that $T_3 = T_6 + 20°F$, and $T_8 = T_7$. It is found that f_1/f_2 must be at least 9.117, or T_6 will drop below 400°F. The results are shown in Table 14.1.

For optimum performance, superheated steam at 480°F is desired, so f_1/f_2 should be at least 9.67. There is a disadvantage in raising f_1/f_2 much higher than this value because, as f_1/f_2 is increased, T_4 increases, and less heat is drawn out of each pound of circulating water. It is generally desirable to have as wide a separation between high and low storage temperatures (T_1 and T_4) as possible.

Another aspect of solar total energy system design is the balance between high temperature and low temperature storage systems. This balance will depend on the patterns of usage of electricity and low-level heat for the load that is applicable. Since high temperature storage is considerably more expensive than low temperature storage, the economics tend to favor less high temperature storage. On the other hand, the pattern of electrical usage dictates the amount of high temperature storage that is desirable.

14.3.4 Air Conditioning with Solar Total Energy Systems

The choice of a method for air conditioning is an important one. There are two basic options. The turbine exhaust temperature can be elevated to drive an absorption chiller with the turbine exhaust, or the turbine exhaust can be kept as cool as possible with a cooling tower and a compressor-type chiller operating from the mechanical output of the turbine. These two options are illustrated schematically in Figure 14.6. It may appear at first glance as if the absorption system makes fuller use of the energy in the steam because the turbine exhaust is used to drive the absorption chiller instead of being cooled in a condenser. Actually, for reversible machines the two approaches have equal efficiency. The elevated temperature of the turbine exhaust with the absorption system reduces the electrical power produced, so that the net effect of cooling plus excess electricity is the same as with the compressor system. Consider the diagrams shown in Figure 14.7. In the absorption system, a reversible turbine operates between temperatures T_1 and T_2 with efficiency η_A. For each unit of heat input to the turbine, $(1 - \eta_A)$ units of heat appears in the exhaust. The heat pumped by the absorption chiller out of the chilled water loop is $(COP_A)(1 - \eta_A)$, and the heat delivered to the cooling tower is $(1 + COP_A)(1 - \eta_A)$. The efficiencies of reversible machines are

$$\eta_A = \frac{T_1 - T_2}{T_1} \tag{14.4}$$

$$COP_A = \frac{T_3(T_2 - T_4)}{T_2(T_4 - T_3)} \tag{14.5}$$

TABLE 14.1 Enthalpies (Btu/lb) and temperatures (°F) in heat exchanger

f_1/f_2	h_1	h_2	h_3	h_4	h_5	h_6	h_7	h_8	T_1	T_2	T_3	T_4	T_5	T_6	T_7	T_8
20	487.7	485.2	443.9	430.5	108	375.1	1202	1252	500	497	463	450	140	400	400	400
15	487.7	484.4	429.3	411.5	108	375.1	1202	1252	500	496	449	445	140	400	400	480
10	487.7	482.7	400.0	373.3	108	375.1	1202	1252	500	494	424	398	140	400	400	480
9.67*	487.7	482.5	397.0	369.4	108	375.1	1202	1252	500	494	420	395	140	400	400	480
9.5	487.7	484.0	397.0	368.9	108	375.1	1202	1237	500	496	420	394	140	400	400	455
9.117†	487.7	487.7	397.0	367.7	108	375.1	1202	1202	500	500	420	393	140	400	400	400

*Critical ratio such that when $f_1/f_2 > 9.67$, $T_8 = T_1 - 20°F$.
†Critical ratio such that when $9.67 > f_1/f_2 > 9.117$, $T_3 = T_6 + 20°F$.

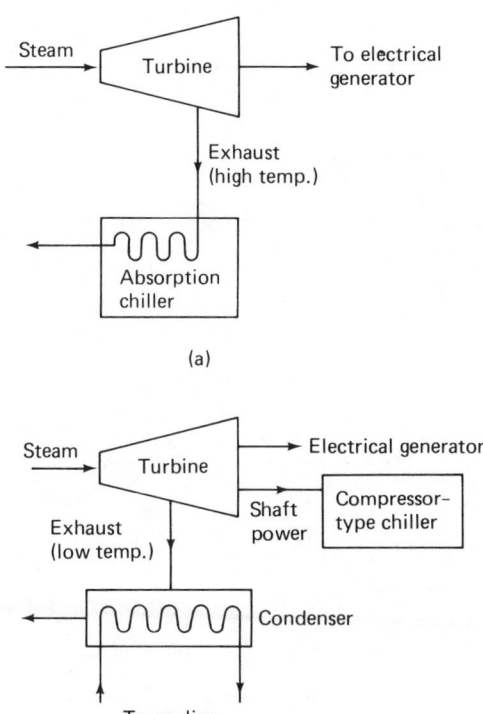

(a)

(b)

Figure 14.6 Comparison of methods of space cooling with a solar total energy system.

Thus, the heat pumped is

$$(1 - \eta_A)(\mathrm{COP}_A) = \frac{T_3(T_2 - T_4)}{T_1(T_4 - T_3)} \tag{14.6}$$

In the compressor system, the mechanical output is divided into one part, η_E, that drives an electrical generator, and η_c to drive the compressor chiller. The COP of the compressor chiller is denoted as COP_c. Let the heat pumped out of the chilled water in the compressor system be the same as the heat pumped with the absorption system. Thus,

$$Q_c = \eta_c(\mathrm{COP}_c) = \eta_c \frac{T_3}{T_4 - T_3} = \frac{T_3}{T_1} \frac{(T_2 - T_4)}{T_4 - T_3} \tag{14.7}$$

and, therefore,

$$\eta_c = \frac{T_2 - T_4}{T_1} \tag{14.8}$$

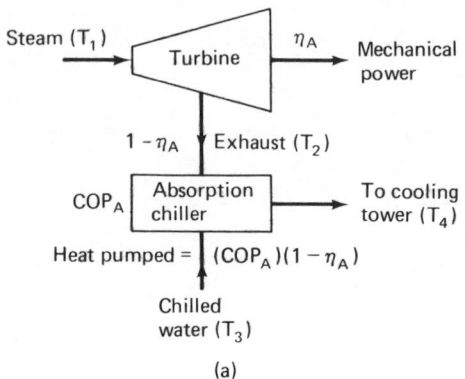

(a)

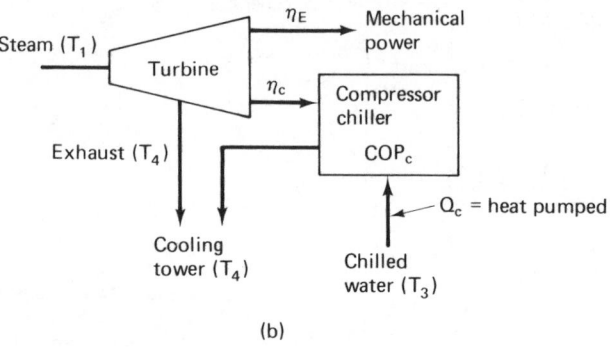

(b)

Figure 14.7 Thermodynamic comparison of absorption and compression chilling in total energy systems: (a) absorption system; (b) compressor system.

Now, if η_E is evaluated from

$$\eta_E = \eta - \eta_c \tag{14.9}$$

where the efficiency of the turbine is

$$\eta = \frac{T_1 - T_4}{T_1} \tag{14.10}$$

it follows that

$$\eta_E = \frac{T_1 - T_2}{T_1} = \eta_A \tag{14.11}$$

Thus, the same net amount of electricity and heat pumped is produced with either system if the machines are reversible. In reality, the machines are not reversible. The degree to which each machine approximates reversibility depends on the application. It is difficult to lay down general rules for choosing between the options. One factor that favors the compressor-type approach is that the use of auxiliary make-up energy is more efficient. If the absorption chiller approach is used, then when make-up energy is required it will be supplied by on-site burning of natural gas or fuel oil in a boiler to produce either steam or hot water to drive the chiller. Since gas and oil are scarce fuels, this is an undesirable feature of the absorption chiller approach. With a compression chiller, make-up energy can be used in the form of electricity purchased from a central power plant which uses (at least in part) uranium or coal as fuels. If the central power plant has 33% cycle efficiency and the COP_c of the compressor chiller is about 4, the expenditure of 1 Btu of fuel at the central power plant pumps about 1.32 Btu out of the chilled water loop when make-up energy is used. In the absorption system, when enough fuel is burned to liberate 1 Btu of heat, only about 0.75 Btu is usable in the form of steam due to the $\sim 75\%$ efficiency of the boiler. This 0.75 Btu of heat drives the turbine and produces perhaps 0.10 Btu equivalent of electricity and 0.65 Btu of heat for the absorption chiller. Since the COP of a single stage absorption chiller is perhaps 0.6, only 0.39 Btu of heat is pumped out of the chilled water. Valuing the electricity produced as being capable of pumping $0.10 \times 4 = 0.40$ Btu out of the chilled water, the total value of the absorption system is equivalent to about 0.79/1.32 of the value of the compression system for utilizing auxiliary make-up energy. There is a clear advantage in using a compression chiller running on electricity from a central power plant as a make-up source of air conditioning.

14.3.5 Other Aspects

The choice of turbine fluid (water or organic) in a solar total energy system depends on the operating temperatures. The advantages and disadvantages of each have been described in Chapter 8. In general, at working temperatures greater than 400°F, water is superior to organics because the parasitic pumping losses are much lower, pump seals are simpler, and any leakage is much less of a problem.

The type and orientation of solar collectors is an important option. When parabolic troughs are used, the east-west orientation offers the advantages of simple support structures and piping runs and fairly even performance across the seasons of the year. Parabolic troughs aligned in the north-south direction are most easily used if placed horizontally on the ground. Such an orientation leads to very poor winter performance, although summer performance is superb. For certain types of loads this may be appropriate, but generally the winter performance is not adequate to justify this orientation. The ideal orientation is north-south with the troughs tilted at the latitude angle, or slightly less. Unfortunately, the costs of substructure and piping are increased substantially, and this arrangement may be too expensive. "Dish"-type collectors offer the advantage of higher working temperatures than linear arrays but are not yet developed adequately. Small central receiver systems might be very useful in conjunction with solar total energy systems.

The choice of working fluid for the collectors and high temperature storage is an important one. Some controversy has developed over the use of pressurized water vs. heat transfer oils. In the solar collector field, the use of pressurized water is an effective choice because of the good heat transfer characteristics of water and its high heat capacity. As a result, there is good heat transfer from the walls of the receiver to the liquid, and the mass throughput is low for any fixed amount of heat transferred. Heat transfer oils have the advantage of lower working pressure but require greater mass throughput and therefore lead to greater parasitic pumping losses. Heat transfer oils also present great dangers from fire or explosion if leaks develop. For high temperature storage, the use of pressurized water requires a very expensive storage tank. Heat transfer oils, on the other hand, can be used in conjunction with rock-filled storage tanks at much lower working pressures. If adequate sites are available, pumped reservoir storage is probably the best approach for solar total energy systems.

Worked Examples

1. Consider the vapor generator shown in Figure 14.3. If 10,000 lb/hr of super-heated steam at 700°F and 600 psi are to be provided to the turbine, and condensate return is at 200°F, what minimum flow rate of heat transfer oil at 730°F (assume density and specific heat are 0.8 that of water at room temperature) must be pumped through the primary loop?

Since the oil is cooled by the water stream, the oil temperature can never drop below the water temperature at any point in the heat exchanger. At 600 psi, the temperature of the boiler $T_6 = 486°F$. For a "pinch" difference of 20°F, we require $T_3 = 486 + 20 = 506°F$. The enthalpy balance across the upper part of the heat exchanger is

$$f_2(h_8 - h_6) = f_1 \times 0.8 \times (T_1 - T_3)$$

$$\frac{(1351.1 - 471)}{0.8(730 - 506)} = \frac{f_1}{f_2}$$

$$\frac{f_1}{f_2} = 4.91$$

The minimum flow rate required is, therefore,

$$f_1 = 49,100 \text{ lb/hr} = 16.4 \text{ ft}^3/\text{min}$$

The minimum final exit temperature T_4 is determined from

$$f_2(h_8 - h_5) = f_1 \times 0.8 \times (730 - T_4)$$

$$T_4 = \frac{-(1351.1 - 168.1)}{0.8 \times 4.91} + 730$$

$$T_4 = 428.8°F$$

2. What cycle efficiency can be expected for the turbine described in Example 1?

Assuming a 200°F exhaust temperature, we have

$$\eta = \frac{1351.1 - h_{ex}}{1351.1 - 168.1}$$

For an isentropic expansion in the turbine,

$$1.5875 = x (1.7760) + (1 - x)(0.2940)$$

$$x = 0.873 = \text{fraction vapor in turbine exhaust}$$

$$h_{ex} = 0.873(1.7760) + 0.127(168.1)$$

$$h_{ex} = 1021.6 \text{ Btu/lb}$$

$$\eta = 0.279$$

This is the theoretical efficiency of a reversible cycle. In practice, perhaps 60% of this value can be achieved, so the actual cycle efficiency would be about

$$0.6 \times 0.279 = 0.167$$

15 | Rights to Solar Access*

Guaranteed access to sunlight is currently an important aspect of solar energy market economics due to its potential impact on sales and use of solar collection equipment. The legal ramifications of solar rights have only recently begun to be reviewed by the courts[1] but have already been the subject of symposia, workshops, and articles in numerous journals.[2] In this chapter, some of the suggested methods for implementing solar rights are described. Though in no way exhaustive, the review will acquaint the reader with terminology and basic concepts currently in use. Chapters 2 and 3 on solar geometry and intensities will be helpful in understanding and applying the theories described here.

---

*The author of this chapter is Ian Whitlock, graduate student at The University of Texas at Dallas. This chapter has been included with his permission.

[1]The courts will not have long to wait. At the time of writing, lawsuits are threatened against a proposed San Diego, CA, ordinance which would require the installation of solar water heaters in new homes. *Solar Energy Intelligence Report*, June 11, 1979, p. 236.

[2]Some examples not referred to specifically in the text are:
Proceedings of the Workshop on Solar Energy and the Law, American Bar Federation, 1975.

R. Becker, Jr., "Common Law Sunrights: An Obstacle to Solar Heating and Cooling?" *Journal of Contemporary Law*, **3**, 19, 1976.

D. Bersohn, "Securing Solar Rights: Easements, Nuisance, or Zoning?" *Columbia Journal of Environmental Law*, **3**, 112, 1976.

D. Moskowitz, "Legal Access to Sunlight: The Solar Energy Imperative." *Natural Resources Lawyer*, **9**, 177, 1976.

R. Zillman and R. Deeny, "Legal Aspects of Solar Energy Development," *Arizona State Law Journal 1976*, 25, 1976.

15.1 Introduction

The legal status of solar access, or a property owner's right to receive unimpeded sunlight from all parts of the sky, is becoming a point of some interest to designers, manufacturers, and buyers of solar collection equipment. If the solar industry is to penetrate a cautious market, solar access needs reinforcement of a kind that will instill confidence in buyers' minds. Assurance is needed that their active collection equipment or passive architectural design features will not be reduced in value by construction or tree planting that occludes the sun. Banks and lending institutions, including the government, will be wary of financing solar collector plants that might be shaded before their useful lifetimes have expired. And it may also be considered within the national interest to ensure that investments in fossil-fuel saving apparatus and design techniques are encouraged and protected.

Solar access is a concept that has been under some scrutiny in the legal world for several years now, even though it is difficult to find a case specifically addressing a solar user's rights to sunlight that has been heard in the courts. Other legal impediments to the development of a solar market have received attention, such as the need for modified building codes, materials performance standards for solar equipment, tax and finance incentives to improve marketing conditions, and the problem of determining electric utility rates when large numbers of solar-powered buildings are interfaced with conventional power grids.[3] But these questions arise after an assumption of guaranteed solar access and are therefore secondary.

15.2 The Need for Legislation

Various methods have been proposed for the creation and implementation of rights to solar access. The record of litigation over issues involving access to light, view, and the free passage of air is generally thought to be insufficient basis for proceedings in defense of solar rights. Stated briefly, although the right to light and air directly above property tends to be absolute, the light which passes over adjoining property first is not guaranteed. Obviously, this is crucial to the collection of solar energy at latitudes distant from the equator. In order to reinforce the judicial basis of solar rights, and given the judiciary direction in this relatively novel field of law, a statement of policy by state or federal legislators seems necessary. The form and content of such statements is now being debated in the literature and in state governments.

The power of Congress to exert federal control over the question of solar access may be construed from past activity under the Commerce Clause

---

[3]W. H. Lawrence and J. H. Minan, "Solar Energy and Public Utility Rate Regulation," *U.C.L.A. Law Review*, **26**, 550, 1979.

of the Constitution.[4] Under this broadly interpreted clause, Congress is empowered to regulate interstate commerce and any activities which significantly affect that commerce. Because the sale and use of solar collection equipment is projected to have an impact on the interstate sale of electrical energy and compete with all forms of fossil fuel currently traded between states, solar access appears to be an issue for consideration by Congress. However, given the activity of state legislatures in this area, some of which have already moved to institute solar rights, the necessity and wisdom of federal action has been questioned, and it seems unlikely that Congress will need to act at all.[5] Already, the passage of the Solar Energy Research, Development and Demonstration Act and the Solar Heating and Cooling Demonstration Act in 1974, and the more recent passage of The Energy Act of 1978, constitute a major statement of intent on the part of the federal government to protect and encourage solar energy usage.

In order for legislative action to effectively guide judicial enforcement of solar rights, many ambiguities that now complicate the concept must be removed. For instance,

— *Should a property owner be entitled to full sunlight over the entire area of his lot, or should he be guaranteed only the light falling on the walls of his house, the roof alone, or only portions of the roof bearing collection equipment?*

— *Is it necessary that solar access be guaranteed from sunrise to sunset, or can shading be allowed at certain times of the day over limited parts of a home or collector?*

— *Should rules of solar access vary over the seasons as the solar azimuth and elevation vary?*

These questions are important if passive architectural features are to be protected as well as active roof top or detached systems. Differences in collector design suggest the need for different regulations. A fixed flat plate or trough collector has a fixed "window" through which it receives sunlight, while a tracking collector has a constantly moving one. Fractional shading of a large flat plate assembly may not affect overall performance greatly, while partial shading of a narrowly focused parabolic concentrator may greatly reduce its efficiency. The law will have to take notice of these variables.

[4]U. S. Const., Article I, Sec. 8, cl. 3. For alternative justifications of federal intervention, see *Application of Solar Technology to Today's Energy Needs*, Vol. 1, Office of Technology Assessment, Washington, DC, June 1978, pp. 78–79.

[5]M. Schiflett and J. Zuckerman, "Solar Heating and Cooling: State and Municipal Impediments and Incentives," *Natural Resources Journal*, **18**, 313, 1978.

Solar energy has several attributes which are of legal importance.[6] It is variable and diffuse, changing with the time of day and year, atmospheric conditions, latitude, and altitude. A prerequisite to the establishment of solar energy as a property right is a strict economic definition. Until recently, sunlight has been thought of as a source of illumination rather than an energy resource or commodity, which is a significant legal difference. Eisenstadt and Utton suggest defining "solar energy" as all the wavelengths of the solar spectrum and "solar light" as only that portion in the visible range.[7] Computer simulations can yield reliable estimates of energy lost through partial reduction of solar input due to shading. With such estimates, it is possible to allocate solar property rights and guarantee them in law.

There are many ways by which state lawmakers could institute guarantees of solar access. In the following sections, four methods will be discussed: easements, nuisance law, land use controls, and water law. Any of these devices could be made available to the public by legislative incorporation into statutory law.

15.3 Easements

An easement can be thought of as one person's right to use another's property in a particular way for a particular reason. City governments and public utilities have easements in private property to allow the construction of sidewalks or the placement of telephone poles. The owner of an enclosed lot may have an easement across a neighbor's land to allow access to his property. The term *solar easement* is generally conceived as a specific type of negative easement, or one

> "... where the owner of the servient estate is prohibited from doing something otherwise lawful upon his estate, because it will affect the dominant estate (as interrupting the light and air from the latter by building on the former.)"[8]

A solar easement obligates the owner of the servient estate not to obstruct the sunlight passing over his land to the dominant (solar user's) estate. Establishing solar rights by easement law is discussed under three categories in the literature: prescriptive, implied, and express easements.

[6]D. Goble, "Solar Rights: Guaranteeing a Place in the Sun," *Oregon Law Review*, **57**, 94, 1977.

[7]M. Eisenstadt, and A. Utton, "Solar Rights and Their Effect on Heating and Cooling." *Natural Resources Journal*, **16**, 363, 1976.

[8]*Black's Law Dictionary*, 4th ed., West Publishing Co., St. Paul, Minn., 1968, p. 600.

Where the authority does not already exist for establishing solar easements, state legislation can standardize procedure and notify the public of the mechanism's existence.[9]

15.3.1 Prescriptive Rights

An activity that has been exercised from time immemorial (or for some period stated in law) can become established in law and is called a *prescriptive right*. The "doctrine of ancient lights," resurrected in most discussions of solar access, is a provision of English common law which states that if light, air, and view have been enjoyed from a window for more than twenty years, the right to those amenities becomes "absolute and indefeasible," a right by prescription.[10] Problems with interpretation of the doctrine and its treatment in American courts have led most commentators to the conclusion that a grant of solar rights based purely on prescription would never pass judicial muster in this country.

The case of *Parker v. Foote* in 1838 is generally cited as the reason for the doctrine's abandonment in American jurisprudence.[11] In that New York State case, a suit was brought in objection to the obstruction of light falling on one property by the construction of a building on another. In a decision that was as much a statement of public policy as law, the court found the doctrine inapplicable to contemporary economic conditions and stated that upholding it would unfairly stifle urban development.

More recently, the doctrine was rejected in *Fontainbleu Hotel Corporation v. Forty-five Twenty-five Incorporated*.[12] In this case, the Fontainbleu Hotel in Miami, Florida, sued the owners of the Eden Roc Hotel to the south when the latter built an extension which shaded the Fontainbleu swimming pool after 2:00 in the afternoon. The court cited the judicial history of the ancient lights doctrine in its decision, concluding that "in the absence of some contractual or statutory obligation, a landowner has no legal right to the free flow of light and air across the adjoining land of his neighbor."[13] The court went on to say that the proper method for ensuring against shading was a

[9]P. Pollock, "The Implementation of State Solar Incentives: Land-Use Planning to Ensure Solar Access," *Solar Energy Research Institute Rept. No. TR-51-163*, March, 1979, p. 8.

[10]D. Dawes, "Urban Sun Rights," in *New Energy Technologies for Building* by R. Schoen, A. Hirshberg, and J. Weingart, Ballinger Publishing, Cambridge, MA, 1975, pp. 201–204.

[11]19 Wendell 309 (NY, 1838).

[12]114 So. 2d 35F (Fla. App. 1959).

[13]*Ibid.*, at 359.

specific amendment of the planning and zoning ordinances of the city.[14] Though the doctrine is still in force in other parts of the English-speaking world, it has been lost to American law.[15]

In English law, the doctrine has traditionally been applied to solar light rather than solar energy. Were it not for other disadvantages, a redefinition might have made the doctrine useful. However, many drawbacks make prescriptive solar rights impractical. First, there is the question of lead time before the right is established. Twenty years may be too long a period to offer any protection against shading, yet "retroactive" prescription would be objected to in many cases as constituting an unfair burden on servient property holders. In general, prescriptive rights are seen to be too inflexible to be successfully applied.

15.3.2 Implied Easements

An implied easement is one thought to be necessary for the reasonable use of a property, such as the use of a pathway for access to and from a building. It differs from an express easement in that it need not be written but is assumed from the circumstances. An example of such an easement of relevance to our subject is the right recognized in all the states of a property abutting a public street to light, air, and view from that street.[16] Litigation over the loss of these rights arose first in urban areas when level and overground railroads were constructed, resulting in the shading of street front property.[17] Both lower courts and the Supreme Court have upheld suits against the taking of these rights and have found such takings to be compensable. Eisenstadt and Utton list three dominant characteristics of the implied easement:

— *The easement is considered a property right, or property in itself.*

— *The monetary value of the easement can be determined.*

— *The easement can be sold and conveyed in cases of condemnation under the power of eminent domain.*

One suggested method for instituting solar rights is the legislative creation of universal easements to solar energy similar to those now existing for street front properties. This enactment would be followed by public acqui-

[14]It is ironic that this prescient comment should have gone unnoticed in Florida, where solar water heaters enjoyed early popularity, before the availability of cheap natural gas.

[15]W. Greene, "The Right to Light," *Solicitor's Journal* **112**, 576, 1968.

[16]Eisenstadt and Utton, *op. cit.*, p. 369.

[17]See *Muhlker v. New York and Harlem R. R. Co.*, 197 U.S. 544 (1905).

sition, through eminent domain, of these easements, making the protection of solar access defensible in the method already established for "light, air, and view." The primary disadvantages of the public acquisition proposal are its administrative complexity and its large cost to the public. Its outstanding advantages are that it would provide immediate and uniform protection and could be incorporated into existing land-use control agencies.[18]

15.3.3 Express Easements

Easements may be created by grant, covenant, or reservation.[19] When an easement for unimpeded solar energy is created in such a contract, it is perhaps the most specific and unimpeachable assurance of solar access because the boundaries and conditions of access are clearly stated, not assumed or implied. In many states, express easements pertaining to property are binding on landholders who follow the original parties to the contract. Elsewhere, statutes would have to be modified to assure enforcement of solar easements through changes in ownership over time.[20]

The use of express easements to ensure solar access avoids the expense of public acquisition mentioned above, places the responsibility of action only on interested individuals, and makes a more permanent insurance than zoning laws, which are susceptible to amendment. Express easements are site-specific and adaptable to individual preferences and bargaining. However, they place the cost of solar access on the solar energy user, who already has the burden of large capital investment to confront and the expense of enforcing the new easements in court is uncertain. Because express easements can be considered property they can be leased as well as taxed.[21] Ancillary legislation may be required to control leasing rates and hopefully to eliminate taxes altogether.

15.4 Nuisance Laws

An entity or activity that causes substantial interference with public health, safety, or welfare is actionable under public nuisance law. By declaring the shading of solar equipment a public nuisance, jurisdictions could grant solar energy users protection similar to that offered by public acquisition of solar easements, injunction to halt shading, or compensation for energy lost. Although this solution requires a relatively simple legislative statement, its

[18]Pollock, *op. cit.*, p. 9.

[19]Annot., 142 ALR 467.

[20]Office of Technology Assessment Report, *op. cit.*, p. 177.

[21]Eisenstadt and Utton, *op. cit.*, pp. 376–379.

efficacy has been questioned.[22] Following such a legislative declaration, lawsuits may still be required to prove nuisance in individual cases, property owners burdened by nuisance restrictions may demand compensation, and suits brought to protect solar access before collectors have been installed may be dismissed as premature. Injunctive relief, which would require offenders to halt any shading activity is not likely to be awarded in most cases. Moreover, only a public official may bring suit to enforce such a statute, unless the language of the statute provides specifically for attack by an individual. Otherwise, if an individual is to bring suit he must prove damage different in kind, rather than degree, to the public damage.

Private nuisance law is largely ignored in the literature.[23] This may be due to the judicial standard used in deciding private nuisance cases, wherein the plaintiff must show a greater hardship from enduring the nuisance than would be caused by enjoining the defendant's activity.[24] In a solar access case, such a standard might be met in a homeowner's action against the planting of a tree, which is of lesser economic importance to his neighbor,[25] or in cases of large solar arrays whose functioning saves large quantities of energy. However, in a case where a commercial establishment wished to construct an addition of some sort, the expected additional profits from the extension of its physical plant could usually be expected to outweigh an individual homeowner's fuel savings.

15.5 Land Use Controls

Ensuring solar access is fundamentally a question of land use, specifically, activity on the land which might inhibit surrounding properties' sunlight. Zoning and subdivision law has evolved for the express purpose of regulating land use to promote the public health and well-being. For this reason, some commentators view zoning as the most suitable device for guaranteeing solar access.[26]

[22]Goble, *op. cit.*, pp. 128, 129.

[23]See generally, F. Gevurtz, "Obstruction of Sunlight as a Private Nuisance," *California Law Review*, **65**, 94, 1977.

[24]Annot., 40 ALR 3d 601.

[25]It is doubtful that even these facts would favor the plaintiff. In *Bliss v. Ball*, 99 Mass. 938 (1868) shading by defendant's tree, alleged to be causing dampness in home of plaintiff, was not found to be a violation of plaintiff's rights. Annot., 83 ALR 2d 936, 938.

[26]See generally:
D. Phelps and R. Yoxall, "Solar Energy: An Analysis of the Implementation of Solar Zoning," *Washburn Law Journal*, **17**, 146, 1977.
G. Reitze, "A Solar Rights Zoning Guarantee: Seeking New Law in Old Concepts," *Washington University Law Quarterly*, **1976**, 375, 1976.

In 1924 Congress drafted a Standard States Zoning Enabling Act. This exemplary Act, or modifications of it, was adopted by most state governments, and empowered municipalities desirous of this form of land use control to draft and implement zoning ordinances. Since that time, zoning and subdivision ordinances have been designed that regulate such aspects of community design as homogenous land use zones, street width, lot size and setback requirements for buildings and fences. Architectural features such as roofing materials, maximum building height, and permissible outbuilding design are also regulated in many ordinances. More recently regulations have been developed that limit construction of new housing units by yearly quotas, require construction to be compatible with terrain to prevent soil erosion, and allow higher than average residential density in exchange for donations of land to be preserved as open space.[27] The constitutional issues surrounding the "taking" of property rights by such land use controls have been extensively argued in the courts,[28] and the function of planning and zoning boards is a familiar aspect of American urban government. In these respects, zoning has already accomplished many of the tasks a more novel approach to solar access would have to undertake afresh.

Several modifications could be made to existing zoning ordinances to protect solar access.[29] Street layout along east-west lines could be stipulated, affording property southern orientation of adjoining lots. Similar requirements for roof orientation would allow retrofitting of solar equipment in later years. Setback regulations forcing the placement of buildings near the northern lot boundaries would reduce the problems of shadows cast from the south. Height restrictions could be defined more specifically to control the shadows cast by buildings.

A significant advantage of the zoning approach is that each municipality drafts its own regulations, enabling the regional flexibility needed to accommodate local solar geometry. The familiar political climate of the city planning department may also be more favorable than a state agency for the implementation and protection of solar access. A zoning ordinance incorporating many of these modifications has been adopted in Albuquerque, New Mexico, and has received much attention in the literature.[30]

Despite the adaptability of zoning regulations, there are criticisms of this approach.[31] The redesigning of zoning ordinances and official maps is a

[27]See generally, R. W. Scott, ed., *Management and Control of Growth*, Vols. 1 and 2. Urban Land Institute, Washington, DC, 1975.

[28]F. Bosselman, D. Callies, and J. Banta, *The Taking Issue*. Report to the Conucil on Environmental Quality, Washington, DC, 1973.

[29]Schiflett and Zuckerman, *op. cit.*, p. 317. Eisenstadt and Utton, *op. cit.*, p. 434 *et seq.*

[30]Eisenstadt and Utton provide a summary. *Op. cit.*, pp. 387, 388.

[31]M. White, "The Allocation of Sunlight: Solar Rights and the Prior Appropriation Doctrine," *University of Colorado Law Review*, **47**, 421, 423, 1976.

complex and costly task, which many planning boards will be unable or unwilling to do. Expertise in this field is lacking even in vigorous and up-to-date planning agencies. New zoning ordinances will not have any effect on existing neighborhoods, so other methods of assuring solar access will still be needed. Currently, zoning ordinances in many areas are an impediment to the use of solar equipment and will have to be altered. Specifications outlawing the use of private property for the production of power, certain types of nonconforming structures (such as detached collectors), or novel roofing materials will need modification. Orientation of streets and buildings to expedite passive collection or retrofitting of collection equipment may conflict with landscaping necessary to reduce soil erosion, rainwater runoff, or noise.

Zoning for solar access may find its best application in the planned unit development (PUD). A concept begun in 1965, PUDs are an option for community planners who wish to combine residential and commercial activity within thoroughly designed, large-scale communities. PUD ordinances could contain all the stipulations mentioned above, and covenants protecting solar access could be included in deeds to the properties within the developments.[32] Other land use planning techniques have been suggested for application to solar access, such as transferable development rights (TDR) and special "solar zones,"[33] but their limited application in conventional planning circles casts doubt on their success elsewhere.

15.6 Water Rights

By incorporating the principles of solar geometry into zoning ordinances or express easements, the body of experience already established in the law could be used to implement the novel concept of guaranteed solar access. By making an analogy between sunlight and some other natural resource with a proven history in law, another vehicle could be developed. Oil and gas law is a possibility, but its emphasis on ownership, complex leasing arrangements, and the exploitation of finite areas and quantities are not characteristics shared with solar energy. Because solar energy is used, not extracted for sale, and because its availability is not diminished by use, water is a better analogy for sunlight. White suggests that the law of water rights be used as a model for guaranteeing solar access.[34]

[32]See generally:
R. Burchell, ed., *Frontiers of Planned Unit Development*, Center for Urban Policy Research, Rutgers University, New Brunswick, NJ, 1973.
C. Harwood, *Using Land to Save Energy*. Ballinger Publishing, Cambridge, MA, 1977, pp. 107–116.

[33]Schiflett and Zuckerman, *op. cit.*, p. 319.

[34]White, *op. cit.*, p. 428, 429, 435.

American water law is divided into two schools of theory. In many of the moist states, the doctrine of riparian rights is followed. Here, the right to use water flowing in a stream is incidental to the ownership of land abutting the stream. In drier western states, rights to water are apportioned according to the doctrine of prior appropriation, where right is established on the basis of who first extracted the waters for a beneficial use. For a number of reasons, including a weaker judicial history, lack of flexibility, and administrative disfavor, the riparian doctrine has been passed over in favor of the prior appropriation system.

Basing her analogy on the Colorado system of water rights allocation, White has described a plan wherein permits must be obtained to establish solar use under the protection of law. Permits would be issued by an administrative agency which would define the requirements of issuance. One requirement would be a clear demonstration of intent to use the allocated sunlight.[35] Such intent might be established by the applicant's purchase of solar equipment, his request for a loan to install a solar system, or his consultation with a builder over the installation. Following the application, the applicant would be required to notify adjacent property holders of his intentions. A review and hearing by the administrative board, where any opposition to the proposed allocation could be voiced, would conclude the process. Presumably, appeal of the board's decision by either party could be heard in the courts.

The proposal has several advantages. It provides an equitable framework for the allotment of solar rights without great public expense or complex administration, as might be required under zoning and public acquisition of solar easement schemes. Those who want assurance of solar access can apply for it; those who don't are not affected by the system in any way. The system provides for the assessment of individual cases in an equitable, efficient, and inexpensive manner: except in cases of re-hearing or appeal, a lawyer's assistance would not be required by applicants or those in opposition. According to White, the legislation needed to erect the system is quite simple: it need only instruct the people and the courts to recognize the analogy between prior appropriation and solar access and provide for the hearing board.[36]

15.7 Contents of State Legislation

Consideration of the advantages and disadvantages of the models for ensuring solar access yields an appreciation of the troubles that accompany the establishment of a previously unrecognized right. Potential conflict generated in a transaction between only two adjacent property holders multiplies as

[35]Ibid., p. 437.

[36]White includes a "Proposed Solar Rights Act,"Ibid., at p. 443.

more become involved. Problems 2.10 and 2.11 in Chapter 2 illustrate the type of calculations that will need to be done to determine "windows" for solar access. Variations in latitude, topography, and time of day can mean that single buildings will disrupt an entire neighborhood's solar access, possibly at critical hours in the operation of solar plant.

The key to success in the early stages of solar access implementation will be successful negotiation between solar energy users and the property owners yielding rights to sections of their airspace. It can be argued that property owners aren't losing an old right so much as gaining a new one, but this will not impress those who hadn't planned to use solar energy in any case or those whose homes are already shaded by buildings or vegetation. Standards for tradeoffs and negotiation must be developed for the many instances where complete solar access will not be available due to existing circumstances.

One suggested tradeoff is the allowance of a 10% maximum loss in total daily insolation from shading.[37] As can be seen from Figure 3.3, comparatively less energy is collected near sunrise and sunset than at near-noon hours. A 10% shading allowance would permit greater shading at hours of lower solar elevation, while only brief periods of shading would be tolerated at midday. A percentage shading allowance could be designed on a daily, seasonal, or yearly basis, depending on the locale and the type of solar plant in question. The placement of collectors on higher portions of roofs or detached structures would be another tradeoff more preferable than litigation.

In his review of current state legislation, Pollock mentions five recommendations for legislative action ensuring solar access:[38]

— *Legislation should not be a simple demonstration of concern or good intention, but should constitute a practical response to the problems of solar access as they exist in the state.*

— *Methods chosen to ensure solar access should be selected on the basis of compatibility with state politics and attitudes.*

— *Legislation should clearly indicate the hierarchy of administrative authority and the basis upon which administrative rulings must be made.*

— *Legislation should coordinate solar access with land use and energy agencies, where they exist, to aid the transfer of technical and procedural information.*

— *For the same reason, coordination and communication should be maintained between state and local governmental organs.*

[37]Eisenstadt and Utton, *op. cit.*, p. 397 *et seq.*

[38]Pollock, *op. cit.*, p. 25.

As of mid-1979, fourteen states had enacted some form of solar access guarantees.[39] The most popular approach has been the creation of solar easement mechanisms, which have been adopted by 11 states. New Mexico has adopted a "sun rights" provision based loosely on concepts of water rights, and California has instituted nuisance provisions for the shading of solar collectors by vegetation. Oregon, a state with a history of progressive land use control measures, has a comprehensive state planning law, including provisions for renewable energy sources.

15.8 Conclusion

Summarized briefly, the problem of solar access in America stems from a judicial history which has established and reaffirmed the priority of individual freedom of action within the bounds of private property. The assumption that freedom is limited when its exercise causes harm or loss to others has not in the past encompassed direct solar energy. Lawmakers in federal and state governments must therefore enunciate their desired reform of judicial opinion and raise the priority of solar access in statutory law. This statement cannot take the simple form of a universal grant to all property owners of full solar access, for such a grant would unfairly burden many and waste resources by protecting others who do not require it. Reform must be tailored to afford equitable solutions to problems where they arise, implementing the guarantee of sunlight in a manner which will bring the greatest public benefit. The most effective method for achieving these goals is to follow closely the tested and established principles of common and civil law.

[39]The states were CA, CO, CT, FL, GA, ID, KS, MD, MN, NJ, NM, ND, OR, and VA. *Ibid.*, p. 12.

Appendix I Units and conversion factors

The units primarily used in this book are common English engineering units. In many instances, units are also illustrated in terms of standard international units (SI). The SI units utilize the basic reference units:

Length:	meter (m)
Mass:	kilogram (kg)
Time:	second (sec)
Temperature:	degrees Kelvin (°K)

A number of derived units can be obtained from these basic units:

Energy:	joule (1 N·m) (J)
Force:	newton (1 kg-m/sec^2) (N)
Power:	watt (1 J/sec) (W)
Pressure:	pascal (1 N/m^2) (P)

Prefixes for the basic units and derived units include:

Multiplier	Symbol	Prefix
10^6	M	mega
10^3	k	kilo
10^{-1}	d	deci
10^{-2}	c	centi
10^{-3}	m	milli
10^{-6}	μ	micro

Some useful conversion factors from common units to SI units are listed below:

Variable	Common Unit	SI Unit	Conversion Factor from Common Units to SI Units
Length	foot	meter	0.3048
Mass	pound	kilogram	0.4536
Time	hour	second	1/3600
Temperature	°F	°K	$(5/9)(T(°F) - 32) + 273.16$
Density	lb/ft^3	kg/m^3	16.02
Force	ft-lb/sec^2	N	4.448
Pressure	lb/in^2	P	6895
Energy	Btu	J	1055
Power	Btu/hr	W	0.2931

Variable	Common Unit	SI Unit	Conversion Factor from Cnmmon Units to SI Units
Specific heat	Btu/lb-°F	J/kg-°K	4187
Energy content	Btu/lb	J/kg	2326
Power density	Btu/hr-ft^2	W/m^2	3.155
Area	ft^2	m^2	0.0929
Volume	ft^3	m^3	0.0283
	gal	m^3	0.00378
Temperature difference	°F	°K	0.5555

Appendix II

Input Data for Solar Systems Abstracted from "Input Data for Solar Systems" by V. Cinquemani, J. R. Owenby, Jr., and R. G. Baldwin, National Oceanic and Atmospheric Administration, Asheville, NC, November, 1978.

```
**************************************************************************
              STATION:  FAIRBANKS                      STATE:  AK
              -----------------------------------      ----------
STATION NUMBER:  26411   LATITUDE:  6449N  LONGITUDE:  14752W  ELEVATION:    138
-----------------------  -----------------  -------------------  ------------------
                                    NORMAL DEGREE         TOTAL HEMISPHERIC
              NORMAL TEMPERATURE (DEG F)*       DAYS*      MEAN DAILY SOLAR RADIATION#

              DAILY     DAILY            BASE 65 DEG F
        MONTH MAXIMUM   MINIMUM  MONTHLY HEATING COOLING   BTU/FT2   KJ/M2   LANGLEYS

        JAN    -2.2     -21.6    -11.9    2384      0        30.1    342.0     8.2
        FEB     9.3     -14.3     -2.5    1890      0       221.4   2513.0    60.1
        MAR    23.3      -4.3      9.5    1720      0       674.2   7651.0   182.9
        APR    40.4      17.3     28.9    1083      0      1193.9  13549.0   323.8
        MAY    58.8      35.7     47.3     549      0      1603.6  18199.0   435.0
        JUN    70.7      47.2     59.0     211     31      1751.9  19882.0   475.2
        JUL    71.8      49.6     60.7     148     15      1542.5  17506.0   418.4
        AUG    65.8      44.9     55.4     304      6      1118.0  12688.0   303.3
        SEP    54.4      34.4     44.4     618      0       709.4   8051.0   192.4
        OCT    33.5      16.9     25.2    1234      0       292.6   3321.0    79.4
        NOV    11.7      -6.2      2.8    1866      0        74.1    841.0    20.1
        DEC    -1.5     -19.3    -10.4    2337      0         2.5     28.0     0.7
        ANN    36.3      15.0     25.7   14344     52       767.8   8714.0   208.3

    * BASED ON 1941-1970 PERIOD                 # AS NOTED IN SOLMET VOLUME 1
    **************************************************************************



              **************************************************************
              STATION:  BIRMINGHAM                     STATE:  AL
              -----------------------------------      ----------
STATION NUMBER:  13876   LATITUDE:  3334N  LONGITUDE:  8645W   ELEVATION:    192
-----------------------  -----------------  -------------------  ------------------
                                    NORMAL DEGREE         TOTAL HEMISPHERIC
              NORMAL TEMPERATURE (DEG F)*       DAYS*      MEAN DAILY SOLAR RADIATION#

              DAILY     DAILY            BASE 65 DEG F
        MONTH MAXIMUM   MINIMUM  MONTHLY HEATING COOLING   BTU/FT2   KJ/M2   LANGLEYS

        JAN    54.3      34.1     44.2     654      9       706.6   8019.0   191.7
        FEB    57.7      36.1     46.9     517     10       967.1  10975.0   262.3
        MAR    64.8      41.8     53.3     389     26      1296.1  14709.0   351.6
        APR    75.3      51.0     63.2     116     62      1673.5  18993.0   453.9
        MAY    82.5      58.4     70.5      20    190      1856.9  21074.0   503.7
        JUN    88.4      66.4     77.4       0    372      1918.5  21773.0   520.4
        JUL    90.3      69.5     79.9       0    462      1809.8  20539.0   490.9
        AUG    89.7      68.7     79.2       0    440      1723.8  19563.0   467.6
        SEP    84.7      63.0     73.9       6    273      1454.6  16508.0   394.6
        OCT    75.8      50.8     63.3     137     84      1210.8  13741.0   328.4
        NOV    64.0      40.1     52.1     391      0       857.9   9736.0   232.7
        DEC    55.5      34.9     45.2     614      0       661.4   7506.0   179.4
        ANN    73.6      51.2     62.4    2844   1928      1344.7  15261.0   364.7

    * BASED ON 1941-1970 PERIOD                 # AS NOTED IN SOLMET VOLUME 1
    **************************************************************************
```

```
********************************************************************************
                    STATION:  MOBILE                    STATE:  AL
          ----------------------------------------      ----------
  STATION NUMBER:  13894  LATITUDE:  3041N  LONGITUDE:   8815W  ELEVATION:    67
  ----------------------  -----------------  --------------------  ----------------

                                        NORMAL DEGREE      TOTAL HEMISPHERIC
            NORMAL TEMPERATURE (DEG F)*       DAYS*       MEAN DAILY SOLAR RADIATION#

            DAILY    DAILY                BASE 65 DEG F
  MONTH  MAXIMUM  MINIMUM   MONTHLY   HEATING   COOLING    BTU/FT2    KJ/M2    LANGLEYS

   JAN    61.1     41.3      51.2       451        23       828.2    9399.0    224.6
   FEB    64.1     43.9      54.0       337        29      1099.6   12479.0    298.3
   MAR    69.5     49.2      59.4       221        47      1407.5   15974.0    381.8
   APR    78.0     57.7      67.9        40       127      1721.7   19540.0    467.0
   MAY    85.0     64.5      74.8         0       304      1872.1   21246.0    507.8
   JUN    89.8     70.7      80.3         0       459      1868.5   21205.0    506.8
   JUL    90.5     72.6      81.6         0       515      1715.3   19467.0    465.3
   AUG    90.6     72.3      81.5         0       512      1641.5   18629.0    445.2
   SEP    86.5     68.4      77.5         0       375      1449.4   16449.0    393.1
   OCT    79.7     58.0      68.9        39       160      1298.7   14739.0    352.3
   NOV    69.5     47.5      58.5       211        16       955.1   10839.0    259.1
   DEC    63.0     42.8      52.9       385        10       759.2    8616.0    205.9
   ANN    77.3     57.4      67.4      1684      2577      1384.7   15715.0    375.6

  * BASED ON 1941-1970 PERIOD                    # AS NOTED IN SOLMET VOLUME 1
  ******************************************************************************

  ******************************************************************************
                    STATION:  LITTLE ROCK               STATE:  AR
          ----------------------------------------      ----------
  STATION NUMBER:  13963  LATITUDE:  3444N  LONGITUDE:   9214W  ELEVATION:    81
  ----------------------  -----------------  --------------------  ----------------

                                        NORMAL DEGREE      TOTAL HEMISPHERIC
            NORMAL TEMPERATURE (DEG F)*       DAYS*       MEAN DAILY SOLAR RADIATION#

            DAILY    DAILY                BASE 65 DEG F
  MONTH  MAXIMUM  MINIMUM   MONTHLY   HEATING   COOLING    BTU/FT2    KJ/M2    LANGLEYS

   JAN    50.1     28.9      39.5       791         0       731.3    8299.0    198.4
   FEB    53.8     31.9      42.9       619         0      1002.8   11381.0    272.0
   MAR    61.8     38.7      50.3       470        14      1312.7   14898.0    356.1
   APR    73.5     49.9      61.7       139        40      1610.7   18280.0    436.9
   MAY    81.4     58.1      69.8        21       169      1929.3   21895.0    523.3
   JUN    89.3     66.8      78.1         0       393      2106.5   23907.0    571.4
   JUL    92.6     70.1      81.4         0       508      2032.3   23064.0    551.2
   AUG    92.6     68.6      80.6         0       484      1860.5   21115.0    504.7
   SEP    85.8     60.8      73.3         5       254      1518.0   17228.0    411.8
   OCT    76.0     48.7      62.4       143        63      1228.3   13940.0    333.2
   NOV    62.4     38.1      50.3       441         0       847.2    9615.0    229.8
   DEC    52.1     31.1      41.6       725         0       673.7    7646.0    182.7
   ANN    72.6     49.3      61.0      3354      1925      1404.4   15939.0    381.0

  * BASED ON 1941-1970 PERIOD                    # AS NOTED IN SOLMET VOLUME 1
  ******************************************************************************

  ******************************************************************************
                    STATION:  PHOENIX                   STATE:  AZ
          ----------------------------------------      ----------
  STATION NUMBER:  23183  LATITUDE:  3326N  LONGITUDE:  11201W  ELEVATION:   339
  ----------------------  -----------------  --------------------  ----------------

                                        NORMAL DEGREE      TOTAL HEMISPHERIC
            NORMAL TEMPERATURE (DEG F)*       DAYS*       MEAN DAILY SOLAR RADIATION#

            DAILY    DAILY                BASE 65 DEG F
  MONTH  MAXIMUM  MINIMUM   MONTHLY   HEATING   COOLING    BTU/FT2    KJ/M2    LANGLEYS

   JAN    64.8     37.6      51.2       428         0      1021.3   11591.0    277.0
   FEB    69.3     40.8      55.1       292        14      1374.1   15595.0    372.7
   MAR    74.5     44.8      59.7       185        21      1814.1   20588.0    492.1
   APR    83.6     51.8      67.7        60       141      2354.8   26725.0    638.7
   MAY    92.9     59.6      76.3         0       355      2676.5   30375.0    726.0
   JUN   101.5     67.7      84.6         0       588      2739.2   31087.0    743.0
   JUL   104.8     77.5      91.2         0       812      2486.5   28219.0    674.5
   AUG   102.2     76.0      89.1         0       747      2292.6   26019.0    621.9
   SEP    98.4     69.1      83.8         0       564      2015.4   22873.0    546.7
   OCT    87.6     56.8      72.2        17       240      1576.5   17892.0    427.6
   NOV    74.7     44.8      59.8       182        26      1150.5   13057.0    312.1
   DEC    66.4     38.5      52.5       388         0       932.0   10577.0    252.8
   ANN    85.1     55.4      70.3      1552      3508      1869.4   21216.0    507.1

  * BASED ON 1941-1970 PERIOD                    # AS NOTED IN SOLMET VOLUME 1
  ******************************************************************************
```

```
*****************************************************************************
                    STATION:  LOS ANGELES                 STATE:  CA
                ---------------------------------------  ----------
     STATION NUMBER:  23174  LATITUDE:  3356N  LONGITUDE:  11824W  ELEVATION:    32
     --------------------------  ------------------  --------------------  ------------------

                                          NORMAL DEGREE          TOTAL HEMISPHERIC
               NORMAL TEMPERATURE (DEG F)*     DAYS*         MEAN DAILY SOLAR RADIATION#

                 DAILY    DAILY             BASE 65 DEG F
         MONTH MAXIMUM  MINIMUM   MONTHLY  HEATING  COOLING    BTU/FT2    KJ/M2    LANGLEYS

          JAN    63.5     45.4      54.5      331       5       926.1    10510.0    251.2
          FEB    64.1     47.0      55.6      270       7      1214.0    13778.0    329.3
          MAR    64.3     48.6      56.5      267       0      1618.7    18370.0    439.1
          APR    65.9     51.7      58.8      195       9      1950.9    22141.0    529.2
          MAY    68.4     55.3      61.9      114      17      2059.6    23374.0    558.7
          JUN    70.3     58.6      64.5       71      56      2119.1    24049.0    574.8
          JUL    74.8     62.1      68.5       19     127      2307.5    26188.0    625.9
          AUG    75.8     63.2      69.5       15     154      2079.5    23600.0    564.1
          SEP    75.7     61.6      68.7       23     134      1681.4    19082.0    456.1
          OCT    72.9     57.5      65.2       77      83      1317.0    14946.0    357.2
          NOV    69.6     51.3      60.5      158      23      1003.9    11393.0    272.3
          DEC    66.5     47.3      56.9      279       0       848.5     9630.0    230.2
          ANN    69.2     54.1      61.7     1819     615      1593.8    18088.0    432.3

       * BASED ON 1941-1970 PERIOD                 # AS NOTED IN SOLMET VOLUME 1
     *****************************************************************************


     *****************************************************************************
                     STATION:  OAKLAND                     STATE:  CA
                ---------------------------------------  ----------
     STATION NUMBER:  23230  LATITUDE:  3744N  LONGITUDE:  12212W  ELEVATION:     2
     --------------------------  ------------------  --------------------  ------------------

                                          NORMAL DEGREE          TOTAL HEMISPHERIC
               NORMAL TEMPERATURE (DEG F)*     DAYS*         MEAN DAILY SOLAR RADIATION#

                 DAILY    DAILY             BASE 65 DEG F
         MONTH MAXIMUM  MINIMUM   MONTHLY  HEATING  COOLING    BTU/FT2    KJ/M2    LANGLEYS

          JAN    54.5     42.7      48.6      508       0       707.8     8033.0    192.0
          FEB    58.0     45.7      51.9      367       0      1017.5    11547.0    276.0
          MAR    60.2     47.2      53.7      350       0      1456.3    16528.0    395.0
          APR    62.8     49.4      56.1      270       0      1922.1    21814.0    521.4
          MAY    65.4     52.4      58.9      193       0      2211.3    25096.0    599.8
          JUN    68.5     55.2      61.9      114      21      2350.0    26670.0    637.4
          JUL    69.7     56.4      63.1       80      21      2322.5    26358.0    630.0
          AUG    70.2     56.8      63.5       74      28      2052.6    23295.0    556.8
          SEP    72.3     56.6      64.5       59      44      1701.1    19306.0    461.4
          OCT    68.7     53.4      61.1      135      14      1212.0    13755.0    328.8
          NOV    62.0     48.5      55.3      291       0       822.1     9330.0    223.0
          DEC    55.5     44.2      49.9      468       0       647.0     7343.0    175.5
          ANN    64.0     50.7      57.4     2909     128      1535.2    17423.0    416.4

       * BASED ON 1941-1970 PERIOD                 # AS NOTED IN SOLMET VOLUME 1
     *****************************************************************************


     *****************************************************************************
                     STATION:  SACRAMENTO                  STATE:  CA
                ---------------------------------------  ----------
     STATION NUMBER:  23232  LATITUDE:  3831N  LONGITUDE:  12130W  ELEVATION:     8
     --------------------------  ------------------  --------------------  ------------------

                                          NORMAL DEGREE          TOTAL HEMISPHERIC
               NORMAL TEMPERATURE (DEG F)*     DAYS*         MEAN DAILY SOLAR RADIATION#

                 DAILY    DAILY             BASE 65 DEG F
         MONTH MAXIMUM  MINIMUM   MONTHLY  HEATING  COOLING    BTU/FT2    KJ/M2    LANGLEYS

          JAN    53.0     37.1      45.1      617       0       596.9     6774.0    161.9
          FEB    59.1     40.4      49.8      426       0       939.4    10661.0    254.8
          MAR    64.1     41.9      53.0      372       0      1458.4    16551.0    395.6
          APR    71.3     45.3      58.3      227      26      2003.6    22739.0    543.5
          MAY    78.8     49.8      64.3      120      98      2434.8    27632.0    660.4
          JUN    86.4     54.6      70.5       20     185      2683.8    30458.0    728.0
          JUL    92.9     57.5      75.2        0     316      2688.0    30506.0    729.1
          AUG    91.3     56.9      74.1        0     286      2368.3    26878.0    642.4
          SEP    87.7     55.3      71.5        5     200      1906.7    21639.0    517.2
          OCT    77.1     49.5      63.3      101      48      1314.9    14923.0    356.7
          NOV    63.6     42.4      53.0      360       0       731.9     8874.0    212.1
          DEC    53.3     38.3      45.8      595       0       538.4     6110.0    146.0
          ANN    73.2     47.4      60.3     2843    1159      1642.9    18645.0    445.6

       * BASED ON 1941-1970 PERIOD                 # AS NOTED IN SOLMET VOLUME 1
     *****************************************************************************
```

```
****************************************************************************
                  STATION:  SAN DIEGO                 STATE:  CA
       ----------------------------------------------    ----------
  STATION NUMBER:  23188  LATITUDE:  3244N  LONGITUDE:  11710U  ELEVATION:      9
  ----------------------    ------------------    ------------------    ------------------


                                      NORMAL DEGREE         TOTAL HEMISPHERIC
         NORMAL TEMPERATURE (DEG F)*       DAYS*        MEAN DAILY SOLAR RADIATION#

           DAILY     DAILY                  BASE 65 DEG F
  MONTH  MAXIMUM   MINIMUM    MONTHLY   HEATING   COOLING    BTU/FT2    KJ/M2     LANGLEYS

    JAN    64.6      45.8       55.2       314       10       975.7    11073.0     264.7
    FEB    65.6      47.8       56.7       237        0      1266.3    14371.0     343.5
    MAR    66.0      50.1       58.0       219        0      1631.6    18517.0     442.6
    APR    67.6      53.8       60.7       144       15      1936.7    21980.0     525.3
    MAY    69.4      57.2       63.3        79       26      2002.8    22730.0     543.3
    JUN    71.1      59.9       65.5        52       67      2062.2    23404.0     559.4
    JUL    75.3      63.9       69.6         6      149      2186.5    24814.0     593.1
    AUG    77.3      65.4       71.4         0      201      2057.3    23348.0     558.0
    SEP    76.5      63.2       69.9        16      163      1717.4    19491.0     465.8
    OCT    73.8      58.4       66.1        43       77      1373.3    15586.0     372.5
    NOV    70.1      51.5       60.8       140       14      1062.7    12060.0     288.2
    DEC    66.1      47.2       56.7       257        0       903.8    10257.0     245.1
    ANN    70.3      55.4       62.9      1507      722      1598.0    18136.0     433.5

  * BASED ON 1941-1970 PERIOD                    # AS NOTED IN SOLMET VOLUME 1
  ****************************************************************************


  ****************************************************************************
                  STATION:  SAN FRANCISCO             STATE:  CA
       ----------------------------------------------    ----------
  STATION NUMBER:  23234  LATITUDE:  3737N  LONGITUDE:  12223W  ELEVATION:      5
  ----------------------    ------------------    ------------------    ------------------


                                      NORMAL DEGREE         TOTAL HEMISPHERIC
         NORMAL TEMPERATURE (DEG F)*       DAYS*        MEAN DAILY SOLAR RADIATION#

           DAILY     DAILY                  BASE 65 DEG F
  MONTH  MAXIMUM   MINIMUM    MONTHLY   HEATING   COOLING    BTU/FT2    KJ/M2     LANGLEYS

    JAN    55.3      41.2       48.3       518        0       707.6     8031.0     191.9
    FEB    58.6      43.8       51.2       386        0      1009.3    11454.0     273.8
    MAR    61.0      44.9       53.0       372        0      1455.1    16514.0     394.7
    APR    63.5      47.0       55.3       291        0      1920.0    21790.0     520.8
    MAY    66.6      49.9       58.3       210        0      2225.6    25258.0     603.7
    JUN    70.2      53.0       61.6       120       18      2376.9    26975.0     644.7
    JUL    70.9      54.0       62.5        93       16      2391.6    27142.0     648.7
    AUG    71.6      54.3       63.0        84       22      2116.5    24020.0     574.1
    SEP    73.6      54.5       64.1        66       39      1742.0    19770.0     472.5
    OCT    70.3      51.6       61.0       137       13      1226.1    13915.0     332.6
    NOV    63.3      47.2       55.3       291        0       821.4     9322.0     222.8
    DEC    56.3      42.9       49.7       474        0       642.4     7290.0     174.2
    ANN    65.1      48.7       56.9      3042      108      1552.8    17623.0     421.2

  * BASED ON 1941-1970 PERIOD                    # AS NOTED IN SOLMET VOLUME 1
  ****************************************************************************


  ****************************************************************************
                  STATION:  DENVER                    STATE:  CO
       ----------------------------------------------    ----------
  STATION NUMBER:  23062  LATITUDE:  3945N  LONGITUDE:  10452W  ELEVATION:   1625
  ----------------------    ------------------    ------------------    ------------------


                                      NORMAL DEGREE         TOTAL HEMISPHERIC
         NORMAL TEMPERATURE (DEG F)*       DAYS*        MEAN DAILY SOLAR RADIATION#

           DAILY     DAILY                  BASE 65 DEG F
  MONTH  MAXIMUM   MINIMUM    MONTHLY   HEATING   COOLING    BTU/FT2    KJ/M2     LANGLEYS

    JAN    43.5      16.2       29.9      1088        0       840.1     9534.0     227.9
    FEB    46.2      19.4       32.8       902        0      1127.0    12790.0     305.7
    MAR    50.1      23.8       37.0       868        0      1530.4    17368.0     415.1
    APR    61.0      33.9       47.5       525        0      1879.3    21328.0     509.8
    MAY    70.3      43.6       57.0       253        0      2134.9    24229.0     579.1
    JUN    80.1      51.9       66.0        80      110      2350.7    26678.0     637.6
    JUL    87.4      58.6       73.0         0      248      2272.6    25792.0     616.4
    AUG    85.8      57.4       71.6         0      208      2044.1    23198.0     554.4
    SEP    77.7      47.8       62.8       120       54      1726.8    19597.0     468.4
    OCT    66.8      37.2       52.0       408        5      1300.5    14759.0     352.7
    NOV    53.3      25.4       39.4       768        0       883.5    10027.0     239.7
    DEC    46.2      18.9       32.6      1004        0       731.8     8305.0     198.5
    ANN    64.0      36.2       50.1      6016      625      1568.4    17800.0     425.4

  * BASED ON 1941-1970 PERIOD                    # AS NOTED IN SOLMET VOLUME 1
  ****************************************************************************
```

```
*****************************************************************************
                  STATION:  HARTFORD                    STATE:  CT
         ----------------------------------------    ----------
STATION NUMBER:  14740  LATITUDE:  4156N  LONGITUDE:  7241W  ELEVATION:    55
-----------------------  -----------------  --------------------  -----------------

                                        NORMAL DEGREE        TOTAL HEMISPHERIC
             NORMAL TEMPERATURE (DEG F)*     DAYS*      MEAN DAILY SOLAR RADIATION#

             DAILY    DAILY                BASE 65 DEG F
      MONTH MAXIMUM  MINIMUM  MONTHLY  HEATING  COOLING    BTU/FT2    KJ/M2    LANGLEYS

      JAN    33.4     16.1     24.8     1246       0        477.5    5419.0     129.5
      FEB    35.7     17.9     26.8     1070       0        714.7    8111.0     193.9
      MAR    44.6     26.6     35.6      911       0        978.5   11105.0     265.4
      APR    58.9     36.5     47.7      519       0       1315.0   14924.0     356.7
      MAY    70.3     46.2     58.3      226      18       1568.5   17801.0     425.5
      JUN    79.5     56.0     67.8       24     108       1685.6   19130.0     457.2
      JUL    84.1     61.2     72.7        0     239       1649.0   18714.0     447.3
      AUG    81.9     58.9     70.4       12     179       1421.7   16135.0     385.6
      SEP    74.5     51.0     62.8      106      40       1154.5   13102.0     313.1
      OCT    64.3     40.8     52.6      384       0        852.9    9679.0     231.3
      NOV    50.6     31.9     41.3      711       0        497.3    5644.0     134.9
      DEC    36.8     19.6     28.2     1141       0        385.1    4370.0     104.4
      ANN    59.6     38.6     49.1     6350     584       1058.3   12011.0     287.1

      * BASED ON 1941-1970 PERIOD              # AS NOTED IN SOLMET VOLUME 1
*****************************************************************************


*****************************************************************************
                  STATION:  WILMINGTON                  STATE:  DE
         ----------------------------------------    ----------
STATION NUMBER:  13781  LATITUDE:  3940N  LONGITUDE:  7536W  ELEVATION:    24
-----------------------  -----------------  --------------------  -----------------

                                        NORMAL DEGREE        TOTAL HEMISPHERIC
             NORMAL TEMPERATURE (DEG F)*     DAYS*      MEAN DAILY SOLAR RADIATION#

             DAILY    DAILY                BASE 65 DEG F
      MONTH MAXIMUM  MINIMUM  MONTHLY  HEATING  COOLING    BTU/FT2    KJ/M2    LANGLEYS

      JAN    40.2     23.8     32.0     1023       0        571.4    6485.0     155.0
      FEB    42.2     24.9     33.6      879       0        827.0    9386.0     224.3
      MAR    51.1     32.0     41.6      725       0       1149.2   13042.0     311.7
      APR    63.0     41.5     52.3      381       0       1480.1   16798.0     401.5
      MAY    73.1     51.6     62.4      128      48       1710.2   19409.0     463.9
      JUN    81.6     61.1     71.4        0     196       1882.6   21365.0     510.6
      JUL    85.5     66.1     75.8        0     335       1822.8   20687.0     494.4
      AUG    83.9     64.3     74.1        0     282       1614.6   18324.0     438.0
      SEP    78.2     57.6     67.9       32     119       1317.7   14955.0     357.4
      OCT    67.8     46.5     57.2      254      12        983.9   11166.0     266.9
      NOV    55.2     36.2     45.7      579       0        644.6    7315.0     174.8
      DEC    43.0     26.3     34.7      939       0        488.6    5545.0     132.5
      ANN    63.7     44.3     54.0     4940     992       1207.7   13706.0     327.6

      * BASED ON 1941-1970 PERIOD              # AS NOTED IN SOLMET VOLUME 1
*****************************************************************************


*****************************************************************************
                  STATION:  JACKSONVILLE                STATE:  FL
         ----------------------------------------    ----------
STATION NUMBER:  13889  LATITUDE:  3030N  LONGITUDE:  8142W  ELEVATION:     9
-----------------------  -----------------  --------------------  -----------------

                                        NORMAL DEGREE        TOTAL HEMISPHERIC
             NORMAL TEMPERATURE (DEG F)*     DAYS*      MEAN DAILY SOLAR RADIATION#

             DAILY    DAILY                BASE 65 DEG F
      MONTH MAXIMUM  MINIMUM  MONTHLY  HEATING  COOLING    BTU/FT2    KJ/M2    LANGLEYS

      JAN    64.6     44.5     54.6      348      25        899.9   10213.0     244.1
      FEB    66.9     45.7     56.3      282      38       1164.3   13214.0     315.8
      MAR    72.2     50.1     61.2      176      58       1521.7   17270.0     412.8
      APR    79.0     57.1     68.1       24     117       1855.7   21060.0     503.3
      MAY    84.6     63.9     74.3        0     288       1956.3   22202.0     530.6
      JUN    88.3     70.0     79.2        0     426       1885.2   21395.0     511.4
      JUL    90.0     72.0     81.0        0     496       1802.0   20451.0     488.8
      AUG    89.7     72.3     81.0        0     496       1694.2   19227.0     459.5
      SEP    86.0     70.4     78.2        0     396       1442.3   16369.0     391.2
      OCT    79.2     61.7     70.5       19     190       1223.1   13881.0     331.8
      NOV    71.4     51.0     61.2      161      47        996.0   11303.0     270.1
      DEC    65.6     45.1     55.4      317      19        817.6    9279.0     221.8
      ANN    78.1     58.7     68.4     1327    2596       1438.2   16322.0     390.1

      * BASED ON 1941-1970 PERIOD              # AS NOTED IN SOLMET VOLUME 1
*****************************************************************************
```

```
**************************************************************************
                    STATION:  MIAMI                      STATE:  FL
                    ------------------------------------ -----------
    STATION NUMBER:  12839  LATITUDE:  2548N  LONGITUDE:  8016W  ELEVATION:      2
    ------------------------  ------------------  --------------------  ------------------


                                         NORMAL DEGREE           TOTAL HEMISPHERIC
              NORMAL TEMPERATURE (DEG F)*      DAYS*        MEAN DAILY SOLAR RADIATION#

              DAILY     DAILY                BASE 65 DEG F
    MONTH  MAXIMUM   MINIMUM   MONTHLY   HEATING   COOLING     BTU/FT2    KJ/M2    LANGLEYS

     JAN     75.6      58.7      67.2       53       121       1057.4   12000.0    286.8
     FEB     76.6      59.0      67.8       67       145       1314.0   14912.0    356.4
     MAR     79.5      63.0      71.3       17       212       1603.3   18196.0    434.9
     APR     82.7      67.3      75.0        0       300       1859.0   21098.0    504.3
     MAY     85.3      70.7      78.0        0       403       1843.6   20923.0    500.1
     JUN     88.0      73.9      81.0        0       480       1707.9   19383.0    463.3
     JUL     89.1      75.5      82.3        0       536       1763.4   20013.0    478.3
     AUG     89.9      75.8      82.9        0       555       1629.8   18497.0    442.1
     SEP     88.3      75.0      81.7        0       501       1456.3   16527.0    395.0
     OCT     84.6      71.0      77.8        0       397       1302.7   14784.0    353.3
     NOV     79.9      64.5      72.2       13       229       1118.6   12695.0    303.4
     DEC     76.6      60.0      68.3       56       159       1019.1   11566.0    276.4
     ANN     83.0      67.9      75.5      206      4038       1472.9   16716.0    399.5

    * BASED ON 1941-1970 PERIOD                  # AS NOTED IN SOLMET VOLUME 1
    **************************************************************************

    **************************************************************************
                    STATION:  TAMPA                      STATE:  FL
                    ------------------------------------ -----------
    STATION NUMBER:  12842  LATITUDE:  2758N  LONGITUDE:  8232W  ELEVATION:      3
    ------------------------  ------------------  --------------------  ------------------


                                         NORMAL DEGREE           TOTAL HEMISPHERIC
              NORMAL TEMPERATURE (DEG F)*      DAYS*        MEAN DAILY SOLAR RADIATION#

              DAILY     DAILY                BASE 65 DEG F
    MONTH  MAXIMUM   MINIMUM   MONTHLY   HEATING   COOLING     BTU/FT2    KJ/M2    LANGLEYS

     JAN     70.6      50.1      60.4      203        60       1010.7   11470.0    274.1
     FEB     71.9      51.7      61.8      176        87       1259.4   14293.0    341.6
     MAR     76.1      55.9      66.0       90       121       1593.7   18087.0    432.3
     APR     82.4      61.6      72.0        9       219       1908.5   21660.0    517.7
     MAY     87.5      66.9      77.2        0       378       1998.2   22677.0    542.0
     JUN     89.9      72.0      81.0        0       480       1847.4   20966.0    501.1
     JUL     90.1      73.7      81.9        0       524       1752.7   19891.0    475.4
     AUG     90.4      74.0      82.2        0       533       1653.1   18761.0    448.4
     SEP     89.0      72.6      80.8        0       474       1492.0   16933.0    404.7
     OCT     83.9      65.5      74.7        0       301       1346.4   15280.0    365.2
     NOV     77.1      56.4      66.8       71       125       1107.8   12572.0    300.5
     DEC     72.0      51.2      61.6      169        64        935.4   10616.0    253.7
     ANN     81.7      62.6      72.2      718      3366       1492.1   16934.0    404.7

    * BASED ON 1941-1970 PERIOD                  # AS NOTED IN SOLMET VOLUME 1
    **************************************************************************

    **************************************************************************
                    STATION:  ATLANTA                    STATE:  GA
                    ------------------------------------ -----------
    STATION NUMBER:  13874  LATITUDE:  3339N  LONGITUDE:  8426W  ELEVATION:    315
    ------------------------  ------------------  --------------------  ------------------


                                         NORMAL DEGREE           TOTAL HEMISPHERIC
              NORMAL TEMPERATURE (DEG F)*      DAYS*        MEAN DAILY SOLAR RADIATION#

              DAILY     DAILY                BASE 65 DEG F
    MONTH  MAXIMUM   MINIMUM   MONTHLY   HEATING   COOLING     BTU/FT2    KJ/M2    LANGLEYS

     JAN     51.4      33.4      42.4      701         0        717.6    8144.0    194.6
     FEB     54.5      35.5      45.0      560         0        968.9   10996.0    262.8
     MAR     61.1      41.1      51.1      443        12       1303.6   14795.0    353.6
     APR     71.4      50.7      61.1      144        27       1686.2   19136.0    457.4
     MAY     79.0      59.2      69.1       27       154       1853.8   21039.0    502.8
     JUN     84.6      66.6      75.6        0       321       1913.8   21720.0    519.1
     JUL     86.5      69.4      78.0        0       403       1812.2   20566.0    491.5
     AUG     86.4      68.6      77.5        0       388       1708.5   19390.0    463.4
     SEP     81.2      63.4      72.3        8       227       1422.0   16138.0    385.7
     OCT     72.5      52.3      62.4      137        57       1199.9   13618.0    325.5
     NOV     61.9      40.8      51.4      408         0        882.9   10020.0    239.5
     DEC     52.7      34.3      43.5      667         0        674.2    7652.0    182.9
     ANN     70.3      51.3      60.8     3095      1589       1345.3   15268.0    364.9

    * BASED ON 1941-1970 PERIOD                  # AS NOTED IN SOLMET VOLUME 1
    **************************************************************************
```

```
****************************************************************************
                STATION:  HONOLULU                   STATE:  HI
          -------------------------------------    -----------
STATION NUMBER:  22521  LATITUDE:  2120N  LONGITUDE:  15755W  ELEVATION:      5
-----------------------  -----------------  -------------------  ------------------


                                      NORMAL DEGREE          TOTAL HEMISPHERIC
          NORMAL TEMPERATURE (DEG F)*      DAYS*       MEAN DAILY SOLAR RADIATION#

          DAILY    DAILY                BASE 65 DEG F
MONTH  MAXIMUM  MINIMUM   MONTHLY  HEATING  COOLING    BTU/FT2    KJ/M2    LANGLEYS

  JAN    79.3     65.3      72.3      0       226       1179.8   13390.0    320.0
  FEB    79.2     65.3      72.3      0       204       1396.3   15847.0    378.8
  MAR    79.7     66.3      73.0      0       248       1621.7   18404.0    439.9
  APR    81.4     68.1      74.8      0       294       1795.8   20380.0    487.1
  MAY    83.6     70.2      76.9      0       369       1949.3   22123.0    528.8
  JUN    85.6     72.2      78.9      0       417       2004.4   22748.0    543.7
  JUL    86.8     73.4      80.1      0       468       2002.2   22723.0    543.1
  AUG    87.4     74.0      80.7      0       487       1966.5   22318.0    533.4
  SEP    87.4     73.4      80.4      0       462       1810.1   20543.0    491.0
  OCT    85.8     72.0      78.9      0       431       1540.3   17481.0    417.8
  NOV    83.2     69.8      76.5      0       345       1266.1   14369.0    343.4
  DEC    80.3     67.1      73.7      0       270       1132.5   12853.0    307.2
  ANN    83.3     69.8      76.6      0      4221       1638.7   18598.0    444.5

  * BASED ON 1941-1970 PERIOD                  # AS NOTED IN SOLMET VOLUME 1
****************************************************************************

****************************************************************************
                STATION:  DES MOINES                  STATE:  IA
          -------------------------------------    -----------
STATION NUMBER:  14933  LATITUDE:  4132N  LONGITUDE:  9339W  ELEVATION:    294
-----------------------  -----------------  -------------------  ------------------


                                      NORMAL DEGREE          TOTAL HEMISPHERIC
          NORMAL TEMPERATURE (DEG F)*      DAYS*       MEAN DAILY SOLAR RADIATION#

          DAILY    DAILY                BASE 65 DEG F
MONTH  MAXIMUM  MINIMUM   MONTHLY  HEATING  COOLING    BTU/FT2    KJ/M2    LANGLEYS

  JAN    27.5     11.3      19.4    1414       0        580.7     6590.0    157.5
  FEB    32.5     15.8      24.2    1142       0        860.7     9768.0    233.5
  MAR    42.5     25.2      33.9     964       0       1180.5    13397.0    320.2
  APR    59.7     39.2      49.5     465       0       1556.6    17666.0    422.2
  MAY    70.9     50.9      60.9     186      59       1867.5    21194.0    506.5
  JUN    79.8     61.1      70.5      26     191       2124.8    24114.0    576.3
  JUL    84.9     65.3      75.1       0     317       2096.8    23796.0    568.7
  AUG    83.2     63.4      73.3      13     270       1827.9    20745.0    495.8
  SEP    74.6     54.0      64.3      94      73       1433.9    16273.0    388.9
  OCT    64.9     43.6      54.3     350      18       1067.8    12118.0    289.6
  NOV    46.4     29.2      37.8     816       0        658.3     7471.0    178.6
  DEC    32.8     17.2      25.0    1240       0        486.9     5526.0    132.1
  ANN    58.3     39.7      49.0    6710     928       1311.8    14888.0    355.8

  * BASED ON 1941-1970 PERIOD                  # AS NOTED IN SOLMET VOLUME 1
****************************************************************************

****************************************************************************
                STATION:  BOISE                       STATE:  ID
          -------------------------------------    -----------
STATION NUMBER:  24131  LATITUDE:  4334N  LONGITUDE:  11613W  ELEVATION:    874
-----------------------  -----------------  -------------------  ------------------


                                      NORMAL DEGREE          TOTAL HEMISPHERIC
          NORMAL TEMPERATURE (DEG F)*      DAYS*       MEAN DAILY SOLAR RADIATION#

          DAILY    DAILY                BASE 65 DEG F
MONTH  MAXIMUM  MINIMUM   MONTHLY  HEATING  COOLING    BTU/FT2    KJ/M2    LANGLEYS

  JAN    36.5     21.4      29.0    1116       0        485.3     5508.0    131.6
  FEB    43.8     27.2      35.5     826       0        839.7     9530.0    227.8
  MAR    51.6     30.5      41.1     741       0       1304.1    14800.0    353.7
  APR    61.4     36.5      49.0     480       0       1826.9    20733.0    495.5
  MAY    70.6     44.1      57.4     252      17       2276.7    25838.0    617.5
  JUN    78.3     51.2      64.8      97      91       2463.2    27955.0    668.1
  JUL    90.5     58.5      74.5       0     295       2612.7    29651.0    708.7
  AUG    87.6     56.7      72.2      12     235       2196.5    24928.0    595.8
  SEP    77.6     48.5      63.1     127      70       1737.2    19715.0    471.2
  OCT    64.7     39.4      52.1     406       6       1137.8    12913.0    308.6
  NOV    48.9     30.7      39.8     756       0        628.3     7130.0    170.4
  DEC    39.1     25.0      32.1    1020       0        437.2     4962.0    118.6
  ANN    62.6     39.1      50.9    5833     714       1495.5    16972.0    405.6

  * BASED ON 1941-1970 PERIOD                  # AS NOTED IN SOLMET VOLUME 1
****************************************************************************
```

```
******************************************************************************
                    STATION:  CHICAGO                    STATE:  IL
                    ----------------------------------   ----------
STATION NUMBER:  14819  LATITUDE:  4147N  LONGITUDE:   8745W  ELEVATION:    190
----------------------  -----------------  ------------------  ------------------


                                      NORMAL DEGREE         TOTAL HEMISPHERIC
           NORMAL TEMPERATURE (DEG F)*    DAYS*       MEAN DAILY SOLAR RADIATION#


           DAILY    DAILY              BASE 65 DEG F
     MONTH MAXIMUM  MINIMUM  MONTHLY  HEATING  COOLING   BTU/FT2   KJ/M2   LANGLEYS

     JAN    31.5    17.0     24.3     1262       0        507.0    5754.0    137.5
     FEB    34.6    20.2     27.4     1053       0        759.5    8620.0    206.0
     MAR    44.6    29.0     36.8      874       0       1106.9   12562.0    300.2
     APR    59.3    40.4     49.9      453       0       1459.0   16558.0    395.7
     MAY    70.3    49.7     60.0      208      53       1788.9   20302.0    485.2
     JUN    80.6    60.3     70.5       26     191       2007.0   22777.0    544.4
     JUL    84.4    65.0     74.7        0     301       1943.8   22060.0    527.2
     AUG    83.3    64.1     73.7        8     277       1719.4   19513.0    466.4
     SEP    75.8    56.0     65.9       57      84       1353.9   15365.0    367.2
     OCT    65.0    45.6     55.4      316      19        968.9   10996.0    262.8
     NOV    48.1    32.6     40.4      738       0        565.6    6419.0    153.4
     DEC    35.3    21.6     28.5     1132       0        401.5    4557.0    108.9
     ANN    59.4    41.8     50.6     6127     925       1215.1   13790.0    329.6

    * BASED ON 1941-1970 PERIOD                    # AS NOTED IN SOLMET VOLUME 1
******************************************************************************


******************************************************************************
                    STATION:  INDIANAPOLIS              STATE:  IN
                    ----------------------------------   ----------
STATION NUMBER:  93819  LATITUDE:  3944N  LONGITUDE:   8617W  ELEVATION:    246
----------------------  -----------------  ------------------  ------------------


                                      NORMAL DEGREE         TOTAL HEMISPHERIC
           NORMAL TEMPERATURE (DEG F)*    DAYS*       MEAN DAILY SOLAR RADIATION#


           DAILY    DAILY              BASE 65 DEG F
     MONTH MAXIMUM  MINIMUM  MONTHLY  HEATING  COOLING   BTU/FT2   KJ/M2   LANGLEYS

     JAN    36.0    19.7     27.9     1150       0        495.6    5624.0    134.4
     FEB    39.3    22.1     30.7      960       0        746.9    8477.0    202.6
     MAR    49.0    30.3     39.7      784       0       1037.4   11773.0    281.4
     APR    62.8    41.8     52.3      387       6       1398.4   15870.0    379.3
     MAY    72.9    51.5     62.2      159      72       1688.0   19157.0    457.9
     JUN    82.3    61.1     71.7       11     212       1868.1   21201.0    506.7
     JUL    85.4    64.6     75.0        0     310       1806.3   20500.0    490.0
     AUG    84.0    62.4     73.2        5     259       1643.5   18652.0    445.8
     SEP    77.7    54.9     66.3       63     102       1324.0   15026.0    359.1
     OCT    67.0    44.3     55.7      302      13        977.0   11088.0    265.0
     NOV    50.5    32.8     41.7      699       0        579.1    6572.0    157.1
     DEC    38.7    23.1     30.9     1057       0        416.6    4728.0    113.0
     ANN    62.2    42.4     52.3     5577     974       1165.0   13222.0    316.0

    * BASED ON 1941-1970 PERIOD                    # AS NOTED IN SOLMET VOLUME 1
******************************************************************************


******************************************************************************
                    STATION:  TOPEKA                    STATE:  KS
                    ----------------------------------   ----------
STATION NUMBER:  13996  LATITUDE:  3904N  LONGITUDE:   9538W  ELEVATION:    270
----------------------  -----------------  ------------------  ------------------


                                      NORMAL DEGREE         TOTAL HEMISPHERIC
           NORMAL TEMPERATURE (DEG F)*    DAYS*       MEAN DAILY SOLAR RADIATION#


           DAILY    DAILY              BASE 65 DEG F
     MONTH MAXIMUM  MINIMUM  MONTHLY  HEATING  COOLING   BTU/FT2   KJ/M2   LANGLEYS

     JAN    38.3    17.7     28.0     1147       0        680.9    7728.0    184.7
     FEB    44.1    22.7     33.4      885       0        941.0   10679.0    255.2
     MAR    52.6    29.7     41.2      745       8       1256.9   14264.0    340.9
     APR    66.3    42.6     54.5      329      14       1641.6   18630.0    445.3
     MAY    75.8    53.2     64.5      118     103       1915.4   21738.0    519.6
     JUN    84.0    63.0     73.5       13     268       2126.4   24132.0    576.8
     JUL    89.2    67.2     78.2        0     409       2127.9   24149.0    577.2
     AUG    88.5    65.9     77.2        0     378       1910.0   21676.0    518.1
     SEP    80.4    56.0     68.2       55     151       1516.4   17210.0    411.3
     OCT    70.3    44.8     57.6      259      30       1146.6   13013.0    311.0
     NOV    54.3    31.5     42.9      663       0        771.6    8757.0    209.3
     DEC    41.8    21.8     31.8     1029       0        583.5    6622.0    158.3
     ANN    65.5    43.0     54.3     5243    1361       1384.8   15716.0    375.6

    * BASED ON 1941-1970 PERIOD                    # AS NOTED IN SOLMET VOLUME 1
******************************************************************************
```

```
*****************************************************************************
                  STATION:  LOUISVILLE              STATE:  KY
            ------------------------------------    ----------
STATION NUMBER:  93821  LATITUDE:  3811N LONGITUDE:  8544W ELEVATION:   149
----------------------  -----------------  --------------------  -----------------

                                  NORMAL DEGREE          TOTAL HEMISPHERIC
          NORMAL TEMPERATURE (DEG F)*      DAYS*      MEAN DAILY SOLAR RADIATION#

           DAILY    DAILY                BASE 65 DEG F
MONTH  MAXIMUM   MINIMUM  MONTHLY  HEATING  COOLING  BTU/FT2    KJ/M2    LANGLEYS

 JAN     42.0     24.5     33.3      983       0      545.5    6191.0    148.0
 FEB     45.0     26.5     35.8      818       0      789.3    8958.0    214.1
 MAR     54.0     34.0     44.0      661      10     1102.0   12506.0    298.9
 APR     66.9     44.8     55.9      286      13     1466.7   16646.0    397.8
 MAY     75.6     53.9     64.8      105      99     1719.8   19518.0    466.5
 JUN     83.7     62.9     73.3        5     254     1903.5   21603.0    516.3
 JUL     87.3     66.4     76.9        0     369     1837.5   20854.0    498.4
 AUG     86.8     64.9     75.9        0     338     1680.2   19069.0    455.8
 SEP     80.5     57.7     69.1       35     158     1361.2   15448.0    369.2
 OCT     70.3     45.9     58.1      241      27     1042.2   11828.0    282.7
 NOV     54.9     35.1     45.0      600       0      652.8    7409.0    177.1
 DEC     44.1     27.1     35.6      911       0      487.9    5537.0    132.3
 ANN     65.9     45.3     55.6     4645    1268     1215.7   13797.0    329.8

* BASED ON 1941-1970 PERIOD                 # AS NOTED IN SOLMET VOLUME 1
*****************************************************************************

*****************************************************************************
                  STATION:  NEW ORLEANS             STATE:  LA
            ------------------------------------    ----------
STATION NUMBER:  12916  LATITUDE:  2959N LONGITUDE:  9015W ELEVATION:     3
----------------------  -----------------  --------------------  -----------------

                                  NORMAL DEGREE          TOTAL HEMISPHERIC
          NORMAL TEMPERATURE (DEG F)*      DAYS*      MEAN DAILY SOLAR RADIATION#

           DAILY    DAILY                BASE 65 DEG F
MONTH  MAXIMUM   MINIMUM  MONTHLY  HEATING  COOLING  BTU/FT2    KJ/M2    LANGLEYS

 JAN     62.3     43.5     52.9      403      28      834.6    9472.0    226.4
 FEB     65.1     46.0     55.6      299      35     1111.9   12619.0    301.6
 MAR     70.4     50.9     60.7      188      55     1414.8   16056.0    383.7
 APR     78.4     58.8     68.6       29     137     1780.3   20204.0    482.9
 MAY     84.9     65.3     75.1        0     313     1967.7   22331.0    533.7
 JUN     89.6     71.2     80.4        0     462     2003.8   22741.0    543.5
 JUL     90.4     73.3     81.9        0     524     1813.5   20581.0    494.9
 AUG     90.6     73.1     81.9        0     524     1716.6   19482.0    465.6
 SEP     86.6     69.7     78.2        0     396     1513.6   17178.0    410.6
 OCT     79.9     59.6     69.8       40     189     1335.0   15151.0    362.1
 NOV     70.3     49.8     60.1      179      32      972.6   11038.0    263.8
 DEC     64.2     45.3     54.8      327      11      779.4    8845.0    211.4
 ANN     77.7     58.9     68.3     1465    2706     1437.4   16308.0    389.8

* BASED ON 1941-1970 PERIOD                 # AS NOTED IN SOLMET VOLUME 1
*****************************************************************************

*****************************************************************************
                  STATION:  BOSTON                  STATE:  MA
            ------------------------------------    ----------
STATION NUMBER:  94701  LATITUDE:  4222N LONGITUDE:  7102W ELEVATION:     5
----------------------  -----------------  --------------------  -----------------

                                  NORMAL DEGREE          TOTAL HEMISPHERIC
          NORMAL TEMPERATURE (DEG F)*      DAYS*      MEAN DAILY SOLAR RADIATION#

           DAILY    DAILY                BASE 65 DEG F
MONTH  MAXIMUM   MINIMUM  MONTHLY  HEATING  COOLING  BTU/FT2    KJ/M2    LANGLEYS

 JAN     35.9     22.5     29.2     1110       0      475.5    5396.0    129.0
 FEB     37.5     23.3     30.4      969       0      709.6    8053.0    192.5
 MAR     44.6     31.5     38.1      834       0     1016.4   11535.0    275.7
 APR     56.3     40.8     48.6      492       0     1325.8   15046.0    359.6
 MAY     67.1     50.1     58.6      218      20     1620.5   18391.0    439.6
 JUN     76.6     59.3     68.0       27     117     1817.1   20622.0    492.9
 JUL     81.4     65.1     73.3        0     260     1749.2   19852.0    474.5
 AUG     79.3     63.3     71.3        8     203     1486.5   16870.0    403.2
 SEP     72.2     56.7     64.5       76      61     1259.9   14298.0    341.7
 OCT     63.2     47.5     55.4      301       0      889.6   10096.0    241.3
 NOV     51.7     38.7     45.2      594       0      502.9    5707.0    136.4
 DEC     39.3     26.6     33.0      992       0      403.0    4574.0    109.3
 ANN     58.7     43.8     51.3     5621     661     1104.7   12537.0    299.6

* BASED ON 1941-1970 PERIOD                 # AS NOTED IN SOLMET VOLUME 1
*****************************************************************************
```

```
*****************************************************************************
                  STATION:  PORTLAND                       STATE:  ME
                  ------------------------------------     -----------
  STATION NUMBER:   14764  LATITUDE:  4339N  LONGITUDE:   7019W  ELEVATION:     19
  ---------------------   ------------------   --------------------   -------------------


                                            NORMAL DEGREE          TOTAL HEMISPHERIC
        NORMAL TEMPERATURE (DEG F)*             DAYS*         MEAN DAILY SOLAR RADIATION#

          DAILY    DAILY                   BASE 65 DEG F
  MONTH MAXIMUM  MINIMUM   MONTHLY   HEATING  COOLING    BTU/FT2    KJ/M2    LANGLEYS

    JAN   31.2    11.7      21.5       1349      0        450.3     5110.0    122.1
    FEB   33.3    12.5      22.9       1179      0        681.9     7739.0    185.0
    MAR   40.8    22.8      31.8       1029      0        969.6    11004.0    263.0
    APR   52.8    32.5      42.7        669      0       1303.9    14798.0    353.7
    MAY   63.6    41.7      52.7        381      0       1567.4    17788.0    425.1
    JUN   73.2    51.1      62.2        106     22       1711.6    19425.0    464.3
    JUL   79.1    56.9      68.0         27    120       1659.1    18829.0    450.0
    AUG   77.6    55.2      66.4         55     99       1460.9    16580.0    396.3
    SEP   69.9    47.4      58.7        200     11       1157.8    13140.0    314.1
    OCT   60.2    38.0      49.1        493      0        822.4     9333.0    223.1
    NOV   47.5    29.7      38.6        792      0        459.3     5212.0    124.6
    DEC   34.9    16.4      25.7       1218      0        362.9     4118.0     98.4
    ANN   55.3    34.7      45.0       7498    252       1050.6    11923.0    285.0

  * BASED ON 1941-1970 PERIOD                    # AS NOTED IN SOLMET VOLUME 1
  *****************************************************************************

  *****************************************************************************
                  STATION:  DETROIT                         STATE:  MI
                  ------------------------------------     -----------
  STATION NUMBER:   14822  LATITUDE:  4225N  LONGITUDE:   8301W  ELEVATION:    191
  ---------------------   ------------------   --------------------   -------------------


                                            NORMAL DEGREE          TOTAL HEMISPHERIC
        NORMAL TEMPERATURE (DEG F)*             DAYS*         MEAN DAILY SOLAR RADIATION#

          DAILY    DAILY                   BASE 65 DEG F
  MONTH MAXIMUM  MINIMUM   MONTHLY   HEATING  COOLING    BTU/FT2    KJ/M2    LANGLEYS

    JAN   31.7    19.2      25.5       1225      0        417.4     4737.0    113.2
    FEB   33.7    20.1      26.9       1067      0        680.4     7722.0    184.6
    MAR   43.1    27.6      35.4        918      0       1000.2    11351.0    271.3
    APR   57.6    38.6      48.1        507      0       1399.0    15877.0    379.5
    MAY   68.5    48.3      58.4        238     33       1715.9    19474.0    465.4
    JUN   79.1    59.1      69.1         26    149       1866.1    21178.0    506.2
    JUL   83.1    63.4      73.3          0    261       1835.4    20830.0    497.8
    AUG   81.6    62.1      71.9         11    225       1575.5    17880.0    427.3
    SEP   74.2    54.8      64.5         80     65       1253.2    14223.0    339.9
    OCT   63.4    45.2      54.3        342     10        876.1     9943.0    237.6
    NOV   47.7    34.4      41.1        717      0        477.8     5422.0    129.6
    DEC   35.4    23.8      29.6       1097      0        343.5     3898.0     93.2
    ANN   58.3    41.4      49.9       6228    743       1120.0    12711.0    303.8

  * BASED ON 1941-1970 PERIOD                    # AS NOTED IN SOLMET VOLUME 1
  *****************************************************************************

  *****************************************************************************
                  STATION:  MINNEAPOLIS-ST. PAUL            STATE:  MN
                  ------------------------------------     -----------
  STATION NUMBER:   14922  LATITUDE:  4453N  LONGITUDE:   9313W  ELEVATION:    255
  ---------------------   ------------------   --------------------   -------------------


                                            NORMAL DEGREE          TOTAL HEMISPHERIC
        NORMAL TEMPERATURE (DEG F)*             DAYS*         MEAN DAILY SOLAR RADIATION#

          DAILY    DAILY                   BASE 65 DEG F
  MONTH MAXIMUM  MINIMUM   MONTHLY   HEATING  COOLING    BTU/FT2    KJ/M2    LANGLEYS

    JAN   21.2     3.2      12.2       1637      0        464.0     5266.0    125.9
    FEB   25.9     7.1      16.5       1358      0        763.9     8669.0    207.2
    MAR   36.9    19.6      28.3       1138      0       1103.5    12524.0    299.3
    APR   55.5    34.7      45.1        597      0       1441.9    16364.0    391.1
    MAY   67.9    46.3      57.1        271     26       1737.3    19716.0    471.2
    JUN   77.1    56.7      66.9         65    122       1927.5    21875.0    522.8
    JUL   82.4    61.4      71.9         11    225       1970.0    22357.0    534.3
    AUG   80.8    59.6      70.2         21    182       1687.0    19146.0    457.6
    SEP   70.7    49.3      60.0        173     23       1254.7    14239.0    340.3
    OCT   60.7    39.2      50.0        472      7        859.6     9756.0    233.2
    NOV   40.6    24.2      32.4        978      0        480.4     5452.0    130.3
    DEC   26.6    10.6      18.6       1438      0        353.3     4010.0     95.8
    ANN   53.8    34.3      44.1       8159    585       1170.2    13281.0    317.4

  * BASED ON 1941-1970 PERIOD                    # AS NOTED IN SOLMET VOLUME 1
  *****************************************************************************
```

```
********************************************************************************
                   STATION:   ST. LOUIS                 STATE:   MO
          ------------------------------------------      ----------
  STATION NUMBER:  13994  LATITUDE:  3845N  LONGITUDE:   9023W  ELEVATION:    172
  ---------------------   -----------------  ------------------  ------------------


                                        NORMAL DEGREE          TOTAL HEMISPHERIC
             NORMAL TEMPERATURE (DEG F)*     DAYS*       MEAN DAILY SOLAR RADIATION#

             DAILY     DAILY                BASE 65 DEG F
     MONTH MAXIMUM   MINIMUM    MONTHLY  HEATING   COOLING    BTU/FT2    KJ/M2    LANGLEYS

     JAN     39.9      22.6      31.3      1045       0        627.4    7120.0     170.2
     FEB     44.2      26.0      35.1       837       0        885.6   10051.0     240.2
     MAR     53.0      33.5      43.3       682       9       1204.7   13672.0     326.8
     APR     67.0      46.0      56.5       272      17       1564.2   17752.0     424.3
     MAY     76.0      55.5      65.8       103     128       1871.3   21237.0     507.6
     JUN     84.9      64.8      74.9        10     307       2092.5   23748.0     567.6
     JUL     88.4      68.8      78.6         0     422       2049.5   23260.0     555.9
     AUG     87.2      67.1      77.2         0     378       1816.5   20615.0     492.7
     SEP     80.1      59.1      69.6        35     173       1459.2   16560.0     395.8
     OCT     69.8      48.4      59.1       224      41       1099.8   12481.0     298.3
     NOV     54.1      35.9      45.0       600       0        718.3    8152.0     194.8
     DEC     42.7      26.5      34.6       942       0        530.6    6022.0     143.9
     ANN     65.6      46.2      55.9      4750    1475       1326.6   15056.0     359.8

  * BASED ON 1941-1970 PERIOD                 # AS NOTED IN SOLMET VOLUME 1
********************************************************************************

********************************************************************************
                   STATION:   JACKSON                   STATE:   MS
          ------------------------------------------      ----------
  STATION NUMBER:   3940  LATITUDE:  3219N  LONGITUDE:   9005W  ELEVATION:    101
  ---------------------   -----------------  ------------------  ------------------


                                        NORMAL DEGREE          TOTAL HEMISPHERIC
             NORMAL TEMPERATURE (DEG F)*     DAYS*       MEAN DAILY SOLAR RADIATION#

             DAILY     DAILY                BASE 65 DEG F
     MONTH MAXIMUM   MINIMUM    MONTHLY  HEATING   COOLING    BTU/FT2    KJ/M2    LANGLEYS

     JAN     58.4      35.8      47.1       569      14        753.5    8551.0     204.4
     FEB     61.7      37.8      49.8       442      17       1026.4   11648.0     278.4
     MAR     68.7      43.4      56.1       313      37       1369.1   15538.0     371.4
     APR     78.2      53.1      65.7        74      95       1708.4   19388.0     463.4
     MAY     85.0      60.4      72.7         6     245       1940.8   22026.0     526.4
     JUN     91.0      67.7      79.4         0     432       2024.2   22973.0     549.1
     JUL     92.7      70.6      81.7         0     518       1909.0   21665.0     517.8
     AUG     92.6      69.8      81.2         0     502       1780.5   20207.0     483.0
     SEP     88.0      64.0      76.0         0     330       1509.2   17128.0     409.4
     OCT     80.1      51.5      65.8        91     116       1271.4   14429.0     344.9
     NOV     68.5      42.0      55.3       301      10        901.6   10232.0     244.6
     DEC     60.5      37.3      48.9       504       5        708.8    8044.0     192.3
     ANN     77.1      52.8      65.0      2300    2321       1408.6   15986.0     382.1

  * BASED ON 1941-1970 PERIOD                 # AS NOTED IN SOLMET VOLUME 1
********************************************************************************
********************************************************************************
                   STATION:   BILLINGS                  STATE:   MT
          ------------------------------------------      ----------
  STATION NUMBER:  24033  LATITUDE:  4548N  LONGITUDE:  10832W  ELEVATION:   1088
  ---------------------   -----------------  ------------------  ------------------


                                        NORMAL DEGREE          TOTAL HEMISPHERIC
             NORMAL TEMPERATURE (DEG F)*     DAYS*       MEAN DAILY SOLAR RADIATION#

             DAILY     DAILY                BASE 65 DEG F
     MONTH MAXIMUM   MINIMUM    MONTHLY  HEATING   COOLING    BTU/FT2    KJ/M2    LANGLEYS

     JAN     31.2      12.5      21.9      1336       0        486.0    5516.0     131.8
     FEB     37.1      17.7      27.4      1053       0        763.2    8661.0     207.0
     MAR     42.1      23.1      32.6      1004       0       1189.5   13499.0     322.6
     APR     55.8      33.4      44.6       612       0       1526.3   17322.0     414.0
     MAY     65.7      43.3      54.5       333       8       1912.8   21708.0     518.8
     JUN     73.7      51.5      62.6       131      59       2173.7   24669.0     589.6
     JUL     85.6      58.0      71.8        10     220       2383.7   27053.0     646.6
     AUG     83.8      56.3      70.1        15     173       2022.4   22952.0     548.6
     SEP     71.3      46.5      58.9       221      38       1470.0   16683.0     398.7
     OCT     61.0      37.5      49.3       487       0        986.8   11199.0     267.7
     NOV     45.0      26.4      35.7       879       0        561.4    6371.0     152.3
     DEC     35.8      17.7      26.8      1184       0        421.2    4780.0     114.2
     ANN     57.3      35.3      46.3      7265     498       1324.7   15034.0     359.3

  * BASED ON 1941-1970 PERIOD                 # AS NOTED IN SOLMET VOLUME 1
********************************************************************************
```

```
*********************************************************************************
                    STATION:  ASHEVILLE                  STATE:  NC
                    -----------------------------------   ----------
STATION NUMBER:   3812  LATITUDE:  3526N  LONGITUDE:   8232W  ELEVATION:    661
-----------------------   ------------------   --------------------   ------------------


                                     NORMAL DEGREE          TOTAL HEMISPHERIC
          NORMAL TEMPERATURE (DEG F)*      DAYS*       MEAN DAILY SOLAR RADIATION#

          DAILY    DAILY              BASE 65 DEG F
     MONTH MAXIMUM  MINIMUM  MONTHLY  HEATING  COOLING    BTU/FT2   KJ/M2    LANGLEYS

     JAN   48.4    27.3      37.9      840        0        721.7    8190.0    195.7
     FEB   50.6    28.2      39.4      717        0        971.4   11024.0    263.5
     MAR   58.3    33.5      45.9      592        0       1306.0   14822.0    354.3
     APR   69.4    42.4      55.9      279        6       1667.6   18925.0    452.3
     MAY   76.8    50.6      63.7      100       60       1804.4   20478.0    489.4
     JUN   82.5    58.7      70.6       14      182       1854.5   21047.0    503.0
     JUL   84.3    62.6      73.5        0      264       1776.1   20157.0    481.8
     AUG   83.8    61.8      72.8        0      244       1626.7   18461.0    441.2
     SEP   78.0    55.4      66.7       50      101       1360.8   15444.0    369.1
     OCT   69.1    44.5      56.8      269       15       1147.4   13022.0    311.2
     NOV   58.2    34.3      46.3      561        0        848.8    9633.0    230.2
     DEC   49.3    28.1      38.7      815        0        657.6    7463.0    178.4
     ANN   67.4    44.0      55.7     4237      872       1311.9   14889.0    355.9

* BASED ON 1941-1970 PERIOD                    # AS NOTED IN SOLMET VOLUME 1
*********************************************************************************


*********************************************************************************
                    STATION:  CAPE HATTERAS               STATE:  NC
                    -----------------------------------   ----------
STATION NUMBER:  93729  LATITUDE:  3516N  LONGITUDE:   7533W  ELEVATION:      2
-----------------------   ------------------   --------------------   ------------------


                                     NORMAL DEGREE          TOTAL HEMISPHERIC
          NORMAL TEMPERATURE (DEG F)*      DAYS*       MEAN DAILY SOLAR RADIATION#

          DAILY    DAILY              BASE 65 DEG F
     MONTH MAXIMUM  MINIMUM  MONTHLY  HEATING  COOLING    BTU/FT2   KJ/M2    LANGLEYS

     JAN   52.3    38.2      45.3      611        0        685.6    7781.0    186.0
     FEB   53.1    38.5      45.8      538        0        952.2   10806.0    258.3
     MAR   57.9    43.2      50.6      458       12       1326.4   15053.0    359.8
     APR   66.3    51.5      58.9      188        5       1773.9   20132.0    481.2
     MAY   73.8    60.2      67.0       47      109       1961.8   22264.0    532.1
     JUN   80.5    68.1      74.3        0      283       2035.9   23105.0    552.2
     JUL   83.8    72.1      78.0        0      403       1920.6   21797.0    521.0
     AUG   83.4    71.5      77.5        0      388       1705.4   19355.0    462.6
     SEP   79.5    67.8      73.7        0      261       1470.4   16688.0    398.9
     OCT   71.3    59.1      65.2       76       82       1136.6   12899.0    308.3
     NOV   63.1    48.8      56.0      277        7        872.9    9906.0    236.8
     DEC   54.8    40.5      47.7      536        0        658.7    7475.0    178.7
     ANN   68.3    55.0      61.7     2731     1550       1375.0   15605.0    373.0

* BASED ON 1941-1970 PERIOD                    # AS NOTED IN SOLMET VOLUME 1
*********************************************************************************


*********************************************************************************
                    STATION:  BISMARCK                    STATE:  ND
                    -----------------------------------   ----------
STATION NUMBER:  24011  LATITUDE:  4646N  LONGITUDE:  10045W  ELEVATION:    502
-----------------------   ------------------   --------------------   ------------------


                                     NORMAL DEGREE          TOTAL HEMISPHERIC
          NORMAL TEMPERATURE (DEG F)*      DAYS*       MEAN DAILY SOLAR RADIATION#

          DAILY    DAILY              BASE 65 DEG F
     MONTH MAXIMUM  MINIMUM  MONTHLY  HEATING  COOLING    BTU/FT2   KJ/M2    LANGLEYS

     JAN   19.1    -2.8       8.2     1761        0        466.8    5298.0    126.6
     FEB   24.5     2.4      13.5     1442        0        775.7    8803.0    210.4
     MAR   35.4    14.7      25.1     1237        0       1168.1   13257.0    316.8
     APR   54.8    31.1      43.0      660        0       1459.3   16562.0    395.8
     MAY   67.1    41.7      54.4      339       11       1848.1   20974.0    501.3
     JUN   75.8    51.8      63.8      122       86       2059.8   23376.0    558.7
     JUL   84.3    57.3      70.8       18      198       2183.6   24782.0    592.3
     AUG   83.5    54.9      69.2       35      165       1876.7   21298.0    509.0
     SEP   71.3    43.7      57.5      252       27       1354.5   15372.0    367.4
     OCT   60.3    33.2      46.8      564        0        907.8   10302.0    246.2
     NOV   39.4    18.3      28.9     1083        0        507.3    5757.0    137.6
     DEC   26.0     5.2      15.6     1531        0        372.9    4232.0    101.1
     ANN   53.5    29.3      41.4     9044      487       1248.4   14168.0    338.6

* BASED ON 1941-1970 PERIOD                    # AS NOTED IN SOLMET VOLUME 1
*********************************************************************************
```

```
*******************************************************************************
                STATION:  NORTH OMAHA                    STATE:  NE
                ----------------------------------       ----------
STATION NUMBER:  94918  LATITUDE:  4122N  LONGITUDE:   9601W  ELEVATION:    404
-----------------------  ------------------  --------------------  ------------------


                                      NORMAL DEGREE            TOTAL HEMISPHERIC
        NORMAL TEMPERATURE (DEG F)*      DAYS*          MEAN DAILY SOLAR RADIATION#

          DAILY    DAILY                 BASE 65 DEG F
    MONTH MAXIMUM  MINIMUM  MONTHLY  HEATING  COOLING    BTU/FT2    KJ/M2    LANGLEYS

     JAN   29.1    11.2     20.2     1389      0         634.0     7195.0    172.0
     FEB   34.8    16.1     25.5     1106      0         892.1    10124.0    242.0
     MAR   44.1    25.1     34.6      942      0        1222.5    13874.0    331.6
     APR   61.0    38.9     50.0      456      6        1558.4    17686.0    422.7
     MAY   71.4    50.4     60.9      186     59        1872.6    21252.0    507.9
     JUN   80.2    60.2     70.2       33    189        2122.5    24088.0    575.7
     JUL   85.4    64.8     75.1        7    320        2106.5    23906.0    571.4
     AUG   84.0    63.4     73.7       10    280        1858.5    21092.0    504.1
     SEP   75.2    53.6     64.4       99     81        1373.2    15584.0    372.5
     OCT   65.9    42.8     54.4      342     14        1049.8    11914.0    284.8
     NOV   47.4    28.3     37.9      813      0         644.1     7310.0    174.7
     DEC   34.3    17.0     25.7     1218      0         511.2     5802.0    138.7
     ANN   59.4    39.3     49.4     6601    949        1320.5    14986.0    358.2

  * BASED ON 1941-1970 PERIOD                   # AS NOTED IN SOLMET VOLUME 1
*******************************************************************************

*******************************************************************************
                STATION:  CONCORD                        STATE:  NH
                ----------------------------------       ----------
STATION NUMBER:  14745  LATITUDE:  4312N  LONGITUDE:   7130W  ELEVATION:    105
-----------------------  ------------------  --------------------  ------------------


                                      NORMAL DEGREE            TOTAL HEMISPHERIC
        NORMAL TEMPERATURE (DEG F)*      DAYS*          MEAN DAILY SOLAR RADIATION#

          DAILY    DAILY                 BASE 65 DEG F
    MONTH MAXIMUM  MINIMUM  MONTHLY  HEATING  COOLING    BTU/FT2    KJ/M2    LANGLEYS

     JAN   31.3     9.9     20.6     1376      0         459.5     5215.0    124.6
     FEB   33.8    11.3     22.6     1187      0         686.1     7786.0    186.1
     MAR   42.4    22.1     32.3     1014      0         973.6    11049.0    264.1
     APR   56.7    31.7     44.2      624      0        1317.1    14948.0    357.3
     MAY   68.6    41.5     55.1      315      8        1582.2    17956.0    429.2
     JUN   77.7    51.6     64.7       58     49        1704.6    19345.0    462.4
     JUL   82.6    56.7     69.7       16    162        1674.6    19005.0    454.2
     AUG   80.1    54.2     67.2       45    113        1455.3    16516.0    394.7
     SEP   72.4    46.5     59.5      182     17        1140.2    12940.0    309.3
     OCT   62.3    36.3     49.3      487      0         817.1     9273.0    221.6
     NOV   47.9    28.1     38.0      810      0         462.7     5251.0    125.5
     DEC   34.6    14.9     24.8     1246      0         362.1     4110.0     98.2
     ANN   57.5    33.7     45.6     7360    349        1053.0    11950.0    285.6

  * BASED ON 1941-1970 PERIOD                   # AS NOTED IN SOLMET VOLUME 1
*******************************************************************************

*******************************************************************************
                STATION:  NEWARK                         STATE:  NJ
                ----------------------------------       ----------
STATION NUMBER:  14734  LATITUDE:  4042N  LONGITUDE:   7410W  ELEVATION:      9
-----------------------  ------------------  --------------------  ------------------


                                      NORMAL DEGREE            TOTAL HEMISPHERIC
        NORMAL TEMPERATURE (DEG F)*      DAYS*          MEAN DAILY SOLAR RADIATION#

          DAILY    DAILY                 BASE 65 DEG F
    MONTH MAXIMUM  MINIMUM  MONTHLY  HEATING  COOLING    BTU/FT2    KJ/M2    LANGLEYS

     JAN   38.5    24.3     31.4     1042      0         551.7     6261.0    149.6
     FEB   40.2    24.9     32.6      907      0         793.0     9000.0    215.1
     MAR   48.8    32.4     40.6      756      0        1108.7    12582.0    300.7
     APR   61.2    42.2     51.7      399      0        1448.6    16440.0    392.9
     MAY   71.6    52.1     61.9      143     47        1687.1    19147.0    457.6
     JUN   81.1    61.6     71.4        0    197        1795.3    20375.0    487.0
     JUL   85.6    67.2     76.4        0    353        1759.9    19973.0    477.4
     AUG   83.7    65.5     74.6        0    298        1564.8    17759.0    424.5
     SEP   77.0    58.6     67.8       34    118        1272.9    14446.0    345.3
     OCT   66.9    48.1     57.5      243     11         950.9    10792.0    257.9
     NOV   54.2    38.2     46.2      564      0         596.2     6766.0    161.7
     DEC   41.5    27.4     34.5      946      0         454.4     5157.0    123.3
     ANN   62.5    45.2     53.9     5034   1024        1165.3    13225.0    316.1

  * BASED ON 1941-1970 PERIOD                   # AS NOTED IN SOLMET VOLUME 1
*******************************************************************************
```

```
*********************************************************************************
                 STATION:  ALBUQUERQUE                    STATE:  NM
                 ----------------------------------------  -----------
STATION NUMBER:  23050  LATITUDE:  3503N  LONGITUDE:  10637W  ELEVATION:  1619
----------------------  -----------------  --------------------  -------------------

                                       NORMAL DEGREE         TOTAL HEMISPHERIC
        NORMAL TEMPERATURE (DEG F)*       DAYS*         MEAN DAILY SOLAR RADIATION#

            DAILY    DAILY                BASE 65 DEG F
     MONTH MAXIMUM  MINIMUM  MONTHLY  HEATING  COOLING   BTU/FT2   KJ/M2   LANGLEYS

      JAN   46.9     23.5     35.2     924       0       1016.5   11536.0   275.7
      FEB   52.6     27.4     40.0     700       0       1342.0   15230.0   364.0
      MAR   59.2     32.3     45.8     595       0       1767.6   20060.0   479.4
      APR   70.1     41.4     55.8     282       6       2228.4   25290.0   604.4
      MAY   79.9     50.7     65.3      58      67       2538.1   28805.0   688.5
      JUN   89.5     59.7     74.6       0     291       2678.9   30403.0   726.6
      JUL   92.2     65.2     78.7       0     425       2488.6   28243.0   675.0
      AUG   89.7     63.4     76.6       0     360       2290.1   25990.0   621.2
      SEP   83.4     56.7     70.1       7     160       1971.7   22377.0   534.8
      OCT   71.7     44.7     58.2     218       7       1546.7   17553.0   419.5
      NOV   57.1     31.8     44.5     615       0       1133.7   12866.0   307.5
      DEC   47.5     24.9     36.2     893       0        927.7   10528.0   251.6
      ANN   70.0     43.5     56.8    4292    1316       1827.5   20740.0   495.7

     * BASED ON 1941-1970 PERIOD                   # AS NOTED IN SOLMET VOLUME 1
*********************************************************************************
*********************************************************************************
                 STATION:  LAS VEGAS                       STATE:  NV
                 ----------------------------------------  -----------
STATION NUMBER:  23169  LATITUDE:  3605N  LONGITUDE:  11510W  ELEVATION:   664
----------------------  -----------------  --------------------  -------------------

                                       NORMAL DEGREE         TOTAL HEMISPHERIC
        NORMAL TEMPERATURE (DEG F)*       DAYS*         MEAN DAILY SOLAR RADIATION#

            DAILY    DAILY                BASE 65 DEG F
     MONTH MAXIMUM  MINIMUM  MONTHLY  HEATING  COOLING   BTU/FT2   KJ/M2   LANGLEYS

      JAN   55.7     32.6     44.2     645       0        978.0   11099.0   265.3
      FEB   61.3     36.9     49.1     451       6       1339.5   15202.0   363.3
      MAR   67.8     41.7     54.8     324       8       1823.5   20695.0   494.6
      APR   77.5     50.0     63.8     126      90       2319.0   26318.0   629.0
      MAY   87.5     59.0     73.3      10     268       2646.3   30033.0   717.8
      JUN   97.2     67.4     82.3       0     519       2777.8   31525.0   753.5
      JUL  103.9     75.3     89.6       0     763       2588.4   29376.0   702.1
      AUG  101.5     73.3     87.4       0     694       2354.8   26725.0   638.7
      SEP   94.8     65.4     80.1       0     453       2037.3   23121.0   552.6
      OCT   81.0     53.1     67.1      74     139       1539.8   17475.0   417.7
      NOV   65.7     40.8     53.3     357       6       1085.5   12319.0   294.4
      DEC   56.7     33.7     45.2     614       0        880.5    9993.0   238.8
      ANN   79.2     52.4     65.8    2601    2946       1864.2   21157.0   505.7

     * BASED ON 1941-1970 PERIOD                   # AS NOTED IN SOLMET VOLUME 1
*********************************************************************************
*********************************************************************************
                 STATION:  RENO                            STATE:  NV
                 ----------------------------------------  -----------
STATION NUMBER:  23185  LATITUDE:  3930N  LONGITUDE:  11947W  ELEVATION:  1341
----------------------  -----------------  --------------------  -------------------

                                       NORMAL DEGREE         TOTAL HEMISPHERIC
        NORMAL TEMPERATURE (DEG F)*       DAYS*         MEAN DAILY SOLAR RADIATION#

            DAILY    DAILY                BASE 65 DEG F
     MONTH MAXIMUM  MINIMUM  MONTHLY  HEATING  COOLING   BTU/FT2   KJ/M2   LANGLEYS

      JAN   45.4     18.3     31.9    1026       0        800.4    9084.0   217.1
      FEB   51.1     23.0     37.1     781       0       1149.9   13050.0   311.9
      MAR   56.0     24.6     40.3     766       0       1649.4   18719.0   447.4
      APR   64.0     29.6     46.8     546       0       2159.3   24506.0   585.7
      MAY   72.2     37.0     54.6     328       6       2523.1   28635.0   684.4
      JUN   80.4     42.5     61.5     145      40       2701.4   30658.0   732.7
      JUL   91.1     47.4     69.3      17     150       2692.1   30552.0   730.2
      AUG   89.0     44.8     66.9      50     109       2405.7   27302.0   652.5
      SEP   81.8     38.6     60.2     168      24       1997.7   22672.0   541.9
      OCT   70.0     30.5     50.3     456       0       1431.0   16240.0   388.1
      NOV   56.3     23.9     40.1     747       0        912.3   10354.0   247.5
      DEC   46.4     19.6     33.0     992       0        705.5    8007.0   191.4
      ANN   67.0     31.7     49.4    6022     329       1760.7   19982.0   477.6

     * BASED ON 1941-1970 PERIOD                   # AS NOTED IN SOLMET VOLUME 1
*********************************************************************************
```

```
*****************************************************************************
                 STATION:  NEW YORK CITY (CENTRAL PARK)   STATE:   NY
                 -------------------------------------     ----------
STATION NUMBER:  94728  LATITUDE:  4047N  LONGITUDE:   7358W  ELEVATION:      57
-----------------------  -----------------  ---------------------  -----------------
```

| | NORMAL TEMPERATURE (DEG F)* | | | NORMAL DEGREE DAYS* | | TOTAL HEMISPHERIC MEAN DAILY SOLAR RADIATION# | | |
| | DAILY | DAILY | | BASE 65 DEG F | | | | |
MONTH	MAXIMUM	MINIMUM	MONTHLY	HEATING	COOLING	BTU/FT2	KJ/M2	LANGLEYS
JAN	38.5	25.9	32.2	1017	0	500.4	5679.0	135.7
FEB	40.2	26.5	33.4	885	0	721.0	8183.0	195.6
MAR	48.4	33.7	41.1	741	0	1037.1	11770.0	281.3
APR	60.7	43.5	52.1	387	0	1363.9	15479.0	370.0
MAY	71.4	53.1	62.3	137	54	1636.2	18569.0	443.8
JUN	80.5	62.6	71.6	0	202	1710.3	19410.0	463.9
JUL	85.2	68.0	76.6	0	360	1687.8	19155.0	457.8
AUG	83.4	66.4	74.9	0	307	1483.3	16834.0	402.3
SEP	76.8	59.9	68.4	29	131	1213.7	13774.0	329.2
OCT	66.8	50.6	58.7	209	14	895.3	10161.0	242.9
NOV	54.0	40.8	47.4	528	0	532.9	6048.0	144.6
DEC	41.4	29.5	35.5	915	0	404.0	4585.0	109.6
ANN	62.3	46.7	54.5	4848	1068	1098.9	12471.0	293.1

```
* BASED ON 1941-1970 PERIOD                    # AS NOTED IN SOLMET VOLUME 1
*****************************************************************************
*****************************************************************************
                 STATION:  SYRACUSE                      STATE:   NY
                 -------------------------------------    ----------
STATION NUMBER:  14771  LATITUDE:  4307N  LONGITUDE:   7607W  ELEVATION:     124
-----------------------  -----------------  ---------------------  -----------------
```

| | NORMAL TEMPERATURE (DEG F)* | | | NORMAL DEGREE DAYS* | | TOTAL HEMISPHERIC MEAN DAILY SOLAR RADIATION# | | |
| | DAILY | DAILY | | BASE 65 DEG F | | | | |
MONTH	MAXIMUM	MINIMUM	MONTHLY	HEATING	COOLING	BTU/FT2	KJ/M2	LANGLEYS
JAN	31.4	15.8	23.6	1283	0	385.1	4370.0	104.4
FEB	32.7	16.5	24.6	1131	0	571.3	6484.0	155.0
MAR	41.5	24.8	33.2	986	0	890.4	10105.0	241.5
APR	56.5	36.4	46.5	555	0	1323.9	15025.0	359.1
MAY	67.6	46.0	56.8	272	18	1577.9	17908.0	428.0
JUN	77.7	56.1	66.9	46	103	1777.9	20177.0	482.2
JUL	82.0	61.0	71.5	11	212	1757.7	19948.0	476.3
AUG	80.2	59.2	69.7	18	164	1503.6	17064.0	407.8
SEP	73.3	52.3	62.8	120	54	1165.3	13225.0	316.1
OCT	62.4	42.5	52.5	392	0	777.3	8822.0	210.7
NOV	48.3	33.6	41.0	720	0	398.7	4525.0	108.2
DEC	35.0	21.2	28.1	1144	0	285.3	3238.0	77.4
ANN	57.4	38.8	48.1	6678	551	1034.5	11741.0	280.6

```
* BASED ON 1941-1970 PERIOD                    # AS NOTED IN SOLMET VOLUME 1
*****************************************************************************
*****************************************************************************
                 STATION:  CINCINNATI (COVINGTON, KY)    STATE:   OH
                 -------------------------------------    ----------
STATION NUMBER:  93814  LATITUDE:  3904N  LONGITUDE:   8440W  ELEVATION:     271
-----------------------  -----------------  ---------------------  -----------------
```

| | NORMAL TEMPERATURE (DEG F)* | | | NORMAL DEGREE DAYS* | | TOTAL HEMISPHERIC MEAN DAILY SOLAR RADIATION# | | |
| | DAILY | DAILY | | BASE 65 DEG F | | | | |
MONTH	MAXIMUM	MINIMUM	MONTHLY	HEATING	COOLING	BTU/FT2	KJ/M2	LANGLEYS
JAN	39.7	22.4	31.1	1051	0	500.5	5680.0	135.8
FEB	42.7	23.8	33.3	888	0	738.4	8380.0	200.3
MAR	51.8	31.6	41.7	722	0	1027.3	11659.0	278.7
APR	65.0	42.7	53.9	341	8	1398.5	15872.0	379.3
MAY	74.4	51.9	63.2	138	82	1672.4	18980.0	453.6
JUN	83.2	61.0	72.1	9	222	1837.1	20849.0	498.3
JUL	86.5	64.6	75.6	0	329	1770.9	20098.0	480.4
AUG	85.8	63.0	74.4	0	294	1634.4	18549.0	443.3
SEP	79.7	55.9	67.8	44	128	1311.6	14885.0	355.8
OCT	68.5	45.0	56.8	271	17	989.8	11233.0	268.5
NOV	53.2	34.3	43.8	636	0	588.5	6679.0	159.6
DEC	42.0	25.3	33.7	970	0	432.5	4908.0	117.3
ANN	64.4	43.5	54.0	5070	1080	1158.5	13148.0	314.2

```
* BASED ON 1941-1970 PERIOD                    # AS NOTED IN SOLMET VOLUME 1
*****************************************************************************
```

```
*****************************************************************************
                    STATION:  OKLAHOMA CITY                STATE:   OK
                    ------------------------------------   ----------
      STATION NUMBER:  13967  LATITUDE:  3524N  LONGITUDE:  9736W  ELEVATION:    397
      -----------------------  -----------------   -------------------  -------------------

                                           NORMAL DEGREE          TOTAL HEMISPHERIC
              NORMAL TEMPERATURE (DEG F)*        DAYS*       MEAN DAILY SOLAR RADIATION#

                  DAILY     DAILY                   BASE 65 DEG F
            MONTH MAXIMUM  MINIMUM  MONTHLY  HEATING  COOLING    BTU/FT2    KJ/M2    LANGLEYS

            JAN     47.6     26.0     36.8     874        0        800.9    9089.0     217.2
            FEB     52.6     30.0     41.3     664        0       1055.0   11973.0     286.2
            MAR     59.8     36.5     48.2     532       11       1400.1   15890.0     379.8
            APR     71.6     49.1     60.4     180       42       1725.4   19581.0     468.0
            MAY     78.7     57.9     68.3      36      138       1918.1   21768.0     520.3
            JUN     87.0     66.6     76.8       0      354       2143.9   24331.0     581.5
            JUL     92.6     70.4     81.5       0      512       2128.4   24155.0     577.3
            AUG     92.5     69.6     81.1       0      499       1950.3   22134.0     529.0
            SEP     84.7     61.3     73.0      12      252       1554.2   17638.0     421.6
            OCT     74.2     50.6     62.4     148       68       1232.6   13989.0     334.3
            NOV     60.9     37.4     49.2     474        0        901.0   10225.0     244.4
            DEC     50.7     29.2     40.0     775        0        725.4    8233.0     196.8
            ANN     71.1     48.7     59.9    3695     1876       1461.3   16584.0     396.4

      * BASED ON 1941-1970 PERIOD                        # AS NOTED IN SOLMET VOLUME 1
      *****************************************************************************
      *****************************************************************************
                    STATION:  PORTLAND                     STATE:   OR
                    ------------------------------------   ----------
      STATION NUMBER:   24229  LATITUDE:  4536N  LONGITUDE:  12236W  ELEVATION:     12
      -----------------------  -----------------   -------------------  -------------------

                                           NORMAL DEGREE          TOTAL HEMISPHERIC
              NORMAL TEMPERATURE (DEG F)*        DAYS*       MEAN DAILY SOLAR RADIATION#

                  DAILY     DAILY                   BASE 65 DEG F
            MONTH MAXIMUM  MINIMUM  MONTHLY  HEATING  COOLING    BTU/FT2    KJ/M2    LANGLEYS

            JAN     43.6     32.5     38.1     834        0        310.0    3518.0      84.1
            FEB     50.1     35.5     42.8     622        0        554.1    6289.0     150.3
            MAR     54.3     37.0     45.7     598        0        895.0   10157.0     242.8
            APR     60.3     40.8     50.6     432        0       1307.7   14841.0     354.7
            MAY     67.0     46.3     56.7     264        7       1663.2   18876.0     451.1
            JUN     72.1     51.8     62.0     128       38       1772.5   20116.0     480.8
            JUL     79.0     55.2     67.1      48      114       2037.3   23121.0     552.6
            AUG     78.1     55.0     66.6      56      106       1673.7   18995.0     454.0
            SEP     73.9     50.5     62.2     119       35       1216.7   13808.0     330.0
            OCT     62.9     44.7     53.8     347        0        723.6    8212.0     196.3
            NOV     52.1     38.5     45.3     591        0        387.5    4398.0     105.1
            DEC     46.0     35.3     40.7     753        0        259.8    2949.0      70.5
            ANN     61.6     43.6     52.6    4792      300       1066.8   12107.0     289.4

      * BASED ON 1941-1970 PERIOD                        # AS NOTED IN SOLMET VOLUME 1
      *****************************************************************************
      *****************************************************************************
                    STATION:  PHILADELPHIA                 STATE:   PA
                    ------------------------------------   ----------
      STATION NUMBER:  13739  LATITUDE:  3953N  LONGITUDE:   7515W  ELEVATION:      9
      -----------------------  -----------------   -------------------  -------------------

                                           NORMAL DEGREE          TOTAL HEMISPHERIC
              NORMAL TEMPERATURE (DEG F)*        DAYS*       MEAN DAILY SOLAR RADIATION#

                  DAILY     DAILY                   BASE 65 DEG F
            MONTH MAXIMUM  MINIMUM  MONTHLY  HEATING  COOLING    BTU/FT2    KJ/M2    LANGLEYS

            JAN     40.1     24.4     32.3    1014        0        555.3    6302.0     150.6
            FEB     42.2     25.5     33.9     871        0        794.5    9017.0     215.5
            MAR     51.2     32.5     41.9     716        0       1108.2   12577.0     300.6
            APR     63.5     42.3     52.9     367        0       1433.9   16273.0     388.9
            MAY     74.1     52.3     63.2     122       67       1659.9   18838.0     450.2
            JUN     83.0     61.6     72.3       0      223       1811.2   20555.0     491.3
            JUL     86.8     66.7     76.8       0      366       1758.1   19953.0     476.9
            AUG     84.8     64.7     74.8       0      304       1574.5   17869.0     427.1
            SEP     78.4     57.8     68.1      38      131       1281.4   14542.0     347.6
            OCT     67.9     46.9     57.4     249       13        958.5   10878.0     260.0
            NOV     55.5     36.9     46.2     564        0        619.3    7028.0     168.0
            DEC     43.2     27.2     35.2     924        0        470.4    5339.0     127.6
            ANN     64.2     44.9     54.6    4865     1104       1168.7   13264.0     317.0

      * BASED ON 1941-1970 PERIOD                        # AS NOTED IN SOLMET VOLUME 1
      *****************************************************************************
```

```
**********************************************************************
               STATION:  PITTSBURGH              STATE:  PA
           --------------------------------------   ----------
STATION NUMBER:  94823  LATITUDE:  4030N  LONGITUDE:  8013W  ELEVATION:    373
-----------------------  -----------------   --------------------   ------------------


                                  NORMAL DEGREE          TOTAL HEMISPHERIC
           NORMAL TEMPERATURE (DEG F)*      DAYS*       MEAN DAILY SOLAR RADIATION#

              DAILY    DAILY               BASE 65 DEG F
       MONTH MAXIMUM  MINIMUM   MONTHLY  HEATING  COOLING    BTU/FT2    KJ/M2    LANGLEYS

        JAN   35.3     20.8      28.1     1144       0        424.4    4817.0    115.1
        FEB   37.3     21.3      29.3     1000       0        625.3    7096.0    169.6
        MAR   47.2     29.0      38.1      834       0        942.6   10697.0    255.7
        APR   60.9     39.4      50.2      444       0       1316.6   14942.0    357.1
        MAY   70.8     48.7      59.8      208      46       1601.7   18178.0    434.5
        JUN   79.5     57.7      68.6       26     134       1761.6   19992.0    477.8
        JUL   82.5     61.3      71.9        7     221       1689.2   19171.0    458.2
        AUG   80.9     59.4      70.2       16     177       1510.4   17141.0    409.7
        SEP   74.9     52.7      63.8       98      62       1208.9   13720.0    327.9
        OCT   63.9     42.4      53.2      372       7        895.0   10157.0    242.8
        NOV   49.3     33.3      41.3      711       0        504.7    5728.0    136.9
        DEC   37.3     23.6      30.5     1070       0        346.8    3936.0     94.1
        ANN   60.0     40.8      50.4     5930     647       1068.9   12131.0    289.9

    * BASED ON 1941-1970 PERIOD                    # AS NOTED IN SOLMET VOLUME 1
**********************************************************************


**********************************************************************
               STATION:  SAN JUAN                STATE:  PR
           --------------------------------------   ----------
STATION NUMBER:  11641  LATITUDE:  1826N  LONGITUDE:  6600W  ELEVATION:     19
-----------------------  -----------------   --------------------   ------------------


                                  NORMAL DEGREE          TOTAL HEMISPHERIC
           NORMAL TEMPERATURE (DEG F)*      DAYS*       MEAN DAILY SOLAR RADIATION#

              DAILY    DAILY               BASE 65 DEG F
       MONTH MAXIMUM  MINIMUM   MONTHLY  HEATING  COOLING    BTU/FT2    KJ/M2    LANGLEYS

        JAN   81.9     68.8      75.4        0     322       1325.6   15044.0    359.6
        FEB   82.1     68.4      75.3        0     288       1535.9   17431.0    416.6
        MAR   83.6     68.9      76.3        0     350       1787.9   20291.0    485.0
        APR   84.4     70.6      77.5        0     375       1890.9   21460.0    512.9
        MAY   85.6     72.8      79.2        0     440       1812.9   20575.0    491.8
        JUN   87.0     74.0      80.5        0     465       1817.0   20621.0    492.9
        JUL   87.0     74.8      80.9        0     493       1873.7   21265.0    508.2
        AUG   87.5     75.1      81.3        0     505       1838.2   20862.0    498.6
        SEP   87.6     74.6      81.1        0     483       1674.5   19004.0    454.2
        OCT   87.4     73.7      80.6        0     484       1515.3   17197.0    411.0
        NOV   85.0     72.3      78.7        0     411       1367.7   15522.0    371.0
        DEC   83.1     70.5      76.8        0     366       1235.8   14025.0    335.2
        ANN   85.2     72.0      78.6        0    4982       1639.6   18608.0    444.7

    * BASED ON 1941-1970 PERIOD                    # AS NOTED IN SOLMET VOLUME 1
**********************************************************************


**********************************************************************
               STATION:  PROVIDENCE              STATE:  RI
           --------------------------------------   ----------
STATION NUMBER:  14765  LATITUDE:  4144N  LONGITUDE:  7126W  ELEVATION:     19
-----------------------  -----------------   --------------------   ------------------


                                  NORMAL DEGREE          TOTAL HEMISPHERIC
           NORMAL TEMPERATURE (DEG F)*      DAYS*       MEAN DAILY SOLAR RADIATION#

              DAILY    DAILY               BASE 65 DEG F
       MONTH MAXIMUM  MINIMUM   MONTHLY  HEATING  COOLING    BTU/FT2    KJ/M2    LANGLEYS

        JAN   36.2     20.6      28.4     1135       0        506.2    5745.0    137.3
        FEB   37.6     21.2      29.4      997       0        738.5    8381.0    200.3
        MAR   44.7     29.0      36.9      871       0       1031.8   11710.0    279.9
        APR   56.7     37.8      47.3      531       0       1373.9   15592.0    372.7
        MAY   66.8     46.9      56.9      259       8       1655.1   18784.0    448.9
        JUN   76.3     56.5      66.4       36      78       1775.5   20150.0    481.6
        JUL   81.1     63.0      72.1        0     224       1695.4   19241.0    459.9
        AUG   79.8     61.0      70.4       10     177       1498.6   17007.0    406.5
        SEP   73.1     53.6      63.4       93      45       1208.8   13719.0    327.9
        OCT   63.9     43.4      53.7      350       0        906.7   10290.0    245.9
        NOV   52.0     34.6      43.3      651       0        537.5    6100.0    145.8
        DEC   39.6     23.4      31.5     1039       0        418.5    4750.0    113.5
        ANN   59.0     40.9      50.0     5972     532       1112.2   12622.0    301.7

    * BASED ON 1941-1970 PERIOD                    # AS NOTED IN SOLMET VOLUME 1
**********************************************************************
```

```
*****************************************************************************
                 STATION:  CHARLESTON                 STATE:  SC
    --------------------------------------------------  -----------
 STATION NUMBER:  13880  LATITUDE:  3254N  LONGITUDE:  8002W  ELEVATION:     12
 -----------------------  ------------------  --------------------  -----------------

                                      NORMAL DEGREE          TOTAL HEMISPHERIC
         NORMAL TEMPERATURE (DEG F)*      DAYS*        MEAN DAILY SOLAR RADIATION#

          DAILY    DAILY           BASE 65 DEG F
    MONTH MAXIMUM  MINIMUM  MONTHLY  HEATING  COOLING    BTU/FT2    KJ/M2    LANGLEYS

     JAN   59.8    37.3     48.6      521      12        744.2     8446.0    201.9
     FEB   61.9    39.0     50.5      419      13        995.3    11296.0    270.0
     MAR   67.8    45.1     56.5      300      36       1338.6    15192.0    363.1
     APR   76.2    53.0     64.6       69      57       1732.3    19660.0    469.9
     MAY   83.1    61.1     72.1        5     225       1860.2    21111.0    504.6
     JUN   87.7    68.1     77.9        0     387       1843.9    20926.0    500.1
     JUL   89.1    71.2     80.2        0     471       1798.9    20416.0    488.0
     AUG   88.6    70.6     79.6        0     453       1585.3    17991.0    430.0
     SEP   84.5    65.9     75.2        0     306       1394.1    15822.0    378.2
     OCT   77.1    55.1     66.1       74     108       1192.7    13536.0    323.5
     NOV   68.4    44.1     56.3      271      10        934.1    10601.0    253.4
     DEC   60.8    37.7     49.3      487       0        720.7     8179.0    195.5
     ANN   75.4    54.0     64.7     2146    2078       1345.1    15265.0    364.8

 * BASED ON 1941-1970 PERIOD                    # AS NOTED IN SOLMET VOLUME 1
 *****************************************************************************


 *****************************************************************************
                 STATION:  RAPID CITY                 STATE:  SD
    --------------------------------------------------  -----------
 STATION NUMBER:  24090  LATITUDE:  4403N  LONGITUDE:  10304W  ELEVATION:    966
 -----------------------  ------------------  --------------------  -----------------

                                      NORMAL DEGREE          TOTAL HEMISPHERIC
         NORMAL TEMPERATURE (DEG F)*      DAYS*        MEAN DAILY SOLAR RADIATION#

          DAILY    DAILY           BASE 65 DEG F
    MONTH MAXIMUM  MINIMUM  MONTHLY  HEATING  COOLING    BTU/FT2    KJ/M2    LANGLEYS

     JAN   34.2     9.6     21.9     1336       0        542.3     6154.0    147.1
     FEB   37.6    13.9     25.8     1098       0        826.5     9380.0    224.2
     MAR   42.7    19.7     31.2     1048       0       1228.8    13946.0    333.3
     APR   57.2    32.0     44.6      612       0       1589.1    18035.0    431.0
     MAY   67.4    42.9     55.2      319      15       1887.0    21415.0    511.8
     JUN   76.3    52.0     64.2      134     110       2131.2    24187.0    578.1
     JUL   86.3    58.8     72.6       13     249       2223.0    25229.0    603.0
     AUG   85.9    57.2     71.6       17     222       1962.7    22275.0    532.4
     SEP   74.7    46.3     60.5      191      56       1517.9    17227.0    411.7
     OCT   63.6    36.4     50.0      474       9       1063.6    12071.0    288.5
     NOV   47.5    23.2     35.4      888       0        646.7     7339.0    175.4
     DEC   38.0    14.9     26.5     1194       0        476.4     5407.0    129.2
     ANN   59.3    33.9     46.5     7324     661       1341.3    15222.0    363.8

 * BASED ON 1941-1970 PERIOD                    # AS NOTED IN SOLMET VOLUME 1
 *****************************************************************************


 *****************************************************************************
                 STATION:  MEMPHIS                 STATE:  TN
    --------------------------------------------------  -----------
 STATION NUMBER:  13893  LATITUDE:  3503N  LONGITUDE:  8959W  ELEVATION:     87
 -----------------------  ------------------  --------------------  -----------------

                                      NORMAL DEGREE          TOTAL HEMISPHERIC
         NORMAL TEMPERATURE (DEG F)*      DAYS*        MEAN DAILY SOLAR RADIATION#

          DAILY    DAILY           BASE 65 DEG F
    MONTH MAXIMUM  MINIMUM  MONTHLY  HEATING  COOLING    BTU/FT2    KJ/M2    LANGLEYS

     JAN   49.4    31.6     40.5      760       0        682.7     7748.0    185.2
     FEB   53.1    34.4     43.8      594       0        944.8    10722.0    256.3
     MAR   60.8    41.1     51.0      457      23       1278.1    14505.0    346.7
     APR   72.7    52.3     62.5      131      56       1638.7    18598.0    444.5
     MAY   81.2    60.6     70.9       22     205       1884.9    21392.0    511.3
     JUN   88.7    68.5     78.6        0     408       2044.6    23204.0    554.6
     JUL   91.6    71.5     81.6        0     515       1972.0    22380.0    534.9
     AUG   90.6    70.1     80.4        0     477       1824.0    20700.0    494.7
     SEP   84.3    62.8     73.6        7     265       1470.9    16693.0    399.0
     OCT   74.9    51.1     63.0      142      80       1204.5    13670.0    326.7
     NOV   61.5    40.3     50.9      423       0        816.7     9269.0    221.5
     DEC   51.7    33.7     42.7      691       0        628.6     7134.0    170.5
     ANN   71.7    51.5     61.6     3227    2029       1365.9    15501.0    370.5

                                                # AS NOTED IN SOLMET VOLUME 1
 * BASED ON 1941-1970 PERIOD
 *****************************************************************************
```

```
****************************************************************************
                STATION:   DALLAS                        STATE:   TX
                -----------------------------------      -----------
 STATION NUMBER:  13960  LATITUDE:  3251N  LONGITUDE:   9651W  ELEVATION:    149
 ----------------------  ------------------  --------------------  -----------------

                                     NORMAL DEGREE            TOTAL HEMISPHERIC
        NORMAL TEMPERATURE (DEG F)*      DAYS*         MEAN DAILY SOLAR RADIATION#

           DAILY    DAILY              BASE 65 DEG F
 MONTH   MAXIMUM  MINIMUM   MONTHLY  HEATING  COOLING    BTU/FT2    KJ/M2    LANGLEYS

  JAN     55.1     35.7      45.4      608        0        821.5    9323.0    222.8
  FEB     59.2     39.5      49.4      437        0       1071.1   12156.0    290.5
  MAR     66.4     45.2      55.8      314       29       1421.8   16136.0    385.7
  APR     76.3     56.4      66.4       71      113       1626.8   18463.0    441.3
  MAY     83.1     64.4      73.8        0      273       1888.5   21433.0    512.3
  JUN     90.6     72.6      81.6        0      498       2134.9   24229.0    579.1
  JUL     95.1     76.3      85.7        0      642       2122.1   24083.0    575.6
  AUG     95.7     75.9      85.8        0      645       1950.2   22133.0    529.0
  SEP     88.0     68.3      78.2        0      396       1587.1   18012.0    430.5
  OCT     78.4     57.5      68.0       55      148       1276.1   14482.0    346.1
  NOV     66.4     45.4      55.9      284       11        936.4   10627.0    254.0
  DEC     57.8     38.6      48.2      521        0        780.1    8853.0    211.6
  ANN     76.0     56.3      66.2     2290     2755       1468.1   16661.0    398.2

 * BASED ON 1941-1970 PERIOD                  # AS NOTED IN SOLMET VOLUME 1
 ****************************************************************************

 ****************************************************************************
                STATION:   BROWNSVILLE                   STATE:   TX
                -----------------------------------      -----------
 STATION NUMBER:  12919  LATITUDE:  2554N  LONGITUDE:   9726W  ELEVATION:      6
 ----------------------  ------------------  --------------------  -----------------

                                     NORMAL DEGREE            TOTAL HEMISPHERIC
        NORMAL TEMPERATURE (DEG F)*      DAYS*         MEAN DAILY SOLAR RADIATION#

           DAILY    DAILY              BASE 65 DEG F
 MONTH   MAXIMUM  MINIMUM   MONTHLY  HEATING  COOLING    BTU/FT2    KJ/M2    LANGLEYS

  JAN     69.5     51.0      60.3      225       79        912.8   10359.0    247.6
  FEB     72.7     54.1      63.4      151      106       1135.4   12886.0    308.0
  MAR     76.6     58.8      67.7       89      173       1457.8   16545.0    395.4
  APR     83.1     66.7      74.9        0      297       1737.2   19715.0    471.2
  MAY     87.1     71.4      79.3        0      443       1927.1   21870.0    522.7
  JUN     90.6     75.0      82.8        0      534       2115.3   24006.0    573.8
  JUL     92.8     75.9      84.4        0      601       2212.5   25109.0    600.1
  AUG     93.0     75.7      84.4        0      601       2027.3   23008.0    549.9
  SEP     89.9     73.2      81.6        0      498       1693.9   19224.0    459.5
  OCT     84.7     66.6      75.7        5      337       1438.9   16330.0    390.3
  NOV     77.5     58.7      68.1       35      128       1054.5   11968.0    286.0
  DEC     72.3     53.3      62.8      145       77        862.4    9787.0    233.9
  ANN     82.5     65.0      73.8      650     3874       1547.9   17567.0    419.9

 * BASED ON 1941-1970 PERIOD                  # AS NOTED IN SOLMET VOLUME 1
 ****************************************************************************

 ****************************************************************************
                STATION:   HOUSTON                       STATE:   TX
                -----------------------------------      -----------
 STATION NUMBER:  12960  LATITUDE:  2959N  LONGITUDE:   9522W  ELEVATION:     33
 ----------------------  ------------------  --------------------  -----------------

                                     NORMAL DEGREE            TOTAL HEMISPHERIC
        NORMAL TEMPERATURE (DEG F)*      DAYS*         MEAN DAILY SOLAR RADIATION#

           DAILY    DAILY              BASE 65 DEG F
 MONTH   MAXIMUM  MINIMUM   MONTHLY  HEATING  COOLING    BTU/FT2    KJ/M2    LANGLEYS

  JAN     62.6     41.5      52.1      416       16        772.4    8766.0    209.5
  FEB     66.0     44.6      55.3      294       22       1034.2   11737.0    280.5
  MAR     71.8     49.8      60.8      189       59       1297.4   14724.0    351.9
  APR     79.4     59.3      69.4       23      155       1522.3   17277.0    412.9
  MAY     85.9     65.6      75.8        0      335       1774.9   20143.0    481.4
  JUN     91.3     70.9      81.1        0      483       1898.1   21541.0    514.8
  JUL     93.8     72.8      83.3        0      567       1828.1   20747.0    495.9
  AUG     94.3     72.4      83.4        0      570       1686.2   19137.0    457.4
  SEP     90.1     68.2      79.2        0      426       1471.0   16694.0    399.0
  OCT     83.5     58.3      70.9       24      207       1275.6   14477.0    346.0
  NOV     73.0     49.1      61.1      155       38        924.0   10486.0    250.6
  DEC     65.8     43.4      54.6      333       11        729.6    8280.0    197.9
  ANN     79.8     58.0      68.9     1434     2889       1351.1   15334.0    366.5

 * BASED ON 1941-1970 PERIOD                  # AS NOTED IN SOLMET VOLUME 1
 ****************************************************************************
```

```
*******************************************************************************
                  STATION:  EL PASO                    STATE:  TX
                  ----------------------------------   ----------
   STATION NUMBER:  23044  LATITUDE:  3148N  LONGITUDE:  10624W  ELEVATION:  1194
   ---------------------  ------------------  -------------------  -----------------


                                      NORMAL DEGREE           TOTAL HEMISPHERIC
            NORMAL TEMPERATURE (DEG F)*      DAYS*          MEAN DAILY SOLAR RADIATION#

              DAILY    DAILY                BASE 65 DEG F
        MONTH MAXIMUM  MINIMUM  MONTHLY  HEATING  COOLING   BTU/FT2   KJ/M2    LANGLEYS

         JAN   57.0    30.2     43.6      663       0       1125.1   12769.0    305.2
         FEB   62.5    34.3     48.4      465       0       1480.1   16798.0    401.5
         MAR   68.9    40.3     54.6      328       6       1909.3   21668.0    517.9
         APR   78.5    49.3     63.9       89      56       2363.5   26823.0    641.1
         MAY   87.2    57.2     72.2        0     223       2600.6   29514.0    705.4
         JUN   94.9    65.7     80.3        0     459       2682.5   30443.0    727.6
         JUL   94.6    69.9     82.3        0     536       2450.1   27806.0    664.6
         AUG   92.8    68.2     80.5        0     481       2284.5   25927.0    619.7
         SEP   87.4    61.0     74.2        0     276       1987.1   22552.0    539.0
         OCT   78.5    49.5     64.0       92      61       1639.0   18601.0    444.6
         NOV   66.1    37.0     51.6      402       0       1243.7   14115.0    337.4
         DEC   57.8    30.9     44.4      639       0       1030.7   11697.0    279.6
         ANN   77.2    49.5     63.4     2678    2098       1899.7   21559.0    515.3

   * BASED ON 1941-1970 PERIOD                    # AS NOTED IN SOLMET VOLUME 1
*******************************************************************************


*******************************************************************************
                  STATION:  SALT LAKE CITY            STATE:  UT
                  ----------------------------------  ----------
   STATION NUMBER:  24127  LATITUDE:  4046N  LONGITUDE:  11158W  ELEVATION:  1288
   ---------------------  ------------------  -------------------  -----------------


                                      NORMAL DEGREE           TOTAL HEMISPHERIC
            NORMAL TEMPERATURE (DEG F)*      DAYS*          MEAN DAILY SOLAR RADIATION#

              DAILY    DAILY                BASE 65 DEG F
        MONTH MAXIMUM  MINIMUM  MONTHLY  HEATING  COOLING   BTU/FT2   KJ/M2    LANGLEYS

         JAN   37.4    18.5     28.0     1147       0        639.1    7253.0    173.4
         FEB   43.4    23.3     33.4      885       0        988.7   11221.0    268.2
         MAR   50.8    28.3     39.6      787       0       1454.3   16505.0    394.5
         APR   61.8    36.6     49.2      474       0       1894.3   21498.0    513.8
         MAY   72.4    44.2     58.3      237      30       2362.4   26811.0    640.8
         JUN   81.3    51.1     66.2       88     124       2560.9   29063.0    694.6
         JUL   92.8    60.5     76.7        0     363       2590.1   29395.0    702.6
         AUG   90.2    58.7     74.5        5     300       2253.6   25576.0    611.3
         SEP   80.3    49.3     64.8      105      99       1843.3   20920.0    500.0
         OCT   66.4    38.4     52.4      402      11       1293.3   14678.0    350.8
         NOV   50.0    28.1     39.1      777       0        787.9    8942.0    213.7
         DEC   39.0    21.5     30.3     1076       0        569.8    6467.0    154.6
         ANN   63.8    38.2     51.0     5983     927       1603.1   18194.0    434.8

   * BASED ON 1941-1970 PERIOD                    # AS NOTED IN SOLMET VOLUME 1
*******************************************************************************


*******************************************************************************
                  STATION:  RICHMOND                   STATE:  VA
                  ----------------------------------   ----------
   STATION NUMBER:  13740  LATITUDE:  3730N  LONGITUDE:  7720W  ELEVATION:  50
   ---------------------  ------------------  -------------------  -----------------


                                      NORMAL DEGREE           TOTAL HEMISPHERIC
            NORMAL TEMPERATURE (DEG F)*      DAYS*          MEAN DAILY SOLAR RADIATION#

              DAILY    DAILY                BASE 65 DEG F
        MONTH MAXIMUM  MINIMUM  MONTHLY  HEATING  COOLING   BTU/FT2   KJ/M2    LANGLEYS

         JAN   47.4    27.6     37.5      853       0        631.9    7171.0    171.4
         FEB   49.9    28.8     39.4      717       0        877.1    9954.0    237.9
         MAR   58.2    35.5     46.9      569       8       1210.4   13737.0    328.3
         APR   70.3    45.2     57.8      226      10       1566.0   17772.0    424.8
         MAY   78.4    54.5     66.5       64     111       1762.0   19997.0    477.9
         JUN   85.4    62.9     74.2        0     276       1872.4   21250.0    507.9
         JUL   88.2    67.5     77.9        0     400       1774.4   20138.0    481.3
         AUG   86.6    65.9     76.3        0     350       1600.6   18165.0    434.2
         SEP   80.9    59.0     70.0       21     171       1347.9   15297.0    365.6
         OCT   71.2    47.4     59.3      203      27       1032.7   11720.0    280.1
         NOV   60.6    37.3     49.0      480       0        733.0    8319.0    198.8
         DEC   49.1    28.8     39.0      806       0        566.7    6432.0    153.7
         ANN   68.8    46.7     57.8     3939    1353       1248.0   14163.0    338.5

   * BASED ON 1941-1970 PERIOD                    # AS NOTED IN SOLMET VOLUME 1
*******************************************************************************
```

```
**********************************************************************
                 STATION:  BURLINGTON                STATE:  VT
          -----------------------------------------   ----------
STATION NUMBER:   14742  LATITUDE:  4428N  LONGITUDE:   7309W  ELEVATION:   104
----------------------  -----------------  -------------------  ------------------


                                        NORMAL DEGREE            TOTAL HEMISPHERIC
        NORMAL TEMPERATURE (DEG F)*         DAYS*          MEAN DAILY SOLAR RADIATION#

          DAILY    DAILY                BASE 65 DEG F
   MONTH MAXIMUM  MINIMUM  MONTHLY  HEATING  COOLING     BTU/FT2    KJ/M2    LANGLEYS

    JAN   25.9      7.6     16.8     1494       0         385.3    4373.0     104.5
    FEB   28.2      8.9     18.6     1299       0         606.8    6887.0     164.6
    MAR   38.0     20.1     29.1     1113       0         940.2   10670.0     255.0
    APR   53.3     32.6     43.0      660       0        1296.2   14711.0     351.6
    MAY   66.1     43.5     54.8      331      15        1574.1   17864.0     427.0
    JUN   76.5     53.9     65.2       63      69        1728.9   19621.0     469.0
    JUL   81.0     58.5     69.8       20     169        1721.1   19533.0     466.8
    AUG   78.3     56.4     67.4       49     123        1475.0   16740.0     400.1
    SEP   70.0     48.6     59.3      191      20        1122.2   12736.0     304.4
    OCT   58.7     38.8     48.8      502       0         740.5    8404.0     200.9
    NOV   44.3     29.7     37.0      840       0         374.6    4251.0     101.6
    DEC   30.3     14.8     22.6     1314       0         283.2    3214.0      76.8
    ANN   54.2     34.5     44.4     7876     396        1020.7   11584.0     276.9

* BASED ON 1941-1970 PERIOD                  # AS NOTED IN SOLMET VOLUME 1
**********************************************************************

**********************************************************************
                 STATION:  SEATTLE-TACOMA            STATE:  WA
          -----------------------------------------   ----------
STATION NUMBER:   24233  LATITUDE:  4727N  LONGITUDE:  12218W  ELEVATION:   122
----------------------  -----------------  -------------------  ------------------


                                        NORMAL DEGREE            TOTAL HEMISPHERIC
        NORMAL TEMPERATURE (DEG F)*         DAYS*          MEAN DAILY SOLAR RADIATION#

          DAILY    DAILY                BASE 65 DEG F
   MONTH MAXIMUM  MINIMUM  MONTHLY  HEATING  COOLING     BTU/FT2    KJ/M2    LANGLEYS

    JAN   43.4     33.0     38.2      831       0         261.7    2970.0      71.0
    FEB   48.5     36.0     42.3      636       0         495.0    5618.0     134.3
    MAR   51.5     36.6     44.1      648       0         849.4    9640.0     230.4
    APR   57.0     40.3     48.7      489       0        1293.5   14680.0     350.9
    MAY   64.1     45.6     54.9      313       0        1713.9   19451.0     464.9
    JUN   69.0     50.6     59.8      167      11        1801.8   20449.0     488.7
    JUL   75.1     53.8     64.5       80      65        2248.2   25515.0     609.8
    AUG   73.8     53.7     63.8       82      45        1616.3   18343.0     438.4
    SEP   68.7     50.4     59.6      170       8        1147.7   13025.0     311.3
    OCT   59.4     44.9     52.2      397       0         656.2    7447.0     178.0
    NOV   50.4     38.8     44.6      612       0         337.2    3827.0      91.5
    DEC   45.4     35.5     40.5      760       0         211.1    2396.0      57.3
    ANN   58.8     43.3     51.1     5185     129        1052.7   11947.0     285.5

* BASED ON 1941-1970 PERIOD                  # AS NOTED IN SOLMET VOLUME 1
**********************************************************************

**********************************************************************
                 STATION:  MADISON                   STATE:  WI
          -----------------------------------------   ----------
STATION NUMBER:   14837  LATITUDE:  4308N  LONGITUDE:   8920W  ELEVATION:   262
----------------------  -----------------  -------------------  ------------------


                                        NORMAL DEGREE            TOTAL HEMISPHERIC
        NORMAL TEMPERATURE (DEG F)*         DAYS*          MEAN DAILY SOLAR RADIATION#

          DAILY    DAILY                BASE 65 DEG F
   MONTH MAXIMUM  MINIMUM  MONTHLY  HEATING  COOLING     BTU/FT2    KJ/M2    LANGLEYS

    JAN   25.4      8.2     16.8     1494       0         515.2    5847.0     139.7
    FEB   29.5     11.1     20.3     1252       0         804.0    9125.0     218.1
    MAR   39.2     21.2     30.2     1079       0        1136.0   12892.0     308.1
    APR   56.0     34.6     45.3      591       0        1398.4   15870.0     379.3
    MAY   67.3     44.6     56.0      297      18        1743.2   19784.0     472.8
    JUN   76.9     54.6     65.8       72      96        1947.9   22107.0     528.4
    JUL   81.4     58.8     70.1       14     172        1934.4   21953.0     524.7
    AUG   80.0     57.3     68.7       39     154        1708.1   19385.0     463.3
    SEP   70.9     48.5     59.7      173      14        1299.4   14747.0     352.5
    OCT   60.9     38.9     49.9      474       6         910.9   10338.0     247.1
    NOV   43.0     26.4     34.7      909       0         504.2    5722.0     136.8
    DEC   29.8     14.0     21.9     1336       0         388.9    4414.0     105.5
    ANN   55.0     34.8     44.9     7730     460        1190.9   13515.0     323.0

* BASED ON 1941-1970 PERIOD                  # AS NOTED IN SOLMET VOLUME 1
**********************************************************************
```

Appendix III

TABLE A3-1 Thermodynamic properties of moist air^a (standard atmospheric pressure, 29.921 in. Hg)

Fahr. Temp. t(F)	Humidity Ratio $W_s \times 10^3$	Volume cu ft/lb dry air			Enthalpy Btu/lb dry air			Entropy Btu per (°F)(lb dry air)			Condensed Water			Fahr. Temp. t(F)
		v_a	v_{as}	v_s	h_a	h_{as}	h_s	s_a	s_{as}	s_s	Enthalpy Btu/Lb h_w	Entropy Btu per (°F)(Lb) s_w	Vap. Press In. Hg $p_s \times 10^2$	
7	1.130	11.756	0.021	11.777	1.681	1.202	2.883	0.00364	0.00271	0.00635	−155.61	−0.3172	5.4022	7
8	1.189	11.781	0.022	11.803	1.922	1.266	3.188	0.00415	0.00285	0.00700	−155.13	−0.3162	5.6832	8
9	1.251	11.806	0.024	11.830	2.162	1.332	3.494	0.00467	0.00299	0.00766	−154.65	−0.3152	5.9776	9
10	1.315	11.831	0.025	11.856	2.402	1.401	3.803	0.00518	0.00314	0.00832	−154.17	−0.3141	6.2858	10
11	1.383	11.857	0.026	11.883	2.642	1.474	4.116	0.00569	0.00330	0.00899	−153.69	−0.3131	6.6085	11
12	1.454	11.882	0.028	11.910	2.882	1.550	4.432	0.00620	0.00346	0.00966	−153.21	−0.3121	6.9462	12
13	1.528	11.907	0.029	11.936	3.123	1.630	4.753	0.00671	0.00363	0.01034	−152.73	−0.3111	7.2997	13
14	1.606	11.933	0.030	11.963	3.363	1.713	5.076	0.00721	0.00380	0.01101	−152.24	−0.3100	7.6696	14
15	1.687	11.958	0.032	11.990	3.603	1.800	5.403	0.00772	0.00399	0.01171	−151.76	−0.3090	8.0565	15
16	1.772	11.983	0.034	12.017	3.843	1.892	5.735	0.00822	0.00418	0.01240	−151.27	−0.3080	8.4612	16
17	1.861	12.009	0.035	12.044	4.083	1.988	6.071	0.00873	0.00438	0.01311	−150.78	−0.3070	8.8543	17
18	1.953	12.034	0.038	12.072	4.324	2.088	6.412	0.00923	0.00459	0.01382	−150.29	−0.3059	9.3267	18
19	2.051	12.059	0.040	12.099	4.564	2.192	6.756	0.00973	0.00481	0.01454	−149.80	−0.3049	9.7889	19
20	2.152	12.084	0.042	12.126	4.804	2.302	7.106	0.01023	0.00504	0.01527	−149.31	−0.3039	10.272	20
21	2.258	12.110	0.044	12.151	5.044	2.416	7.460	0.01073	0.00528	0.01601	−148.82	−0.3029	10.777	21
22	2.369	12.135	0.046	12.181	5.284	2.536	7.820	0.01123	0.00553	0.01676	−148.33	−0.3018	11.305	22
23	2.485	12.160	0.049	12.209	5.525	2.661	8.186	0.01173	0.00579	0.01752	−147.84	−0.3008	11.856	23
24	2.606	12.186	0.051	12.237	5.765	2.792	8.557	0.01223	0.00607	0.01830	−147.34	−0.2998	12.431	24
25	2.733	12.211	0.054	12.265	6.005	2.929	8.934	0.01273	0.00635	0.01908	−146.85	−0.2988	13.032	25
26	2.865	12.236	0.057	12.293	6.245	3.072	9.317	0.01322	0.00665	0.01987	−146.35	−0.2977	13.659	26
27	3.003	12.262	0.059	12.321	6.485	3.221	9.706	0.01372	0.00696	0.02068	−145.85	−0.2967	14.313	27
28	3.147	12.287	0.062	12.349	6.726	3.377	10.103	0.01421	0.00728	0.02149	−145.36	−0.2957	14.966	28
29	3.297	12.312	0.065	12.377	6.966	3.540	10.506	0.01470	0.00761	0.02231	−144.86	−0.2947	15.709	29
30	3.454	12.338	0.068	12.406	7.206	3.709	10.915	0.01519	0.00796	0.02315	−144.36	−0.2936	16.452	30
31	3.617	12.363	0.071	12.434	7.446	3.887	11.333	0.01568	0.00832	0.02400	−143.86	−0.2926	17.227	31
32	3.788	12.388	0.075	12.463	7.686	4.072	11.758	0.01617	0.00870	0.02487	−143.36	−0.2916	18.035	32
32*	3.788	12.388	0.075	12.463	7.686	4.072	11.758	0.01617	0.00870	0.02487	0.04	0.0000	18.037	32*
33	3.944	12.413	0.079	12.492	7.927	4.242	12.169	0.01666	0.00904	0.02570	1.05	0.0020	18.778	33
34	4.107	12.438	0.082	12.520	8.167	4.418	12.585	0.01715	0.00940	0.02655	2.06	0.0041	19.546	34
35	4.275	12.464	0.085	12.549	8.407	4.601	13.008	0.01764	0.00977	0.02741	3.06	0.0061	20.342	35
36	4.450	12.489	0.089	12.578	8.647	4.791	13.438	0.01812	0.01016	0.02828	4.07	0.0081	21.166	36
37	4.631	12.514	0.093	12.607	8.887	4.987	13.874	0.01861	0.01056	0.02917	5.07	0.0102	22.020	37

^{a}Reprinted with permission from the 1977 Fundamentals Volume, ASHRAE Handbook and Product Directory.

TABLE A3-1 Thermodynamic properties of moist air[a] (standard atmospheric pressure, 29.921 in. Hg) (continued)

Fahr. Temp. $t(F)$	Humidity Ratio $W_s \times 10^3$	Volume cu ft/lb dry air			Enthalpy Btu/lb dry air			Entropy Btu per (°F) (lb dry air)			Condensed Water			Fahr. Temp. $t(F)$
		v_a	v_{as}	v_s	h_a	h_{as}	h_s	s_a	s_{as}	s_s	Enthalpy Btu/lb h_w	Entropy Btu per (°F)(Lb) s_w	Vap. Press In. Hg p_s	
38	4.818	12.540	0.097	12.637	9.128	5.191	14.319	0.01909	0.01097	0.03006	6.08	0.0122	0.22904	38
39	5.012	12.565	0.101	12.666	9.368	5.403	14.771	0.01957	0.01139	0.03096	7.08	0.0142	0.23819	39
40	5.213	12.590	0.105	12.695	9.608	5.622	15.230	0.02005	0.01183	0.03188	8.09	0.0162	0.24767	40
41	5.421	12.616	0.109	12.725	9.848	5.849	15.697	0.02053	0.01228	0.03281	9.09	0.0182	0.25748	41
42	5.638	12.641	0.114	12.755	10.088	6.084	16.172	0.02101	0.01275	0.03376	10.09	0.0202	0.26763	42
43	5.860	12.666	0.119	12.785	10.329	6.328	16.657	0.02149	0.01323	0.03472	11.10	0.0222	0.27813	43
44	6.091	12.691	0.124	12.815	10.569	6.580	17.149	0.02197	0.01373	0.03570	12.10	0.0242	0.28899	44
45	6.331	12.717	0.129	12.846	10.809	6.841	17.650	0.02245	0.01425	0.03670	13.10	0.0262	0.30023	45
46	6.578	12.742	0.134	12.876	11.049	7.112	18.161	0.02293	0.01478	0.03771	14.10	0.0282	0.31185	46
47	6.835	12.767	0.140	12.907	11.289	7.391	18.680	0.02340	0.01534	0.03874	15.11	0.0302	0.32386	47
48	7.100	12.792	0.146	12.938	11.530	7.681	19.211	0.02387	0.01591	0.03978	16.11	0.0321	0.33629	48
49	7.374	12.818	0.151	12.969	11.770	7.981	19.751	0.02434	0.01650	0.04084	17.11	0.0341	0.34913	49
50	7.658	12.843	0.158	13.001	12.010	8.291	20.301	0.02481	0.01711	0.04192	18.11	0.0361	0.36240	50
51	7.952	12.868	0.164	13.032	12.250	8.612	20.862	0.02528	0.01774	0.04302	19.11	0.0381	0.37611	51
52	8.256	12.894	0.170	13.064	12.491	8.945	21.436	0.02575	0.01839	0.04414	20.11	0.0400	0.39028	52
53	8.569	12.919	0.178	13.097	12.731	9.289	22.020	0.02622	0.01906	0.04528	21.12	0.0420	0.40492	53
54	8.894	12.944	0.185	13.129	12.971	9.644	22.615	0.02669	0.01976	0.04645	22.12	0.0439	0.42004	54
55	9.229	12.970	0.192	13.162	13.211	10.01	23.22	0.02716	0.02047	0.04763	23.12	0.0459	0.43565	55
56	9.575	12.995	0.200	13.195	13.452	10.39	23.84	0.02762	0.02121	0.04883	24.12	0.0478	0.45176	56
57	9.934	13.020	0.208	13.228	13.692	10.79	24.48	0.02809	0.02197	0.05006	25.12	0.0497	0.46840	57
58	10.30	13.045	0.216	13.261	13.932	11.19	25.12	0.02855	0.02276	0.05131	26.12	0.0517	0.48558	58
59	10.69	13.071	0.224	13.295	14.172	11.61	25.78	0.02902	0.02357	0.05259	27.12	0.0536	0.50330	59
60	11.08	13.096	0.233	13.329	14.413	12.05	26.46	0.02948	0.02441	0.05389	28.12	0.0555	0.52159	60
61	11.49	13.121	0.242	13.363	14.653	12.50	27.15	0.02994	0.02527	0.05521	29.12	0.0574	0.54047	61
62	11.91	13.147	0.251	13.398	14.893	12.96	27.85	0.03040	0.02616	0.05656	30.12	0.0594	0.55994	62
63	12.35	13.172	0.261	13.433	15.134	13.44	28.57	0.03086	0.02708	0.05794	31.12	0.0613	0.58002	63
64	12.80	13.197	0.271	13.468	15.374	13.94	29.31	0.03132	0.02803	0.05935	32.12	0.0632	0.60073	64
65	13.26	13.222	0.282	13.504	15.614	14.45	30.06	0.03177	0.02901	0.06078	33.11	0.0651	0.62209	65
66	13.74	13.247	0.292	13.539	15.855	14.98	30.83	0.03223	0.03002	0.06225	34.11	0.0670	0.64411	66
67	14.21	13.273	0.303	13.576	16.095	15.53	31.62	0.03269	0.03106	0.06375	35.11	0.0689	0.66681	67
68	14.75	13.298	0.315	13.613	16.335	16.09	32.42	0.03314	0.03213	0.06527	36.11	0.0708	0.69019	68
69	15.28	13.323	0.327	13.650	16.576	16.67	33.25	0.03360	0.03323	0.06683	37.11	0.0727	0.71430	69

a Compiled by John A. Goff and S. Gratch.

* Extrapolated to represent metastable equilibrium with undercooled liquid.

[a]Reprinted with permission from the 1977 Fundamentals Volume, ASHRAE Handbook and Product Directory.

TABLE A3-1 Thermodynamic properties of moist air^a (standard atmospheric pressure, 29.921 in. Hg) (continued)

Fahr. Temp. t(F)	Humidity Ratio $W_s \times 10^2$	Volume cu ft/lb dry air			Enthalpy Btu/lb dry air			Entropy Btu per (°F)(lb dry air)			Condensed Water			Fahr. Temp. t(F)
		v_a	v_{as}	v_s	h_a	h_{as}	h_s	s_a	s_{as}	s_s	Enthalpy Btu/lb h_w	Entropy Btu per (°F)(Lb) s_w	Vap. Press In. Hg p_s	
70	1.582	13.348	0.339	13.687	16.816	17.27	34.09	0.03405	0.03437	0.06842	38.11	0.0746	0.73915	70
71	1.639	13.373	0.351	13.724	17.056	17.89	34.95	0.03450	0.03554	0.07004	39.11	0.0765	0.76475	71
72	1.697	13.398	0.364	13.762	17.297	18.53	35.83	0.03495	0.03675	0.07170	40.11	0.0784	0.79112	72
73	1.757	13.424	0.377	13.801	17.537	19.20	36.74	0.03540	0.03800	0.07340	41.11	0.0803	0.81828	73
74	1.819	13.449	0.392	13.841	17.778	19.88	37.66	0.03585	0.03928	0.07513	42.10	0.0821	0.84624	74
75	1.882	13.474	0.407	13.881	18.018	20.59	38.61	0.03630	0.04060	0.07690	43.10	0.0840	0.87504	75
76	1.948	13.499	0.422	13.921	18.259	21.31	39.57	0.03675	0.04197	0.07872	44.10	0.0859	0.90470	76
77	2.016	13.525	0.437	13.962	18.499	22.07	40.57	0.03720	0.04337	0.08057	45.10	0.0877	0.93523	77
78	2.086	13.550	0.453	14.003	18.740	22.84	41.58	0.03765	0.04482	0.08247	46.10	0.0896	0.96665	78
79	2.158	13.575	0.470	14.045	18.980	23.64	42.62	0.03810	0.04631	0.08441	47.10	0.0914	0.99899	79
80	2.233	13.601	0.486	14.087	19.221	24.47	43.69	0.03854	0.04784	0.08638	48.10	0.0933	1.0323	80
81	2.310	13.626	0.504	14.130	19.461	25.32	44.78	0.03899	0.04942	0.08841	49.09	0.0952	1.0665	81
82	2.389	13.651	0.523	14.174	19.702	26.20	45.90	0.03943	0.05105	0.09048	50.09	0.0970	1.1017	82
83	2.471	13.676	0.542	14.218	19.942	27.10	47.04	0.03987	0.05273	0.09260	51.09	0.0989	1.1379	83
84	2.555	13.702	0.560	14.262	20.183	28.04	48.22	0.04031	0.05446	0.09477	52.09	0.1007	1.1752	84
85	2.642	13.727	0.581	14.308	20.423	29.01	49.43	0.04075	0.05624	0.09699	53.09	0.1025	1.2135	85
86	2.731	13.752	0.602	14.354	20.663	30.00	50.66	0.04119	0.05807	0.09926	54.08	0.1043	1.2529	86
87	2.824	13.777	0.624	14.401	20.904	31.03	51.93	0.04163	0.05995	0.10158	55.08	0.1062	1.2934	87
88	2.919	13.803	0.645	14.448	21.144	32.09	53.23	0.04207	0.06189	0.10396	56.08	0.1080	1.3351	88
89	3.017	13.828	0.668	14.496	21.385	33.18	54.56	0.04251	0.06389	0.10640	57.08	0.1098	1.3779	89
90	3.118	13.853	0.692	14.545	21.625	34.31	55.93	0.04295	0.06596	0.10890	58.08	0.1116	1.4219	90
91	3.223	13.879	0.716	14.595	21.865	35.47	57.33	0.04339	0.06807	0.11146	59.07	0.1135	1.4671	91
92	3.330	13.904	0.741	14.645	22.106	36.67	58.78	0.04382	0.07025	0.11407	60.07	0.1153	1.5135	92
93	3.441	13.929	0.768	14.697	22.346	37.90	60.25	0.04426	0.07249	0.11675	61.07	0.1171	1.5612	93
94	3.556	13.954	0.795	14.749	22.587	39.18	61.77	0.04469	0.07480	0.11949	62.07	0.1188	1.6102	94
95	3.673	13.980	0.822	14.802	22.827	40.49	63.32	0.04513	0.07718	0.12231	63.07	0.1206	1.6606	95
96	3.795	14.005	0.851	14.856	23.068	41.85	64.92	0.04556	0.07963	0.12519	64.06	0.1224	1.7123	96
97	3.920	14.030	0.881	14.911	23.308	43.24	66.55	0.04600	0.08215	0.12815	65.06	0.1242	1.7654	97
98	4.049	14.056	0.911	14.967	23.548	44.68	68.23	0.04643	0.08474	0.13117	66.06	0.1260	1.8199	98
99	4.182	14.081	0.942	15.023	23.789	46.17	69.96	0.04686	0.08741	0.13427	67.06	0.1278	1.8759	99
100	4.319	14.106	0.975	15.081	24.029	47.70	71.73	0.04729	0.09016	0.13745	68.06	0.1296	1.9333	100
101	4.460	14.131	1.009	15.140	24.270	49.28	73.55	0.04772	0.09299	0.14071	69.05	0.1314	1.9923	101
102	4.606	14.157	1.043	15.200	24.510	50.91	75.42	0.04815	0.09591	0.14406	70.05	0.1332	2.0528	102
103	4.756	14.182	1.079	15.261	24.751	52.59	77.34	0.04858	0.09891	0.14749	71.05	0.1350	2.1149	103
104	4.911	14.207	1.117	15.324	24.991	54.32	79.31	0.04900	0.1020	0.1510	72.05	0.1367	2.1786	104

^{a}Reprinted with permission from the 1977 Fundamentals Volume, ASHRAE Handbook and Product Directory.

TABLE A3-1 Thermodynamic properties of moist air[a] (standard atmospheric pressure, 29.921 in. Hg) (continued)

Fahr. Temp. t(F)	Humidity Ratio $W_s \times 10$	Volume cu ft/lb dry air			Enthalpy Btu/lb dry air			Entropy Btu per (°F) (lb dry air)			Condensed Water			Fahr. Temp. t(F)
		v_a	v_{as}	v_s	h_a	h_{as}	h_s	s_a	s_{as}	s_s	Enthalpy Btu/Lb h_w	Entropy Btu per (°F)(Lb) s_w	Vap. Press In. Hg p_s	
105	0.5070	14.232	1.155	15.387	25.232	56.11	81.34	0.04943	0.1052	0.1546	73.04	0.1385	2.2439	105
106	0.5234	14.258	1.194	15.452	25.472	57.95	83.42	0.04985	0.1085	0.1584	74.04	0.1403	2.3109	106
107	0.5404	14.283	1.235	15.518	25.713	59.85	85.56	0.05028	0.1118	0.1621	75.04	0.1421	2.3797	107
108	0.5578	14.308	1.278	15.586	25.953	61.80	87.76	0.05070	0.1153	0.1660	76.04	0.1438	2.4502	108
109	0.5758	14.333	1.321	15.654	26.194	63.82	90.03	0.05113	0.1189	0.1700	77.04	0.1456	2.5225	109
110	0.5944	14.359	1.365	15.724	26.434	65.91	92.34	0.05155	0.1226	0.1742	78.03	0.1472	2.5966	110
111	0.6135	14.384	1.412	15.796	26.675	68.05	94.72	0.05197	0.1264	0.1784	79.03	0.1491	2.6726	111
112	0.6333	14.409	1.460	15.869	26.915	70.27	97.18	0.05239	0.1302	0.1826	80.03	0.1508	2.7505	112
113	0.6536	14.435	1.509	15.944	27.156	72.55	99.71	0.05281	0.1342	0.1870	81.03	0.1525	2.8304	113
114	0.6746	14.460	1.560	16.020	27.397	74.91	102.31	0.05323	0.1384	0.1916	82.03	0.1543	2.9123	114
115	0.6962	14.485	1.613	16.098	27.637	77.34	104.98	0.05365	0.1426	0.1963	83.02	0.1560	2.9962	115
116	0.7185	14.510	1.668	16.178	27.878	79.85	107.73	0.05407	0.1470	0.2011	84.02	0.1577	3.0821	116
117	0.7415	14.536	1.723	16.259	28.119	82.43	110.55	0.05449	0.1515	0.2060	85.02	0.1595	3.1701	117
118	0.7652	14.561	1.782	16.343	28.359	85.10	113.46	0.05490	0.1562	0.2111	86.02	0.1612	3.2603	118
119	0.7897	14.586	1.842	16.428	28.600	87.86	116.46	0.05532	0.1610	0.2163	87.02	0.1629	3.3527	119
120	0.8149	14.611	1.905	16.516	28.841	90.70	119.54	0.05573	0.1659	0.2216	88.01	0.1646	3.4474	120
121	0.8410	14.637	1.968	16.605	29.082	93.64	122.72	0.05615	0.1710	0.2272	89.01	0.1664	3.5443	121
122	0.8678	14.662	2.034	16.696	29.322	96.66	125.98	0.05656	0.1763	0.2329	90.01	0.1681	3.6436	122
123	0.8955	14.687	2.103	16.790	29.563	99.79	129.35	0.05698	0.1817	0.2387	91.01	0.1698	3.7452	123
124	0.9242	14.712	2.174	16.886	29.804	103.0	132.8	0.05739	0.1872	0.2446	92.01	0.1715	3.8493	124
125	0.9537	14.738	2.247	16.985	30.044	106.4	136.4	0.05780	0.1930	0.2508	93.01	0.1732	3.9558	125
126	0.9841	14.763	2.323	17.086	30.285	109.8	140.1	0.05821	0.1989	0.2571	94.01	0.1749	4.0649	126
127	1.016	14.788	2.401	17.189	30.536	113.4	143.9	0.05862	0.2050	0.2636	95.00	0.1766	4.1765	127
128	1.048	14.813	2.482	17.295	30.766	117.0	147.8	0.05903	0.2113	0.2703	96.00	0.1783	4.2907	128
129	1.082	14.839	2.565	17.404	31.007	120.8	151.8	0.05944	0.2178	0.2772	97.00	0.1800	4.4076	129
130	1.116	14.864	2.652	17.516	31.248	124.7	155.9	0.05985	0.2245	0.2844	98.00	0.1817	4.5272	130
131	1.152	14.889	2.742	17.631	31.489	128.8	160.3	0.06026	0.2314	0.2917	99.00	0.1834	4.6495	131
132	1.189	14.915	2.834	17.749	31.729	133.0	164.7	0.06067	0.2386	0.2993	100.00	0.1851	4.7747	132
133	1.227	14.940	2.930	17.870	31.970	137.3	169.3	0.06108	0.2459	0.3070	101.00	0.1868	4.9028	133
134	1.267	14.965	3.029	17.994	32.211	141.8	174.0	0.06148	0.2536	0.3151	102.00	0.1885	5.0337	134

[a] Compiled by John A. Goff and S. Gratch.

[a] Reprinted with permission from the 1977 Fundamentals Volume, ASHRAE Handbook and Product Directory.

TABLE A3-1 Thermodynamic properties of moist air^a (standard atmospheric pressure, 29.921 in. Hg) (continued)

Fahr. Temp. t(F)	Humidity Ratio W_s	Volume cu ft/lb dry air v_a	v_{as}	v_s	Enthalpy Btu/lb dry air h_a	h_{as}	h_s	Entropy Btu per (°F) (lb dry air) s_a	s_{as}	s_s	Condensed Water Enthalpy Btu/lb h_w	Entropy Btu per (°F)(Lb) s_w	Vap. Press In. Hg p_s	Fahr. Temp. t(F)
135	0.1308	14.990	3.132	18.122	32.452	146.4	178.9	0.06189	0.2614	0.3233	103.00	0.1902	5.1676	135
136	0.1350	15.016	3.237	18.253	32.692	151.2	183.9	0.06229	0.2695	0.3318	104.00	0.1918	5.3046	136
137	0.1393	15.041	3.348	18.389	32.933	156.1	189.0	0.06270	0.2778	0.3405	105.00	0.1935	5.4446	137
138	0.1439	15.066	3.462	18.528	33.174	161.2	194.4	0.06310	0.2865	0.3496	106.00	0.1952	5.5878	138
139	0.1485	15.091	3.580	18.671	33.414	166.5	199.9	0.06350	0.2954	0.3589	107.00	0.1969	5.7342	139
140	0.1534	15.117	3.702	18.819	33.655	172.0	205.7	0.06390	0.3047	0.3686	107.99	0.1985	5.8838	140
141	0.1584	15.142	3.829	18.971	33.896	177.7	211.6	0.06430	0.3142	0.3785	108.99	0.2002	6.0367	141
142	0.1636	15.167	3.961	19.128	34.136	183.6	217.7	0.06470	0.3241	0.3888	109.99	0.2018	6.1930	142
143	0.1689	15.192	4.098	19.290	34.377	189.7	224.1	0.06510	0.3343	0.3994	110.99	0.2035	6.3527	143
144	0.1745	15.218	4.239	19.457	34.618	196.0	230.6	0.06549	0.3449	0.4104	111.99	0.2051	6.5160	144
145	0.1803	15.243	4.386	19.629	34.859	202.5	237.4	0.06589	0.3559	0.4218	112.99	0.2068	6.6828	145
146	0.1862	15.268	4.539	19.807	35.099	209.3	244.4	0.06629	0.3672	0.4335	113.99	0.2084	6.8532	146
147	0.1924	15.293	4.698	19.991	35.340	216.4	251.7	0.06669	0.3790	0.4457	114.99	0.2101	7.0273	147
148	0.1989	15.319	4.862	20.181	35.581	223.7	259.3	0.06708	0.3912	0.4583	115.99	0.2117	7.2051	148
149	0.2055	15.344	5.033	20.377	35.822	231.3	267.1	0.06748	0.4038	0.4713	116.99	0.2134	7.3867	149
150	0.2125	15.369	5.211	20.580	36.063	239.2	275.3	0.06787	0.4169	0.4848	117.99	0.2150	7.5722	150
151	0.2197	15.394	5.396	20.790	36.304	247.3	283.6	0.06827	0.4304	0.4987	118.99	0.2167	7.7616	151
152	0.2271	15.420	5.587	21.007	36.545	255.9	292.4	0.06866	0.4445	0.5132	119.99	0.2183	7.9550	152
153	0.2349	15.445	5.788	21.233	36.785	264.7	301.5	0.06906	0.4591	0.5282	120.99	0.2200	8.1525	153
154	0.2430	15.470	5.996	21.466	37.026	273.9	310.9	0.06945	0.4743	0.5438	121.99	0.2216	8.3541	154
155	0.2514	15.496	6.213	21.709	37.267	283.5	320.8	0.06984	0.4901	0.5599	122.99	0.2232	8.5599	155
156	0.2602	15.521	6.439	21.960	37.508	293.5	331.0	0.07023	0.5066	0.5768	123.99	0.2248	8.7701	156
157	0.2693	15.546	6.675	22.221	37.749	303.9	341.7	0.07062	0.5237	0.5943	124.99	0.2265	8.9846	157
158	0.2788	15.571	6.922	22.493	37.990	314.7	352.7	0.07101	0.5415	0.6125	125.99	0.2281	9.2036	158
159	0.2887	15.597	7.178	22.775	38.231	326.0	364.2	0.07140	0.5600	0.6314	127.00	0.2297	9.4271	159
160	0.2990	15.622	7.446	23.068	38.472	337.8	376.3	0.07179	0.5793	0.6511	128.00	0.2313	9.6556	160
161	0.3098	15.647	7.727	23.374	38.713	350.1	388.8	0.07218	0.5994	0.6716	129.00	0.2329	9.8876	161
162	0.3211	15.672	8.020	23.692	38.954	363.0	402.0	0.07257	0.6204	0.6930	130.00	0.2345	10.125	162
163	0.3329	15.698	8.326	24.024	39.195	376.5	415.7	0.07296	0.6423	0.7153	131.00	0.2361	10.367	163
164	0.3452	15.723	8.648	24.371	39.436	390.5	429.9	0.07334	0.6652	0.7385	132.00	0.2377	10.614	164
165	0.3581	15.748	8.985	24.733	39.677	405.3	445.0	0.07373	0.6892	0.7629	133.00	0.2393	10.866	165
166	0.3716	15.773	9.339	25.112	39.918	420.8	460.7	0.07411	0.7142	0.7883	134.00	0.2409	11.123	166
167	0.3858	15.799	9.708	25.507	40.159	437.0	477.2	0.07450	0.7405	0.8150	135.01	0.2426	11.385	167
168	0.4007	15.824	10.098	25.922	40.400	454.0	494.4	0.07488	0.7680	0.8429	136.01	0.2441	11.652	168
169	0.4163	15.849	10.508	26.357	40.641	471.8	512.4	0.07527	0.7969	0.8722	137.01	0.2457	11.925	169

^{a}Reprinted with permission from the 1977 Fundamentals Volume, ASHRAE Handbook and Product Directory.

TABLE A3-1 Thermodynamic properties of moist air^a (standard atmospheric pressure, 29.921 in. Hg) (continued)

Fahr. Temp. t(F)	Humidity Ratio W_s	Volume cu ft/lb dry air			Enthalpy Btu/lb dry air			Entropy Btu per (°F) (lb dry air)			Condensed Water			Fahr. Temp. t(F)
		v_a	v_{as}	v_s	h_a	h_{as}	h_s	s_a	s_{as}	s_s	Enthalpy Btu/Lb h_w	Entropy Btu/(Lb)(°F) s_w	Vap. Press In. Hg p_s	
170	0.4327	15.874	10.938	26.812	40.882	490.6	531.5	0.07565	0.8273	0.9030	138.01	0.2473	12.203	170
171	0.4500	15.900	11.391	27.291	41.123	510.4	551.5	0.07603	0.8592	0.9352	139.01	0.2489	12.486	171
172	0.4682	15.925	11.870	27.795	41.364	531.3	572.7	0.07641	0.8927	0.9691	140.01	0.2505	12.775	172
173	0.4875	15.950	12.376	28.326	41.605	553.3	594.9	0.07680	0.9281	1.0049	141.01	0.2521	13.069	173
174	0.5078	15.975	12.911	28.886	41.846	576.5	618.3	0.07718	0.9654	1.0426	142.02	0.2537	13.369	174
175	0.5292	16.001	13.475	29.476	42.087	601.1	643.2	0.07756	1.005	1.083	143.02	0.2553	13.675	175
176	0.5519	16.026	14.074	30.100	42.328	627.1	669.4	0.07794	1.047	1.125	144.02	0.2568	13.987	176
177	0.5760	16.051	14.710	30.761	42.569	654.7	697.3	0.07832	1.091	1.169	145.02	0.2584	14.304	177
178	0.6016	16.076	15.386	31.462	42.810	684.1	726.9	0.07870	1.137	1.216	146.03	0.2600	14.628	178
179	0.6288	16.102	16.104	32.206	43.051	715.2	758.3	0.07908	1.187	1.266	147.03	0.2616	14.958	179
180	0.6578	16.127	16.870	32.997	43.292	748.5	791.8	0.07946	1.240	1.319	148.03	0.2631	15.294	180
181	0.6887	16.152	17.689	33.841	43.534	783.9	827.4	0.07984	1.296	1.376	149.03	0.2647	15.636	181
182	0.7218	16.177	18.565	34.742	43.775	821.9	865.7	0.08021	1.357	1.437	150.04	0.2662	15.985	182
183	0.7572	16.203	19.504	35.707	44.016	862.5	906.5	0.08059	1.421	1.502	151.04	0.2678	16.340	183
184	0.7953	16.228	20.513	36.741	44.257	906.2	950.5	0.08096	1.490	1.571	152.04	0.2693	16.702	184
185	0.8363	16.253	21.601	37.854	44.498	953.2	997.7	0.08134	1.565	1.646	153.05	0.2709	17.071	185
186	0.8805	16.278	22.775	39.053	44.740	1004	1049	0.08171	1.645	1.727	154.05	0.2724	17.446	186
187	0.9283	16.304	24.047	40.351	44.981	1059	1104	0.08208	1.731	1.813	155.05	0.2740	17.828	187
188	0.9802	16.329	25.427	41.756	45.222	1119	1164	0.08245	1.825	1.907	156.06	0.2755	18.217	188
189	1.037	16.354	26.934	43.288	45.463	1184	1229	0.08283	1.928	2.011	157.06	0.2771	18.614	189
190	1.099	16.379	28.580	44.959	45.704	1255	1301	0.08320	2.039	2.122	158.07	0.2786	19.017	190
191	1.166	16.405	30.385	46.790	45.946	1332	1378	0.08357	2.161	2.245	159.07	0.2802	19.427	191
192	1.241	16.430	32.375	48.805	46.187	1418	1464	0.08394	2.296	2.380	160.07	0.2817	19.845	192
193	1.324	16.455	34.581	51.036	46.428	1513	1559	0.08431	2.444	2.528	161.08	0.2833	20.271	193
194	1.416	16.480	37.036	53.516	46.670	1619	1666	0.08468	2.609	2.694	162.08	0.2848	20.704	194
195	1.519	16.506	39.785	56.291	46.911	1737	1784	0.08505	2.794	2.879	163.09	0.2864	21.145	195
196	1.635	16.531	42.885	59.416	47.153	1871	1918	0.08542	3.002	3.087	164.09	0.2879	21.594	196
197	1.767	16.556	46.402	62.958	47.394	2022	2069	0.08579	3.238	3.324	165.10	0.2895	22.020	197
198	1.917	16.581	50.426	67.007	47.636	2195	2243	0.08616	3.507	3.593	166.10	0.2910	22.514	198
199	2.091	16.607	55.074	71.681	47.877	2395	2443	0.08653	3.817	3.904	167.11	0.2925	22.974	199
200	2.295	16.632	60.510	77.142	48.119	2629	2677	0.08689	4.179	4.266	168.11	0.2940	23.468	200

^a Compiled by John A. Goff and S. Gratch.

^aReprinted with permission from the 1977 Fundamentals Volume, ASHRAE Handbook and Product Directory.

TABLE A3-2 Thermodynamic properties of water at saturation[a]

Fahr. Temp. t(F)	Absolute Pressure P		Specific Volume, cu ft per lb			Enthalpy, Btu per lb			Entropy, Btu per (lb) (°F)			Fahr. Temp. t(F)
	Lb/Sq In.	In. Hg	Sat. Liquid v_f	Evap. v_{fg}	Sat. Vapor v_g	Sat. Liquid h_f	Evap. h_{fg}	Sat. Vapor h_g	Sat. Liquid s_f	Evap. s_{fg}	Sat. Vapor s_g	
72	0.38856	0.79113	0.01606	813.95	813.97	40.07	1052.58	1092.65	0.07834	1.9797	2.0580	72
73	0.40190	0.81829	0.01606	788.38	788.40	41.07	1052.01	1093.08	0.08022	1.9749	2.0551	73
74	0.41564	0.84626	0.01603	763.73	763.75	42.06	1051.46	1093.52	0.08209	1.9701	2.0522	74
75	0.42979	0.87506	0.01606	739.95	739.97	43.06	1050.89	1093.95	0.08396	1.9654	2.0494	75
76	0.44435	0.90472	0.01606	717.01	717.03	44.06	1050.32	1094.38	0.08582	1.9607	2.0465	76
77	0.45935	0.93524	0.01607	694.88	694.90	45.06	1049.76	1094.82	0.08769	1.9560	2.0437	77
78	0.47478	0.96666	0.01607	673.52	673.54	46.06	1049.19	1095.25	0.08954	1.9513	2.0408	78
79	0.49066	0.99900	0.01607	652.91	652.93	47.06	1048.62	1095.68	0.09140	1.9466	2.0380	79
80	0.50701	1.0323	0.01607	633.01	633.03	48.05	1048.07	1096.12	0.09325	1.9419	2.0352	80
81	0.52382	1.0665	0.01608	613.80	613.82	49.05	1047.50	1096.55	0.09510	1.9373	2.0324	81
82	0.54112	1.1017	0.01608	595.25	595.27	50.05	1046.93	1096.98	0.09694	1.9328	2.0297	82
83	0.55892	1.1380	0.01608	577.34	577.36	51.05	1046.37	1097.42	0.09878	1.9281	2.0269	83
84	0.57722	1.1752	0.01608	560.01	560.06	52.05	1045.80	1097.85	0.10062	1.9236	2.0242	84
85	0.59604	1.2136	0.01609	543.33	543.35	53.05	1045.23	1098.28	0.10246	1.9189	2.0214	85
86	0.61540	1.2530	0.01609	527.19	527.21	54.04	1044.67	1098.71	0.10429	1.9144	2.0187	86
87	0.63530	1.2935	0.01609	511.60	511.62	55.04	1044.10	1099.14	0.10611	1.9099	2.0160	87
88	0.65575	1.3351	0.01610	496.52	496.54	56.04	1043.54	1099.58	0.10794	1.9054	2.0133	88
89	0.67678	1.3779	0.01610	481.96	481.98	57.04	1042.97	1100.01	0.10976	1.9008	2.0106	89
90	0.69938	1.4219	0.01610	467.88	467.90	58.04	1042.40	1100.44	0.11158	1.8963	2.0079	90
91	0.72059	1.4671	0.01610	454.26	454.28	59.03	1041.84	1100.87	0.11339	1.8919	2.0053	91
92	0.74340	1.5136	0.01611	441.10	441.12	60.03	1041.27	1101.30	0.11520	1.8874	2.0026	92
93	0.76684	1.5613	0.01611	428.38	428.40	61.03	1040.70	1101.73	0.11701	1.8830	2.0000	93
94	0.79091	1.6103	0.01611	416.07	416.09	62.03	1040.13	1102.16	0.11881	1.8786	1.9974	94
95	0.81564	1.6607	0.01612	404.17	404.19	63.03	1039.56	1102.59	0.12061	1.8741	1.9947	95
96	0.84103	1.7124	0.01612	392.65	392.67	64.02	1039.00	1103.02	0.12241	1.8698	1.9922	96
97	0.86711	1.7655	0.01612	381.51	381.53	65.02	1038.43	1103.45	0.12420	1.8654	1.9896	97
98	0.89388	1.8200	0.01612	370.73	370.75	66.02	1037.86	1103.88	0.12600	1.8610	1.9870	98
99	0.92137	1.8759	0.01613	360.30	360.32	67.02	1037.29	1104.31	0.12778	1.8566	1.9844	99
100	0.94959	1.9334	0.01613	350.20	350.22	68.02	1036.72	1104.74	0.12957	1.8523	1.9819	100
101	0.97854	1.9923	0.01614	340.42	340.44	69.01	1036.16	1105.17	0.13135	1.8480	1.9793	101
102	1.0083	2.0529	0.01614	330.96	330.98	70.01	1035.58	1105.59	0.13313	1.8437	1.9768	102
103	1.0388	2.1149	0.01614	321.80	321.82	71.01	1035.01	1106.02	0.13490	1.8394	1.9743	103
104	1.0700	2.1786	0.01614	312.93	312.95	72.01	1034.44	1106.45	0.13667	1.8351	1.9718	104
105	1.1021	2.2440	0.01615	304.34	304.36	73.01	1033.87	1106.88	0.13844	1.8309	1.9693	105
106	1.1351	2.3110	0.01615	296.02	296.04	74.01	1033.29	1107.30	0.14021	1.8266	1.9668	106
107	1.1688	2.3798	0.01616	287.96	287.98	75.00	1032.73	1107.73	0.14197	1.8224	1.9644	107

[a]Reprinted with permission from the 1977 Fundamentals Volume, ASHRAE Handbook and Product Directory.

TABLE A3-2 Thermodynamic properties of water at saturation^a (continued)

Fahr. Temp. t(F)	Absolute Pressure P_s		Specific Volume, cu ft per lb			Enthalpy, Btu per lb			Entropy, Btu per (lb) (°F)			Fahr. Temp. t(F)
	Lb/Sq In.	In. Hg	Sat. Liquid v_f	Evap. v_{fg}	Sat. Vapor v_g	Sat. Liquid h_f	Evap. h_{fg}	Sat. Vapor h_g	Sat. Liquid s_f	Evap. s_{fg}	Sat. Vapor s_g	
108	1.2035	2.4503	0.01616	280.14	280.16	76.00	1032.16	1108.16	0.14373	1.8182	1.9619	108
109	1.2390	2.5226	0.01616	272.58	272.60	77.00	1031.58	1108.58	0.14549	1.8140	1.9595	109
110	1.2754	2.5968	0.01617	265.24	265.26	78.00	1031.01	1109.01	0.14724	1.8098	1.9570	110
111	1.3128	2.6728	0.01617	258.14	258.16	79.00	1030.44	1109.44	0.14899	1.8056	1.9546	111
112	1.3510	2.7507	0.01617	251.25	251.27	80.00	1029.86	1109.86	0.15074	1.8015	1.9522	112
113	1.3902	2.8306	0.01618	244.57	244.59	80.99	1029.30	1110.29	0.15248	1.7973	1.9498	113
114	1.4305	2.9125	0.01618	238.10	238.12	81.99	1028.72	1110.71	0.15423	1.7932	1.9474	114
115	1.4717	2.9963	0.01618	231.82	231.84	82.99	1028.15	1111.14	0.15596	1.7890	1.9450	115
116	1.5139	3.0823	0.01619	225.73	225.75	83.99	1027.57	1111.56	0.15770	1.7849	1.9426	116
117	1.5571	3.1703	0.01619	219.83	219.85	84.99	1026.99	1111.98	0.15943	1.7809	1.9403	117
118	1.6014	3.2606	0.01620	214.10	214.12	85.99	1026.42	1112.41	0.16116	1.7767	1.9379	118
119	1.6468	3.3530	0.01620	208.54	208.56	86.98	1025.85	1112.83	0.16289	1.7727	1.9356	119
120	1.6933	3.4477	0.01620	203.16	203.18	87.98	1025.28	1113.26	0.16461	1.7687	1.9333	120
121	1.7409	3.5446	0.01621	197.93	197.95	88.98	1024.70	1113.68	0.16634	1.7647	1.9310	121
122	1.7897	3.6439	0.01621	192.85	192.87	89.98	1024.12	1114.10	0.16805	1.7606	1.9286	122
123	1.8396	3.7455	0.01622	187.93	187.95	90.98	1023.54	1114.52	0.16977	1.7566	1.9264	123
124	1.8907	3.8496	0.01622	183.15	183.17	91.98	1022.96	1114.94	0.17148	1.7526	1.9241	124
125	1.9430	3.9561	0.01622	178.51	178.53	92.98	1022.39	1115.37	0.17319	1.7486	1.9218	125
126	1.9966	4.0651	0.01623	174.00	174.02	93.98	1021.81	1115.79	0.17490	1.7446	1.9195	126
127	2.0514	4.1768	0.01623	169.63	169.65	94.97	1021.24	1116.21	0.17660	1.7407	1.9173	127
128	2.1075	4.2910	0.01624	165.38	165.40	95.97	1020.66	1116.63	0.17830	1.7367	1.9150	128
129	2.1649	4.4078	0.01624	161.26	161.28	96.97	1020.08	1117.05	0.18000	1.7328	1.9128	129
130	2.2237	4.5274	0.01625	157.25	157.27	97.97	1019.50	1117.47	0.18170	1.7289	1.9106	130
131	2.2838	4.6498	0.01625	153.36	153.38	98.97	1018.92	1117.89	0.18339	1.7250	1.9084	131
132	2.3452	4.7750	0.01626	149.58	149.60	99.97	1018.34	1118.31	0.18508	1.7211	1.9062	132
133	2.4081	4.9030	0.01626	145.91	145.93	100.97	1017.76	1118.73	0.18676	1.7172	1.9040	133
134	2.4725	5.0340	0.01626	142.34	142.36	101.97	1017.18	1119.15	0.18845	1.7134	1.9018	134
135	2.5382	5.1679	0.01627	138.87	138.89	102.97	1016.59	1119.56	0.19013	1.7095	1.8996	135
136	2.6055	5.3049	0.01627	135.50	135.52	103.97	1016.01	1119.98	0.19181	1.7056	1.8974	136
137	2.6743	5.4450	0.01628	132.22	132.24	104.97	1015.43	1120.40	0.19348	1.7018	1.8953	137
138	2.7446	5.5881	0.01628	129.04	129.06	105.97	1014.85	1120.82	0.19516	1.6979	1.8931	138
139	2.8165	5.7345	0.01629	125.94	125.96	106.97	1014.26	1121.23	0.19683	1.6942	1.8910	139
140	2.8900	5.8842	0.01629	122.94	122.96	107.96	1013.69	1121.65	0.19850	1.6903	1.8888	140
141	2.9651	6.0371	0.01630	120.01	120.03	108.96	1013.11	1122.07	0.20016	1.6865	1.8867	141
142	3.0419	6.1934	0.01630	117.16	117.18	109.96	1012.52	1122.48	0.20182	1.6828	1.8846	142
143	3.1204	6.3532	0.01631	114.40	114.42	110.96	1011.94	1122.90	0.20348	1.6790	1.8825	143

^a Compiled by John A. Goff and S. Gratch.

^a Reprinted with permission from the 1977 Fundamentals Volume, ASHRAE Handbook and Product Directory.

ᵃTABLE A3-2 Thermodynamic properties of water at saturationᵃ (continued)

Fahr. Temp. t(F)	Absolute Pressure p Lb/Sq In.	In. Hg	Specific Volume, cu ft per lb Sat. Liquid v_f	Evap. v_{fg}	Sat. Vapor v_g	Enthalpy, Btu per lb Sat. Liquid h_f	Evap. h_{fg}	Sat. Vapor h_g	Entropy, Btu per (lb) (°F) Sat. Liquid s_f	Evap. s_{fg}	Sat. Vapor s_g	Fahr. Temp. t(F)
144	3.2006	6.5164	0.01631	111.70	111.72	111.96	1011.35	1123.31	0.20514	1.6753	1.8804	144
145	3.2825	6.6832	0.01632	109.09	109.11	112.96	1010.77	1123.73	0.20679	1.6715	1.8783	145
146	3.3662	6.8536	0.01632	106.56	106.56	113.96	1010.18	1124.14	0.20845	1.6678	1.8763	146
147	3.4517	7.0277	0.01633	104.06	104.08	114.96	1009.59	1124.55	0.21010	1.6641	1.8742	147
148	3.5390	7.2056	0.01633	101.65	101.67	115.96	1009.01	1124.97	0.21174	1.6604	1.8721	148
149	3.6282	7.3872	0.01634	99.306	99.322	116.96	1008.42	1125.38	0.21339	1.6567	1.8701	149
150	3.7194	7.5727	0.01634	97.022	97.038	117.96	1007.83	1125.79	0.21503	1.6530	1.8680	150
151	3.8124	7.7622	0.01635	94.799	94.815	118.96	1007.24	1126.20	0.21667	1.6493	1.8660	151
152	3.9074	7.9556	0.01635	92.635	92.651	119.96	1006.66	1126.62	0.21830	1.6457	1.8640	152
153	4.0044	8.1532	0.01636	90.528	90.544	120.97	1006.06	1127.03	0.21994	1.6421	1.8620	153
154	4.1035	8.3548	0.01636	88.477	88.493	121.97	1005.47	1127.44	0.22157	1.6384	1.8600	154
155	4.2046	8.5607	0.01637	86.480	86.496	122.97	1004.88	1127.85	0.22320	1.6348	1.8580	155
156	4.3078	8.7708	0.01637	84.536	84.552	123.97	1004.29	1128.26	0.22482	1.6312	1.8560	156
157	4.4132	8.9853	0.01638	82.642	82.658	124.97	1003.70	1128.67	0.22645	1.6276	1.8540	157
158	4.5207	9.2042	0.01638	80.798	80.814	125.97	1003.11	1129.08	0.22807	1.6239	1.8520	158
159	4.6304	9.4276	0.01639	79.001	79.017	126.97	1002.51	1129.48	0.22969	1.6204	1.8501	159
160	4.7424	9.6556	0.01639	77.251	77.267	127.97	1001.92	1129.89	0.23130	1.6168	1.8481	160
161	4.8566	9.8882	0.01640	75.546	75.562	128.97	1001.33	1130.30	0.23292	1.6133	1.8462	161
162	4.9732	10.126	0.01640	73.885	73.901	129.97	1000.74	1130.71	0.23453	1.6097	1.8442	162
163	5.0921	10.368	0.01641	72.267	72.283	130.98	1000.13	1131.11	0.23614	1.6062	1.8423	163
164	5.2134	10.615	0.01642	70.690	70.706	131.98	999.54	1131.52	0.23774	1.6027	1.8404	164
165	5.3372	10.867	0.01642	69.153	69.169	132.98	998.94	1131.92	0.23935	1.5990	1.8384	165
166	5.4634	11.124	0.01643	67.654	67.670	133.98	998.35	1132.33	0.24095	1.5956	1.8365	166
167	5.5921	11.386	0.01643	66.194	66.210	134.98	997.75	1132.73	0.24255	1.5920	1.8346	167
168	5.7233	11.653	0.01644	64.770	64.786	135.98	997.16	1133.14	0.24414	1.5887	1.8328	168
169	5.8572	11.925	0.01644	63.382	63.398	136.99	996.55	1133.54	0.24574	1.5852	1.8309	169
170	5.9936	12.203	0.01645	62.029	62.045	137.99	995.95	1133.94	0.24733	1.5817	1.8290	170
171	6.1328	12.487	0.01645	60.710	60.726	138.99	995.36	1134.35	0.24892	1.5782	1.8271	171
172	6.2746	12.775	0.01646	59.423	59.439	139.99	994.76	1134.76	0.25051	1.5748	1.8253	172
173	6.4192	13.070	0.01647	58.168	58.184	141.00	994.15	1135.15	0.25209	1.5713	1.8234	173
174	6.5666	13.370	0.01647	56.944	56.960	142.00	993.55	1135.55	0.25367	1.5679	1.8216	174
175	6.7168	13.676	0.01648	55.750	55.766	143.00	992.95	1135.95	0.25525	1.5644	1.8197	175
176	6.8699	13.987	0.01648	54.586	54.602	144.00	992.35	1136.35	0.25683	1.5611	1.8179	176
177	7.0259	14.305	0.01649	53.450	53.466	145.00	991.75	1136.75	0.25841	1.5577	1.8161	177

ᵃReprinted with permission from the 1977 Fundamentals Volume, ASHRAE Handbook and Product Directory.

TABLE A3-2 Thermodynamic properties of water at saturation[a] (continued)

Fahr. Temp. t(F)	Absolute Pressure Ps		Specific Volume, cu ft per lb			Enthalpy, Btu per lb			Entropy, Btu per (lb) (°F)			Fahr. Temp. t(F)
	Lb/Sq In.	In. Hg	Sat. Liquid vf	Evap. vfg	Sat. Vapor vg	Sat. Liquid hf	Evap. hfg	Sat. Vapor hg	Sat. Liquid sf	Evap. sfg	Sat. Vapor sg	
178	7.1849	14.629	0.01650	52.341	52.357	146.01	991.14	1137.15	0.25598	1.5543	1.8143	178
179	7.3469	14.959	0.01650	51.260	51.276	147.01	990.54	1137.55	0.26155	1.5508	1.8124	179
180	7.5119	15.295	0.01651	50.203	50.220	148.01	989.93	1137.94	0.26312	1.5475	1.8106	180
181	7.6801	15.637	0.01651	49.173	49.190	149.02	989.32	1138.34	0.26468	1.5442	1.8089	181
182	7.8514	15.986	0.01652	48.168	48.185	150.02	988.72	1138.74	0.26625	1.5408	1.8071	182
183	8.0258	16.341	0.01652	47.187	47.204	151.02	988.12	1139.14	0.26781	1.5375	1.8053	183
184	8.2035	16.703	0.01653	46.229	46.246	152.03	987.50	1139.53	0.26937	1.5341	1.8035	184
185	8.3845	17.071	0.01654	45.294	45.311	153.03	986.89	1139.92	0.27093	1.5308	1.8017	185
186	8.5688	17.446	0.01654	44.381	44.398	154.04	986.28	1140.32	0.27248	1.5275	1.8000	186
187	8.7565	17.829	0.01655	43.489	43.506	155.04	985.67	1140.71	0.27404	1.5242	1.7982	187
188	8.9476	18.218	0.01656	42.619	42.636	156.04	985.07	1141.11	0.27559	1.5209	1.7965	188
189	9.1422	18.614	0.01656	41.769	41.786	157.05	984.45	1141.50	0.27713	1.5176	1.7947	189
190	9.3403	19.017	0.01657	40.939	40.956	158.05	983.84	1141.89	0.27868	1.5143	1.7930	190
191	9.5420	19.428	0.01658	40.128	40.145	159.06	983.22	1142.28	0.28022	1.5111	1.7913	191
192	9.7473	19.846	0.01658	39.337	39.354	160.06	982.61	1142.67	0.28176	1.5078	1.7896	192
193	9.9563	20.271	0.01659	38.563	38.580	161.06	982.00	1143.06	0.28330	1.5045	1.7878	193
194	10.169	20.704	0.01659	37.807	37.824	162.07	981.38	1143.45	0.28484	1.5013	1.7861	194
195	10.386	21.145	0.01660	37.069	37.086	163.08	980.76	1143.84	0.28638	1.4980	1.7844	195
196	10.606	21.594	0.01661	36.348	36.365	164.08	980.15	1144.23	0.28791	1.4949	1.7828	196
197	10.830	22.050	0.01661	35.643	35.660	165.08	979.54	1144.62	0.28944	1.4917	1.7811	197
198	11.058	22.515	0.01662	34.954	34.971	166.09	978.91	1145.00	0.29097	1.4884	1.7794	198
199	11.290	22.987	0.01663	34.281	34.298	167.10	978.29	1145.39	0.29250	1.4852	1.7777	199
200	11.526	23.468	0.01663	33.623	33.640	168.10	977.68	1145.78	0.29402	1.4820	1.7760	200
201	11.767	23.957	0.01664	32.980	32.997	169.11	977.05	1146.16	0.29554	1.4789	1.7744	201
202	12.011	24.455	0.01665	32.351	32.368	170.11	976.43	1146.54	0.29706	1.4756	1.7727	202
203	12.260	24.961	0.01665	31.737	31.754	171.12	975.81	1146.93	0.29858	1.4725	1.7711	203
204	12.513	25.476	0.01666	31.136	31.153	172.12	975.19	1147.31	0.30010	1.4693	1.7694	204
205	12.770	26.000	0.01667	30.549	30.566	173.13	974.56	1147.69	0.30161	1.4662	1.7678	205
206	13.031	26.532	0.01667	29.974	29.991	174.14	973.94	1148.08	0.30312	1.4631	1.7662	206
207	13.297	27.074	0.01668	29.413	29.430	175.14	973.32	1148.46	0.30463	1.4600	1.7646	207
208	13.568	27.625	0.01669	28.863	28.880	176.15	972.69	1148.84	0.30614	1.4568	1.7629	208
209	13.843	28.185	0.01669	28.326	28.343	177.16	972.06	1149.22	0.30765	1.4536	1.7613	209
210	14.123	28.754	0.01670	27.801	27.818	178.17	971.43	1149.60	0.30915	1.4506	1.7597	210
211	14.407	29.333	0.01671	27.287	27.301	179.17	970.81	1149.98	0.31035	1.4474	1.7581	211
212	14.696	29.921	0.01671	26.784	26.801	180.18	970.17	1150.35	0.31215	1.4444	1.7565	212

[a] Compiled by John A. Goff and S. Gratch.
[a] Reprinted with permission from the 1977 Fundamentals Volume, ASHRAE Handbook and Product Directory.

TABLE A3-2 Thermodynamic properties of water at saturation[a] (continued)

Fahr. Temp. t(F)	Abs. Press. Lb/Sq In. p	Specific Volume		Enthalpy			Entropy			Fahr. Temp. t(F)
		Sat. Liquid v_f	Sat. Vapor v_g	Sat. Liquid h_f	Evap. h_{fg}	Sat. Vapor h_g	Sat. Liquid s_f	Evap. s_{fg}	Sat. Vapor s_g	
212	14.698	0.016716	26.80	180.16	970.3	1150.5	0.31213	1.4446	1.7567	212
214	15.291	.016729	25.83	182.17	969.1	1151.2	.31513	1.4384	1.7535	214
216	15.903	.016743	24.90	184.18	967.8	1152.0	.31811	1.4322	1.7504	216
218	16.535	.016758	24.00	186.20	966.5	1152.7	.32109	1.4261	1.7472	218
220	17.188	.016772	23.15	188.22	965.3	1153.5	.32406	1.4201	1.7441	220
222	17.861	.016786	22.33	190.24	964.0	1154.2	.32702	1.4140	1.7410	222
224	18.557	.016801	21.55	192.26	962.7	1154.9	.32998	1.4080	1.7380	224
226	19.275	.016816	20.80	194.28	961.4	1155.7	.33292	1.4020	1.7349	226
228	20.015	.016830	20.08	196.30	960.1	1156.4	.33586	1.3961	1.7319	228
230	20.78	.016845	19.386	198.32	958.8	1157.1	.33880	1.3901	1.7289	230
232	21.57	0.016860	18.723	200.34	957.5	1157.9	0.34172	1.3842	1.7260	232
234	22.38	.016875	18.087	202.37	956.2	1158.6	.34464	1.3784	1.7230	234
236	23.22	.016891	17.476	204.39	954.9	1159.3	.34755	1.3725	1.7201	236
238	24.08	.016906	16.890	206.42	953.6	1160.0	.35045	1.3667	1.7172	238
240	24.97	.016922	16.327	208.44	952.3	1160.7	.35335	1.3609	1.7143	240
242	25.88	.016937	15.786	210.47	950.9	1161.4	.35624	1.3552	1.7114	242
244	26.82	.016953	15.267	212.49	949.6	1162.1	.35912	1.3494	1.7085	244
246	27.79	.016969	14.767	214.52	948.3	1162.8	.36199	1.3437	1.7057	246
248	28.79	.016985	14.287	216.55	947.0	1163.5	.36486	1.3380	1.7029	248
250	29.82	.017001	13.826	218.59	945.6	1164.2	.36772	1.3324	1.7001	250
252	30.88	0.017017	13.382	220.62	944.3	1164.9	0.37058	1.3267	1.6973	252
254	31.97	.017034	12.955	222.65	942.9	1165.6	.37342	1.3211	1.6945	254
256	33.09	.017050	12.544	224.68	941.6	1166.2	.37626	1.3155	1.6918	256
258	34.24	.017067	12.149	226.72	940.2	1166.9	.37910	1.3100	1.6891	258
260	35.42	.017084	11.768	228.76	938.8	1167.6	.38193	1.3044	1.6864	260
262	36.64	.017101	11.402	230.79	937.5	1168.3	.38475	1.2989	1.6837	262
264	37.89	.017118	11.049	232.83	936.1	1168.9	.38756	1.2935	1.6810	264
266	39.17	.017135	10.709	234.87	934.7	1169.6	.39037	1.2880	1.6784	266
268	40.49	.017152	10.382	236.91	933.3	1170.2	.39317	1.2825	1.6757	268
270	41.85	.017170	10.066	238.95	932.0	1170.9	.39597	1.2771	1.6731	270

[a]Reprinted with permission from the 1977 Fundamentals Volume, ASHRAE Handbook and Product Directory.

TABLE A3-2 Thermodynamic properties of water at saturation^a (continued)

272	43.24	0.017187	9.762	241.00	930.6	1171.6	0.39876	1.2717	1.6705	272
274	44.67	.017205	9.469	243.04	929.2	1172.2	.40154	1.2664	1.6679	274
276	46.13	.017223	9.186	245.08	927.8	1172.8	.40432	1.2610	1.6653	276
278	47.64	.017241	8.913	247.13	926.3	1173.5	.40709	1.2557	1.6628	278
280	49.18	.017259	8.650	249.18	924.9	1174.1	.40986	1.2504	1.6602	280
282	50.77	.017277	8.397	251.23	923.5	1174.7	.41262	1.2451	1.6577	282
284	52.40	.017296	8.152	253.28	922.1	1175.4	.41537	1.2398	1.6552	284
286	54.07	.017314	7.915	255.33	920.6	1176.0	.41812	1.2346	1.6527	286
288	55.78	.017333	7.687	257.38	919.2	1176.6	.42086	1.2293	1.6502	288
290	57.53	.017352	7.467	259.44	917.8	1177.2	.42360	1.2241	1.6477	290
292	59.33	0.017371	7.254	261.50	916.3	1177.8	0.42633	1.2189	1.6453	292
294	61.17	.017390	7.048	263.55	914.9	1178.4	.42906	1.2138	1.6428	294
296	63.06	.017409	6.849	265.61	913.4	1179.0	.43178	1.2086	1.6404	296
298	65.00	.017429	6.657	267.67	911.9	1179.6	.43449	1.2035	1.6380	298
300	66.98	.017448	6.472	269.73	910.4	1180.2	.43720	1.1984	1.6356	300
310	77.64	.017548	5.632	280.06	903.0	1183.0	.45067	1.1731	1.6238	310
320	89.60	.017652	4.919	290.43	895.3	1185.8	.46400	1.1483	1.6123	320
330	103.00	.017760	4.312	300.84	887.5	1188.4	.47722	1.1238	1.6010	330
340	117.93	.017872	3.792	311.30	879.5	1190.8	.49031	1.0997	1.5901	340
350	134.53	.017988	3.346	321.80	871.3	1193.1	.50329	1.0760	1.5793	350
360	152.92	0.018108	2.961	332.35	862.9	1195.2	0.51617	1.0526	1.5688	360
370	173.23	.018233	2.628	342.96	854.2	1197.2	.52894	1.0295	1.5585	370
380	195.60	.018363	2.339	353.62	845.4	1199.0	.54163	1.0067	1.5483	380
390	220.2	.018498	2.087	364.34	836.2	1200.6	.55422	0.9841	1.5383	390
400	247.1	.018638	1.8661	375.12	826.8	1202.0	.56672	.9617	1.5284	400
410	276.5	.018784	1.6726	385.97	817.2	1203.1	.57916	.9395	1.5187	410
420	308.5	.018936	1.5024	396.89	807.2	1204.1	.59152	.9175	1.5091	420
430	343.3	.019094	1.3521	407.89	796.9	1204.8	.60381	.8957	1.4995	430
440	381.2	.019260	1.2192	418.98	786.3	1205.3	.61605	.8740	1.4900	440
450	422.1	.019433	1.1011	430.2	775.4	1205.6	.6282	.8523	1.4806	450
460	466.3	0.019614	0.9961	441.4	764.1	1205.5	0.6404	0.8308	1.4712	460
470	514.1	.019803	.9025	452.8	752.4	1205.2	.6525	.8093	1.4618	470
480	565.5	.020002	.8187	464.3	740.3	1204.6	.6646	.7878	1.4524	480
490	620.7	.020211	.7436	475.9	727.8	1203.7	.6767	.7663	1.4430	490
500	680.0	.02043	.6761	487.7	714.8	1202.5	.6888	.7448	1.4335	500

Temperatures in this table follow the Thermodynamic Fahrenheit Scale.

^{a}Reprinted with permission from the 1977 Fundamentals Volume, ASHRAE Handbook and Product Directory.

TABLE A3.3 Thermodynamic properties of steam. Superheated-steam table[a]

Abs. Press., Psia (sat. temp.)		Temp. °F												
		200	300	400	500	600	700	800	900	1000	1100	1200	1400	1600
1 (101.74)	v	392.6	452.3	512.0	571.6	631.2	690.8	750.4	809.9	869.5	929.1	988.7	1107.8	1227.0
	h	1150.4	1195.8	1241.7	1288.3	1335.7	1383.8	1432.8	1482.7	1533.5	1585.2	1637.7	1745.7	1857.5
	s	2.0512	2.1153	2.1720	2.2233	2.2702	2.3137	2.3542	2.3923	2.4283	2.4625	2.4952	2.5566	2.6137
5 (162.24)	v	78.16	90.25	102.26	114.22	126.16	138.10	150.03	161.95	173.87	185.79	197.71	221.6	245.4
	h	1148.8	1195.0	1241.2	1288.0	1335.4	1383.6	1432.7	1482.6	1533.4	1585.1	1637.7	1745.7	1857.4
	s	1.8718	1.9370	1.9942	2.0456	2.0927	2.1361	2.1767	2.2148	2.2509	2.2851	2.3178	2.3792	2.4363
10 (193.21)	v	38.85	45.00	51.04	57.05	63.03	69.01	74.98	80.95	86.92	92.88	98.84	110.77	122.69
	h	1146.6	1193.9	1240.6	1287.5	1335.1	1383.4	1432.5	1482.4	1533.2	1585.0	1637.6	1745.6	1857.3
	s	1.7927	1.8595	1.9172	1.9689	2.0160	2.0596	2.1002	2.1383	2.1744	2.2086	2.2413	2.3028	2.3598
14.696 (212.00)	v		30.53	34.68	38.78	42.86	46.94	51.00	55.07	59.13	63.19	67.25	75.37	83.48
	h		1192.8	1239.9	1287.1	1334.8	1383.2	1432.3	1482.3	1533.1	1584.8	1637.5	1745.5	1857.3
	s		1.8160	1.8743	1.9261	1.9734	2.0170	2.0576	2.0958	2.1319	2.1662	2.1989	2.2603	2.3174
20 (227.96)	v		22.36	25.43	28.46	31.47	34.47	37.46	40.45	43.44	46.42	49.41	55.37	61.34
	h		1191.6	1239.2	1286.6	1334.4	1382.9	1432.1	1482.1	1533.0	1584.7	1637.4	1745.4	1857.2
	s		1.7808	1.8396	1.8918	1.9392	1.9829	2.0235	2.0618	2.0978	2.1321	2.1648	2.2263	2.2834
40 (267.25)	v		11.040	12.628	14.168	15.688	17.198	18.702	20.20	21.70	23.20	24.69	27.68	30.66
	h		1186.8	1236.5	1284.8	1333.1	1381.9	1431.3	1481.4	1532.4	1584.3	1637.0	1745.1	1857.0
	s		1.6994	1.7608	1.8140	1.8619	1.9058	1.9467	1.9850	2.0212	2.0555	2.0883	2.1498	2.2069
60 (292.71)	v		7.259	8.357	9.403	10.427	11.441	12.449	13.452	14.454	15.453	16.451	18.446	20.44
	h		1181.6	1233.6	1283.0	1331.8	1380.9	1430.5	1480.8	1531.9	1583.8	1636.6	1744.8	1856.7
	s		1.6402	1.7135	1.7678	1.8162	1.8605	1.9015	1.9400	1.9762	2.0106	2.0434	2.1049	2.1621
80 (312.03)	v			6.220	7.020	7.797	8.562	9.322	10.077	10.830	11.582	12.332	13.830	15.325
	h			1230.7	1281.1	1330.5	1379.9	1429.7	1480.1	1531.3	1583.4	1636.2	1744.5	1856.5
	s			1.6791	1.7346	1.7836	1.8281	1.8694	1.9079	1.9442	1.9787	2.0115	2.0731	2.1303
100 (327.81)	v			4.937	5.589	6.218	6.835	7.446	8.052	8.656	9.259	9.860	11.060	12.258
	h			1227.6	1279.1	1329.1	1378.9	1428.9	1479.5	1530.8	1582.9	1635.7	1744.2	1856.2
	s			1.6518	1.7085	1.7581	1.8029	1.8443	1.8829	1.9193	1.9538	1.9867	2.0484	2.1056
120 (341.25)	v			4.081	4.636	5.165	5.683	6.195	6.702	7.207	7.710	8.212	9.214	10.213
	h			1224.4	1277.2	1327.7	1377.8	1428.1	1478.8	1530.2	1582.4	1635.3	1743.9	1856.0
	s			1.6287	1.6869	1.7370	1.7822	1.8237	1.8625	1.8990	1.9335	1.9664	2.0281	2.0854
140 (353.02)	v			3.468	3.954	4.413	4.861	5.301	5.738	6.172	6.604	7.035	7.895	8.752
	h			1221.1	1275.2	1326.4	1376.8	1427.3	1478.2	1529.7	1581.9	1634.9	1743.5	1855.7
	s			1.6087	1.6683	1.7190	1.7645	1.8063	1.8451	1.8817	1.9163	1.9493	2.0110	2.0683
160 (363.53)	v			3.008	3.443	3.849	4.244	4.631	5.015	5.396	5.775	6.152	6.906	7.656
	h			1217.6	1273.1	1325.0	1375.7	1426.4	1477.5	1529.1	1581.4	1634.5	1743.2	1855.5
	s			1.5908	1.6519	1.7033	1.7491	1.7911	1.8301	1.8667	1.9014	1.9344	1.9962	2.0535
180 (373.06)	v			2.649	3.044	3.411	3.764	4.110	4.452	4.792	5.129	5.466	6.136	6.804
	h			1214.0	1271.0	1323.5	1374.7	1425.6	1476.8	1528.6	1581.0	1634.1	1742.9	1855.2
	s			1.5745	1.6373	1.6894	1.7355	1.7776	1.8167	1.8534	1.8882	1.9212	1.9831	2.0404
200 (381.79)	v			2.361	2.726	3.060	3.380	3.693	4.002	4.309	4.613	4.917	5.521	6.123
	h			1210.3	1268.9	1322.1	1373.6	1424.8	1476.2	1528.0	1580.5	1633.7	1742.6	1855.0
	s			1.5594	1.6240	1.6767	1.7232	1.7655	1.8048	1.8415	1.8763	1.9094	1.9713	2.0287
220 (389.86)	v			2.125	2.465	2.772	3.066	3.352	3.634	3.913	4.191	4.467	5.017	5.565
	h			1206.5	1266.7	1320.7	1372.0	1424.0	1475.5	1527.5	1580.0	1633.3	1742.3	1854.7
	s			1.5453	1.6117	1.6652	1.7120	1.7545	1.7939	1.8308	1.8656	1.8987	1.9607	2.0181
240 (397.37)	v			1.9276	2.247	2.533	2.804	3.068	3.327	3.584	3.839	4.093	4.597	5.100
	h			1202.5	1264.5	1319.2	1371.5	1423.2	1474.8	1526.9	1579.6	1632.9	1742.0	1854.5
	s			1.5319	1.6003	1.6546	1.7017	1.7444	1.7839	1.8209	1.8558	1.8889	1.9510	2.0084
260 (404.42)	v				2.063	2.330	2.582	2.827	3.067	3.305	3.541	3.776	4.242	4.707
	h				1262.3	1317.7	1370.4	1422.3	1474.2	1526.3	1579.1	1632.5	1741.7	1854.2
	s				1.5897	1.6447	1.6922	1.7352	1.7748	1.8118	1.8467	1.8799	1.9420	1.9995
280 (411.05)	v				1.9047	2.156	2.392	2.621	2.845	3.066	3.286	3.504	3.938	4.370
	h				1260.0	1316.2	1369.4	1421.5	1473.5	1525.8	1578.6	1632.1	1741.4	1854.0
	s				1.5796	1.6354	1.6834	1.7265	1.7662	1.8033	1.8383	1.8716	1.9337	1.9912
300 (417.33)	v				1.7675	2.005	2.227	2.442	2.652	2.859	3.065	3.269	3.674	4.078
	h				1257.6	1314.7	1368.3	1420.6	1472.8	1525.2	1578.1	1631.7	1741.0	1853.7
	s				1.5701	1.6268	1.6751	1.7184	1.7582	1.7954	1.8305	1.8638	1.9260	1.9835
350 (431.72)	v				1.4923	1.7036	1.8980	2.084	2.266	2.445	2.622	2.798	3.147	3.493
	h				1251.5	1310.9	1365.5	1418.5	1471.1	1523.8	1577.0	1630.7	1740.3	1853.1
	s				1.5481	1.6070	1.6563	1.7002	1.7403	1.7777	1.8130	1.8463	1.9086	1.9663
400 (444.59)	v				1.2851	1.4770	1.6508	1.8161	1.9767	2.134	2.290	2.445	2.751	3.055
	h				1245.1	1306.9	1362.7	1416.4	1469.4	1522.4	1575.8	1629.6	1739.5	1852.5
	s				1.5281	1.5894	1.6398	1.6842	1.7247	1.7623	1.7977	1.8311	1.8936	1.9513

[a]Abridged from "Thermodynamic Properties of Steam," by Joseph H. Keenan and Frederick G. Keyes, John Wiley & Sons, Inc., New York, 1937, by permission.

TABLE A3.3 Thermodynamic properties of steam. Superheated-steam table[a] (continued)

Abs. Press., Psia (sat. temp.)		Temp. °F												
		500	550	600	620	640	660	680	700	800	900	1000	1200	1400	1600
450 (456.28)	v	1.1231	1.2155	1.3005	1.3332	1.3652	1.3967	1.4278	1.4584	1.6074	1.7516	1.8928	2.170	2.443	2.714
	h	1238.4	1272.0	1302.8	1314.6	1326.2	1337.5	1348.8	1359.9	1414.3	1467.7	1521.0	1628.6	1738.7	1851.9
	s	1.5095	1.5437	1.5735	1.5845	1.5951	1.6054	1.6153	1.6250	1.6699	1.7108	1.7486	1.8177	1.8803	1.9381
500 (467.01)	v	0.9927	1.0800	1.1591	1.1893	1.2188	1.2478	1.2763	1.3044	1.4405	1.5715	1.6996	1.9504	2.197	2.442
	h	1231.3	1266.8	1298.6	1310.7	1322.6	1334.2	1345.7	1357.0	1412.1	1466.0	1519.6	1627.6	1737.9	1851.3
	s	1.4919	1.5280	1.5588	1.5701	1.5810	1.5915	1.6016	1.6115	1.6571	1.6982	1.7363	1.8056	1.8683	1.9262
550 (476.94)	v	0.8852	0.9686	1.0431	1.0714	1.0989	1.1259	1.1523	1.1783	1.3038	1.4241	1.5414	1.7706	1.9957	2.219
	h	1223.7	1261.2	1294.3	1306.8	1318.9	1330.8	1342.5	1354.0	1409.9	1464.3	1518.2	1626.6	1737.1	1850.6
	s	1.4751	1.5131	1.5451	1.5568	1.5680	1.5787	1.5890	1.5991	1.6452	1.6868	1.7250	1.7946	1.8575	1.9155
600 (486.21)	v	0.7947	0.8753	0.9463	0.9729	0.9988	1.0241	1.0489	1.0732	1.1899	1.3013	1.4096	1.6203	1.8279	2.033
	h	1215.7	1255.5	1289.9	1302.7	1315.2	1327.4	1339.3	1351.1	1407.7	1462.5	1516.7	1625.5	1736.3	1850.0
	s	1.4586	1.4990	1.5323	1.5443	1.5558	1.5667	1.5773	1.5875	1.6343	1.6762	1.7147	1.7846	1.8476	1.9056
700 (503.10)	v		0.7277	0.7934	0.8177	0.8411	0.8639	0.8860	0.9077	1.0108	1.1082	1.2024	1.3853	1.5641	1.7405
	h		1243.2	1280.6	1294.3	1307.5	1320.3	1332.8	1345.0	1403.2	1459.0	1515.9	1623.5	1734.8	1848.8
	s		1.4722	1.5084	1.5212	1.5333	1.5449	1.5559	1.5665	1.6147	1.6573	1.6963	1.7666	1.8299	1.8881
800 (518.23)	v		0.6154	0.6779	0.7006	0.7223	0.7433	0.7635	0.7833	0.8763	0.9633	1.0470	1.2088	1.3662	1.5214
	h		1229.8	1270.7	1285.4	1299.4	1312.9	1325.9	1338.6	1398.6	1455.4	1511.0	1621.4	1733.2	1847.5
	s		1.4467	1.4863	1.5000	1.5129	1.5250	1.5366	1.5476	1.5972	1.6407	1.6801	1.7510	1.8146	1.8729
900 (531.98)	v		0.5264	0.5873	0.6089	0.6294	0.6491	0.6680	0.6863	0.7716	0.8506	0.9262	1.0714	1.2124	1.3509
	h		1215.0	1260.1	1275.9	1290.9	1305.1	1318.8	1332.1	1393.9	1451.8	1508.1	1619.3	1731.6	1846.3
	s		1.4216	1.4653	1.4800	1.4938	1.5066	1.5187	1.5303	1.5814	1.6257	1.6656	1.7371	1.8009	1.8595
1000 (544.61)	v		0.4533	0.5140	0.5350	0.5546	0.5733	0.5912	0.6084	0.6878	0.7604	0.8294	0.9615	1.0893	1.2146
	h		1198.3	1248.8	1265.9	1281.9	1297.0	1311.4	1325.3	1389.2	1448.2	1505.1	1617.3	1730.0	1845.0
	s		1.3961	1.4450	1.4610	1.4757	1.4893	1.5021	1.5141	1.5670	1.6121	1.6525	1.7245	1.7886	1.8474
1100 (556.31)	v			0.4532	0.4738	0.4929	0.5110	0.5281	0.5445	0.6191	0.6866	0.7503	0.8716	0.9885	1.1031
	h			1236.7	1255.3	1272.4	1288.5	1303.7	1318.3	1384.3	1444.5	1502.2	1615.2	1728.4	1843.8
	s			1.4251	1.4425	1.4583	1.4728	1.4862	1.4989	1.5535	1.5995	1.6405	1.7130	1.7775	1.8363
1200 (567.22)	v			0.4016	0.4222	0.4410	0.4586	0.4752	0.4909	0.5617	0.6250	0.6843	0.7967	0.9046	1.0101
	h			1223.5	1243.9	1262.4	1279.6	1295.7	1311.0	1379.3	1440.7	1499.2	1613.1	1726.9	1842.5
	s			1.4052	1.4243	1.4413	1.4568	1.4710	1.4843	1.5409	1.5879	1.6293	1.7025	1.7672	1.8263
1400 (587.10)	v			0.3174	0.3390	0.3580	0.3753	0.3912	0.4062	0.4714	0.5281	0.5805	0.6789	0.7727	0.8640
	h			1193.0	1218.4	1240.4	1260.3	1278.5	1295.5	1369.1	1433.1	1493.2	1608.9	1723.7	1840.0
	s			1.3639	1.3877	1.4079	1.4258	1.4419	1.4567	1.5177	1.5666	1.6093	1.6836	1.7489	1.8083
1600 (604.90)	v				0.2733	0.2936	0.3112	0.3271	0.3417	0.4034	0.4553	0.5027	0.5906	0.6738	0.7545
	h				1187.8	1215.2	1238.7	1259.6	1278.7	1358.4	1425.3	1487.0	1604.6	1720.5	1837.5
	s				1.3489	1.3741	1.3952	1.4137	1.4303	1.4964	1.5476	1.5914	1.6669	1.7328	1.7926
1800 (621.03)	v				0.2407	0.2597	0.2760	0.2907	0.3502	0.3986	0.4421	0.5218	0.5968	0.6693	
	h				1185.1	1214.0	1238.5	1260.3	1347.2	1417.4	1480.8	1600.4	1717.3	1835.0	
	s				1.3377	1.3638	1.3855	1.4044	1.4765	1.5301	1.5752	1.6520	1.7185	1.7788	
2000 (635.82)	v				0.1936	0.2161	0.2337	0.2489	0.3074	0.3532	0.3935	0.4668	0.5352	0.6011	
	h				1145.6	1184.9	1214.8	1240.0	1335.5	1409.2	1474.5	1596.1	1714.1	1832.5	
	s				1.2945	1.3300	1.3564	1.3783	1.4576	1.5139	1.5603	1.6384	1.7055	1.7660	
2500 (668.13)	v					0.1484	0.1686	0.2294	0.2710	0.3061	0.3678	0.4244	0.4784		
	h					1132.3	1176.8	1303.6	1387.8	1458.4	1585.3	1706.1	1826.2		
	s					1.2687	1.3073	1.4127	1.4772	1.5273	1.6088	1.6775	1.7389		
3000 (695.36)	v						0.0984	0.1760	0.2159	0.2476	0.3018	0.3505	0.3966		
	h						1060.7	1267.2	1365.0	1441.8	1574.3	1698.0	1819.9		
	s						1.1966	1.3690	1.4439	1.4984	1.5837	1.6540	1.7163		
3206.2 (705.40)	v							0.1583	0.1981	0.2288	0.2806	0.3267	0.3707		
	h							1250.5	1355.2	1434.7	1569.8	1694.6	1817.2		
	s							1.3508	1.4309	1.4874	1.5742	1.6452	1.7080		
3500	v							0.0306	0.1364	0.1762	0.2058	0.2546	0.2977	0.3381	
	h							780.5	1224.9	1340.7	1424.5	1563.3	1689.8	1813.6	
	s							0.9515	1.3241	1.4127	1.4723	1.5615	1.6336	1.6968	
4000	v							0.0287	0.1052	0.1462	0.1743	0.2192	0.2581	0.2943	
	h							763.8	1174.8	1314.4	1406.8	1552.1	1681.7	1807.2	
	s							0.9347	1.2757	1.3827	1.4482	1.5417	1.6154	1.6795	
4500	v							0.0276	0.0798	0.1226	0.1500	0.1917	0.2273	0.2602	
	h							753.5	1113.9	1286.5	1388.4	1540.8	1673.5	1800.9	
	s							0.9235	1.2204	1.3529	1.4253	1.5235	1.5990	1.6640	
5000	v							0.0268	0.0593	0.1036	0.1303	1.1696	0.2027	0.2329	
	h							746.4	1047.1	1256.5	1369.5	1529.5	1665.3	1794.5	
	s							0.9152	1.1622	1.3231	1.4034	1.5066	1.5839	1.6499	
5500	v							0.0262	0.0463	0.0880	0.1143	0.1516	0.1825	0.2106	
	h							741.3	985.0	1224.1	1349.3	1518.2	1657.0	1788.1	
	s							0.9090	1.1093	1.2930	1.3821	1.4908	1.5699	1.6369	

[a]Abridged from "Thermodynamic Properties of Steam," by Joseph H. Keenan and Frederick G. Keyes, John Wiley & Sons, Inc., New York, 1937, by permission.

TABLE A3-4 Refrigerant 114 (dichlorotetrafluoroethane) properties of liquid and saturated vapor[a]

Temp F	Pressure psia	Pressure psig*	Volume cu ft/lb Vapor v_g	Density lb/cu ft Liquid $1/v_f$	Enthalpy Btu/lb Liquid h_f	Enthalpy Btu/lb Vapor h_g	Entropy Btu/(lb)(°R) Liquid s_f	Entropy Btu/(lb)(°R) Vapor s_g
-135	0.034	29.852	603.346525	109.8970	-18.182	51.282	-0.04898	0.16496
-130	0.045	29.830	463.532655	109.5052	-17.277	51.908	-0.04622	.16363
-125	0.058	29.802	359.388410	109.1114	-16.368	52.540	-0.04348	.16241
-120	0.076	29.767	281.079992	108.7157	-15.453	53.178	-0.04077	.16127
-115	0.098	29.722	221.666914	108.3179	-14.533	53.821	-0.03808	.16023
-110	0.124	29.668	176.201685	107.9182	-13.607	54.470	-0.03541	0.15927
-105	0.158	29.600	141.123750	107.5164	-12.676	55.124	-0.03277	.15839
-100	0.198	29.518	113.847231	107.1124	-11.740	55.783	-0.03015	.15758
-95	0.247	29.418	92.477644	106.7063	-10.797	56.448	-0.02754	.15685
-90	0.306	29.297	75.615391	106.2979	-9.848	57.117	-0.02496	.15618
-85	0.377	29.153	62.218156	105.8873	-8.894	57.792	-0.02240	0.15558
-80	0.462	28.981	51.503710	105.4744	-7.933	58.472	-0.01985	.15504
-75	0.561	28.778	42.880661	105.0591	-6.965	59.156	-0.01732	.15456
-70	0.679	28.539	35.898716	104.6414	-5.991	59.844	-0.01480	.15414
-65	0.817	28.258	30.212674	104.2211	-5.011	60.538	-0.01230	.15377
-60	0.977	27.932	25.556162	103.7984	-4.023	61.235	-0.00982	0.15346
-58	1.048	27.787	23.933843	103.6286	-3.626	61.515	-0.00883	.15334
-56	1.124	27.633	22.431598	103.4583	-3.228	61.796	-0.00784	.15324
-54	1.204	27.471	21.039402	103.2876	-2.828	62.077	-0.00685	.15314
-52	1.288	27.298	19.748155	103.1165	-2.428	62.359	-0.00587	.15305
-50	1.378	27.116	18.549588	102.9450	-2.026	62.642	-0.00488	0.15296
-48	1.472	26.923	17.436181	102.7730	-1.623	62.925	-0.00390	.15289
-46	1.572	26.720	16.401089	102.6006	-1.219	63.209	-0.00292	.15282
-44	1.678	26.505	15.438076	102.4278	-0.814	63.493	-0.00195	.15275
-42	1.789	26.279	14.541455	102.2545	-0.408	63.778	-0.00097	.15270
-40	1.906	26.040	13.706038	102.0807	0.000	64.064	0.00000	0.15265
-38	2.030	25.788	12.927084	101.9065	0.409	64.350	0.00097	.15261
-36	2.160	25.523	12.200262	101.7318	0.819	64.637	0.00194	.15256
-34	2.297	25.244	11.521608	101.5566	1.230	64.924	0.00291	.15253
-32	2.441	24.951	10.887491	101.3810	1.643	65.212	0.00388	.15251
-30	2.593	24.642	10.294588	101.2049	2.057	65.500	0.00484	0.15249
-28	2.752	24.319	9.739848	101.0283	2.472	65.789	0.00580	.15248
-26	2.919	23.979	9.220474	100.8512	2.888	66.078	0.00677	.15247
-24	3.094	23.622	8.733894	100.6737	3.306	66.367	0.00773	.15246
-22	3.277	23.248	8.277746	100.4956	3.725	66.657	0.00868	.15247
-20	3.470	22.857	7.849858	100.3170	4.145	66.948	0.00964	0.15248
-18	3.671	22.446	7.448229	100.1379	4.567	67.239	0.01060	.15249
-16	3.882	22.017	7.071017	99.9583	4.990	67.531	0.01155	.15251

Temp F	Pressure psia	Pressure psig	Volume cu ft/lb Vapor v_g	Density lb/cu ft Liquid $1/v_f$	Enthalpy Btu/lb Liquid h_f	Enthalpy Btu/lb Vapor h_g	Entropy Btu/(lb)(°R) Liquid s_f	Entropy Btu/(lb)(°R) Vapor s_g
70	27.261	12.561	1.142969	91.6512	24.444	80.277	0.05146	0.15687
72	28.282	13.582	1.103851	91.4416	24.926	80.574	.05236	.15703
74	29.331	14.631	1.066355	91.2310	25.409	80.871	.05327	.15719
76	30.410	15.710	1.030402	91.0196	25.893	81.168	.05417	.15735
78	31.519	16.819	0.995918	90.8072	26.379	81.464	.05507	.15752
80	32.659	17.959	0.962831	90.5939	26.865	81.760	0.05597	0.15768
82	33.829	19.129	.931075	90.3796	27.353	82.056	.05687	.15785
84	35.032	20.332	.900588	90.1643	27.843	82.352	.05776	.15802
86	36.266	21.566	.871308	89.9481	28.333	82.647	.05866	.15819
88	37.533	22.833	.843179	89.7308	28.825	82.942	.05955	.15836
90	38.833	24.133	0.816148	89.5125	29.318	83.237	0.06045	0.15854
92	40.167	25.467	.790164	89.2931	29.812	83.531	.06134	.15871
94	41.535	26.835	.765180	89.0727	30.307	83.825	.06223	.15889
96	42.938	28.238	.741149	88.8512	30.804	84.119	.06312	.15907
98	44.377	29.677	.718028	88.6286	31.302	84.413	.06401	.15924
100	45.851	31.151	0.695777	88.4049	31.801	84.706	0.06490	0.15942
102	47.362	32.662	.674356	88.1800	32.301	84.998	.06578	.15960
104	48.910	34.210	.653729	87.9540	32.803	85.290	.06667	.15978
106	50.495	35.795	.633861	87.7268	33.305	85.582	.06755	.15997
108	52.119	37.419	.614718	87.4985	33.809	85.873	.06844	.16015
110	53.782	39.082	0.596269	87.2689	34.314	86.164	0.06932	0.16033
112	55.483	40.783	.578484	87.0380	34.820	86.454	.07020	.16052
114	57.225	42.525	.561335	86.8059	35.327	86.744	.07108	.16070
116	59.007	44.307	.544793	86.5725	35.835	87.033	.07196	.16089
118	60.830	46.130	.528835	86.3379	36.345	87.322	.07283	.16108
120	62.695	47.995	0.513434	86.1018	36.855	87.610	0.07371	0.16126
122	64.602	49.902	.498568	85.8644	37.367	87.897	.07458	.16145
124	66.551	51.851	.484214	85.6257	37.880	88.184	.07546	.16164
126	68.544	53.864	.470351	85.3855	38.394	88.470	.07633	.16183
128	70.581	55.881	.456959	85.1439	38.909	88.755	.07720	.16201
130	72.662	57.962	0.444019	84.9008	39.425	89.040	0.07807	0.16220
132	74.961	60.089	.431512	84.6562	39.943	89.324	.07893	.16239
134	76.961	62.261	.419421	84.4101	40.461	89.608	.07980	.16258
136	79.179	64.479	.407729	84.1624	40.981	89.890	.08067	.16277
138	81.444	66.744	.396420	83.9131	41.501	90.172	.08153	.16296
140	83.757	69.057	0.385479	83.6622	42.023	90.453	0.08239	0.16315
142	86.118	71.418	.374892	83.4096	42.546	90.733	.08326	.16334
144	88.521	73.827	.364644	83.1554	43.070	91.012	.08412	.16353

[a]Reprinted with permission from the 1977 Fundamentals Volume, ASHRAE Handbook and Product Directory.

TABLE A3-4 Refrigerant 114 (dichlorotetrafluoroethane) properties of liquid and saturated vapor[a] (continued)

°F																
-14	4.103	21.568	6.716525	99.7782	5.414	67.822	82.8994	.354723	76.286	146	90.986	.15253	.01250	91.291	.08498	.16372
-12	4.334	21.098	6.383184	99.5976	5.839	68.114	82.6416	.345116	78.795	148	93.495	.15256	.01364	91.568	.08583	.16391
-10	4.575	20.607	6.069551	99.4164	6.266	68.407	82.3820	0.335810	81.354	150	96.054	0.15259	0.01441	91.845	0.08669	0.16410
-8	4.827	20.094	5.774289	99.2347	6.694	68.700	82.1206	.326794	83.965	152	98.665	.15263	.01536	92.120	.08754	.16429
-6	5.090	19.559	5.496164	99.0524	7.124	68.993	81.8572	.318058	86.627	154	101.327	.15267	.01630	92.395	.08840	.16448
-4	5.364	19.000	5.234035	98.8696	7.555	69.287	81.5919	.309590	89.342	156	104.042	.15272	.01725	92.669	.08925	.16467
-2	5.650	18.417	4.986844	98.6862	7.987	69.581	81.3246	.301380	92.110	158	106.810	.15277	.01819	92.941	.09010	.16486
0	5.949	17.810	4.753614	98.5023	8.420	69.875	81.0553	0.293419	94.932	160	109.632	0.15283	0.01914	93.212	0.09095	0.16504
2	6.260	17.177	4.533437	98.3178	8.855	70.170	80.7839	.285697	97.808	162	112.508	.15289	.02008	93.483	.09180	.16523
4	6.583	16.517	4.325473	98.1327	9.291	70.465	80.5103	.278205	100.739	164	115.439	.15295	.02102	93.752	.09265	.16542
6	6.921	15.831	4.128942	97.9471	9.729	70.760	80.2346	.270936	103.726	166	118.426	.15302	.02196	94.019	.09350	.16560
8	7.271	15.117	3.943119	97.7608	10.168	71.056	79.9566	.263880	106.769	168	121.469	.15309	.02290	94.286	.09434	.16579
10	7.636	14.374	3.767332	97.5740	10.608	71.351	79.6763	0.257030	109.870	170	124.570	0.15317	0.02384	94.551	0.09519	0.16597
12	8.016	13.601	3.600955	97.3865	11.050	71.647	79.3936	.250379	113.028	172	127.728	.15325	.02478	94.815	.09603	.16616
14	8.410	12.799	3.443408	97.1985	11.493	71.943	79.1084	.243919	116.244	174	130.944	.15333	.02571	95.078	.09687	.16634
16	8.819	11.965	3.294147	97.0098	11.937	72.240	78.8208	.237643	119.520	176	134.220	.15342	.02665	95.339	.09771	.16652
18	9.244	11.100	3.152670	96.8205	12.383	72.537	78.5306	.231545	122.856	178	137.556	.15351	.02758	95.599	.09855	.16670
20	9.686	10.201	3.018507	96.6306	12.830	72.834	78.2377	0.225618	126.252	180	140.952	0.15360	0.02851	95.857	0.09939	0.16688
22	10.143	9.269	2.891220	96.4400	13.278	73.131	77.9422	.219857	129.710	182	144.410	.15370	.02944	96.113	.10023	.16706
24	10.618	8.303	2.770402	96.2488	13.728	73.428	77.6438	.214254	133.230	184	147.930	.15380	.03037	96.368	.10107	.16723
26	11.110	7.301	2.655670	96.0569	14.179	73.725	77.3425	.208806	136.812	186	151.512	.15390	.03130	96.622	.10190	.16741
28	11.620	6.263	2.546570	95.8644	14.632	74.023	77.0382	.203506	140.459	188	155.159	.15401	.03223	96.873	.10274	.16758
30	12.148	5.188	2.443070	95.6712	15.085	74.320	76.7309	0.198349	144.169	190	158.869	0.15412	0.03316	97.123	0.10357	0.16776
32	12.695	4.074	2.344558	95.4773	15.541	74.618	76.4204	.193330	147.945	192	162.645	.15423	.03408	97.371	.10441	.16793
34	13.261	2.922	2.250844	95.2827	15.997	74.916	76.1066	.188444	151.787	194	166.487	.15435	.03501	97.617	.10524	.16810
36	13.847	1.730	2.161657	95.0874	16.455	75.214	75.7895	.183686	155.696	196	170.396	.15447	.03593	97.861	.10607	.16826
38	14.452	0.496	2.076742	94.8914	16.914	75.512	75.4689	.179053	159.672	198	174.372	.15459	.03685	98.102	.10690	.16843
40	15.078	0.378	1.995861	94.6947	17.375	75.810	75.1446	0.174539	163.717	200	178.417	0.15471	0.03777	98.342	0.10773	0.16859
42	15.726	1.026	1.918791	94.4973	17.837	76.108	74.3173	.163752	174.136	205	188.836	.15484	.03869	98.931	.10980	.16899
44	16.394	1.694	1.845323	94.2991	18.300	76.406	73.4645	.153627	185.002	210	199.702	.15497	.03961	99.505	.11187	.16937
46	17.085	2.385	1.775259	94.1002	18.765	76.704	72.5835	.144107	196.332	215	211.032	.15510	.04053	100.062	.11394	.16973
48	17.798	3.098	1.708417	93.9005	19.231	77.002	71.6714	.135141	208.141	220	222.841	.15524	.04145	100.599	.11601	.17008
50	18.534	3.834	1.644623	93.7001	19.698	77.300	70.7246	0.126680	220.448	225	235.148	0.15538	0.04236	101.115	0.11808	0.17040
52	19.293	4.593	1.583714	93.4989	20.167	77.598	69.7391	.118681	233.271	230	247.971	.15552	.04328	101.607	.12015	.17069
54	20.076	5.376	1.525539	93.2969	20.637	77.896	68.7096	.111101	246.630	235	261.330	.15566	.04419	102.072	.12223	.17095
56	20.884	6.184	1.469953	93.0941	21.108	78.194	67.6300	.103900	260.550	240	275.250	.15580	.04510	102.505	.12432	.17117
58	21.716	7.016	1.416822	92.8905	21.581	78.492	66.4923	.097039	275.054	245	289.754	.15595	.04601	102.901	.12642	.17136
60	22.574	7.874	1.366019	92.6860	22.055	78.790	65.2865	0.090480	290.172	250	304.872	0.15610	0.04693	103.254	0.12855	0.17148
62	23.458	8.758	1.317424	92.4808	22.530	79.087	62.6138	.078107	322.385	260	337.085	.15625	.04783	103.795	.13290	.17154
64	24.368	9.668	1.270924	92.2749	23.007	79.385	59.4330	.066411	357.516	270	372.216	.15640	.04874	104.010	.13748	.17119
66	25.304	10.604	1.226414	92.0677	23.484	79.683	55.3261	.054798	396.039	280	410.739	.15655	.04965	104.629	.14261	.17008
68	26.269	11.569	1.183794	91.8599	23.964	79.980	48.6173	.041364	438.719	290	453.419	.15671	.05056	101.575	.14944	.16682

* Inches of mercury below one standard atmosphere.   ** Based on 0 for the saturated liquid at −40 F.

[a] From published data (1966) of E. I. du Pont de Nemours & Co., Inc. Used by permission.

[a]Reprinted with permission from the 1977 Fundamentals Volume, ASHRAE Handbook and Product Directory.

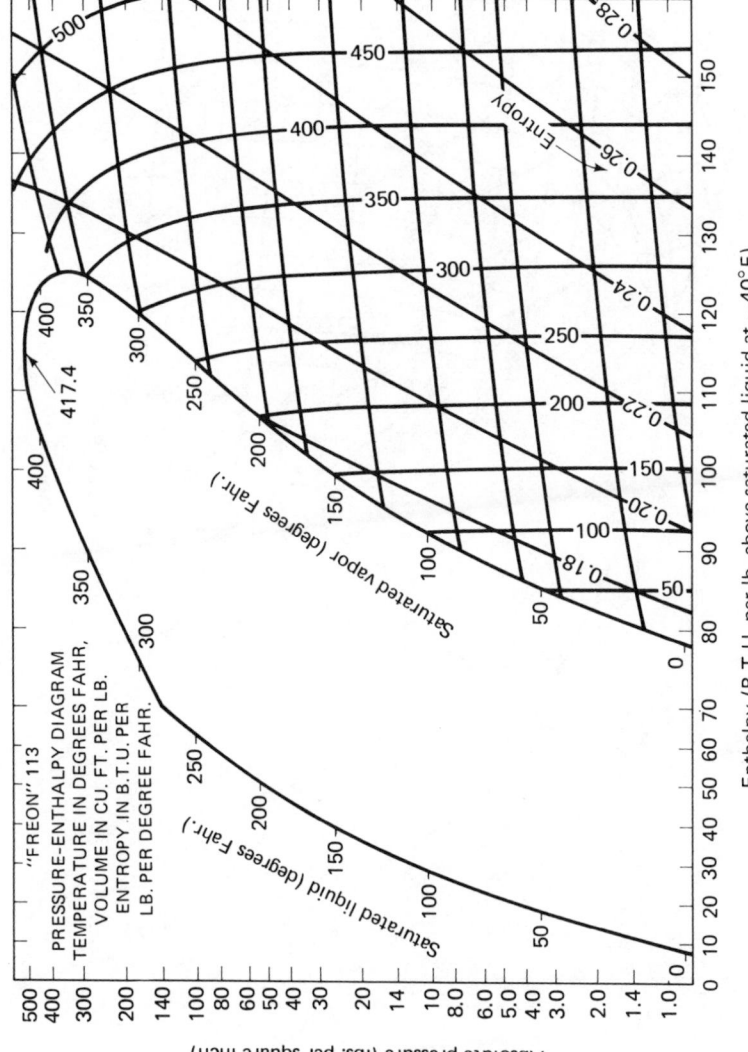

Figure A3.1

509

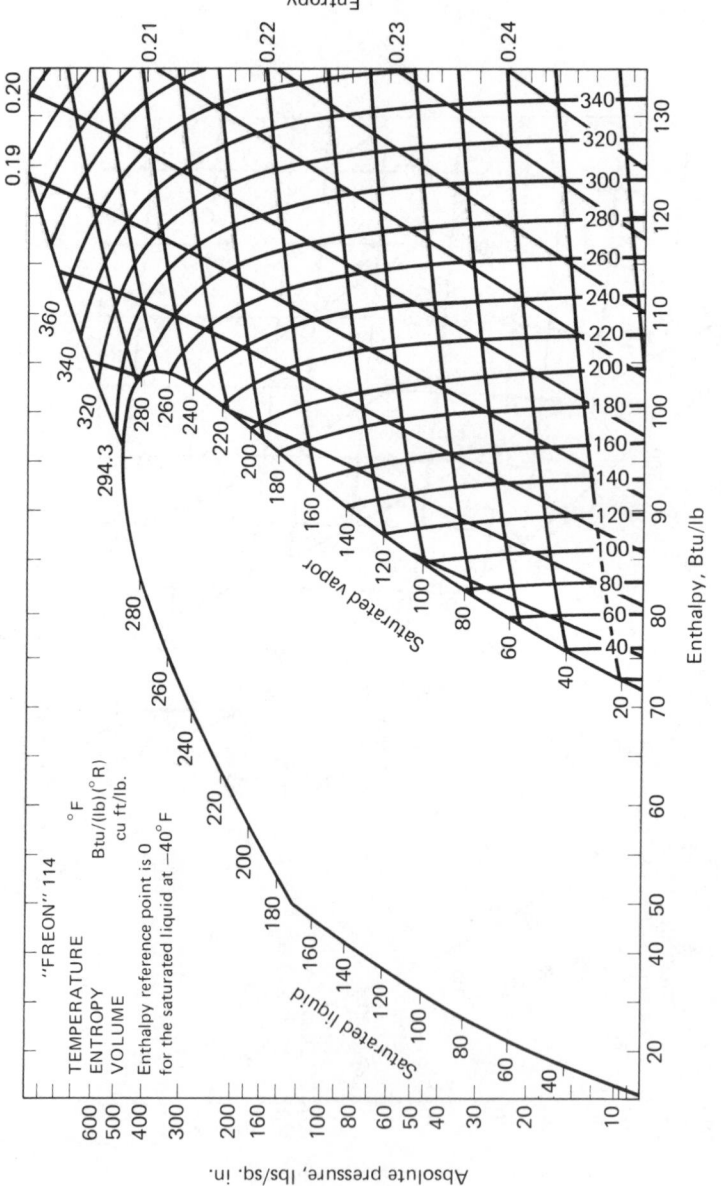

Figure A3.2

"FREON" 114

TEMPERATURE   °F
ENTROPY   Btu/(lb)(°R)
VOLUME   cu ft/lb.

Enthalpy reference point is 0
for the saturated liquid at −40° F

Entropy

Enthalpy, Btu/lb

Absolute pressure, lbs/sq. in.

Saturated vapor

Saturated liquid

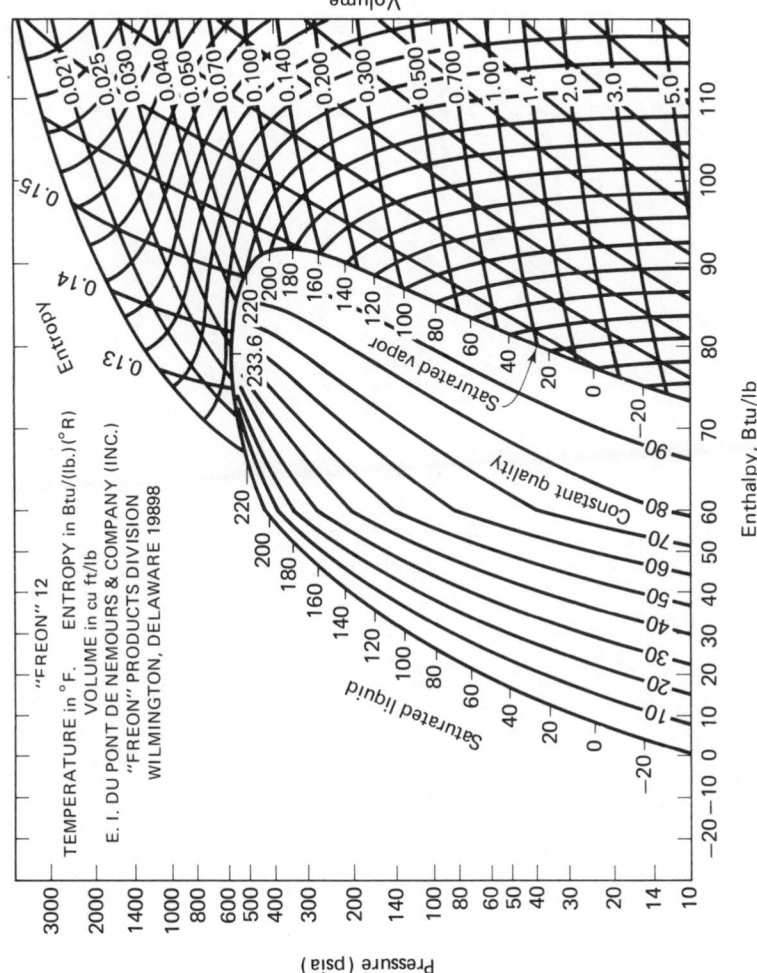

Figure A3.3

Enthalpy, Btu/lb

511

Index